S0-CKK-373

Methods in Enzymology

Volume 190

RETINOIDS

Part B

Cell Differentiation and Clinical Applications

METHODS IN ENZYMOLOGY

Methods in Enzymology

Volume 190

Retinoids

Part B

Cell Differentiation and Clinical Applications

EDITED BY

Lester Packer

DEPARTMENT OF MOLECULAR AND CELL BIOLOGY
UNIVERSITY OF CALIFORNIA, BERKELEY
BERKELEY, CALIFORNIA

ACADEMIC PRESS, INC.
Harcourt Brace Jovanovich, Publishers
San Diego New York Boston
London Sydney Tokyo Toronto

This book is printed on acid-free paper. ∞

Academic Press, Inc.
San Diego, California 92101

United Kingdom Edition published by
Academic Press Limited
24-28 Oval Road, London NW1 7DX

Library of Congress Catalog Card Number: 54-9110

ISBN 0-12-182091-2 (alk. paper)

Printed in the United States of America
90 91 92 93 9 8 7 6 5 4 3 2 1

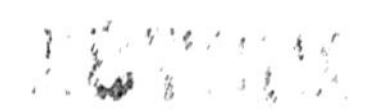

Table of Contents

Section I. Cell Differentiation

A. Normal Cells

B. Cell Lines

C. Tissues and Organ Culture

D. Retinoids as Morphogens and Teratogens

E. Cell Synopsis

Section II. Nutrition, Tissue and Immune Status, and Antioxidant Action

A. Nutrition

B. Tissue and Immune Status

C. Antioxidant Action

Section III. Pharmacokinetics, Pharmacology, and Toxicology

A. Pharmacokinetics

B. Pharmacology

C. Toxicology

Contributors to Volume 190

Article numbers are in parentheses following the names of contributors.
Affiliations listed are current.

SERGIO ADAMO (9), *Department of Experimental Medicine, University of L'Aquila, 67100 L'Aquila, Italy*

BRAD AMOS (23), *Department of Tumor Biology, M. D. Anderson Cancer Center, University of Texas, Houston, Texas 77030*

EVA ANDERSSON (18), *Department of Dermatology, University Hospital, S-581 85 Linköping, Sweden*

ADRIANNE BENDICH (27), *Department of Clinical Nutrition, Hoffmann-La Roche Inc., Nutley, New Jersey 07110*

TROND BERG (6), *Institute for Nutrition Research, University of Oslo, N-0316 Oslo 3, Norway*

RUNE BLOMHOFF (6), *Institute for Nutrition Research, University of Oslo, N-0316 Oslo 3, Norway*

STEVEN A. BOOK (43), *Health and Welfare Agency, State of California, Sacramento, California 95814*

D. LUCILLE BREDBERG (17), *Department of Ophthalmology, University of Washington, Seattle, Washington 98195*

THEODORE R. BREITMAN (13), *Laboratory of Biological Chemistry, Division of Cancer Treatment, National Cancer Institute, National Institutes of Health, Bethesda, Maryland 20892*

C. D. B. BRIDGES (15), *Department of Biological Sciences, Purdue University, West Lafayette, Indiana 47907*

CONSTANCE E. BRINCKERHOFF (19), *Department of Medicine and Biochemistry, Dartmouth Medical School, Hanover, New Hampshire 03756*

ADRIAAN BROUWER (5), *TNO Institute for Experimental Gerontology, 2280 HV Rijswijk, The Netherlands*

GRAEME F. BRYCE (38), *Roche Dermatologics, Hoffmann-La Roche Inc., Nutley, New Jersey 07110*

ROBERT J. CAPETOLA (37), *Research Laboratories, The R. W. Johnson Pharmaceutical Research Institute, Raritan, New Jersey 08869*

ANDREW L. DARROW (12), *Department of Molecular Pharmacology, State University of New York at Stony Brook, Stony Brook, New York 11794*

WILLEM J. DE GRIP (1), *Center of Eye Research Nijmegen, University of Nijmegen, 6500 HB Nijmegen, The Netherlands*

LUIGI M. DE LUCA (9, 10), *National Cancer Institute, Bethesda, Maryland 20892*

THOMAS I. DORAN (35), *Preclinical Research, Roche Dermatologics, Hoffmann-La Roche Inc., Nutley, New Jersey 07110*

R. EHLERT (31), *Department of Dermatology and Venereology, University Medical Center Steglitz, The Free University of Berlin, Berlin, Federal Republic of Germany*

GREGOR EICHELE (21), *Department of Cellular and Molecular Physiology, Harvard Medical School, Boston, Massachusetts 02115*

DENISE A. FAHERTY (27), *Department of Immunopharmacology, Hoffmann-La Roche Inc., Nutley, New Jersey 07110*

SHAO-LING FONG (15), *Departments of Ophthalmology and Biochemistry and Molecular Biology, Indiana University, Indianapolis, Indiana 46202*

ELAINE V. FUCHS (2), *Department of Molecular Genetics and Cell Biology, Howard Hughes Medical Institute, University of Chicago, Chicago, Illinois 60637*

GERARD J. GENDIMENICO (37), *Research Laboratories, The R. W. Johnson Pharmaceutical Research Institute, Raritan, New Jersey 08869*

MARGARET A. GEORGE (4), *Cell Biology Section, Laboratory of Pulmonary Pathobiology, National Institute of Environmental Health Sciences, Research Triangle Park, North Carolina 27709*

GEORGE J. GIUDICE (2), *Department of Dermatology, Medical College of Wisconsin, Milwaukee, Wisconsin 53226*

H. GOLLNICK (31), *Department of Dermatology and Venereology, University Medical Center Steglitz, The Free University of Berlin, Berlin, Federal Republic of Germany*

JOANNE BALMER GREEN (32), *Nutrition Department, Pennsylvania State University, University Park, Pennsylvania 16802*

MICHAEL H. GREEN (32), *Nutrition Department, Pennsylvania State University, University Park, Pennsylvania 16802*

JOSEPH F. GRIPPO (16), *Department of Toxicology and Pathology, Hoffmann-La Roche Inc., Nutley, New Jersey 07110*

MICHAEL D. GRISWOLD (7), *Department of Biochemistry, Washington State University, Pullman, Washington 99164*

GARY L. GROVE (39), *KGL's Skin Study Center, Broomall, Pennsylvania 19008*

MARY JO GROVE (39), *KGL's Skin Study Center, Broomall, Pennsylvania 19008*

LORRAINE J. GUDAS (14), *Department of Biological Chemistry and Molecular Pharmacology, Harvard Medical School and Dana-Farber Cancer Institute, Boston, Massachusetts 02115*

HENK F. J. HENDRIKS (5), *TNO Institute for Experimental Gerontology, 2280 HV Rijswijk, The Netherlands*

MIDORI HIRAMATSU (29), *Department of Neurochemistry, Institute for Neurobiology, Okayama University Medical School, Okayama 700, Japan*

W. BRIAN HOWARD (44, 45, 46), *La Jolla Cancer Research Foundation, La Jolla, California 92037*

FREESIA L. HUANG (10), *National Institute of Child Health and Human Development, Endocrinology and Reproduction Research Branch, National Institutes of Health, Bethesda, Maryland 20892*

JAMES HURLEY (34), *Business Development, Roche Dermatologics, Hoffmann-La Roche Inc., Nutley, New Jersey 07110*

ANTON M. JETTEN (4), *Cell Biology Section, Laboratory of Pulmonary Pathobiology, National Institute of Environmental Health Sciences, Research Triangle Park, North Carolina 27709*

ALICE F. KARL (7), *Department of Biochemistry, Washington State University, Pullman, Washington 99164*

SHIGEMI KATO (9), *St. Marianna University School of Medicine, Miyamae-Ku, Kawasaki 213, Japan*

TERENCE KEALEY (36), *Department of Clinical Biochemistry, Cambridge University, Addenbrooke's Hospital, Cambridge CB2 2QR, England*

ANDREAS KISTLER (44, 45, 46), *Clinical Research, F. Hoffmann-La Roche Ltd., CH-4002 Basel, Switzerland*

LORRAINE H. KLIGMAN (40), *Department of Dermatology, University of Pennsylvania School of Medicine, Philadelphia, Pennsylvania 19104*

DICK L. KNOOK (5), *TNO Institute for Experimental Gerontology, 2280 HV Rijswijk, The Netherlands*

DEVENDRA M. KOCHHAR (33), *Department of Anatomy, Thomas Jefferson University, Philadelphia, Pennsylvania 19107*

DAFNA LOTAN (11), *Department of Tumor Biology, M. D. Anderson Cancer Center, University of Texas, Houston, Texas 77030*

REUBEN LOTAN (11, 23), *Department of Tumor Biology, M. D. Anderson Cancer Center, University of Texas, Houston, Texas 77030*

MALCOLM MADEN (20), *Anatomy and Human Biology Group, Kings College, London WC2R 2LS, England*

LAWRENCE J. MARNETT (30), *Department of Biochemistry, Vanderbilt University Medical Center, Nashville, Tennessee 37232*

JOHN L. MCGUIRE (37), *Research Laboratories, The R. W. Johnson Pharmaceutical Research Institute, Raritan, New Jersey 08869*

RAJENDRA G. MEHTA (42), *Laboratory of Pathophysiology, IIT Research Institute, Chicago, Illinois 60616*

JAMES A. MEZICK (37), *Research Laboratories, The R. W. Johnson Pharmaceutical Research Institute, Raritan, New Jersey 08869*

LEONARD M. MILSTONE (8), *Department of Dermatology, Yale University School of Medicine, New Haven, Connecticut 06510*

RICHARD C. MOON (42), *Laboratory of Pathophysiology, IIT Research Institute, Chicago, Illinois 60616*

HEINZ NAU (47), *Institute of Toxicology and Embryopharmacology, The Free University of Berlin, D-1000 Berlin 33, Federal Republic of Germany*

KATHLEEN M. NAUSS (28), *Health Effects Institute, Cambridge, Massachusetts 02139*

C. E. ORFANOS (31), *Department of Dermatology and Venereology, University Medical Center Steglitz, The Free University of Berlin, Berlin, Federal Republic of Germany*

LESTER PACKER (29), *Department of Molecular and Cell Biology, University of California, Berkeley, Berkeley, California 94720*

MARIA PONEC (3), *Department of Dermatology, University Hospital Leiden, 2300 RC Leiden, The Netherlands*

JAMES I. REARICK (4), *Department of Biochemistry, Kirksville College of Osteopathic Medicine, Kirksville, Missouri 63501*

RICHARD J. RICKLES (12), *Department of Biochemistry and Molecular Biology, Harvard University, Cambridge, Massachusetts 02138*

G. RINCK (31), *Department of Dermatology and Venereology, University Medical Center Steglitz, The Free University of Berlin, Berlin, Federal Republic of Germany*

OLA ROLLMAN (18), *Department of Dermatology, University Hospital, S-751 85 Uppsala, Sweden*

DENNIS R. ROOP (10), *Department of Cell Biology, Baylor College of Medicine, Houston, Texas 77030*

MARVIN E. ROSENTHALE (37), *Research Laboratories, The R. W. Johnson Pharmaceutical Research Institute, Raritan, New Jersey 08869*

A. CATHARINE ROSS (28), *Department of Physiology and Biochemistry, Division of Nutrition, Medical College of Pennsylvania, Philadelphia, Pennsylvania 19129*

JOHN C. SAARI (17), *Department of Ophthalmology, University of Washington, Seattle, Washington 98195*

PETER G. SACKS (11), *Department of Tumor Biology, M. D. Anderson Cancer Center, University of Texas, Houston, Texas 77030*

VICTOR M. SAMOKYSZYN (30), *Department of Pharmaceutical Chemistry, School of Pharmacy, University of California, San Francisco, San Francisco, California 94143*

STANLEY S. SHAPIRO (34, 35, 38), *Preclinical Research, Roche Dermatologics, Hoffmann-La Roche Inc., Nutley, New Jersey 07110*

MICHAEL I. SHERMAN (16), *Department of Cell Biology, Hoffmann-La Roche Inc., Nutley, New Jersey 07110*

KENNETH L. SIMPSON (25), *Food Science and Nutrition Research Center, West Kingston, Rhode Island 02892*

JOHN EDGAR SMITH (24), *Nutrition Department, Pennsylvania State University, University Park, Pennsylvania 16802*

DAVID L. STOCUM (20), *Department of Biology, Indiana University-Purdue University at Indianapolis, Indianapolis, Indiana 46205*

SIDNEY STRICKLAND (12), *Department of Molecular Pharmacology, State University of New York at Stony Brook, Stony Brook, New York 11794*

KAMPE TEELMANN (41), *Pharmaceutical Division, F. Hoffmann-La Roche Ltd., CH-4002 Basel, Switzerland*

CHRISTINA THALLER (21), *Department of Cellular and Molecular Physiology, Harvard Medical School, Boston, Massachusetts 02115*

ADRIAN M. TIMMERS (1), *Department of Ophthalmology, University of Florida, Gainesville, Florida 32610*

HANS TÖRMÄ (18, 22), *Department of Dermatology, University Hospital, S-581 85 Linköping, Sweden*

SALLY S. TWINING (28), *Department of Biochemistry, Medical College of Wisconsin, Milwaukee, Wisconsin 53226*

BARBARA A. UNDERWOOD (26), *National Eye Institute, National Institutes of Health, Bethesda, Maryland 20892*

ANDERS VAHLQUIST (18, 22), *Department of Dermatology, University Hospital, S-581 85 Linköping, Sweden*

SARAH WEDDEN (21), *Department of Anatomy, University Medical School, Edinburgh EH8 9AG, Scotland*

ARIJ WEERHEIM (3), *Department of Dermatology, University Hospital Leiden, 2300 RC Leiden, The Netherlands*

CALVIN C. WILLHITE (43), *Toxic Substances Control Program, Department of Health Services, State of California, Berkeley, California 94710*

Preface

Spectacular progress and unprecedented interest in the field of retinoids prompted us to consider this topic for two volumes in the *Methods in Enzymology* series: Volume 189, Retinoids, Part A: Molecular and Metabolic Aspects and Volume 190, Retinoids, Part B: Cell Differentiation and Clinical Applications.

From a historical perspective we know that studies in the 1930s showed that vitamin A (retinol) and retinal had a role in the visual process. It was also recognized that some link between vitamin A and cancer incidence existed. Several decades ago it was discovered that retinoic acid had a dramatic effect on the chemically induced DMBA mouse skin carcinogenesis model in which enormous reductions in the tumor burden were observed. This led to the realization that retinoids had important effects on cell differentiation. This resulted almost immediately in the synthesis and evaluation of new retinoids. Indeed, the effects of retinoids on cell differentiation appear to be more universal and of greater importance than their light-dependent role in vision and microbial energy transduction.

Progress has been rapid, and the importance of accurate methodology for this field is imperative to its further development. The importance of methodology applies to the use of retinoids in basic research in molecular, cellular, and developmental biology, and in clinical medicine. In medicine, applications have been mainly to cancer and in dermatology to the treatment of skin diseases and skin aging. As new retinoids are being tested in biological models and in clinical medicine, interest in the nutrition and pharmacology of retinoids has arisen. Moreover, the beneficial effects of retinoids in pharmacological treatment have led to a recognition of the "double-edged sword" of toxicity (teratogenicity).

In Section I of this volume, Cell Differentiation, the effects of retinoids in various cell differentiation systems are covered. Many new systems in which retinoids exhibit their effects have been employed. Both normal diploid cells and cell lines *in vitro* have been used, and the methods and systems employed are presented. In addition, tissue and organ culture are important areas for retinoid methodology. The effects of retinoids as morphogens and teratology agents are also included. In Sections II, Nutrition, Tissue and Immune Status, and Antioxidant Action, and III, Pharmacokinetics, Pharmacology, and Toxicology, nutritional and pharmacological methods are presented. Retinoids in the treatment of skin disease and in cancer chemotherapy are probably the most important areas in which methodological developments have occurred. New methodology has also revealed the antioxidant activity of retinoids, and since any antioxidant may also be a pro-oxidant such considerations may be important for clinical pharmacology and therapeutics.

Volume 189 covers structure and analysis, receptors, transport, and binding proteins, and enzymology and metabolism.

I am very grateful to the Advisory Board—Frank Chytil, DeWitt Goodman, Maria A. Livrea, Leonard Milstone, Concetta Nicotra, James A. Olson, and Stanley S. Shapiro—for their unique input, advice, counsel, and encouragement in the planning and organization of this volume. In most instances, I met with every member of the board on one or more occasions to discuss the topics and to identify the most important contributors. Indeed, we found almost universal acceptance, and virtually no one turned down our invitation to contribute to this volume. In fact it was somewhat autocatalytic in that many contributors, realizing the timeliness and significance of having all of the methods dealing with retinoids included, made suggestions for additional contributions which were evaluated by the board. In a few instances we may have been somewhat overzealous, and more than one article on a method has been included. We do apologize for this slight redundancy for the sake of completeness.

LESTER PACKER

METHODS IN ENZYMOLOGY

VOLUME I. Preparation and Assay of Enzymes
Edited by SIDNEY P. COLOWICK AND NATHAN O. KAPLAN

VOLUME II. Preparation and Assay of Enzymes
Edited by SIDNEY P. COLOWICK AND NATHAN O. KAPLAN

VOLUME III. Preparation and Assay of Substrates
Edited by SIDNEY P. COLOWICK AND NATHAN O. KAPLAN

VOLUME IV. Special Techniques for the Enzymologist
Edited by SIDNEY P. COLOWICK AND NATHAN O. KAPLAN

VOLUME V. Preparation and Assay of Enzymes
Edited by SIDNEY P. COLOWICK AND NATHAN O. KAPLAN

VOLUME VI. Preparation and Assay of Enzymes (*Continued*)
Preparation and Assay of Substrates
Special Techniques
Edited by SIDNEY P. COLOWICK AND NATHAN O. KAPLAN

VOLUME VII. Cumulative Subject Index
Edited by SIDNEY P. COLOWICK AND NATHAN O. KAPLAN

VOLUME VIII. Complex Carbohydrates
Edited by ELIZABETH F. NEUFELD AND VICTOR GINSBURG

VOLUME IX. Carbohydrate Metabolism
Edited by WILLIS A. WOOD

VOLUME X. Oxidation and Phosphorylation
Edited by RONALD W. ESTABROOK AND MAYNARD E. PULLMAN

VOLUME XI. Enzyme Structure
Edited by C. H. W. HIRS

VOLUME XII. Nucleic Acids (Parts A and B)
Edited by LAWRENCE GROSSMAN AND KIVIE MOLDAVE

VOLUME XIII. Citric Acid Cycle
Edited by J. M. LOWENSTEIN

VOLUME XIV. Lipids
Edited by J. M. LOWENSTEIN

VOLUME XV. Steroids and Terpenoids
Edited by RAYMOND B. CLAYTON

VOLUME XVI. Fast Reactions
Edited by KENNETH KUSTIN

VOLUME XVII. Metabolism of Amino Acids and Amines (Parts A and B)
Edited by HERBERT TABOR AND CELIA WHITE TABOR

VOLUME XVIII. Vitamins and Coenzymes (Parts A, B, and C)
Edited by DONALD B. MCCORMICK AND LEMUEL D. WRIGHT

VOLUME XIX. Proteolytic Enzymes
Edited by GERTRUDE E. PERLMANN AND LASZLO LORAND

VOLUME XX. Nucleic Acids and Protein Synthesis (Part C)
Edited by KIVIE MOLDAVE AND LAWRENCE GROSSMAN

VOLUME XXI. Nucleic Acids (Part D)
Edited by LAWRENCE GROSSMAN AND KIVIE MOLDAVE

VOLUME XXII. Enzyme Purification and Related Techniques
Edited by WILLIAM B. JAKOBY

VOLUME XXIII. Photosynthesis (Part A)
Edited by ANTHONY SAN PIETRO

VOLUME XXIV. Photosynthesis and Nitrogen Fixation (Part B)
Edited by ANTHONY SAN PIETRO

VOLUME XXV. Enzyme Structure (Part B)
Edited by C. H. W. HIRS AND SERGE N. TIMASHEFF

VOLUME XXVI. Enzyme Structure (Part C)
Edited by C. H. W. HIRS AND SERGE N. TIMASHEFF

VOLUME XXVII. Enzyme Structure (Part D)
Edited by C. H. W. HIRS AND SERGE N. TIMASHEFF

VOLUME XXVIII. Complex Carbohydrates (Part B)
Edited by VICTOR GINSBURG

VOLUME XXIX. Nucleic Acids and Protein Synthesis (Part E)
Edited by LAWRENCE GROSSMAN AND KIVIE MOLDAVE

VOLUME XXX. Nucleic Acids and Protein Synthesis (Part F)
Edited by KIVIE MOLDAVE AND LAWRENCE GROSSMAN

VOLUME XXXI. Biomembranes (Part A)
Edited by SIDNEY FLEISCHER AND LESTER PACKER

VOLUME XXXII. Biomembranes (Part B)
Edited by SIDNEY FLEISCHER AND LESTER PACKER

VOLUME XXXIII. Cumulative Subject Index Volumes I–XXX
Edited by MARTHA G. DENNIS AND EDWARD A. DENNIS

VOLUME XXXIV. Affinity Techniques (Enzyme Purification: Part B)
Edited by WILLIAM B. JAKOBY AND MEIR WILCHEK

VOLUME XXXV. Lipids (Part B)
Edited by JOHN M. LOWENSTEIN

VOLUME XXXVI. Hormone Action (Part A: Steroid Hormones)
Edited by BERT W. O'MALLEY AND JOEL G. HARDMAN

VOLUME XXXVII. Hormone Action (Part B: Peptide Hormones)
Edited by BERT W. O'MALLEY AND JOEL G. HARDMAN

VOLUME XXXVIII. Hormone Action (Part C: Cyclic Nucleotides)
Edited by JOEL G. HARDMAN AND BERT W. O'MALLEY

VOLUME XXXIX. Hormone Action (Part D: Isolated Cells, Tissues, and Organ Systems)
Edited by JOEL G. HARDMAN AND BERT W. O'MALLEY

VOLUME XL. Hormone Action (Part E: Nuclear Structure and Function)
Edited by BERT W. O'MALLEY AND JOEL G. HARDMAN

VOLUME XLI. Carbohydrate Metabolism (Part B)
Edited by W. A. WOOD

VOLUME XLII. Carbohydrate Metabolism (Part C)
Edited by W. A. WOOD

VOLUME XLIII. Antibiotics
Edited by JOHN H. HASH

VOLUME XLIV. Immobilized Enzymes
Edited by KLAUS MOSBACH

VOLUME XLV. Proteolytic Enzymes (Part B)
Edited by LASZLO LORAND

VOLUME XLVI. Affinity Labeling
Edited by WILLIAM B. JAKOBY AND MEIR WILCHEK

VOLUME XLVII. Enzyme Structure (Part E)
Edited by C. H. W. HIRS AND SERGE N. TIMASHEFF

VOLUME XLVIII. Enzyme Structure (Part F)
Edited by C. H. W. HIRS AND SERGE N. TIMASHEFF

VOLUME XLIX. Enzyme Structure (Part G)
Edited by C. H. W. HIRS AND SERGE N. TIMASHEFF

VOLUME L. Complex Carbohydrates (Part C)
Edited by VICTOR GINSBURG

VOLUME LI. Purine and Pyrimidine Nucleotide Metabolism
Edited by PATRICIA A. HOFFEE AND MARY ELLEN JONES

VOLUME LII. Biomembranes (Part C: Biological Oxidations)
Edited by SIDNEY FLEISCHER AND LESTER PACKER

VOLUME LIII. Biomembranes (Part D: Biological Oxidations)
Edited by SIDNEY FLEISCHER AND LESTER PACKER

VOLUME LIV. Biomembranes (Part E: Biological Oxidations)
Edited by SIDNEY FLEISCHER AND LESTER PACKER

VOLUME LV. Biomembranes (Part F: Bioenergetics)
Edited by SIDNEY FLEISCHER AND LESTER PACKER

VOLUME LVI. Biomembranes (Part G: Bioenergetics)
Edited by SIDNEY FLEISCHER AND LESTER PACKER

VOLUME LVII. Bioluminescence and Chemiluminescence
Edited by MARLENE A. DELUCA

VOLUME LVIII. Cell Culture
Edited by WILLIAM B. JAKOBY AND IRA PASTAN

VOLUME LIX. Nucleic Acids and Protein Synthesis (Part G)
Edited by KIVIE MOLDAVE AND LAWRENCE GROSSMAN

VOLUME LX. Nucleic Acids and Protein Synthesis (Part H)
Edited by KIVIE MOLDAVE AND LAWRENCE GROSSMAN

VOLUME 61. Enzyme Structure (Part H)
Edited by C. H. W. HIRS AND SERGE N. TIMASHEFF

VOLUME 62. Vitamins and Coenzymes (Part D)
Edited by DONALD B. MCCORMICK AND LEMUEL D. WRIGHT

VOLUME 63. Enzyme Kinetics and Mechanism (Part A: Initial Rate and Inhibitor Methods)
Edited by DANIEL L. PURICH

VOLUME 64. Enzyme Kinetics and Mechanism (Part B: Isotopic Probes and Complex Enzyme Systems)
Edited by DANIEL L. PURICH

VOLUME 65. Nucleic Acids (Part I)
Edited by LAWRENCE GROSSMAN AND KIVIE MOLDAVE

VOLUME 66. Vitamins and Coenzymes (Part E)
Edited by DONALD B. MCCORMICK AND LEMUEL D. WRIGHT

VOLUME 80. Proteolytic Enzymes (Part C)
Edited by LASZLO LORAND

VOLUME 81. Biomembranes (Part H: Visual Pigments and Purple Membranes, I)
Edited by LESTER PACKER

VOLUME 82. Structural and Contractile Proteins (Part A: Extracellular Matrix)
Edited by LEON W. CUNNINGHAM AND DIXIE W. FREDERIKSEN

VOLUME 83. Complex Carbohydrates (Part D)
Edited by VICTOR GINSBURG

VOLUME 84. Immunochemical Techniques (Part D: Selected Immunoassays)
Edited by JOHN J. LANGONE AND HELEN VAN VUNAKIS

VOLUME 85. Structural and Contractile Proteins (Part B: The Contractile Apparatus and the Cytoskeleton)
Edited by DIXIE W. FREDERIKSEN AND LEON W. CUNNINGHAM

VOLUME 86. Prostaglandins and Arachidonate Metabolites
Edited by WILLIAM E. M. LANDS AND WILLIAM L. SMITH

VOLUME 87. Enzyme Kinetics and Mechanism (Part C: Intermediates, Stereochemistry, and Rate Studies)
Edited by DANIEL L. PURICH

VOLUME 88. Biomembranes (Part I: Visual Pigments and Purple Membranes, II)
Edited by LESTER PACKER

VOLUME 89. Carbohydrate Metabolism (Part D)
Edited by WILLIS A. WOOD

VOLUME 90. Carbohydrate Metabolism (Part E)
Edited by WILLIS A. WOOD

VOLUME 91. Enzyme Structure (Part I)
Edited by C. H. W. HIRS AND SERGE N. TIMASHEFF

VOLUME 92. Immunochemical Techniques (Part E: Monoclonal Antibodies and General Immunoassay Methods)
Edited by JOHN J. LANGONE AND HELEN VAN VUNAKIS

VOLUME 93. Immunochemical Techniques (Part F: Conventional Antibodies, Fc Receptors, and Cytotoxicity)
Edited by JOHN J. LANGONE AND HELEN VAN VUNAKIS

VOLUME 94. Polyamines
Edited by HERBERT TABOR AND CELIA WHITE TABOR

VOLUME 95. Cumulative Subject Index Volumes 61–74, 76–80
Edited by EDWARD A. DENNIS AND MARTHA G. DENNIS

VOLUME 96. Biomembranes [Part J: Membrane Biogenesis: Assembly and Targeting (General Methods; Eukaryotes)]
Edited by SIDNEY FLEISCHER AND BECCA FLEISCHER

VOLUME 97. Biomembranes [Part K: Membrane Biogenesis: Assembly and Targeting (Prokaryotes, Mitochondria, and Chloroplasts)]
Edited by SIDNEY FLEISCHER AND BECCA FLEISCHER

VOLUME 98. Biomembranes (Part L: Membrane Biogenesis: Processing and Recycling)
Edited by SIDNEY FLEISCHER AND BECCA FLEISCHER

VOLUME 99. Hormone Action (Part F: Protein Kinases)
Edited by JACKIE D. CORBIN AND JOEL G. HARDMAN

VOLUME 100. Recombinant DNA (Part B)
Edited by RAY WU, LAWRENCE GROSSMAN, AND KIVIE MOLDAVE

VOLUME 101. Recombinant DNA (Part C)
Edited by RAY WU, LAWRENCE GROSSMAN, AND KIVIE MOLDAVE

VOLUME 102. Hormone Action (Part G: Calmodulin and Calcium-Binding Proteins)
Edited by ANTHONY R. MEANS AND BERT W. O'MALLEY

VOLUME 103. Hormone Action (Part H: Neuroendocrine Peptides)
Edited by P. MICHAEL CONN

VOLUME 104. Enzyme Purification and Related Techniques (Part C)
Edited by WILLIAM B. JAKOBY

VOLUME 105. Oxygen Radicals in Biological Systems
Edited by LESTER PACKER

VOLUME 106. Posttranslational Modifications (Part A)
Edited by FINN WOLD AND KIVIE MOLDAVE

VOLUME 107. Posttranslational Modifications (Part B)
Edited by FINN WOLD AND KIVIE MOLDAVE

VOLUME 108. Immunochemical Techniques (Part G: Separation and Characterization of Lymphoid Cells)
Edited by GIOVANNI DI SABATO, JOHN J. LANGONE, AND HELEN VAN VUNAKIS

VOLUME 109. Hormone Action (Part I: Peptide Hormones)
Edited by LUTZ BIRNBAUMER AND BERT W. O'MALLEY

VOLUME 110. Steroids and Isoprenoids (Part A)
Edited by JOHN H. LAW AND HANS C. RILLING

VOLUME 111. Steroids and Isoprenoids (Part B)
Edited by JOHN H. LAW AND HANS C. RILLING

VOLUME 112. Drug and Enzyme Targeting (Part A)
Edited by KENNETH J. WIDDER AND RALPH GREEN

VOLUME 113. Glutamate, Glutamine, Glutathione, and Related Compounds
Edited by ALTON MEISTER

VOLUME 114. Diffraction Methods for Biological Macromolecules (Part A)
Edited by HAROLD W. WYCKOFF, C. H. W. HIRS, AND SERGE N. TIMASHEFF

VOLUME 115. Diffraction Methods for Biological Macromolecules (Part B)
Edited by HAROLD W. WYCKOFF, C. H. W. HIRS, AND SERGE N. TIMASHEFF

VOLUME 116. Immunochemical Techniques (Part H: Effectors and Mediators of Lymphoid Cell Functions)
Edited by GIOVANNI DI SABATO, JOHN J. LANGONE, AND HELEN VAN VUNAKIS

VOLUME 130. Enzyme Structure (Part K)
Edited by C. H. W. HIRS AND SERGE N. TIMASHEFF

VOLUME 131. Enzyme Structure (Part L)
Edited by C. H. W. HIRS AND SERGE N. TIMASHEFF

VOLUME 132. Immunochemical Techniques (Part J: Phagocytosis and Cell-Mediated Cytotoxicity)
Edited by GIOVANNI DI SABATO AND JOHANNES EVERSE

VOLUME 133. Bioluminescence and Chemiluminescence (Part B)
Edited by MARLENE DELUCA AND WILLIAM D. MCELROY

VOLUME 134. Structural and Contractile Proteins (Part C: The Contractile Apparatus and the Cytoskeleton)
Edited by RICHARD B. VALLEE

VOLUME 135. Immobilized Enzymes and Cells (Part B)
Edited by KLAUS MOSBACH

VOLUME 136. Immobilized Enzymes and Cells (Part C)
Edited by KLAUS MOSBACH

VOLUME 137. Immobilized Enzymes and Cells (Part D)
Edited by KLAUS MOSBACH

VOLUME 138. Complex Carbohydrates (Part E)
Edited by VICTOR GINSBURG

VOLUME 139. Cellular Regulators (Part A: Calcium- and Calmodulin-Binding Proteins
Edited by ANTHONY R. MEANS AND P. MICHAEL CONN

VOLUME 140. Cumulative Subject Index Volumes 102–119, 121–134

VOLUME 141. Cellular Regulators (Part B: Calcium and Lipids)
Edited by P. MICHAEL CONN AND ANTHONY R. MEANS

VOLUME 142. Metabolism of Aromatic Amino Acids and Amines
Edited by SEYMOUR KAUFMAN

VOLUME 143. Sulfur and Sulfur Amino Acids
Edited by WILLIAM B. JAKOBY AND OWEN W. GRIFFITH

VOLUME 194. Guide to Yeast Genetics and Molecular Biology
Edited by CHRISTINE GUTHRIE AND GERALD R. FINK

VOLUME 195. Adenylyl Cyclase, G Proteins, and Guanylyl Cyclase (in preparation)
Edited by ROGER A. JOHNSON AND JACKIE D. CORBIN

VOLUME 196. Molecular Motors and the Cytoskeleton (in preparation)
Edited by RICHARD B. VALLEE

VOLUME 197. Phospholipases (in preparation)
Edited by EDWARD A. DENNIS

VOLUME 198. Peptide Growth Factors (Part C) (in preparation)
Edited by DAVID BARNES, J. P. MATHER, AND GORDON H. SATO

Section I

Cell Differentiation

A. Normal Cells
Articles 1 through 10

B. Cell Lines
Articles 11 through 16

C. Tissues and Organ Culture
Articles 17 through 19

D. Retinoids as Morphogens and Teratogens
Articles 20 through 22

E. Cell Synopsis
Article 23

[1] Analysis of the Visual Cycle by Short-Term Incubation of Isolated Retinal Pigment Epithelial Cells

By ADRIAN M. TIMMERS and WILLEM J. DE GRIP

Introduction

The recycling of retinoids in retina and retinal pigment epithelium (RPE), effectuating the regeneration of bleached visual pigment to the photoactive state, is called the visual cycle. The trigger reaction for the process of vision as well as the visual cycle is photoinduced isomerization (11-cis to all-trans) of the chromophore of vertebrate visual pigments.[1] Major parts of the pathway of the visual cycle remained obscure for decades. Only recently has insight been obtained into some of the long-standing enigmas: which type of retinoid is isomerized where and how and transported in what form to the outer segments to eventually regenerate the visual pigment, rhodopsin.[2] This regeneration pathway is initiated in the RPE where all-*trans*-retinol is enzymatically isomerized to 11-*cis*-retinol followed later by conversion to 11-*cis*-retinaldehyde. The isomerization reaction is driven by the free energy of hydrolysis of the retinyl ester.[3]

Retinoid metabolism in the RPE, with respect to the visual cycle, encompasses several steps: uptake of retinol, intracellular transport, acylation, isomerization, oxidation, and secretion of retinoid. In order to study such a complex set of metabolic pathways in RPE cells, the development of a reliable *in vitro* system would be highly desirable. Investigations on the multifaceted retinoid metabolism in RPE cells *in vitro* require isolated RPE cells in which the complexity of cellular organization is preserved to a high extent and which are viable and metabolically active.[4] In order to meet these requirements of physiological fitness, we optimized the isolation of bovine RPE cells and the *in vitro* incubation conditions by applying a variety of criteria. These criteria included morphology (ultrastructure of the cells), viability (exclusion of viability stains and retention of small cellular proteins), and metabolic activity (energy charge). Here we describe an approach to study retinoid metabolism in isolated bovine RPE cells during short-term *in vitro* incubation.

[1] G. Wald, *Nature (London)* **219,** 800 (1968).

[2] P. S. Bernstein, C. W. Law, and R. R. Rando, *Proc. Natl. Acad. Sci. U.S.A.* **84,** 1849 (1987).

[3] R. R. Rando, J. Canada, P. S. Deigner, and C. W. Law, *Invest. Ophthalmol. Visual Sci.* **30,** 331 (1989).

[4] A. M. M. Timmers, W. J. De Grip, and F. J. M. Daemen, *in* "Proceedings on Retinal Proteins" (N. G. Abdulaev and Y. A. Ovchinnikov, eds.), p. 381. VNU Science Press, Utrecht, 1987.

Isolation of Viable Bovine Retinal Pigment Epithelial Cells

Principle

Isolation of intact RPE cells requires release of cells from their intimate contact with both retina and Brüchs membrane. Strong reduction of the attachment of RPE to retina and Brüchs membrane is achieved by perfusion of the intact bovine eye through the central ophthalmic artery with a divalent cation-free isotonic salt buffer (perfusion buffer; Ca^{2+}, Mg^{2+}-free Hanks-EDTA[5]) kept at 0°.[6] A high recovery of RPE cells which are not contaminated by either red blood cells or rod outer segments is obtained after 12–18 min of perfusion. The yield of $1-2 \times 10^6$ RPE cells per bovine eye represents 20–40% of the total RPE cell population. Furthermore, over 80% of the cells exclude didansylcystine (viability stain),[6] and 85% of the cellular retinol-binding protein (CRBP) is retained in the cells as assayed with the Lipidex 1000 binding assay.[7] The ultrastructure of the isolated cells is very well preserved (Fig. 1). The yield, purity, and integrity of this RPE cell population meet the high standards for *in vitro* studies.

Procedure

Bovine eyes are enucleated 20 min after the death of the animal; the optic nerve is kept at least 2 cm long. The eyes are immediately transported in a light-tight container to the laboratory, where excess fat and muscle are trimmed off under dim red light. The eyes are wrapped in aluminum foil, leaving the optic nerve accessible, and kept in a dark container at 8–10°.

The central ophthalmic artery is cannulated with a blunt 21-gauge needle connected to a reservoir of perfusion buffer (137.8 m*M* NaCl; 5.4 m*M* KCl; 0.3 m*M* Na_2HPO_4; 0.4 m*M* KH_2PO_4; 2 m*M* EDTA; 5.5 m*M* glucose; 10 m*M* HEPES, pH 7.4) positioned 100–120 cm above the eye (Fig. 2). The central ophthalmic artery, which supplies the entire eye and runs along the optic nerve, can be readily identified by its translucent white color and the blood clot at its end. Further differentiation from fat tissue is achieved by pulling it gently with forceps; this ruptures fat tissue but not the artery. The artery is grasped with two small forceps (5SA, Technical Tools, Rotterdam), pulled over the blunt needle, and tied with a suture. The wrapped eye is perfused with ice-cold perfusion buffer for 13–17 min at a flow rate of 0.5–1 ml/min. Routinely, eyes are processed within 2 hr

[5] J. Heller and P. Jones, *Exp. Eye Res.* **30,** 481 (1980).

[6] A. M. M. Timmers, E. A. Dratz, W. J. De Grip, and F. J. M. Daemen, *Invest. Ophthalmol. Visual Sci.* **25,** 1013 (1984).

[7] A. M. M. Timmers, W. A. H. M. van Groningen-Luyben, F. J. M. Daemen, and W. J. De Grip, *J. Lipid Res.* **27,** 979 (1986).

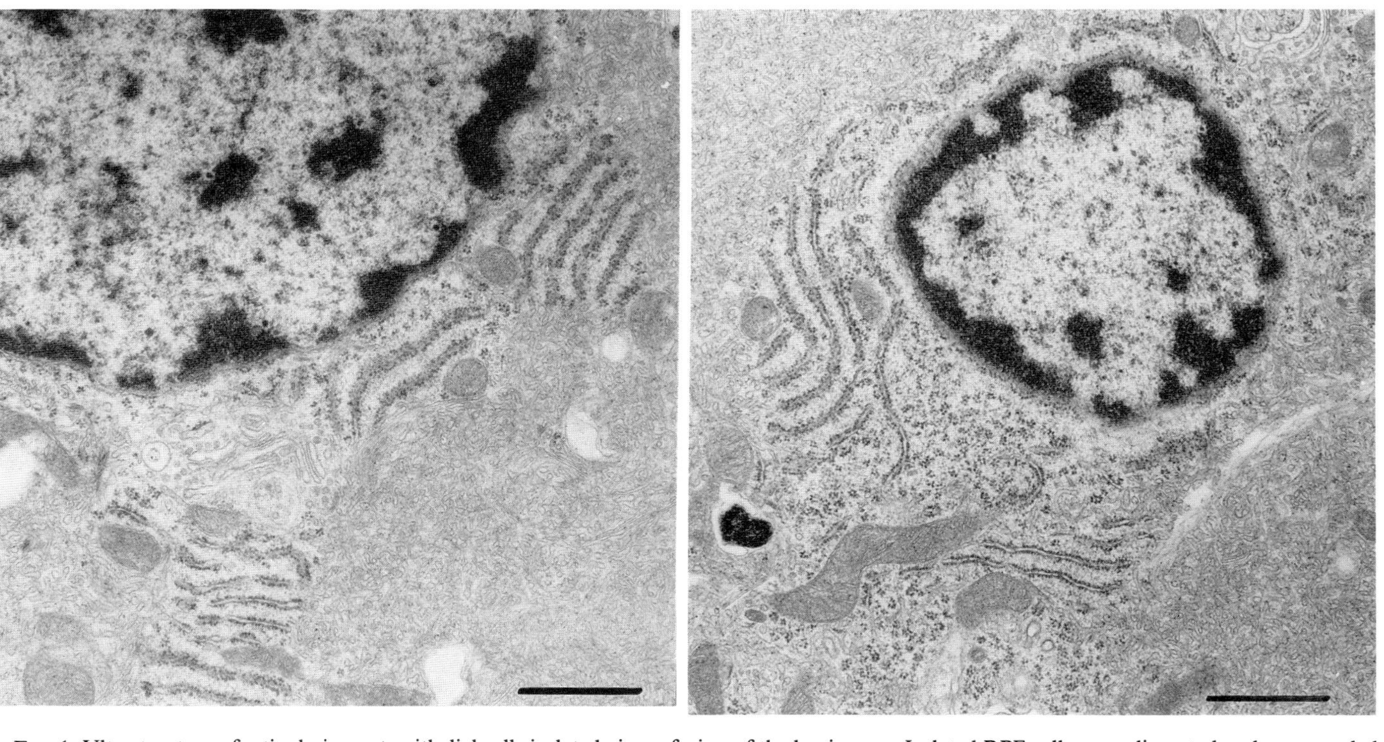

FIG. 1. Ultrastructure of retinal pigment epithelial cells isolated via perfusion of the bovine eye. Isolated RPE cells are sedimented and resuspended for 2 hr in modified Karnovsky fixative (2% paraformaldehyde, 2.5% glutaraldehyde, 0.1 *M* sodium cacodylate, pH 7.4). The cells are embedded in 2% low-temperature gelling agarose, post-fixed with 2% OsO_4 (1 hr), dehydrated, and embedded in Epon 812. Polymerization is carried out at 60°. Ultrathin sections (60–80 nm) are treated with uranyl acetate and lead citrate and examined with a Philips EM 300 or EM 301 transmission microscope. Bar: 1 μm. The mitochondrial ultrastructure, as well as the nuclear membrane and membrane-dense cytoplasm, is highly preserved.

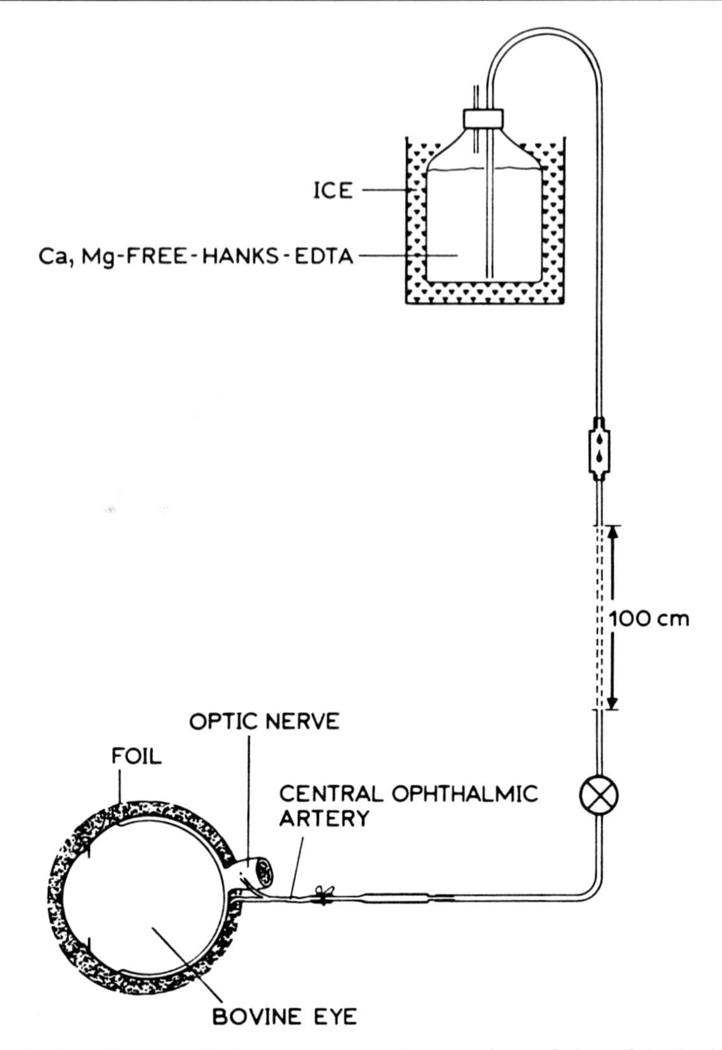

FIG. 2. Semidiagrammatic drawing showing the setup for perfusion of the bovine eye, adapted from a perfusion system for cat eyes (J. M. Thijssen, Department of Ophthalmology, University of Nijmegen, personal communication, 1983).

after the death of the animal. Under these conditions, less than 5% of the eyes fail to perfuse.

The perfusion is ended by disconnecting the artery from the needle, and the anterior part is removed under normal fluorescent light. Special care is taken to remove the retina, which should not slide over the RPE, to avoid considerable loss of RPE cells. This is achieved by carefully detaching the retina at its edge and when it is about to dislodge, the eyecup is turned to allow the retina to release by gravity. Then the retina is cut as close to the optic disc as possible. RPE cells are dislodged from Brüchs membrane using a gentle stream of the appropriate incubation buffer (perfusion buffer, Krebs–Ringer, or RPMI 1640 DM) from a syringe equipped with a hypodermic needle. The cell suspension is centrifuged at 40 *g* for 10 min at 0°. The sedimented cells are washed with a large volume of incubation buffer and resuspended carefully in a small volume. A 50-μl aliquot is taken to determine cell density; the rest is immediately processed for further analysis.

Cell Density. Since isolation of RPE cells via perfusion yields clusters of cells, cell density is determined by means of a DNA assay[8] based on the fluorescence of DAPI (4′,6-diamidino-2-phenylindole-2HCl, Boehringer Mannheim, Mannheim, FRG) complexed with double-stranded DNA.[9] A 50-μl aliquot of cells is centrifuged (10 sec at 8000 *g* at room temperature), and the precipitated cells are lysed in 0.5 ml of 10 m*M* NaCl. DNA is released from the nucleus by sonication (Branson, Danbury, CT, B12 sonifier with microtip, 10 W output) for 2 times for 10 sec each. Following centrifugation (10 sec at 8000 *g* at room temperature), the supernatant is utilized directly in the DNA assay.

Calibration curves (0.5–4.0 μg DNA/ml) of high molecular weight DNA and a dilution series of the RPE extract are prepared in 10 m*M* NaCl. To 50 μl of DNA solution 0.5 ml of DAPI solution is added (25 ng/ml in 10 m*M* Tris-HCl, pH 7.0). Just before use a DAPI working solution is prepared by 1000-fold dilution of stock solution of DAPI in dimethyl sulfoxide, which is stored at −20°. Although the resulting fluorescent complex is stable for 8 hr, fluorescence is routinely measured immediately with a Shimadzu (Kyoto, Japan) Spectrofluorophotometer RF 150 (excitation 360 nm, slit 20 nm; emission 450 nm, slit 40 nm). Cell density is calculated by regression analysis assuming 6 pg DNA/somatic bovine cell.[10] The detection limit is approximately 10,000 cells.

Cell Viability. The 20-fold increase of fluorescence quantum yield of didansylcystine (Sigma, St. Louis, MO) when bound to membranes has been utilized to check the integrity of plasma membranes of photoreceptor cells.[11] Since RPE cells also contain a high cytoplasmic membrane density,[12] we applied didansylcystine as a probe to evaluate the integrity of

[8] P. D. Mier, H. van Rennes, P. E. J. van Erp, and H. Roelfzema, *J. Invest. Dermatol.* **78,** 267 (1982).

[9] J. Kapuscinsky and B. Skoczylas, *Anal. Biochem.* **83,** 252 (1977).

[10] H. A. Sober, "Handbook of Biochemistry." The Chemical Rubber Co., Cleveland, Ohio, 1970.

[11] S. Yoshikami, W. E. Robinson, and W. A. Hagins, *Science* **185,** 1176 (1974).

[12] M. L. Katz, P. J. Farnsworth, K. R. Parker, and E. A. Dratz, unpublished observations (1984).

RPE plasmalemma. An aliquot of 50 μl of RPE cells ($1-2 \times 10^6$ cells/ml) is gently mixed with an equal volume of freshly prepared didansylcystine solution (0.25 mg/ml in perfusion buffer) and examined immediately under a fluorescence microscope (high-pressure mercury lamp excitation light filtered through a Schott, Mainz, FRG, BG 12 filter). Two cell populations are observed: cells that show clear fluorescence above background and cells that do not stain at all. Disruption of the plasmalemma with detergent (0.5% hexadecyltrimethylammonium bromide) or 70% (v/v) ethanol renders 100% of the cells fluorescent within 15 sec.

Short-Term Incubation System for Retinal Pigment Epithelial Cells

Studies on cellular metabolism *in vitro* do not only require viable cells but also require an incubation system in which the cells can be kept metabolically fit for an appropriate time period.[13] Two parameters are applied to evaluate incubation media: (1) retention of CRBP, a small cytoplasmic retinol-binding protein (M_r 14,000), as a criterion for structural viability of the cells, and (2) the energy charge, a measure of chemical energy stores, as an indicator for cellular metabolic viability. The energy charge of a cell is defined as the ratio (ATP + 0.5 ADP)/(ATP + ADP + AMP); in a viable, metabolically active cell it varies between 0.8 and 0.95.[14] The energy charge is calculated from the ribonucleotide content of $5-6 \times 10^6$ RPE cells, extracted with ice-cold $HClO_4$ (final concentration 0.4 *M*) to release the ribonucleotides, which then are separated and quantitated by ion-exchange high-performance liquid chromatography (HPLC) using a ternary elution system.[15]

All utilized media are supplemented with 10 m*M* glutamine and pyruvate. Incubations are carried out at 37° in a humidified atmosphere of 95% O_2/5% CO_2. RPE cells are isolated as described above and incubated at a density of $1-2 \times 10^6$ cells/ml. RPE cells incubated in either RPMI 1640 DM (Flow Labs, Irvine, Scotland) or perfusion buffer (see above) maintain equally well a steady level of CRBP (120 ± 15 pmol/10^6 cells) during at least 8 hr of incubation (data not shown).

The energy charge is measured in RPE cells incubated in RPMI 1640 DM, Krebs–Ringer, or perfusion buffer (Table I). During incubation in perfusion buffer the energy charge plummets within 4 hr, as did the total adenine nucleotide content. On incubation in Krebs–Ringer solution, a slow increase in energy charge is detected to a still suboptimal level, but the adenine nucleotide level actually decreases. On incubation in RPMI 1640

[13] A. M. M. Timmers, Ph.D. Dissertation, University of Nijmegen, 1987.

[14] L. Stryer, "Biochemistry." Freeman, New York, 1988.

[15] R. A. De Abreu, J. M. van Baal, and J. A. J. M. Bakkeren, *J. Chromatogr* **227,** 45 (1982).

TABLE I
METABOLIC FITNESS OF ISOLATED RETINAL PIGMENT EPITHELIAL CELLS DURING SHORT-TERM INCUBATION

Incubation time (hr)	Incubation medium	Metabolic parameters		
		ATP[a]	Total A[a]	Energy charge
0	Perfusion buffer	390	540	0.79
	Krebs–Ringer	370	370	0.58
	RPMI 1640 DM (4)[b]	320 ± 37	600 ± 50	0.69 ± 0.02
2	Perfusion buffer	—	—	—
	Krebs–Ringer	—	—	—
	RPMI 1640 DM	390	720	0.73
4	Perfusion buffer	30	90	0.47
	Krebs–Ringer	260	490	0.66
	RPMI 1640 DM (3)[b]	500 ± 143	730 ± 260	0.82 ± 0.01
8	Perfusion buffer	—	—	—
	Krebs–Ringer	—	—	—
	RPMI 1640 DM (2)[b]	650 ± 90	900 ± 160	0.82 ± 0.01

[a] ATP and total A (total adenine ribonucleotide pool) are expressed as picomoles/10^6 RPE cells.
[b] Values are means ± S.D., with the number of determinations given in parentheses.

DM, a steady increase of both energy charge and total adenine nucleotide level is observed, which levels off to a near steady-state level between 2 and 4 hr of incubation. The cells maintain an energy charge of 0.82 for at least 8 hr. The viability of the cells is further emphasized by the observation that no leakage of ATP or ADP into the incubation media is detected throughout the entire incubation period. Retinoid analysis, performed as described below, indicates that during this period no significant changes in the total retinoid population in the RPMI cells occur, except for a slow and steady reduction in retinol level from about 5 to about 3% of total retinoid.[13] In conclusion, isolated bovine retinal pigment epithelial cells incubated in RPMI 1640 DM meet the requirements for *in vitro* studies, set above.

Quantitative Retinoid Analysis

Chromatographic Separation of Retinoids

A retinoid standard mixture is prepared as described previously[16] and contains retinyl palmitate, 11-*cis*-, 13-*cis*-, all-*trans*-retinaldehyde, and retinol, with *syn*-all-*trans*-retinaloxime added as the internal standard. A

[16] G. W. T. Groenendijk, P. A. A. Jansen, S. L. Bonting, and F. J. M. Daemen, this series, Vol. 67, p. 203.

TABLE II
REPRODUCIBILITY OF GRADIENT CHROMATOGRAPHY AND QUANTITATION OF OCULAR RETINOIDS USING *syn*-all-*trans*-RETINALOXIME AS INTERNAL STANDARD

Retinoid	Quantity[a] (pmol)	Retention time[b] (min)
Retinyl ester	12.1 ± 4%	2.92 ± 0.5%
Retinaldehydes		
13-*cis*-Retinaldehyde	20.8 ± 3%	8.80 ± 0.7%
11-*cis*-Retinaldehyde	7.5 ± 6%	8.90 ± 0.7%
9-*cis*-Retinaldehyde	3.2 ± 5%	10.15 ± 0.9%
all-*trans*-Retinaldehyde	28.1 ± 4%	11.12 ± 0.9%
syn-all-*trans*-Retinaloxime	19	17.67 ± 1.1%
Retinols		
11-*cis*-Retinol	2.2 ± 4%	19.39 ± 2.0%
13-*cis*-Retinol	3.1 ± 3%	20.14 ± 1.1%
all-*trans*-Retinol	9.4 ± 4%	22.86 ± 2.0%

[a] Quantities are expressed as means ± S.E. (%) (per injection), calculated using *syn*-all-*trans*-retinaloxime as the internal standard. Over a period of 2 weeks six analyses were run whereby 15 μl of retinoid standard mixture was injected 6 times.

[b] Retention times are expressed as means ± S.E. (%).

complex gradient elution is developed to allow baseline separation of retinyl esters, retinaldehydes, and retinols in a total analysis time of only 24 min[13] (Table II, Fig. 3). In our setup, an LKB (Bromma, Sweden) 2152 HPLC microprocessor controller commands a LKB 2150 gradient pump at a flow rate of 0.1 ml/min. The elution profile starts with 0.3% dioxane in isooctane for 3 min, then a gradient of dioxane (0.3–2.5%) in isooctane over 5 min, with a steady 2.5% level during the next 4 min, followed by a gradient of dioxane (2.5–7%) from 12 to 16 min and a steady dioxane level of 7% up to 25 min. At the end of each HPLC run the column is equilibrated with 0.3% dioxane in isooctane. Detection is carried out at 328 nm and 0.005 AUFS with a Kratos (Urmston, England) analytical Spectroflow 757 absorbance detector. The detection limit for all-*trans*-retinol is about 50 fmol.

Analysis of Geometric Distribution of Retinyl Ester and Retinaldehyde Fraction

The HPLC procedure described above does not resolve geometric isomers of the retinyl esters; in addition, the 11-cis and 13-cis isomers of retinaldehyde are usually not sufficiently resolved to collect separate radio-

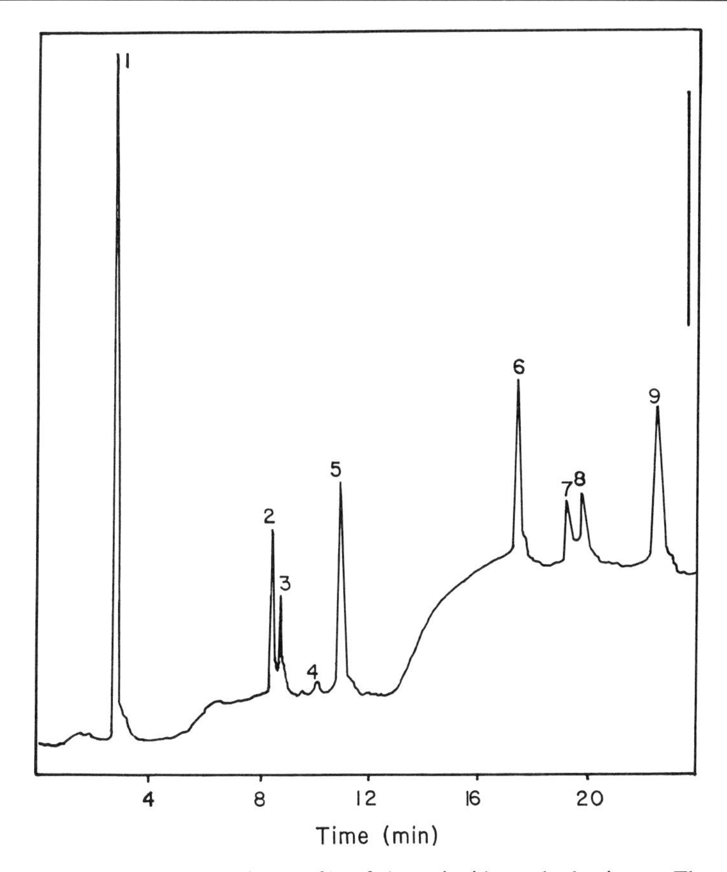

FIG. 3. Gradient HPLC elution profile of the retinoid standard mixture. The elution profile of a standard mixture of retinoids was obtained by straight-phase HPLC in the gradient mode. Column: LiChrosorb Si 60-5 (100 × 3 mm). See text (Quantitative Retinoid Analysis) for details of the gradient profile. The increase in baseline absorption is due to the dioxane gradient. The bar represents 0.0005 absorbance units. Peak 1, retinyl palmitate; peak 2, 13-*cis*-retinaldehyde; peak 3, 11-*cis*-retinaldehyde; peak 4, 9-*cis*-retinaldehyde; peak 5, all-*trans*-retinaldehyde; peak 6, *syn*-all-*trans*-retinaloxime; peak 7, 11-*cis*-retinol; peak 8, 13-*cis*-retinol; peak 9, all-*trans*-retinol. If present, 9-*cis*-retinol would elute at 21.6 min.

labeled fractions. If required, the isomeric composition of the retinaldehyde and retinyl ester fraction can be most simply determined in two steps through conversion by either saponification (retinyl ester)[16] or reduction (retinaldehyde)[17] to the corresponding retinols. The retinols can be analyzed by isocratic HPLC with high sensitivity and with baseline separation of 11-*cis*-, 13-*cis*-, and all-*trans*-retinol.

[17] C. D. B. Bridges, *Exp. Eye Res.* **24**, 571 (1976).

Isomeric Composition of Retinyl Ester Fraction

The retinyl ester fraction is separated from other retinoids by semipreparative straight-phase HPLC (LiChrosorb Si 60-5; 250 × 9 mm) with hexane/diethyl ether (50:50, v/v) as the eluent at a flow rate of 3 ml/min and detection at 328 nm. The retinyl esters elute at 4 min. After evaporation of the eluant, 1.5 ml of 6% KOH (w/v) in methanol is added to the retinyl ester residue. The esters are saponified by a 20-min incubation at room temperature. Then, 2 ml hexane and 1 ml double-distilled water are added, and, after vigorous mixing, the two phases are separated by centrifugation (1–2 min; 7000 *g* at 4°). The organic upper layer is collected, and the lower layer is extracted with 1 ml hexane. Following evaporation of the combined upper layers with a stream of nitrogen, the residue is dissolved in dioxane/hexane (5:95), and the isomeric composition of the resulting retinols is determined by HPLC in an isocratic mode with dioxane/hexane (5:95, v/v).[16] A steady level of 10–15% of 13-cis isomers is detected consistently; this represents a saponification artifact.

Isomeric Composition of Retinaldehyde Fraction

The retinaldehyde fraction can be separated from other retinoids by semipreparative straight-phase HPLC as above. The retinaldehydes elute between 6 and 8 min. Evaporation of the solvent is followed by reduction of aldehyde with excess sodium borohydride in methanol for 3–5 min at room temperature.[17] Retinols are extracted and processed as described for the retinyl esters. Again 13-cis isomers are detected in tissue extracts, but in this case they probably arise as a side effect of the previous extraction, as control experiments with all-*trans*- or 11-*cis*-retinaldehyde do not present any evidence for artificial isomerization during reduction.

Quantitation of Retinoids

syn-all-*trans*-Retinaloxime is selected as an internal standard[13,16] to quantitate the various retinoid species by compensating for experimental variations. The following features demonstrate that *syn*-all-*trans*-retinaloxime, a derivative of all-*trans*-retinal,[16] has excellent credentials for this job: (1) it has physicochemical characteristics that are comparable to the retinoids under investigation, (2) it is geometrically stable under the conditions used for extraction and analysis,[16] and (3) it does not occur naturally in RPE cells. Furthermore, (4) *syn*-all-*trans*-retinaloxime does not interfere with the elution profile of the other retinoids of interest (see Fig. 3), and (5) it can be easily detected at 328 nm.

The total retinoid content in the various peaks is determined via peak integration utilizing a Hewlett Packard (Palo Alto, CA) A/D converter on line with the detector and a modular HP3353 system. The molar absorbance of the various retinoids at 328 nm,[16] in combination with the recovery of the internal standard *syn*-all-*trans*-retinaloxime, then allows quantitation of the retinoid content of the original extract. The amount of labeled component in the various retinoid fractions is quantitated using 20-sec eluate fractions collected in a scintillation vial. Each fraction is mixed with 4 ml of scintillation fluid (Aqua Luma Plus, Lumac:3M BV, Schaesberg, The Netherlands) and counted in a liquid scintillation analyzer (UTPM Tricarb 4000). Counts per minute (cpm) are converted to picomoles using the measured specific activity of the all-*trans*-retinol applied in that particular experiment.

Extraction and Analysis of Cellular Retinoids

Investigation of retinoid metabolism in pigment epithelial cells requires not only reliable separation and quantification of retinoids but certainly also equivalent extraction of the entire retinoid population with preservation of its geometric distribution. An extraction procedure based on dichloromethane/methanol/water (1 : 1 : 1, v/v) has been developed, which will result in quantitative recovery of retinol and retinyl ester with little risk of isomerization.[16] This system, however, yields only partial recovery of protein-bound retinaldehydes,[16] but to our knowledge no equivalent simple alternative is available.

All manipulations are carried out under dim red light to exclude photoisomerization of retinoids. A volume of 0.5 ml of cell suspension (routinely $1-2 \times 10^6$ RPE cells) in a 10-ml stoppered glass tube is mixed with 5 μl of 6% Ammonyx LO (Fluka Chemie AG, Buchs, Switzerland).[13] Subsequently, internal standard is added (0.5 nmol of *syn*-all-*trans*-retinaloxime in 100 μl isooctane), followed by 1.5 ml methanol. The mixture is kept on ice for 10–15 min followed by addition of 1.5 ml dichloromethane. After each addition, the tube contents are vigorously mixed (10–20 sec on a vortex mixer). At this stage a stable single phase should be obtained; otherwise, an additional 200 μl of methanol should be added.[13] Finally, 1.0 ml water is added to induce phase separation, and, after 20 sec of mixing, the organic and aqueous phases are completely separated by centrifugation (10 min, 5000 *g* at 4°). The lower organic phase is collected, and the upper water–methanol layer is back-extracted with 1.5 ml dichloromethane.

To the pooled dichloromethane layers is added 100 μl of the volatile organic base triethylamine,[13] and the organic solvent is evaporated by a stream of nitrogen. The residue is dissolved first in 10 μl of dioxane. Then

10 μl of isooctane is added, followed by addition of 180 μl isooctane. After each addition the tube contents are thoroughly mixed. The samples are either directly analyzed by HPLC or stored at $-20°$ under argon for at most 2 weeks. Cellular retinoids are identified by coinjection of the standard mixture. Supplementation of Ammonyx in the retinoid extraction reduces adherence of retinoids to glass[7] and, in particular, enhances the recovery of 11-*cis*-retinol in our hands. Even after more than 100 HPLC runs, no effect of the detergent on column performance could be observed. The addition of triethylamine appears to be essential in reducing the relative large variation in the recovery of retinaldehydes originally observed (~20%). The slightly alkaline environment prohibits formation of stable protonated Schiff bases of retinaldehyde with amino compounds such as phospholipids. The gradual decrease of the polarity of the solvent, when dissolving the final cellular retinoid extract, in our hands reduces the variation in overall recovery of retinoids. Probably, this protocol more efficiently dissolves polar coextracted components, which otherwise might act as a low solubility "trap" to retard solubilization of retinoids.

The reproducibility of the extraction and quantitation procedure is checked by analyzing a dilution series of RPE cells ($0.3-5 \times 10^6$ cells/extraction tube) to which a fixed amount of *syn*-all-*trans*-retinyloxime is added prior to extraction. Recovery of *syn*-all-*trans*-retinyloxime is a constant $98 \pm 3\%$. Table III shows that the retinoid contents calculated from the various cell dilutions are sufficiently reproducible to allow quantitative analysis. Routinely, extraction for retinoid analysis should be performed on $1-2 \times 10^6$ RPE cells.

TABLE III
REPRODUCIBILITY OF RETINOID EXTRACTION AND QUANTITATION USING *syn*-all-*trans*-RETINALOXIME AS INTERNAL STANDARD

Number of RPE cells per extraction	Retinyl ester[a]	Retinaldehyde[a]		Retinol[a]	
		11-*cis*	all-*trans*	11-*cis*	all-*trans*
5.3×10^6	1603	93	21	52	47
2.6×10^6	1270	77	28	38	32
1.3×10^6	1862	103	49	52	53
0.7×10^6	1794	100	42	47	38
0.3×10^6	1473	83	50	67	38
Mean ± S.E.	1600 ± 110	91 ± 5	38 ± 5	51 ± 5	42 ± 4

[a] Retinoids are expressed as picomoles/10^6 RPE cells.

In Vitro Supply of all-*trans*-[^{3}H]Retinol to Retinal Pigment Epithelial Cells

Retinoid Administration

Illumination of photoreceptor cells results in a flux of all-*trans*-retinol into RPE cells.[18] In order to study retinoid metabolism in RPE relevant to the visual cycle, we administered ^{3}H-labeled all-*trans*-retinol to the isolated RPE cells. However, the very poor solubility of retinol in aqueous media results in aggregation and adherence to whatever surfaces are present.[7] Furthermore, higher vitamin A concentrations become lytic for membranes.[19] These features therefore demand the use of a carrier. Unfortunately, the putative "natural" carrier in the interretinal space, interphotoreceptor matrix retinoid-binding protein (IRBP), is not easily available and is rather unstable. Instead, we opted for an aselective system, phosphatidylcholine vesicles, which have been demonstrated to be reliable and stable retinol carriers, with a large capacity and high transfer rate.[20-22] Furthermore, the geometric configurations of retinoids appeared to be very stable in this system, and vesicles are easily prepared and loaded with all-*trans*-retinol.[13,20] In fact, this system, which performs through passive transfer of retinol, might not be less "natural" than IRBP, for which so far no specific receptor has been demonstrated in RPE.

Preparation of Retinol–Phosphatidylcholine Carrier Vesicles

Soy lecithin (4.5 mg) is dried thoroughly under a stream of nitrogen followed by exposure to 30 min of high vacuum. The lipid residue is resuspended in 0.6 ml RPMI 1640 DM, and liposomes are generated by shaking vigorously. Subsequently, sonication (Branson B12 sonifier with microtip, at full power) for 10–15 min on ice converts the liposome suspension to an opalescent dispersion of vesicles. The vesicles are prepared 1 day before use and kept at 4° under nitrogen.

Before incorporation into phosphatidylcholine vesicles, the all-*trans*-[^{3}H]retinol must be purified from contaminating 13-*cis*-retinol. To prevent photoisomerization all subsequent manipulations should be carried out under red light. ^{3}H-Labeled retinol (all-*trans*-[11,12-$^3H_2$]retinol, 75 μCi, 55 Ci/mmol; Amersham, Amersham, England) is mixed with the desired amount of unlabeled retinol in ethanol. Following evaporation of the

[18] J. E. Dowling, *Nature (London)* **188,** 114 (1960).
[19] J. T. Dingle and J. A. Lucy, *Biol. Rev.* **40,** 422 (1965).
[20] G. W. T. Groenendijk, W. J. De Grip, and F. J. M. Daemen, *Vision Res.* **24,** 1623 (1984).
[21] R. R. Rando and F. W. Bangerter, *Biochem. Biophys. Res. Commun.* **104,** 430 (1982).
[22] S. Yoshikami and G. N. Nöll, this series, Vol. 81, p. 447.

ethanol, the residue is dissolved in hexane/diethyl ether (50 : 50, v/v), and all-*trans*-retinol is purified by preparative HPLC as described above (see isomeric composition of retinyl ester and retinaldehyde fractions). The all-trans peak, which elutes after 25 min, is collected, and the residue remaining after evaporation of the solvent with nitrogen is dissolved in 5 μl of ethanol. To this solution 400 μl of vesicle dispersion is added, and the all-*trans*-retinol is incorporated into the vesicles by carefully but thoroughly mixing under nitrogen for several minutes. Out of this carrier dispersion 30-μl aliquots are transferred to and gently mixed with 0.8 ml of either a preincubated RPE cell suspensions or control incubations in which no cells are present.

Incubation of Retinal Pigment Epithelial Cells with all-trans-[^{3}H]Retinol

These studies as well should be performed in darkness or under red light to prevent photoisomerization of retinoids. Prior to administration of ^{3}H-labeled all-*trans*-retinol in phosphatidylcholine vesicles, the isolated RPE cells ($1-2 \times 10^6$ cells/0.8 ml) should be preincubated in the dark for 2 hr in RPMI 1640 DM at 37° under 95% CO_2/5% O_2 to allow recovery from the isolation shock and to restore the energy charge of the cells. Subsequently, 30 μl of retinol–carrier suspension is added, and, after various time intervals, incubation mixtures are harvested and the RPE cells rapidly sedimented (900 *g;* 3 min; 0°). A 0.5-ml aliquot of the supernatant is set aside for retinoid analysis. Incubation wells and cell pellets are washed once with 1 ml of ice-cold RPMI 1640 DM. The final cell pellet is resuspended in 0.5 ml of ice-cold RPMI 1640 DM. In order to determine the radiolabel recovery, 50-μl aliquots of incubation medium and cell suspension are analyzed by liquid scintillation counting. The remainder is immediately frozen and stored at −20° in the dark for retinoid extraction and analysis; preferably, this is performed within 24 hr.

Retinoid Metabolism in Retinal Pigment Epithelial Cells *in Vitro*

The following results are presented briefly in order to illustrate the potential of the system outlined above. A supply of all-*trans*-[^{3}H]retinol, ranging from 1.1 to 5.8 nmol per 10^6 cells, is administered to bovine RPE cells isolated and incubated as described above. These levels represent a retinol challenge equivalent to a 10–60% bleach of the visual pigment rhodopsin *in vivo*. However, *in vivo* the actual retinol concentration would be much higher owing to the 10- to 100-fold lower ratio of extracellular fluid to cell volume. Hence, this dosage reflects a lower physiological range.

Transfer of all-*trans*-[^{3}H]retinol from the carrier vesicles to the RPE cells occurs rapidly, and after 2 hr the transfer amounts to approximately 30% of the label. No time-dependent loss of label is observed. Surprisingly, the kinetics of this uptake of all-*trans*-[^{3}H]retinol show a first-order dependence on the administered retinol concentration. This raises the question whether uptake is due to retinol transfer (vesicle *to* cell) or to ingestion of entire vesicles (vesicle *in* cell). By marking the vesicles with the nonexchangeable label [^{14}C]cholesteryl oleate,[20] we demonstrated that at least 75% of the uptake of retinol by the cells is due to transfer of retinol from vesicle to cell.

Analysis of the distribution of the internalized label over various retinoid classes is carried out after various incubation times. Since all doses produce a similar trend, we restrict ourselves here to the largest dose. The internalized all-*trans*-retinol is rapidly converted to all-*trans*-retinyl ester (major metabolite). The maximal rates for uptake and acylation calculated under our conditions are close (58 and 54 pmol/min/10^6 RPE cells, respectively), indicating that under these conditions uptake is still rate-limiting. In addition to acylation, a slow but constant oxidation to all-*trans*-retinaldehyde is observed.

Simultaneously with this nonisomerizing metabolism, approximately 10% of the internalized label enters the isomerizing route after 2 hr of incubation and is recovered as 11-*cis*-retinol, 11-*cis*-retinaldehyde.[23] The maximal isomerization rate measured under our conditions is 1–2 pmol/min/10^6 RPE cells.

Our data confirm recent findings[2] that retinal pigment epithelial cells contain the necessary machinery to effect isomerization of all-*trans*-retinoids to 11-*cis*-retinoids. To our knowledge this is the first time that the full scale of metabolic processes (isomerization, acylation, and oxidation) has been observed in a single incubation system which utilizes isolated cells instead of subcellular extracts. The data presented emphasize the unique potential of isolated RPE cells in short-term incubation studies for the investigation of retinoid pathways in this cell layer. The isolated cells are highly viable and metabolically active and immediately accessible for extraction and analysis. In addition, *in vitro* incubation renders them easily accessible to experimental modulation, for example, to study regulatory processes. The described *in vitro* incubation system for RPE cells presents a promising approach for future exploration of the visual cycle.

[23] Of the potential *cis*-retinoids, the 9-cis isomers were never detected, but 13-cis isomers were identified in varying amounts. The formation of 13-cis isomers reflected a nonspecific effect, however, as (1) accumulation of 13-*cis*-retinol is observed in control incubations in the absence of RPE cells (under these conditions no 11-*cis*-retinol was detectable); (2) conversion of all-trans to 13-cis isomers is catalyzed by amino compounds such as aminophospholipids in membranes[24] (this mechanism does not produce 11-cis isomers); and (3) probably owing to the same mechanism, extraction of retinoids from tissues tends to increase 13-cis levels and lower 11-cis levels. [G. W. T. Groenendijk, C. W. M. Jacobs, S. L. Bonting, and F. J. M. Daemen, *Eur. J. Biochem.* **106,** 119 (1980)].

[2] Vitamin A-Mediated Regulation of Keratinocyte Differentiation

By GEORGE J. GIUDICE and ELAINE V. FUCHS

Introduction

The normal structure of the epidermis is the result of a highly regulated process of keratinocyte growth and differentiation. The four morphologically and biochemically distinguishable layers of the epidermis—basal, spinous, granular, and cornified—correspond to successive stages of the programmed cell death. The basal layer consists of actively dividing cells which are firmly anchored to the underlying stroma via a plasma membrane-associated organelle known as the hemidesmosome. This dermal–epidermal association is necessary for maintenance of the normal structure and function of basal cells. The major proteins produced by the basal layer are keratins K5 and K14, which compose a dense cytoplasmic network of intermediate filaments in these cells.[1-3]

The movement of a cell from the basal into the spinous layer signals the onset of the second stage of differentiation. The cell loses its proliferative potential and shifts from a cuboidal shape with a relatively smooth surface to a polygonal shape containing fine, spinelike processes which interdigitate with the processes of adjacent cells. The composition of intermediate filaments of spinous cells changes with the induced synthesis of keratins K1, K2, K10, and K11.[1] As cells move outward through the spinous layer, there is an increase in the density of keratin filaments and in filament bundling.

Cells in the granular layer are characterized by the appearance of intracellular keratohyaline granules, which are made up, in part, of profilaggrin, a histidine-rich, phosphorylated precursor of filaggrin. Filaggrin is a protein which may be involved in the further aggregation of keratin filaments into bundles called macrofibrils. Cells in this layer also generate two other differentiation-specific proteins: transglutaminase, a calcium-activated enzyme which catalyzes the formation of ϵ-(γ-glutamyl)lysine cross-links in proteins deposited on the inner membrane surface, and involucrin, one of the primary substrates of transglutaminase and a component of the submembranous marginal band.

As the epidermal cell enters the terminal stage of keratinization, (1) all

[1] E. Fuchs and H. Green, *Cell (Cambridge, Mass.)* **19,** 1033 (1980).
[2] R. Moll, W. W. Franke, and D. L. Schiller, *Cell (Cambridge, Mass.)* **31,** 11 (1982).
[3] W. Nelson and T.-T. Sun, *J. Cell Biol.* **97,** 244 (1983).

cellular organelles are lost through enzymatic degradation, (2) the contents of the lamellar granules are secreted into the extracellular region, (3) the cell interior becomes densely packed with thick bundles of keratin filaments embedded in an amorphous matrix, and (4) the transglutaminase-catalyzed cross-linked envelope is completed, forming a sacculus for the keratin macrofibrils. The final product of the process of keratinization is a tough and resilient protective barrier, called the stratum corneum, which also functions in fluid and electrolyte homeostasis.

Role of Retinoids in Epithelial Differentiation

The first indication that retinoids may play a role in the control of epithelial differentiation came from observations of the effects of vitamin A deficiency in humans[4] and experimental animals.[5,6] It was demonstrated that the early effects of vitamin A deficiency included the eruption of cutaneous lesions characterized by hyperkeratosis and the transformation of columnar and transitional epithelia into keratinized stratified squamous epithelia. Administration of vitamin A reversed these epithelial changes and resulted in normalization of the affected tissues.[7,8] Vitamin A deficiency was also found to promote epithelial tumorigenesis,[9] although in this case the process seemed to be irreversible.

Early experiments involving the use of cultured cells to study vitamin A-mediated regulation of epithelial differentiation demonstrated that excess vitamin A inhibited keratinization in cultured chick ectoderm, transforming the tissue into a mucus-secreting epithelium.[10] Conversely, removal of vitamin A from the medium of cultured human keratinocytes, by delipidization of the serum, resulted in the induction of terminal differentiation.[11] Since these early studies, *in vitro* culture systems have been used extensively to characterize the morphological and biochemical changes associated with vitamin A regulation of keratinization. In this chapter, we present the two major systems for culturing human epidermal cells and describe some of the analytical tools which have been used to study the

[4] C. E. Bloch, *J. Hyg.* **19,** 283 (1921).
[5] S. Mori, *Bull. Johns Hopkins Hosp.* **33,** 357 (1922).
[6] S. B. Wolbach and P. R. Howe, *J. Exp. Med.* **43,** 753 (1925).
[7] C. N. Frazier and C. K. Hu, *Arch. Intern. Med.* **48,** 507 (1931).
[8] S. B. Wolbach and P. R. Howe, *J. Exp. Med.* **57,** 511 (1933).
[9] D. Burk and R. J. Winzler, *Vitam. Horm.* **2,** 305 (1944).
[10] H. B. Fell and E. Mellanby, *J. Physiol. (London)* **119,** 470 (1953).
[11] E. Fuchs and H. Green, *Cell (Cambridge, Mass.)* **25,** 617 (1981).

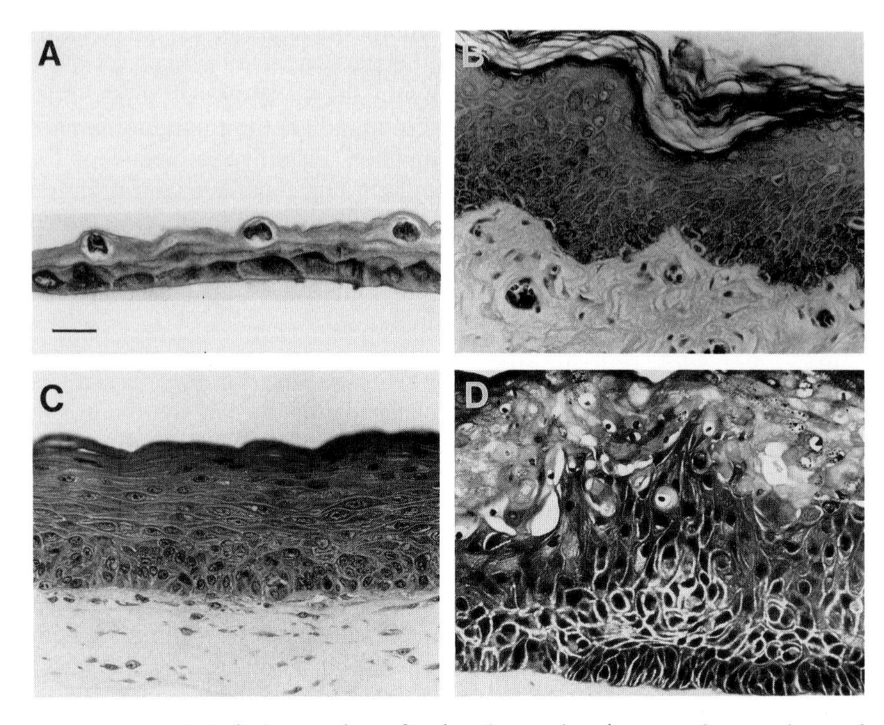

FIG. 1. Morphological comparison of various human keratinocyte cultures and normal human epidermis. All cells were fixed in Carnoy's solution, paraffin-sectioned, and stained with hematoxylin and eosin. (A) Human foreskin keratinocytes grown on a 3T3 feeder layer in a submerged culture system, (B) normal human trunk skin, (C) human foreskin keratinocytes grown at the air–liquid interface, (D) SCC-13 cells grown at the air–liquid interface. Bar: 30 μm. [(A, B, and C) From R. Kopan, G. Traska, and E. Fuchs, *J. Cell Biol.* **105,** 427 (1987). (D) From A. Stoler, R. Kopan, M. Duvic, and E. Fuchs, *J.Cell Biol.* **107,** 427 (1988).]

biosynthetic changes that take place during induction or inhibition of keratinization.

Submerged Keratinocyte Culture

By using a feeder layer of growth-arrested 3T3 mouse fibroblasts, Rheinwald and Green[12] vastly improved methods to serially cultivate both normal human epidermal keratinocytes and keratinocytes from squamous cell carcinomas of the skin. In the presence of feeder cells, keratinocytes attach to the culture plate, grow, and form stratified colonies two to three cell layers thick (Fig. 1A). Although enucleate squames and cornified

[12] J. G. Rheinwald and H. Green, *Cell (Cambridge, Mass.)* **6,** 331 (1975).

envelopes form in the upper layers, these cultures do not produce a gradient of differentiating cell morphologies (compare with human skin, Fig. 1B), nor do they produce the differentiation-specific keratins K1, K2, K10, and K11.[1]

Using both morphological and biochemical criteria, it has been demonstrated that terminal differentiation can be induced in these cultures by the depletion of vitamin A from the medium.[11] Under vitamin A-deficient conditions, stratification and cell adhesiveness increase, cell motility is reduced, and granular cells and squames are produced. Readdition of retinoids suppresses these differentiation-related events. Interestingly, SCC-13, a keratinocyte line cloned from a squamous cell carcinoma of the skin,[13] is even more sensitive to retinoids than its normal epidermal counterpart.[14]

Culture Medium and Solutions

The growth medium for all cultures described here consists of a 3 : 1 mixture of Dulbecco's modified Eagle's medium (DMEM) and Ham's nutrient mixture F12.[12,13] For growth of 3T3 fibroblasts, this medium is supplemented with 10% newborn calf serum. Human keratinocytes are grown in medium supplemented with the following: 15–20% lot-tested fetal calf serum (Sterile Systems, Logan, Utah), 1×10^{-10} *M* cholera toxin, 0.4 μg/ml hydrocortisone, 5 μg/ml insulin, 5 μg/ml transferrin, 2×10^{-11} *M* triiodothyronine, and 5 ng/ml epidermal growth factor (EGF).[12,13]

Preparation of Vitamin A-Depleted Medium

The only significant source of retinoids in vitamin A-depleted culture medium is the fetal calf serum. To remove these retinoids, the fetal calf serum is delipidized according to the procedure of Rothblat *et al.*[15] Briefly, 100 ml of serum is extracted with 1 liter of a 1 : 1 mixture (v/v) of acetone and ethanol at 4° for 4 hr. The precipitate is washed 2 times with 500 ml of cold ether on a Whatman #1 filter in a large Büchner funnel such that the precipitate does not dry out until after the second ether wash. Most of the residual ether is removed by flushing the precipitate with a stream of nitrogen gas for 15 to 30 min. The protein residue is then (1) placed under reduced pressure overnight, (2) ground to a powder using a mortar and pestle, and (3) placed under reduced pressure again (4–12 hr) to eliminate any trace of ether. The final residue is reconstituted in an appropriate

[13] J. G. Rheinwald and M. A. Beckett, *Cancer Res.* **41,** 1657 (1981).

[14] K. H. Kim, F. Schwartz, and E. Fuchs, *Proc. Natl. Acad. Sci. U.S.A.* **81,** 4280 (1984).

[15] G. H. Rothblat, L. Y. Arborgast, L. Ouellet, and B. V. Howard, *In Vitro* **12,** 554 (1976).

volume of phosphate-buffered saline (PBS) such that the original serum protein concentration is restored, as judged by Lowry protein analysis.[11]

Preparation of Fibroblast Feeder Layer

Optimal growth of human epidermal cells is dependent on exogenous dermal factors, which can be supplied by 3T3 mouse fibroblasts.[12] Clones of 3T3 cells to be used as feeder cells must (1) form a contact-inhibited monolayer at saturation density, (2) survive mitomycin C treatment without detaching from the plate, and (3) promote keratinocyte attachment and growth. To prepare growth-arrested 3T3 fibroblasts for use as feeder cells, mitomycin C (8 μg/ml) is added to the medium of a confluent 3T3 culture for 2 hr at 37°. After washing the cells 3 times with PBS, they may be incubated at 37° in fibroblast culture medium until needed. To prepare feeder layers, a confluent culture of growth-arrested 3T3 cells is trypsinized and replated at about one-third saturation density (1.6×10^4 cells/cm^2).

Plating and Growth of Human Keratinocytes on Fibroblast Feeder Layer

Immediately after surgical removal, a neonatal foreskin is washed extensively with serum-free medium to reduce the chance of contamination. Under asceptic conditions, the excess subcutaneous fat and dermis is trimmed from the tissue. The epidermal layer is finely minced with scissors and disaggregated by incubation at 37° for 40 min with constant stirring in 12 ml of a 1:1 mixture (v/v) of 0.25% trypsin in PBS and versene. Large tissue fragments are allowed to settle, and the cell suspension, containing mostly fibroblasts, is removed with a sterile Pasteur pipette and discarded. A fresh aliquot of enzyme solution is added to the remaining tissue for another 40-min incubation. The released cells (mostly keratinocytes) are centrifuged, resuspended in serum-containing epidermal culture medium (minus EGF), and plated onto fibroblast feeder layers at a concentration of $2-6 \times 10^3$ cells/cm^2.[12,13] This trypsinization/plating process is then repeated 3 times. EGF is left out of the plating medium because it has been shown to interfere with keratinocyte colony formation.[16] The cultures are not disturbed for 4 days during which time the keratinocytes attach to the plastic. At the time of the first medium change (on the fifth day of culture), and at subsequent feedings, EGF is added to the cultures at a concentration of 5 ng/ml.[16] These culture conditions are optimized for the growth of epidermal cells,and any contaminating dermal fibroblasts are largely contact-inhibited by the feeder cells. Feeder cells can be selectively removed

[16] J. G. Rheinwald and H. Green, *Nature (London)* **265,** 421 (1977).

from the culture dish by treatment with versene. Pure epidermal cells can then be removed by treatment with a 1 : 1 mixture (v/v) of 0.1% trypsin and versene. Cells can be stored in 20% glycerol under liquid nitrogen until further use. Cultures can be thawed and passaged serially through several hundred cell generations.

Culturing of Keratinocyte at the Air–Liquid Interface

To optimize the system for stratification and terminal differentiation rather than growth, keratinocytes should be cultured at the air–liquid interface.[17-21] Unlike the submerged culture system, this floating culture system does not require the use of vitamin A-depleted medium to induce the differentiation process (Fig. 1C). Vitamin A-mediated suppression of differentiation still occurs under these conditions, but at a higher retinoid concentration.[21] Using these culture methods, SCC-13 cells also stratify and differentiate (see Fig. 1D);[22] however, the differentiation program is clearly abnormal and is completely suppressed by retinoids.[23]

The following protocol for culturing keratinocytes at the air–liquid interface is essentially as described by Asselineau *et al.*[20] Type I collagen (Seikagaku America, Inc., St. Petersburg, FL) is combined with keratinocyte growth medium as described by the manufacturer. A confluent culture of 3T3 fibroblasts is trypsinized and washed once in growth medium. The dissociated cell suspension is pelleted and resuspended in the collagen solution at 4° at a cell density of 1.5×10^5/ml. The cell suspension is pipetted into 35-mm culture plates (2 ml/plate) and allowed to gel by incubating at 37° for 2–3 hr. Gelled lattices are stored submerged in growth medium at 37° until ready for use (usually 12–72 hr after gelation). Keratinocytes are applied to the surface of the submerged collagen lattice at a density of $2-6 \times 10^3$ cells/cm^2 and cultured submerged for 7 days. At this point, the lattice is lifted with a spatula and placed onto a stainless steel grid whose edges have been bent such that the culture is suspended in the dish. Growth medium is added until the undersurface of the grid is in contact with the medium, and nutrients are fed to the epidermis by diffusion through the artificial dermis.[20,21]

[17] M. A. Karasek and M. E. Charlton, *J. Invest. Dermatol.* **56,** 205 (1971).
[18] J. H. Lillie, D. K. MacCallum, and A. Jepsen, *Exp. Cell Res.* **125,** 153 (1980).
[19] J. Yang and S. Nandi, *Int. Rev. Cytol.* **81,** 249 (1983).
[20] D. Asselineau, B. Bernhard, C. Bailly, and M. Darmon, *Exp. Cell Res.* **159,** 536 (1985).
[21] R. Kopan, G. Traska, and E. Fuchs, *J. Cell Biol.* **105,** 427 (1987).
[22] A. Stoler, R. Kopan, M. Duvic, and E. Fuchs, *J. Cell Biol.* **107,** 427 (1988).
[23] R. Kopan and E. Fuchs, *J.Cell Biol.* **109,** 295 (1989).

Analysis of Vitamin A-Regulated Shifts in Keratin Expression

When grown as submerged cultures, epidermal and SCC-13 cells synthesize keratins K5 and K14, characteristic of basal epidermal cells, as well as keratins K6, K16, and K17, typically associated with abnormal differentiation in the suprabasal cells of psoriatic skin and squamous cell carcinomas.[22] Keratinization, induced either by using vitamin A-depleted medium in the submerged culture system[11] or by growing the keratinocytes at the air–liquid interface using normal growth medium,[17–21] results in the suprabasal expression of the terminal differentiation-specific keratins.[11,20,21] Addition of retinoids to the growth medium causes an increase in cell proliferation, the disappearance of most morphological features of terminal differentiation, and the corresponding inhibition of expression of all differentiation-associated keratins.[23,24] The techniques which have been used to analyze these vitamin A-mediated changes are described below.

Immunolocalization of Keratin Proteins

Production of Monospecific Antikeratin Antibodies. The two major technical problems associated with the immunolocalization of keratins are (1) cross-reactivity of antikeratin antibodies with more than one member of the keratin protein family and (2) masking of antigenic determinants recognized by antikeratin antibodies.[25] Because of these problems, it has been extremely difficult to obtain reliable localization profiles for specific keratins. One strategy designed to minimize these problems involves the production of polyclonal antisera against synthetic peptides corresponding to those portions of keratin polypeptides which (1) have amino acid sequences that are highly divergent from other keratins and (2) are present on the surface of keratin filaments, thereby reducing the chance that the antigenic determinants will be masked. Using this strategy, a number of monospecific antikeratin antisera have been prepared against the carboxy-terminal regions of various keratins.[22,26]

Light Microscopic Localization of Keratin Proteins Using Immunogold–Silver Enhancement Technique. For use with monoclonal antibodies, freshly isolated tissue should be frozen in isopentane at −120°. If polyclonal antisera are to be used, tissues can often be fixed in 4% paraformaldehyde and embedded in paraffin. Five-micrometer tissue sections are affixed to glass slides, rehydrated, and treated for 30 min at room temperature with 2% bovine serum albumin (BSA) in PBS to reduce

[24] R. L. Eckert and H. Green, *Proc. Natl. Acad. Sci. U.S.A.* **81,** 4321 (1984).

[25] R. Eichner, P. Bonitz, and T.-T. Sun, *J. Cell Biol.* **98,** 1388 (1984).

[26] D. R. Roop, H. Huitfeldt, A. Kilkenny, and S. H. Yuspa, *Differentiation* **35,** 143 (1987).

background staining. An antikeratin antibody, diluted as necessary in BSA–PBS, is incubated with the sections for 1 hr at room temperature. After three 10-min washes in PBS, sections are incubated overnight at room temperature with gold-conjugated secondary antibodies (15-nm gold particles; Janssen Life Science Products, Piscataway, NJ) diluted in BSA–PBS. Sections are then washed in PBS (6 times, 10 min each), fixed in 2% glutaraldehyde–PBS (v/v) for 15 min, and washed again, first in PBS (3 times, 10 min each) and finally in glass distilled water (3 times, 10 min each). The gold label is silver-enhanced using the IntenSE silver enhancement kit (Janssen Life Science Products) according to the manufacturer's instructions. Figure 2A shows a section of a paraformaldehyde-fixed SCC-13 raft culture stained with a monospecific polyclonal peptide antiserum to the carboxy terminus of human K6.[23]

Combining Autoradiography and Immunohistochemistry

Immunogold localization can also be performed in combination with autoradiographic analysis. This combined approach was recently utilized[23] to investigate the relation between keratinocyte proliferation and the expression of the keratin pair K6/K16, proteins which had previously been shown to be associated with skin diseases involving abnormal differentiation and hyperproliferation.[22,27] Exposure of rapidly proliferating SCC-13 raft cultures to retinoic acid (1×10^{-6} *M*) resulted in the inhibition of synthesis of keratins K6 and K16 but, surprisingly, was not accompanied by a decrease in the rate of incorporation of [^{3}H]thymidine.[23] These findings demonstrate that, at least in one case, the processes of K6/K16 expression and hyperproliferation can be uncoupled.

In this investigation, keratinocytes were cultured at the air–liquid interface. For the final 2 hr of culture, [^{3}H]thymidine (2 μCi/ml, >90 Ci/mmol) was included in the growth medium. The cultures were fixed, sectioned, labeled by the immunogold method described above. After stopping the silver enhancement process, sections were dehydrated by passage through a series of ethanol washes and then prepared for autoradiography using Kodak NTB2 liquid nuclear track emulsion (Eastman Kodak Co., Rochester, NY) as described by the manufacturer. Figure 2B shows a section of an SCC-13 raft culture which was subjected to both immunohistochemistry, to localize keratin K6, and autoradiography, to identify cells that had incorporated [^{3}H]thymidine.[23]

[27] R. A. Weiss, R. Eichner, and T.-T. Sun, *J. Cell Biol.* **98,** 1397 (1984).

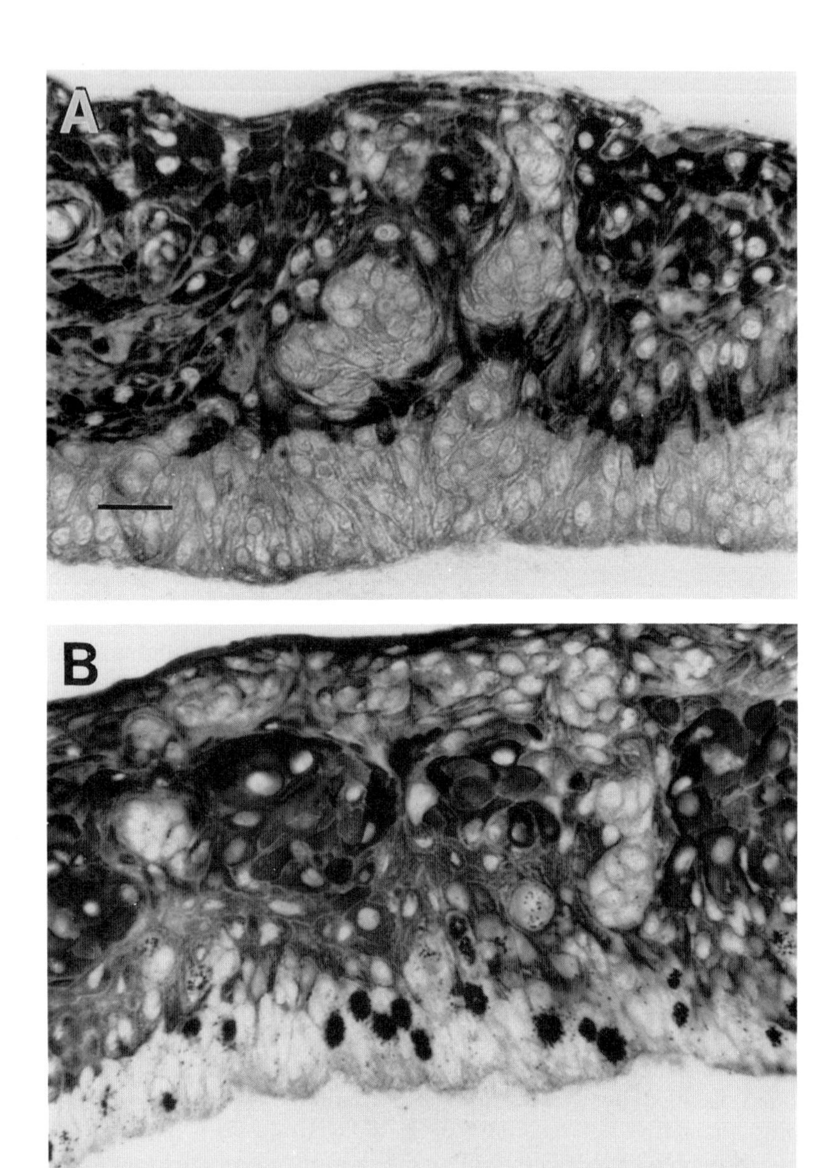

FIG. 2. (A) Immunolocalization of keratin K6 in an SCC-13 raft culture. SCC-13 cells were grown at the air–liquid interface for 10 days. Two hours prior to harvesting, [^{3}H]thymidine was added to the culture medium. The tissue was then embedded in paraffin and sectioned (5 μm), and the sections were labeled with a polyclonal monospecific anti-human K6 antisera using the immunogold–silver enhancement technique. Note that anti-K6 staining is confined to the suprabasal, spinouslike cells of the culture. (B) SCC-13 rafts were prepared and labeled with anti-K6 antisera as in (A), followed by autoradiography to identify cells which incorporated [^{3}H]thymidine. Note that autoradiographic silver grains are primarily over the basal cells. Bar: 30 μm. [From R. Kopan and E. Fuchs, *J. Cell Biol.* **109,** 295 (1989).

Localization of Keratin mRNA

Tissue Preparation. For *in situ* hybridization analysis of normal skin or cultured keratinocytes, tissue is fixed in 4% paraformaldehyde in phosphate-buffered saline (PBS), embedded in paraffin, and sectioned at 5 μm. Optimal fixation time depends on cell type and sample thickness. In general, cultured keratinocytes (either raft or submerged) require 0.5–1.0 hr whereas skin samples require 3–6 hr. In order to minimize detachment of the tissue sections from the glass slides during this procedure, the slides should be pretreated with aminoalkylsilane. To increase probe accessibility, skin sections are treated with 0.5 μg/ml proteinase K for 30 min at 37° before hybridization. Proteinase K treatment is not necessary for cultured cells.[28]

Tissue sections are briefly rinsed in water and then in 0.1 *M* triethanolamine, pH 8.0, at room temperature. Treatment of sections with 0.25% (v/v) acetic anhydride in 0.1 *M* triethanolamine for 10 min at room temperature results in the reduction of background hybridization. After two brief rinses in 2 × saline sodium citrate (SSC; 1 × SSC is 0.15 *M* NaCl, 15 m*M* sodium citrate, pH 7.0) and dehydration through an ethanol series from 70 to 100%, sections are ready for hybridization.

Probe Preparation. ^{35}S-Labeled UTP-containing cRNA antisense probes are prepared by *in vitro* transcription using a construct containing a bacteriophage promoter (the promoters most commonly used for this purpose are derived from SP6, T3, and T7) and the appropriate polymerase.[22] Prior to use, probes are hydrolyzed to an average size of 150 base pairs (bp) by sodium carbonate treatment at 60°.

Hybridization and Washes. The hybridization mixture consists of 50% formamide, 0.3 *M* NaCl, 10 m*M* Tris, pH 8.0, 1 m*M* EDTA, 1 × Denhardt's solution (0.02% each of BSA, Ficoll, and polyvinylpyrrolidone), 500 μg/ml yeast tRNA, 10% dextran sulfate, and 100 m*M* dithiothreitol (DTT)[22,29] The cRNA probe is denatured by incubating at 80° for 3 min in low salt, and it is then added to the hybridization solution at a concentration of 0.2–0.3 μg/ml. A small volume of hybridization mix (3 μl/cm^2) is added to the pretreated tissue sections and covered with a coverslip chip. The slide is immersed in prewarmed mineral oil and incubated at 42° overnight.

Mineral oil is removed with chloroform washes, and the coverslip is floated off in 4× SSC. Tissue sections are then washed at 37° for 30 min in an aqueous solution containing 20 μg/ml RNase A, 0.5 *M* NaCl, 1 m*M* EDTA, and 10 m*M* Tris, pH 8.0.[22] This is followed by two 30-min washes

[28] A. L. Tyner and E. Fuchs, *J. Cell Biol.* **103,** 1945 (1986).

[29] K. H. Cox, D. V. DeLeon, L. M. Angerer, and R. C. Angerer, *Dev. Biol.* **101,** 485 (1984).

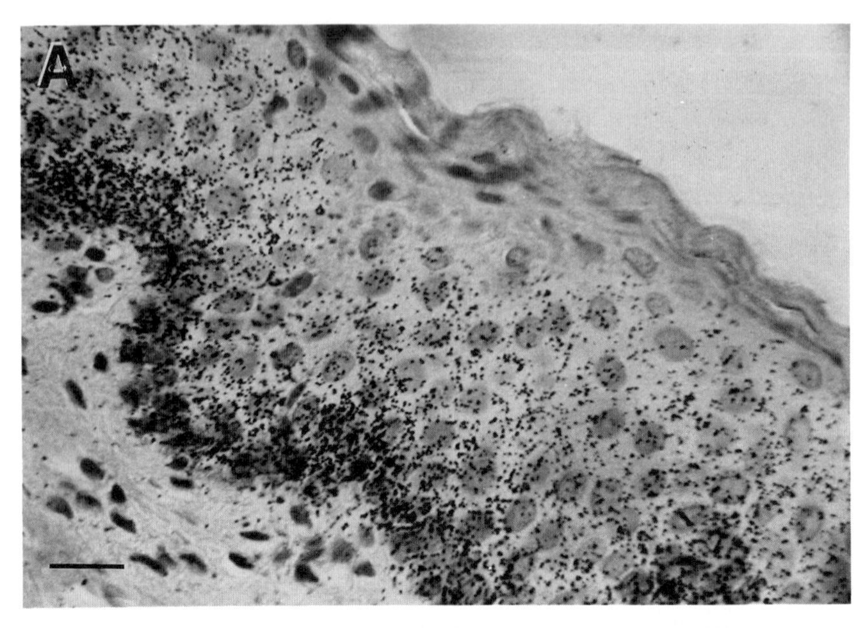

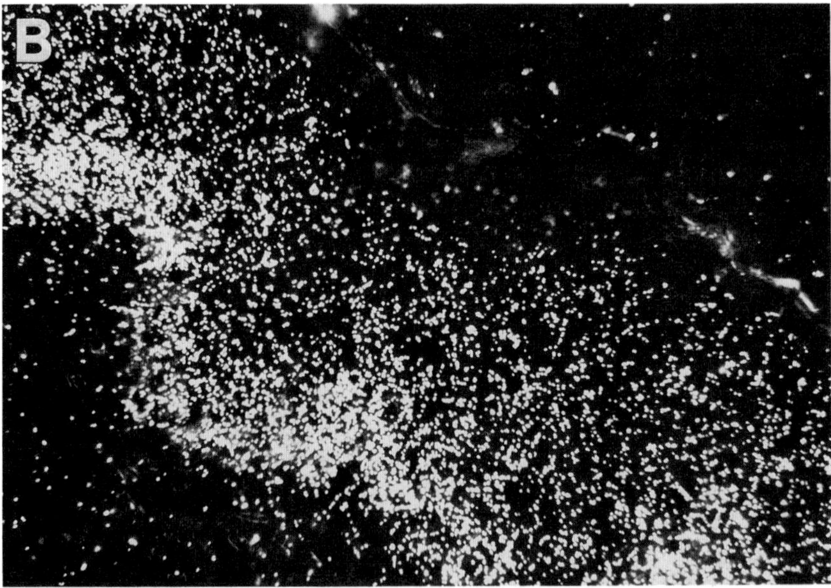

FIG. 3. *In situ* localization of keratin K14 mRNA in normal human epidermis. Human skin was fixed in 4% paraformaldehyde, paraffin-sectioned, and hybridized with a ^{35}S-labeled UTP-containing cRNA probe complementary to human K14 mRNA. Washed sections were exposed to Kodak NTB-2 autoradiographic emulsion for 3 days, developed, and counterstained with hematoxylin and eosin to visualize tissue morphology. Microscopic examination using bright-field (A) and dark-field optics (B) shows that most of the silver grains are localized to the basal epidermal layer. 30 μm. [From A. Stoler, R. Kopan, M. Duvic, and E. Fuchs, *J. Cell Biol.* **107,** 427 (1988).]

at room temperature in 2X SSC and two 60-min washes at 45–50° in 0.1X SSC containing 10 mM DTT. To alter the stringency of the wash conditions, the temperature of the final two washes can be adjusted accordingly. Tissue sections are subsequently dehydrated through an ethanol series, air-dried, and analyzed by autoradiography using Kodak NTB2 liquid nuclear track emulsion (Eastman Kodak Co.) as specified by the manufacturer. Figure 3 shows an example of a section of human skin hybridized with a radiolabeled cRNA probe specific for human K14 mRNA.[22]

Gene Transfection

Gene transfer technology has been used to investigate the regulation of keratin expression and the mechanism of keratin filament assembly.[30-32] Future analyses of vitamin A regulation in keratinocytes will undoubtedly be facilitated by this technology.

Transfection of primary keratinocyte cultures presents technical problems not encountered with most established cell lines. Stratification and terminal differentiation interfere with the introduction and expression of foreign DNA. To circumvent these problems, keratinocytes to be transfected can be grown in a low-calcium medium, which promotes growth and inhibits stratification and differentiation.[33] In cases where it is desirable to maintain monolayer cultures following transfection, the conventional calcium phosphate transfection procedure should not be used. The transfection protocol outlined below, which is a modification of the DEAE-dextran transfection procedure described by Gorman,[34] has been successfully used for the transient transfection of keratin cDNA constructs into cultured keratinocytes.[31]

Human keratinocytes are grown in low-calcium medium containing calcium-depleted serum as described previously.[33] Preconfluent cultures are washed 4 times with serum-free low-calcium medium. Four milliliters of serum-free low-calcium medium containing 5 μg/ml supercoiled DNA, 150 μg/ml DEAE-dextran, and 10 μM chloroquine are then added to each 100-mm plate of cells. After a 3-hr incubation at 37°, cells are treated with 10% dimethyl sulfoxide for 1–2 min, washed twice with low-calcium medium, and returned to the 37° CO_2 incubator in serum-containing low-calcium medium. Cells are typically analyzed 65 hr posttransfection.

[30] G. J. Giudice and E. Fuchs, *Cell (Cambridge, Mass.)* **48,** 453 (1987).

[31] K. Albers and E. Fuchs, *J. Cell Biol.* **105,** 791 (1987).

[32] R. Lersch, V. Stellmach, C. Stocks, G. J. Giudice, and E. Fuchs, *Mol. Cell. Biol.* **9,** 3685 (1989).

[33] H. Hennings, D. Michael, C. Cheng, P. Steinert, K. Holbrook, and S. H. Yuspa, *Cell (Cambridge, Mass.)* **19,** 245 (180).

[34] C. Gorman, *in* "DNA Cloning, Volume II" (D. M. Glover, ed.), p. 143. IRL Press, Oxford, 1985.

[3] Retinoids and Lipid Changes in Keratinocytes

By MARIA PONEC and ARIJ WEERHEIM

Introduction

Cultured keratinocytes have proved to be a useful model to study the regulation of epidermal differentiation, because the degree of their differentiation can be modified experimentally.[1-3] Especially when the "organotypic" culture systems are used, the reconstructed epidermis exhibits morphological and biochemical features close to those seen *in vivo*. In these systems the keratinocytes are attached to a biological matrix, such as deepidermized dermis (DED) with preserved lamina densa, and they are lifted to the air–liquid interface so that the upper layers are exposed to air.[4] Such a three-dimensional culture system provides an attractive model to study the modulation of epidermal differentiation by drugs, such as retinoids, used in the treatment of skin disorders.

The purpose of this chapter is to provide a guide to the use of *in vitro* reconstructed epidermis for the study of drug-induced modulation of the epidermal lipid composition. Epidermal differentiation is known to be accompanied by marked changes in lipid composition.[5-9] A progressive depletion of phospholipids coupled with an increase of sterols and of certain classes of sphingolipids was found to occur during differentiation of both human and animal epidermis.[5,10] The final product of epidermal differentiation is the stratum corneum, the lipids of which, predominantly ceramides and nonpolar lipids, play an important role in cohesion and desquamation of the stratum corneum as well as in the maintenance of normal barrier function.[5,11-17]

[1] M. Prunieras, M. Regnier, and D. Woodley, *J. Invest. Dermatol.* **81s,** 28 (1983).
[2] N. E. Fusening, *in* "Biology of the Integument" (J. Beriter-Hahn, A. G. Matoltsy, and K. S. Richards, eds.), Vol. 2, p. 409, Springer-Verlag, Berlin, 1986.
[3] K. A. Holbrook and H. Hennings, *J. Invest. Dermatol.* **81s,** 28 (1983).
[4] M. Regnier, M. Prunieras, and D. Woodley, *Front. Matrix Biol.* **9,** 4 (1981).
[5] P. M. Elias, *J. Invest. Dermatol.* **80s,** 44 (1983).
[6] P. M. Elias and D. S. Friend, *J. Cell Biol.* **65,** 180 (1975).
[7] M. A. Lampe, M. L. Williams, and P. M. Elias, *J. Lipid Res.* **24,** 131 (1983).
[8] M. A. Lampe, A. L. Burlingame, J. Whitney, M. L. Williams, B. E. Brown, E. Roitman, and P. M. Elias, *J. Lipid Res.* **24,** 120 (1983).
[9] H. J. Yardley and R. Summerly, *Pharmacol. Ther.* **13,** 357 (1981).
[10] G. M. Gray and H. J. Yardley, *J. Lipid Res.* **16,** 441 (1977).
[11] M. L. Williams and P. M. Elias, *Arch. Dermatol.* **121,** 477 (1985).
[12] P. W. Wertz and D. T. Downing, *J. Lipid Res.* **24,** 753 (1983).
[13] P. W. Wertz and D. T. Downing, *J. Lipid Res.* **26,** 761 (1985).

In this chapter attention is focused on the comparison of lipid composition of human epidermis with that of the *in vitro* reconstructed epidermis, the latter cultured either in the absence or presence of retinoic acid.

Cell Culture

Submerged Culture. Juvenile foreskin keratinocytes derived from donors (aged 1–2 years) are cultured together with irradiated mouse 3T3 fibroblast feeder cells in Dulbecco–Vogt and Ham's F12 (3:1) media supplemented with 5% (v/v) fetal calf serum (FCS), 0.4 μg/ml hydrocortisone, 1 μM isoproterenol, and 10 ng/ml epidermal growth factor (EGF).[18]

Air-Exposed Culture. The deepidermized dermis (DED) for air-exposed cultures is prepared as described by Regnier *et al.*[4] Briefly, cadaver skin (stored at 4° in 85% (v/v) glycerol) is carefully washed with phosphate-buffered saline (PBS) and incubated for 3–5 days in PBS at 37°. Subsequently, the epidermis is scraped off and the remaining dermis irradiated (3000 R) and washed several times with culture medium. The dermis is then placed on the stainless steel grid, and 0.5×10^6 normal human keratinocytes (second or third passage) are inoculated inside a stainless steel ring (diameter 1 cm) placed on the top of the dermis. After 24 hr the ring is removed, and the level of culture medium is adjusted to just reach the height of the grid. This method ensures that the cells are exposed to air throughout the remaining period of culture. The medium used for air-exposed cultures is Dulbecco–Vogt and Ham's F12 (3:1) media supplemented with 5% FCS or 5% (v/v) delipidized FCS (DLS),[19] 1 μM isoproterenol, and 10 ng/ml EGF. In experiments in which the effect of retinoic acid on lipid composition is studied, the cultures are refed on days 3 and 7 with media supplemented with 2 μM retinoic acid. One-half microliter per milliliter medium of freshly prepared stock solution in absolute ethanol is used. Addition of retinoic acid is performed under yellow light, and the cultures are maintained in the dark. Controls received 0.5 μl ethanol/ml medium only.

[14] P. W. Wertz, M. C. Miethke, S. A. Long, J. S. Strauss, and D. T. Downing, *J. Invest. Dermatol.* **84,** 253 (1985).
[15] P. W. Wertz, E. S. Cho, and D. T. Downing, *Biochim. Biophys. Acta* **753,** 350 (1986).
[16] P. A. Bowser, D. H. Nugteren, R. J. White, U. M. T. Houtsmuller, and C. Prottey, *Biochim. Biophys. Acta* **834,** 419 (1985).
[17] P. A. Bowser, R. J. White, and D. H. Nugteren, *Int. J. Cosmet. Sci.* **8,** 125 (1986).
[18] J. G. Rheinwald, *in* "Methods in Cell Biology" (C. Harris, B. F. Trump, and G. Stoner, eds.), Vol. 21A, p. 229. Academic Press, New York, 1980.
[19] G. H. Rothblat, L. Y. Arbogast, L. Ovellett, and B. V. Howard, *In Vitro* **12,** 554 (1976).

Lipid Extraction

After 10 days of culture, the reconstructed epidermis separated from DED by heating for 1 min at 60° is washed in PBS and collected in 2 ml chloroform–methanol (1:2, v/v) (Pyrex test tube with a screw containing a Teflon inlay) and stored at −20° until use. Subsequently, the lipids are extracted according to Bligh and Dyer,[20] with the addition of 0.25 *M* KCl to ensure extraction of polar species. Briefly, the following procedure is followed:

1. After defrosting, extract the tissue in chloroform–methanol (1:2) for 1 hr at room temperature, centrifuge 10 min at 900 *g* at room temperature, and transfer supernatant to a "collect tube."
2. Extract the pellet in 2 ml chloroform–methanol–water (1:2:0.5, v/v/v) for 1 hr at 37°, centrifuge, and transfer the supernatant to the collect tube.
3. Extract the pellet at room temperature in 2 ml chloroform–methanol (1:2, v/v), centrifuge, and transfer the supernatant to the collect tube.
4. Extract the pellet in 2 ml chloroform–methanol (2:1), centrifuge at room temperature, and transfer supernatant to the collect tube.
5. Extract the pellet in 2 ml chloroform, centrifuge at room temperature, and transfer the supernatant to the collect tube.
6. Add 200 μl of 2.5% KCl to the collected supernatants, vortex, add 2 ml water, centrifuge 5 min at 900 *g* at room temperature, and transfer the underlying fluid to a second collect tube.
7. Wash the remaining upper layer with 4 ml chloroform, centrifuge 5 min at 900 *g* at room temperature, and transfer the underlying fluid to the second collect tube.
8. Evaporate organic solvents from the second collect tube to dryness at 50° under a stream of nitrogen.
9. Dissolve the residue in 1 ml chloroform–methanol (2:1, v/v) and keep the lipid extract until use at −20° in a closed tube (with Teflon inlay).

The pellet (obtained after Step 5), after drying, is lysed in 1 *N* NaOH and used thereafter for protein determination. For determination of the total lipid content, aliquots of lipid extract are weighed.

Lipid Separation

Separation of extracted lipids is achieved by means of either one- (1D-) or two-dimensional (2D-) high-performance thin-layer chromatography

[20] E. G. Bligh and W. J. Dyer, *Can. J. Biochem. Physiol.* **37**, 911 (1959).

(HPTLC). The advantage of 2D-HPTLC is the high resolution and the relative ease of identification of a great variety of epidermal lipid fractions, such as various sphingolipids, lanosterol, A-acid, and α-hydroxy acids (Fig. 1).[15,16] However, the use of 2D-HPTLC for quantitative determination of individual lipid fractions is not practical. For this purpose a 1D-HPTLC system (Fig. 2), in which a mixture of lipid standards can be applied next to the investigated lipid samples on a single plate, is the more suitable approach and enables a rapid screening and quantification of epidermal lipids.

Cleaning of Thin-Layer Plates

In order to remove impurities that may interfere with lipid separation, the HPTLC plates (Merck, Darmstad, FRG) are washed first in methanol–ethyl acetate (60:40), followed by chloroform–ethyl acetate–

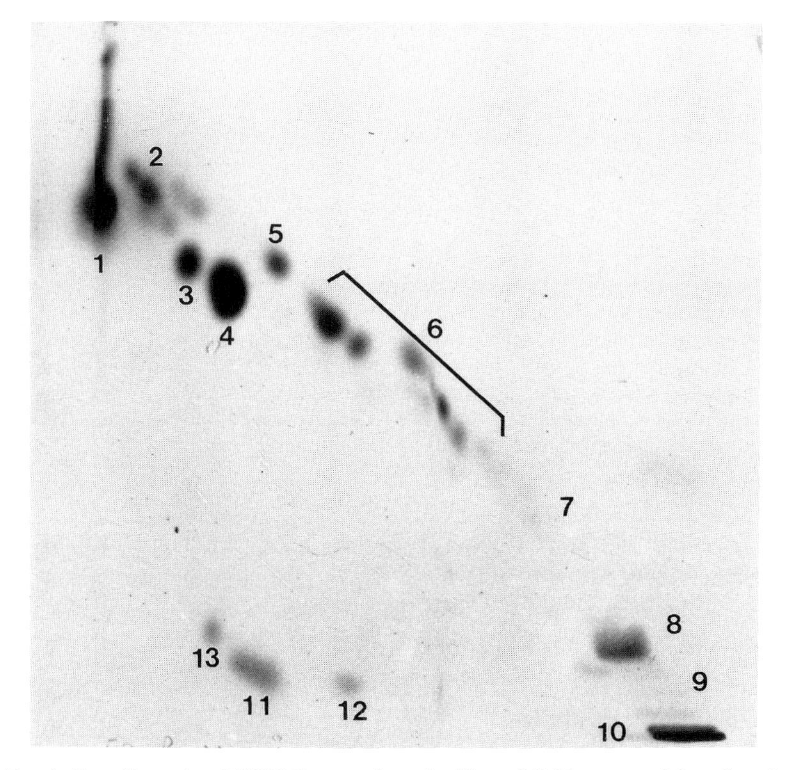

FIG. 1. Two-dimensional HPTLC separation of epidermal lipids extracted from keratinocytes cultured for 14 days at the air–liquid interface on deepidermized dermis. 1, Neutral lipids; 2, tri- and diglycerides; 3, lanosterol; 4, cholesterol; 5, *O*-acylceramide; 6, ceramides; 7, *O*-acylglycosylceramide; 8, cholesterol sulfate; 9, cerebrosides; 10, polar lipids; 11, free fatty acids; 12, α-hydroxy fatty acids; 13, *O*-acyl fatty acids.

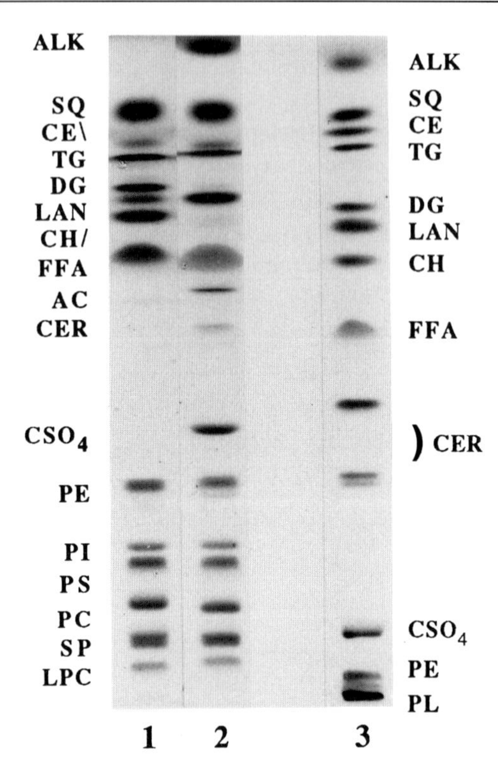

FIG. 2. Separation of lipid standards by means of one-dimensional HPTLC using the total lipid development system (lanes 1,2) or the ceramide development system (lane 3). LPC, Lysophosphatidylcholine; SP, sphingomyelin; PC, phosphatidylcholine; PS, phosphatidylserine; PI, phosphatidylinositol; PE, phosphatidylethanolamine; CSO_4, cholesterol sulfate; CER, ceramides; AC, acylceramides; FFA, free fatty acids; CH, cholesterol; TG, triglycerides; DG, diglycerides; LAN, lanosterol; SQ, squalene; ALK, alkanes; PL, phospholipids; CE, cholesterol esters. FFA: lanes 1, 3, oleic acid; lane 2, palmitic acid.

diethyl ether (30:20:50). After evaporation of all solvents, the HPTLC plates are activated for 15 min at 130°.

Sample Application

One-Dimensional HPTLC. Increasing volumes of lipid extracts, containing up to 50 μg lipids, are applied as narrow bands (0.5 cm broad) on a precleaned and activated HPTLC plate (20 × 10 cm), at a constant distance of 0.5 cm (*x* coordinate), starting at a height of 0.5 cm (*y* coordinate) above the bottom edge of the HPTLC plate. In our experiments Linomat IV (CAMAG, Muttenz, Switzerland) has been used throughout. Standard mixtures containing 0.2 to 10 μg of each lipid are applied for calibration purposes.

Two-Dimensional HPTLC. An aliquot of lipid extract is applied as a narrow band (0.5–1 cm broad) on an HPTLC plate (10 × 10 cm) at a distance of 5 mm in the x direction and 5 mm in the y direction, starting at the right-hand edge of the plate.

High-Performance Thin-Layer Chromatography

For lipid separation CAMAG horizontal development chambers are used. The HPTLC plate is placed in such a way that the silica gel side of the HPTLC plate is facing down. All solvents used for the separation of lipids are of analytical grade (Merck). All developments are carried out at 4°. Following each development step, the HPTLC plate is dried under stream of air at 40° on a heat block (Thermoplate, DESAGA, Heidelberg, FRG) for approximately 10 min.

One-Dimensional HPTLC. Lipids are fractionated using two different development systems.

Total lipid development system. For the analysis of total lipids, 5–50 μg of the total lipid extract is applied to HPTLC plates and developed at 4° sequentially from the bottom edge of the plate as follows:

1. 15 mm: chloroform
2. 10 mm: chloroform–diethyl ether–acetone–methanol (72:8:10:10)
3. 60 mm: chloroform–acetone–methanol–33% ammonia (76:16:8:1)
4. 70 mm: chloroform–diethyl ether–acetone–methanol (72:8:10:10)
5. 30 mm: chloroform–ethyl acetate–ethyl methyl ketone–2-propanol–ethanol–methanol–water–acetic acid (34:4:4:6: 20:28:4:1)
6. 40 mm: chloroform–ethyl acetate–ethyl methyl ketone–2-propanol–ethanol–methanol–water (46:4:4:6:28:6)
7. 80 mm: chloroform–diethyl ether–acetone–methanol (76:4:8:12)
8. 90 mm: hexane–diethyl ether–ethyl acetate (80:16:4)

Ceramide development system. For the analysis of sphingolipids, 5–50 μg of the total lipid extract is applied to HPTLC plates and developed sequentially from the bottom edge of the plate as follows:

1. 15 mm: chloroform
2. 10 mm: chloroform–acetone–methanol (76:8:16)
3. 70 mm: chloroform–hexyl acetate–acetone–methanol (86:1:10:4)
4. 20 mm: chloroform–acetone–methanol (76:4:20)

5. 75 mm: chloroform–diethyl ether–hexyl acetate–ethyl acetate–acetone–methanol (72:4:1:4:16:4)
6. 90 mm: hexane–diethyl ether–ethyl acetate (80:16:4)

Two-Dimensional HPTLC. For 2D-HPTLC the atmosphere inside the development chamber is saturated with solvent vapor by placing a piece of filter paper saturated with the appropriate solvent mixture underneath the HPTLC plate. The development systems for 2D-HPTLC are shown below.

Dimension	Distance	Solvent system
First	10	Chloroform–acetone–methanol (90:5:5)
Second	85	Chloroform–acetone–methanol (90:5:5)
First	85	Chloroform–acetone–methanol–33% ammonia (60:30:8:2)
Second	85	Chloroform–acetone–methanol–acetic acid–water (78:16:4:2:1)

Staining and Charring

Quantitative analysis should proceed further according to the following prescription:

1. After accomplishing the lipid separation, dry the HPTLC plate at 40° under an air stream and subsequently heat the plate for 5 min at 130°.
2. Pipette 5 ml (10 × 10 cm plate) or 10 ml (20 × 10 cm plate) of a staining mixture containing acetic acid–H_2SO_4–H_3PO_4–water (5:1:1:95) and 0.5% $CuSO_4$.
3. Clean the glass side of the plate with a tissue to prevent the corrosion of the heat block.
4. Place the HPTLC plate on the heat block (preheated to 60°) and wait until the staining mixture evaporates.
5. Increase the temperature of the heat block from 60 to 160°.

In the temperature range between 80 and 90° the cholesterol, cholesterol sulfate, and cholesterol esters turn red and subsequently blue; squalene, lathosterol, and lanosterol stain brown/red and subsequently brown.

Standards

For identification and quantification of lipids the following standards are used: lysophosphatidylcholine (Sigma, St. Louis, MO, L-4129); sphingomyelin (Sigma, S-7004); phosphatidylcholine (Sigma, P-7524); phosphatidylserine (Sigma, P-7769); phosphatidylinositol (Sigma, P-0639); phosphatidylethanolamine (Sigma, P-7523); ceramides, bovine, mixture (Applied Science Alltech Assoc., State College, PA, 22100); cerebrosides, type I (Sigma, C-8752); cerebrosides, type II (Sigma, C-1516); lipid standards containing oleic acid, cholesterol, triolein, and cholesterol oleate (Sigma, 178-4); lanosterol (Sigma, L-5678); lathosterol (Sigma, C-3652); cholesterol sulfate (Sigma, C-9523); 1,2-diolein and 1,3-diolein (Sigma, D-8894); squalene (Sigma, S-3626). Although the bovine ceramide standards do not comigrate with all human ceramide bands, they are used for mass calibration purposes. Acylglucosylceramides (AGC) and acylceramides (AC), kindly provided by Dr. D. Nugteren (Unilever Research Laboratory, Vlaardingen, The Netherlands), are used for qualitative purposes only.

Quantification

After 1D-HPTLC development and charring, the HPTLC plates are scanned using the Densitometer CD 60 (DESAGA), which is equipped with suitable software for automatic measurements and integration of samples and evaluation of standards. Scanning profiles of the lipids extracted from human epidermis and reconstructed epidermis are shown in Fig. 3.

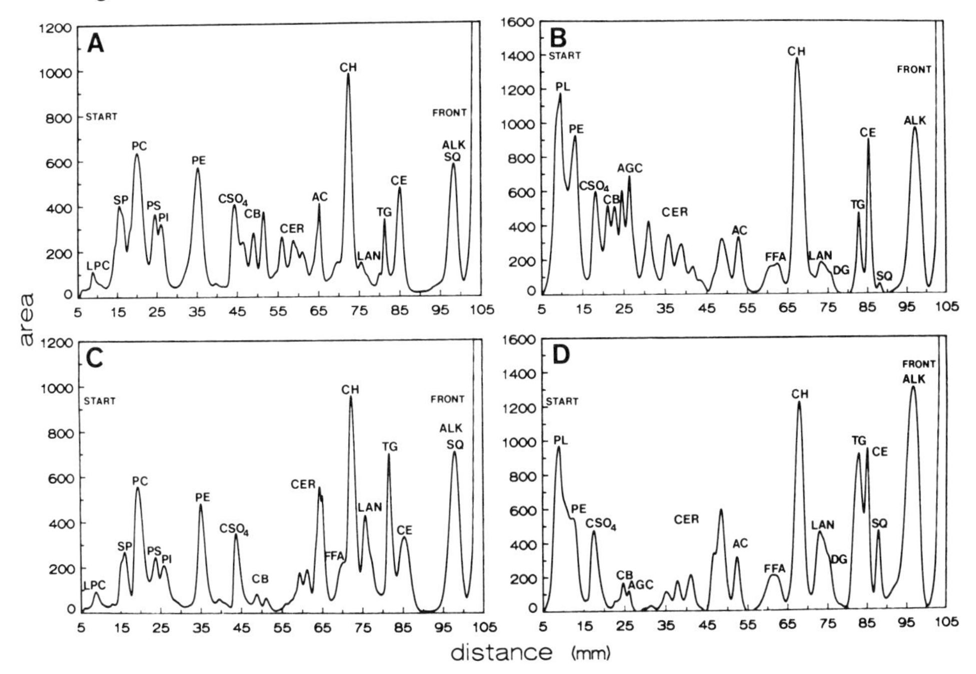

FIG. 3. Scanning profile of lipids extracted from human epidermis (A, B) and reconstructed epidermis (C, D) separated using the total lipid development system (A, C) or the ceramide development system (B, D). CB, Cerebrosides; AGC, acylglucosylceramides. For other abbreviations, see legend to Fig. 2.

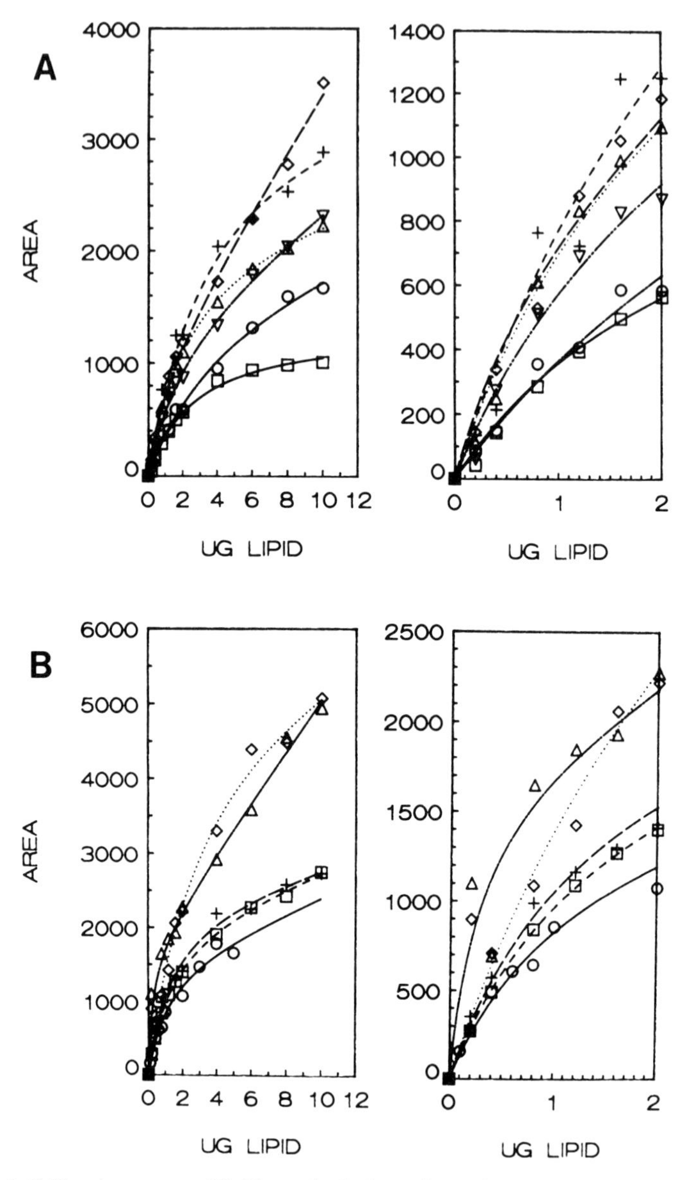

FIG. 4. Calibration curves of lipid standards determined after charring and photodensitometric scanning. The lipids were separated using the total lipid development system. (A) Phospholipids: LPC (□), SP (+), PC (◇), PI (○), PS (△), PE (▽). (B) Neutral lipids: cholesterol sulfate (○), cholesterol (□), lanosterol (+), cholesterol ester (◇), squalene (△). The right-hand sides show calibration curves for lipid amounts up to 2 μg.

Mass determinations of lipids are performed with corresponding standard series of the cochromatographed lipid standards on each HPTLC plate. Examples of calibration curves of lipid standards are presented in Fig. 4. Since large differences exist in the mass of individual epidermal fractions, for example, between cholesterol and cholesterol sulfate or ceramides, it is necessary to apply increasing quantities of the total lipid extracts. When this is done for each sample, the precise mass determination of a given lipid fraction can be performed.

Composition of Epidermal Lipids: *In Vivo* and *in Vitro* Comparison, Effect of Retinoic Acid

The use of the air-exposed culture system, using deepidermized dermis as the supporting substrate, offers an attractive model for studying the differentiation-related changes in lipid composition of keratinocytes. Epidermis cultured in this way is very much like the epidermis *in vivo,* although minor differences in lipid composition between the reconstructed epidermis and epidermis *in vivo* still exist, such as lower contents of certain sphingolipids (showing the same TLC mobility as cerebrosides), lower linoleic and arachidonic acid contents, and a higher content of triglycerides[21] (Fig. 5). This might be due to differences in available nutrients, which have to diffuse to differentiating cells through the dermal equivalent and the underlying basal cells. This impaired diffusion may also explain why, in contrast to submerged cultures, the presence of serum lipids only marginally affects the morphology and keratin and lipid patterns and why relatively high concentrations of retinoic acid (micromolar range) have to be administrated to the *in vitro* reconstructed epidermis for the induction of significant effects. At these concentrations, changes not only in the tissue morphology and the keratin pattern[22,23] but also in lipid composition[23] (Fig. 5) have been reported. Retinoic acid induced a 3- to 4-fold increase in phospholipids, a 3-fold decrease in total sphingolipids, a 6- to 8-fold decrease in acylceramides and lanosterol, and an approximately 2-fold decrease in cholesterol and cholesterol sulfate (Table I).[23]

It can be concluded that the presence and/or content of ceramides, especially acylceramides and lanosterol, can serve as markers of keratinocyte differentiation. Under culture conditions where differentiation is impaired, for example, when normal human keratinocytes are cultured under

[21] M. Ponec, A. Weerheim, J. Kempenaar, A.-M. Mommaas, and D. H. Nugteren, *J. Lipid Res.* **29,** 949 (1988).

[22] R. Kopan, S. Traska, and E. Fuchs, *J. Cell Biol.* **105,** 427 (1987).

[23] M. Ponec, *in* "Pharmacology of the Skin," (U. Reichert, and B. Shroot, eds.) Vol. 3, p. 45. Karger, Basel, 1989.

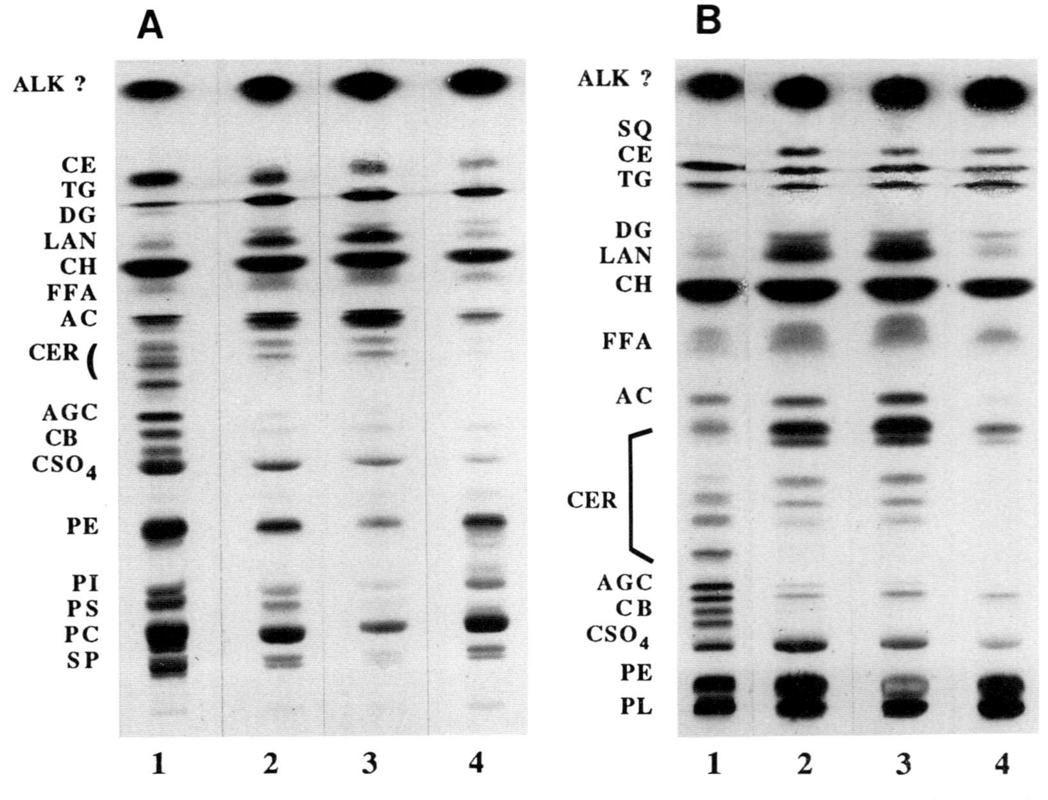

FIG. 5. One-dimensional HPTLC separation of lipids extracted from noncultured epidermis (lane 1), keratinocytes cultured at the air–liquid interface in the presence of 5% fetal calf serum (FCS) (lane 2), in the presence of 5% delipidized serum (DLS) (lane 3), or in the presence of 5% DLS and 2 μM retinoic acid (lane 4), using the total lipid development system (A) or the ceramide development system (B).

TABLE I
EFFECT OF RETINOIC ACID ON LIPID COMPOSITION OF KERATINOCYTES CULTURED AT AIR–MEDIUM INTERFACE[a]

Lipid	Composition: Full serum	Delipidized serum	Delipidized serum + 2 μM RA
Phospholipids	98.2	33.3	123.7
Cholesterol sulfate	5.3	4.6	2.7
Sphingolipids	24.8	36.3	14.9
Glycosphingolipids	2.2	1.1	2.2
Acylglucosylceramides	0.8	0.2	0.8
Acylceramides	3.1	4.3	0.5
Other ceramides	18.7	30.6	11.3
Neutral lipids	160.2	166.9	139.9
Cholesterol	35.4	39.2	28.1
Lanosterol	10.0	8.1	1.4
Free fatty acids	3.1	3.2	1.1
Triglycerides	94.0	103.8	98.3
Cholesterol esters	8.7	12.1	11.0

[a] Normal keratinocytes cultured first for 3 days in the presence of 5% full serum were subsequently refed on days 3 and 7 with media supplemented with 5% full serum, 5% delipidized serum, or 5% delipidized serum to which 2 μM retinoic acid (RA) was added. On day 10 the cells were harvested. Lipid extraction, separation, and quantification were performed as described in the text. Data are expressed as μg lipid/mg protein.

submerged conditions[21] or at the air–liquid interface in the presence of retinoic acid, these lipids are either absent or present in very low concentrations.

[4] Down-Regulation of Squamous Cell-Specific Markers by Retinoids: Transglutaminase Type I and Cholesterol Sulfotransferase

By ANTON M. JETTEN, MARGARET A. GEORGE, and JAMES I. REARICK

Introduction

Epithelial cells from a variety of tissues are able to undergo squamous cell differentiation. In tissues such as the epidermis this is the normal pathway of differentiation, whereas in epithelial cells from other tissues such as the tracheobronchial epithelium this differentiation is only observed under certain pathological conditions such as vitamin A deficiency.[1] During the process of differentiation cells stratify, obtain a squamous morphology, and ultimately cornify. Several biochemical and molecular markers of the squamous differentiated phenotype have been established. In particular, the induction of type I (epidermal) transglutaminase (EC 2.3.2.13) activity and increased levels of cholesterol sulfate and cholesterol sulfotransferase activity have been shown to be general and convenient markers of this differentiation process.[2-6] The transglutaminase type I is thought to catalyze the formation of transpeptide bonds between specific protein precursors of the cross-linked envelope, a layer of cross-linked protein deposited just beneath the plasma membrane of cornifying cells. The function of cholesterol sulfate has not yet been fully established.

Retinoids have been shown to inhibit the expression of the squamous differentiated phenotype and are effective inhibitors of the increase in transglutaminase activity, cholesterol sulfate, and sulfotransferase activity.[3-6] This inhibition occurs at nanomolar concentrations. Moreover, specific structural requirements are critical for the activity of retinoids.[7] The suppression of the squamous differentiated phenotype by retinoids might be mediated by nuclear retinoic acid receptors.[8]

[1] A. M. Jetten, *Environ. Health Perspect.* **80,** 149 (1989).

[2] U. Lichti, T. Ben, and S. H. Yuspa, *J. Biol. Chem.* **260,** 1422 (1985).

[3] S. M. Thacher and R. H. Rice, *Cell (Cambridge, Mass.)* **40,** 685 (1985).

[4] A. M. Jetten and J. E. Shirley, *J. Biol. Chem.* **261,** 15097 (1986).

[5] J. I. Rearick and A. M. Jetten, *J. Biol. Chem.* **261,** 13898 (1986).

[6] J. I. Rearick, P. W. Albro, and A. M. Jetten, *J. Biol. Chem.* **262,** 13069 (1987).

[7] A. M. Jetten, K. Anderson, M. A. Deas, H. Kagechika, R. Lotan, J. I. Rearick, and K. Shudo, *Cancer Res.* **47,** 3523 (1987).

[8] A. M. Jetten, *in* "Mechanisms of Differentiation" (P. B. Fisher, ed.), in press. CRC Press, Boca Raton, Florida, 1990.

Assay of Transglutaminase Activity

Transglutaminases are a family of enzymes which catalyze the covalent linking between peptide glutamine residues and primary amines, resulting in either the covalent incorporation into proteins of small, primary amines, such as putrescine, or in the cross-linking of proteins by the formation of ϵ-(γ-glutamyl)lysine transpeptide bonds.[9] Epithelial cells can synthesize two different types of transglutaminase: type I (epidermal) transglutaminase, which is largely membrane-bound and plays a role in protein cross-linking during cornification, and type II (tissue) transglutaminase, which is localized in the cytosol and has a still unknown function.

Materials

[^{3}H]Putrescine (18.6 Ci/mmol)
Casein hydrolyzate (Sigma, St. Louis, MO)
Mono- Q (HR 5/5) (Pharmacia, Piscataway, NJ)

Procedure

Preparation of Cellular Fractions. Cells grown in 60-mm tissue culture dishes are washed twice in phosphate-buffered saline (PBS) and then placed on ice. Cells can either be collected in 200 μl ice-cold PBS containing 10 U/ml leupeptin, 10 U/ml aprotinin, and 2 m*M* phenylmethylsulfonyl fluoride (PMSF) by scraping with a rubber policeman and assayed for transglutaminase activity immediately, or the dishes can be stored at −70° until the assay is performed.[3,4] The cell suspension is transferred to an Eppendorf tube and briefly sonicated (Branson Sonifier Model 200, equipped with microtip, 10–15 sec at an output setting of 5). Then 10 or 20 μl is taken from the homogenate and set aside for protein determination. Dithiotreitol (DTT, 1 *M*) is added to the remaining homogenate to a final concentration of 10 m*M*. One-half of this homogenate is set aside to measure total transglutaminase activity. The other half is centrifuged in an Eppendorf centrifuge at 10,000 *g* for 5 min at 4°. The resulting supernatant, containing the type II transglutaminase activity, is removed, transferred to another Eppendorf tube, and set on ice. The pellet, containing the type I transglutaminase activity, is resuspended in an equal volume of ice-cold PBS/DTT buffer. Triton X-100 or Nonidet P-40 (NP-40) (1% final concentration) is then added to each fraction. The addition of detergent enhances the transglutaminase activity as well as the reproducibility of the assay.

Measurement of Transglutaminase Activity. Transglutaminase activity

[9] J. E. Folk, *Annu. Rev. Biochem.* **49**, 517 (1980).

in the various cell fractions is determined by measuring the incorporation of [^{3}H]putrescine into casein hydrolyzate (Sigma). Aliquots (50 μl) of the samples are added to 13 × 100mm glass tubes placed on ice and containing 150 μl of the following, freshly prepared buffer: 50 m*M* sodium borate (pH 9.5), 1 m*M* EDTA, 10 m*M* $CaCl_2$, 0.13 *M* NaCl, 10 m*M* DTT, 0.25% Triton X-100, 2 mg/ml of casein hydrolyzate, and 2.5 μM [^{3}H]putrescine (18.6 Ci/mmol). Control tubes receive buffer instead of sample or casein. Determinations are carried out in duplicate. The mixtures are incubated at 25° for 60 min and then placed on ice. Aliquots of 80 μl are removed and spotted onto Whatman 3MM paper disks (2.3 mm diameter), which have been previously washed with 100 m*M* EGTA/100 m*M* EDTA, air-dried, and attached to paper clips. The latter prevent filters from sticking together and optimize washing. The disks are then washed 3 times for 30 min in cold 5% (w/v) trichloroacetic acid containing 1% (w/v) putrescine and once for 10 min in 100% ethanol. Paper clips are removed, and filters are air-dried and then transferred to scintillation vials. After the addition of Hydrofluor (National Diagnostics, Manville, NJ) the radioactivity is determined in a scintillation counter.

Chromatographic Analysis. Transglutaminase types I and II can be separated via anion-exchange chromatography using either a DEAE-cellulose column or a Mono Q (HR 5/5) (Pharmacia) high-performance liquid chromatography (HPLC) column.[2,3] Total cell homogenates are prepared in a buffer consisting of 10 m*M* Tris-HCl (pH 7.4), 10 m*M* DTT, and 0.5 m*M* EDTA as described above. After the addition of Triton X-100 (1% final concentration) the homogenate is centrifuged at 100,000 *g* for 30 min at 4°, and the supernatant is applied to a Mono Q column which has been equilibrated with the same buffer containing 0.1% Triton X-100. After a 10-ml wash with the same buffer, the transglutaminases are eluted from the column with a linear gradient of NaCl (from 0.0 to 0.5 *M*; total 50 ml). Samples of 0.5 ml are collected and analyzed for transglutaminase activity as described above. Type I transglutaminase is eluted at 0.25 *M* NaCl whereas type II elutes at 0.4 *M* NaCl.

Comments

The expression of transglutaminase types I and II can be further analyzed using monoclonal antibodies and cDNA probes specific for each of the transglutaminases.[3-13] Both type I and type II transglutaminase activity

[10] P. J. Birckbichler, H. F. Upchurch, J. K. Patterson, Jr., and E. Conway, *Hybridoma* **4,** 179 (1985).

[11] W. T. Moore, M. P. Murtaugh, and P. J. A. Davies, *J. Biol. Chem.* **259,** 12794 (1984).

[12] K. Ikure, T. Nasu, H. Yokota, Y. Tsuchiya, R. Sasaki, and H. Chiba, *Biochemistry* **27,** 2898 (1988).

[13] E. E. Floyd and A. M. Jetten, *Mol. Cell. Biol.* **9,** 4846 (1989).

can be modulated by retinoids. Retinoids inhibit transglutaminase type I activity during squamous differentiation in several cell types, including epidermal keratinocytes and tracheobronchial epithelial cells.[2-4] This inhibition has been shown to be related to a decrease in the level of corresponding mRNA.[13] In mouse peritoneal macrophages and myeloblastic HL60 cells, and under certain conditions in epidermal keratinocytes, retinoids have been shown to induce type II transglutaminase activity.[2,11] In macrophages retinoids appear to regulate type II transglutaminase at the transcriptional level.

Estimation of Relative Cholesterol Sulfate Content

Cells in culture take up radioactive sulfate from the culture medium and incorporate sulfate into a variety of cellular components which include cholesterol sulfate. Isolation and quantitation of cholesterol [^{35}S]sulfate provides a measure of cholesterol sulfate content which can be compared between cells grown under various culture conditions.[5,14]

Materials

$Na_2{}^{35}SO_4$ or $H_2{}^{35}SO_4$, carrier-free, 1200-1400 Ci/mmol
Chloroform/methanol, 2:1 (v/v)
0.1 *M* KCl
0.1 *M* KCl/methanol, 1:1 (v/v)
Silica gel G plates (Analtech, Newark, DE)

Procedure

Incorporation of [^{35}S]Sulfate into Cells. Cells are incubated with medium containing 10 to 50 μCi [^{35}S]sulfate per milliliter of medium. The lower amount is adequate for cell types grown in low-sulfate medium such as Ham's Nutrient Mixture F12 ([SO_4^{-2}] 6 μM). For cells grown in other media (Dulbecco's modified Eagle's, RPMI 1640, MCDB 151, and Hanks' or Earle's salts) containing higher amounts of sulfate (0.4 to 1.0 mM), the 50 μCi/ml concentration is more appropriate. After an incorporation period (6-24 hr) at 37°, cell monolayers are washed twice with Ca^{2+},Mg^{2+}-free PBS and removed from the dish by scraping or by standard trypsinization procedures. Scraping or trypsinization give identical results in terms of cholesterol [^{35}S]sulfate quantitation. The radiolabeled cells are centrifuged at 4° in 15-ml polypropylene conical tubes at 500 *g* for 5 min. After

[14] A. M. Jetten, M. A. George, C. Nervi, L.D. Boone, and J. I. Rearick, *J. Invest. Dermatol.* **92,** 203 (1989).

resuspension in PBS and recentrifugation, the supernatant is removed, and the cell pellets are either extracted immediately or stored at −70°.

Extraction. Four milliliters of chloroform/methanol (2:1, v/v) is added to each pellet obtained above, which in general should comprise the cells from one 60-mm dish, or 10^5 to 10^7 cells. Brief sonication (Branson Sonifier model 200 or equivalent probe-type sonicator, equipped with microtip, 5–15 sec at an output setting of 5) results in pellet disruption and generation of a cloudy, single-phase suspension. Vortexing (rather than sonication) at this point is counterproductive as clumps of cells adhere to the upper walls of the centrifuge tube and are difficult to dislodge. Occasionally two phases will appear at this point, with a small aqueous layer above a much larger organic layer. This results from attempting to extract too large a pellet or from leaving too much residual supernatant (>100 μl) over the pellet. Should two phases develop, more chloroform/methanol (2:1, v/v) can be added until a single phase suspension is obtained. All subsequent volumes (0.1 *M* KCl and 0.1 *M* KCl/methanol) would be increased proportionally.

At this point an aliquot (5%) may be taken for the determination of protein. The suspension is then centrifuged (1000 *g,* 10 min at room temperature), the resulting supernatant (CM extract) is transferred to a fresh 15-ml polypropylene centrifuge tube, and the pellet (CM-insoluble material) is dried under an air stream while vortexing.

Partitioning. To the 4-ml CM extract is added 1.0 ml of 0.1 *M* KCl; the mixture is vigorously vortexed, then centrifuged (500 *g,* 5 min) to separate the phases. The upper aqueous phase is discarded, and 2 ml of 0.1 *M* KCl/methanol (1:1, v/v) is added, and the mixture is vortexed and centrifuged as before. The upper aqueous phase is discarded, and the final lower phase or an aliquot thereof is transferred to a liquid scintillation vial, dried under an air stream, and radioactivity determined after adding a standard liquid scintillation cocktail.

The CM-insoluble material is dissolved in 10% (w/v) sodium dodecyl sulfate with heating (100°, 30 min). Aliquots (50 μl) are used for the determination of radioactivity and protein. The latter value for protein is used as basis for calculation of the incorporation of radioactivity [disintegrations per minute (dpm) per microgram of protein], recognizing that significant amounts of cellular protein actually are soluble in chloroform/methanol (2:1, v/v).

Identification of Cholesterol Sulfate. Material (10^3 to 10^4 dpm) from the final lower phase as described above can be analyzed by thin-layer chromatography. Authentic cholesterol 3-sulfate [10 μg of a stock solution of 2 mg/ml in chloroform/methanol (2:1, v/v)] may be included in the sample. Useful separation systems include silica gel G plates developed with chloroform/methanol/acetone/glacial acetic acid/water (8 : 2 : 4 : 2 : 1 v/v) or with benzene/methyl ethyl ketone/ethanol/water (3 : 3 : 3 : 1, v/v). After the solvent reaches the top of the plate, the plate is dried, sprayed with acidic iron trichloride (50 mg $FeCl_3 \cdot 6H_2O$ dissolved in 90 ml water, 5 ml glacial acetic acid, and 5 ml concentrated sulfuric acid), and heated at 100° until the colored spots appear (2–10 min). Radioactivity is determined by scraping 0.5-cm sections of each lane into a vial and performing liquid scintillation spectrometry using a standard cocktail. More rigorous and exhaustive techniques are available[5] to identify cholesterol sulfate utilizing double-label procedures, solvolysis, gas chromatography–mass spectrometry, and fast atom bombardment mass spectrometry.

Comments

The partitioning of the CM extract against 0.1 *M* KCl rather than water necessary to ensure that the large majority (95–98%) of the cholesterol [^{35}S]sulfate partitions into the lower phase. Under these conditions, virtually 100% of any sulfate or radiolabeled hydrophilic cellular metabolites (such as 3′-phosphoadenosine 5′-phosphosulfate, PAPS) partition into the aqueous phase. By contrast, partitioning against water without KCl leads to the appearance of a significant amount (10–30%) of cholesterol [^{35}S]sulfate in the upper aqueous phase. Addition of KCl to the aqueous phase results not only in higher recoveries but also in decreased variability between duplicate determinations, which variability never exceeds 15%.

The results of the incorporation of [^{35}S]sulfate into CM-insoluble material serves as an important control on the specificity of any changes in cholesterol [^{35}S]sulfate content. In general, no changes are observed in CM-insoluble radioactivity under conditions where cholesterol [^{35}S]sulfate amounts increase 20- to 100-fold. Changes are thus specific to cholesterol sulfate and not due to increased uptake [^{35}S]sulfate by the cells or increased synthesis of 3′-phosphoadenosine 5′-phosphosulfate.

With regard to the identification of cholesterol sulfate by thin-layer chromatography (TLC) or other techniques, it has been demonstrated that cholesterol sulfate is the sole sulfated species in the CM extract of several cell types: rabbit tracheal epithelial cells, human bronchial cells, and human epidermal keratinocytes. Presumably, it is necessary to demonstrate this fact only once for each cell type or, at most, for each treatment where dramatic increases in [^{35}S]sulfate incorporation into the final lower phase are found, to rule out that the treatment results in the synthesis of a novel sulfated lipophile. In the case of some, but not all, human squamous cell carcinoma cell lines, it has been found that a significant fraction of the organic-soluble [^{35}S]sulfate radioactivity migrates on thin-layer systems much slower that cholesterol sulfate, consistent with the chromatographic behavior of sulfatides. In cell types such as these, TLC identification of the species altered by any experimental treatment would be a necessary adjunct to the experimental procedure.

Assay of Cholesterol Sulfotransferase Activity

Sulfotransferases in cell extracts will transfer [^{35}S]sulfate from the donor substrate, 3′-phosphoadenosine 5′-phospho[^{35}S]sulfate, to acceptor substrates possessing free hydroxyl groups, including cholesterol. The cholesterol [^{35}S]sulfate thus formed is extracted and quantitated in a manner analogous to that described above.[6]

Materials

3′-Phosphoadenosine 5′-phospho[^{35}S]sulfate (New England Nuclear Research Products, Boston, MA), 1–2 Ci/mmol.

Procedure

Cells to be assayed for cholesterol sulfotransferase activity are removed from dishes by scaping or trypsinization and pelleted by centrifugation (500 *g*, 5 min at 4°). After removal of the supernatant solution, cells are either used immediately or stored at −70°.

Cells are resuspended in 50 m*M* HEPES buffer, pH 7.0, at a concentration of about 10^6 cells/ml. The cell suspension is lysed either by sonication (3 times for 15 sec each, Branson Sonifier 200 with microtip at an output setting of 5′) or by freeze–thawing 3 times, cycling between a dry ice/ethanol bath and a 40° water bath. The lysate is centrifuged (13,000 *g*, 15 min, 4°), and the supernatant solution is used as the enzyme source.

Enzyme assay mixtures contain 50 m*M* HEPES, pH 7.0, 200 m*M* NaCl, 0.01% (w/v) Triton X-100, 1.8 μM PAPS (about 10^6 dpm), 2.6 m*M* cholesterol, and up to 240 μg of cellular protein. The cholesterol is present in the assay in suspension as a result of preparing a stock solution containing HEPES, NaCl, Triton X-100, and cholesterol at 5-fold their final concentration and sonicating this solution exhaustively (same instrument as above, 10 to 20 times for 30 sec each burst over a 30-min period, with cooling on ice). Assays are initiated by the addition of enzyme, and the mixtures are incubated at 37° for 1 hr. The incubation is terminated by the addition of 4 ml chloroform/methanol (2:1, v/v) and the cholesterol [^{35}S]sulfate formed is quantitated as described above, except the pelleting of CM-insoluble material, if any, is omitted and the CM extract of the assay mixture is immediately partitioned against 1 ml of 0.1 *M* KCl.

Comments

The conditions of the assay (pH, buffer salt, concentrations of NaCl and Triton X-100) were initially optimized for the enzyme from rabbit tracheal cells.[6] Comparison with the work of Epstein *et al.*[15] suggests that species and/or tissue differences in optimal conditions for cholesterol sulfotransferases may exist. On the other hand, we have seen that optima for the cholesterol sulfotransferase of human epidermal keratinocytes[14] are virtually identical to those of rabbit tracheal cells.[16]

If exogenously added cholesterol is omitted from the reaction mixture, the radioactive incorporation into the final lower phase is reduced by 25 to 80% from that seen in the presence of added cholesterol. The radioactive substance formed in the absence of added cholesterol appears to be cholesterol sulfate by the criterion of thin-layer chromatography. These results indicate that an endogenous acceptor (probably cholesterol) is present in the 13,000 *g* supernatant solution used as an enzyme source. This endogenous acceptor will generally affect the results of quantitative assays because exogenously added cholesterol will bring the total cholesterol concentration close to saturation and the differences in enzyme amounts measured between differentiated and undifferentiated cells are usually quite large. For some types of studies, including substrate specificity and detailed enzyme kinetics, it may be necessary to separate the enzyme from its endogenous acceptor. This can be accomplished by centrifugation at 470,000 *g* for 4.5 hr at 4°, which causes the acceptor to pellet whereas a large majority of the enzyme remains in the supernatant solution.

[15] E. H. Epstein, Jr., J. M. Bonifas, T. C. Barber, and M. Haynes, *J. Invest. Dermatol.* **83**, (1984).
[16] A. M. Jetten, M. A. George, and J. I. Rearick, unpublished observations (1988).

[5] Isolation, Purification, and Characterization of Liver Cell Types

By HENK F. J. HENDRIKS, ADRIAAN BROUWER, and DICK L. KNOOK

Introduction

The liver consists of four main cell types: the parenchymal, Kupffer, endothelial, and fat-storing cells. In the rat, parenchymal cells represent 60% of the total cell number in the liver, occupy 70% of the liver volume, and make up 92.5% of the liver protein content.[1,2] The total nonparenchymal cells represent 40% of the cell number, but together they occupy only 6.5% of the liver volume.[1,2] The remainder of the liver volume is formed by the extracellular spaces.[1,2] Based on estimates made for the BN/BiRij rat strain, a rat liver consists of an average of 108×10^6 parenchymal cells, 9.5×10^6 Kupffer cells, 19.5×10^6 endothelial cells, and about 16×10^6 fat-storing cells per gram of liver.[3]

The liver plays a central role in the uptake, storage, and mobilization of retinol (vitamin A) in the body. The metabolism of retinoids in the liver, where over 95% of the retinoids in the body is found, is both complex and highly regulated.[4] Specific functions in retinoid metabolism have been described for parenchymal and fat-storing cells.[5,6] Possibly, Kupffer cells may have a function in retinoid metabolism as well.[7] Liver cell isolation procedures have been widely applied to study cell-specific functions in liver retinoid metabolism. In this chapter, methods available for the isolation, purification, and characterization of parenchymal, fat-storing, Kupffer, and endothelial cells are described. Isolation and purification of other minor cell populations present in liver cell isolates, such as pit cells, smooth muscle cells, bile duct cells, and several types of blood cells, are not discussed.

Isolation Procedures

Isolation of Rat Liver Cells Using Collagenase

The isolation procedure for rat cells using collagenase is applicable for the isolation of both parenchymal cells and nonparenchymal cells (Table I). The procedure consists of perfusion and incubation of the liver with collagenase to degrade the extracellular matrix, combined with Ca^{2+}-free perfusion.[8] Details concerning the experimental procedures vary among

[1] A. Blouin, R. P. Bolender, and E. R. Weibel, *J. Cell Biol.* **72,** 441 (1977).

[2] A. C. Munthe-Kaas, T. Berg, and R. Sjeljelid, *Exp. Cell Res.* **99,** 146 (1976).

[3] H. F. J. Hendriks, W. A. M. M. Verhoofstad, A. Brouwer, A. M. de Leeuw, and D. L. Knook, *Exp. Cell Res.* **160,** 138 (1985).

[4] D. S. Goodman, *in* "The Retinoids" (M. B. Sporn, A. B. Roberts, D. A. Goodman, eds.), Vol. 2, p. 41. Academic Press, New York, 1984.

[5] H. F. J. Hendriks, A. Brouwer, and D. L. Knook, *Hepatology* **7,** 1368 (1987).

[6] D. L. Knook, W. S. Blaner, A. Brouwer, and H. F. J. Hendriks, *in* "Cells of the Hepatic Sinusoid" (E. Wisse, D. L. Knook, and K. Decker, eds.), Vol. 2, p. 16. Kupffer Cell Foundation, Rijswijk, The Netherlands, 1989.

[7] H. F. J. Hendriks, E. Elhanany, A. Brouwer, A. M. de Leeuw, and D. L. Knook, *Hepatology* **8,** 276 (1988).

[8] M. N. Berry and D. S. Friend, *J. Cell Biol.* **43,** 506 (1969).

TABLE I
METHODS FOR ISOLATING RAT LIVER CELLS

Cell type and method				Composition[a]				
Perfusion	Incubation	Temperature	Yield[a]	P	E	K	FSC	Refs.
Parenchymal cells								
Collagenase	Collagenase	37°	++	++	+	±	− [b–e]	*f*
None	None	<8°	±	±	−	−	− [g]	*h, i*
Nonparenchymal cells								
Pronase	Pronase	37°	+	−	+	+	− [b,j]	*k, l*
Pronase–collagenase	Pronase–collagenase	37°	++	−	+	+	+ [b,j]	*m*
Collagenase	Collagenase	37°	++	++	+	±	− [b–d]	*f, n*
Collagenase	Collagenase; pronase	37°	+	−	+	±	− [b,j]	*o*
Collagenase	Collagenase; enterotoxin	37°	+	−	+	+	− [b,c]	*p*
Pronase	None	<8°	+	−	+	±	− [c,g]	*q*
Collagenase	Collagenase	<8°	±	−	+	±	− [c,d,g,r]	*f*

[a] ++, Excellent; +, good; ±, insufficient; −, very poor. P, Parenchymal cells; E, endothelial cells; K, Kupffer cells; FSC, fat-storing cells.
[b] Uptake of cellular debris during isolation.
[c] Retention of membrane receptors and endocytic capacity.
[d] Significant contamination with parenchymal cells and cytoplasmic particles derived from these cells (blebs).
[e] Kupffer and fat-storing cell yield can be improved using centrifugal elutriation instead of differential centrifugation as the first step in the separation of parenchymal cells from nonparenchymal cells.
[f] J. F. Nagelkerke, K. P. Barto, and T. J. C. van Berkel, *Exp. Cell Res.* **138,** 183 (1982).
[g] Inhibition of cellular metabolism and of degradation of endocytosed ligands during isolation.
[h] H. F. J. Hendriks, W. A. M. M. Verhoofstad, A. Brouwer, A. M. de Leeuw, and D. L. Knook, *Exp. Cell Res.* **160,** 138 (1985).
[i] W. S. Blaner, H. F. J. Hendriks, A. Brouwer, A. M. de Leeuw, D. L. Knook, and D. S. Goodman, *J. Lipid Res.* **26,** 1241 (1985).
[j] Loss of membrane receptors and endocytic capacity.
[k] D. L. Knook, N. Blansjaar, and E. C. Sleyster, *Exp. Cell Res.* **109,** 317 (1977).
[l] D. L. Knook and E. C. Sleyster, *Biochem. Biophys. Res. Commun.* **96,** 250 (1980).
[m] D. L. Knook, A. M. Seffelaar, and A. M. de Leeuw, *Exp. Cell Res.* **139,** 468 (1982).
[n] D. L. Knook, D. P. Praaning-van Daalen, and A. M. de Leeuw, *in* "Cells of the Hepatic Sinusoid" (A. Kirn, D. L. Knook, and E. Wisse, eds.), p. 445. Kupffer Cell Foundation, Rijswijk, The Netherlands, 1986.
[o] T. Berg and D. Boman, *Biochim. Biophys. Acta* **321,** 585 (1973).
[p] R. Blomhoff, B. Smedsrod, W. Eskild, P. E. Granum, and T. Berg, *Exp. Cell Res.* **150,** 194 (1984).
[q] D. P. Praaning-van Daalen, and D. L. Knook, *FEBS Lett.* **141,** 229 (1982).
[r] Exact cellular composition has not been reported.

laboratories. Some of the differences are trivial, but some may have important consequences for the quality of the cell preparation. The protocol described below is employed in our laboratory for the isolation of cell preparations used in biochemical and cell culture experiments.

A perfusion medium having the following composition is used: 0.9 m*M* glutamine, 0.1 m*M* aspartic acid, 0.2 m*M* threonine, 0.3 m*M* serine, 0.5 m*M* glycine, 0.9 m*M* glutamic acid, 0.6 m*M* alanine, 10.0 m*M* lactate, 1.0 m*M* sodium pyruvate, 3.0 m*M* KCl, 0.5 m*M* $NaH_2PO_4 \cdot H_2O$, 20.0 m*M* glucose, 25.0 m*M* HEPES, 0.5 m*M* $MgCl_2$, 24.0 m*M* $NaHCO_3$. All components except $NaHCO_3$ are dissolved, the pH is adjusted to 7.4, and then $NaHCO_3$ is added. The osmolality is adjusted to 308 mOsm with NaCl.

During the isolation procedure, perfusates are constantly saturated through an air stone with a mixture of oxygen and carbon dioxide (95 : 5, v/v) to meet oxygen needs of the metabolically active parenchymal cells. The temperature of the perfusates is maintained at 37° for optimal collagenase activity. It is important to perfuse the liver with a perfusion pressure of less than 40 torr, because parenchymal cells are highly sensitive to hydrostatic pressure. In BN/BiRij rats with an average liver weight of about 4 g, a rate of approximately 10 ml/min is used.

The isolation procedure is as follows. First, the liver is linearly perfused *in situ* through the portal vein using a G18 Luer Braunüle (Melsungen, FRG) and perfusion medium containing 1 m*M* $CaCl_2$ and 0.05% collagenase type I (Sigma, St. Louis, MO) for 2 min to rinse blood from the tissue. Second, the liver is linearly perfused with a Ca^{2+}-free perfusion medium for 15–20 min. Ca^{2+}-free conditions will disrupt the tight junctions, which intimately connect the parenchymal cells. This linear perfusion is followed by a third, recirculation perfusion with perfusion medium containing 0.05% collagenase and 1 m*M* $CaCl_2$ for 30 min. $CaCl_2$ is added to activate collagenase. After the perfusions, the Glisson's capsule is disrupted. At this stage the liver should have the consistency of a thick paste.

Optimal perfusion is critical for a good yield and viability of liver cells. Dead cells will lose their DNA into solution, which may lead to clumping. Clumping may be circumvented to some extent by adding 2% (w/v) bovine serum albumin (BSA) to the incubate.

The liver paste is agitated for 30 min in a 0.05% collagenase solution at 37°, which is constantly saturated with oxygen and carbon dioxide (95 : 5, v/v). Parenchymal cells are harvested by centrifugation at 50 *g* for 4 min, and nonparenchymal cells are harvested by centrifugation at 450 *g* for 10 min. The resulting total liver cell suspension contains about 100×10^6 parenchymal cells, 6×10^6 Kupffer cells, 15×10^6 endothelial cells, and less than 1×10^6 fat-storing cells per gram liver. Parenchymal cells can be easily harvested by differential centrifugation at 50 *g* for 4 min or even shorter, but this leads to substantial impurities of nonparenchymal cells. These include fat-storing cells which often remain attached to parenchymal cells. Centrifugal elutriation is the method of choice for obtaining both parenchymal and the different types of nonparenchymal cells in high purity. Contamination of nonparenchymal with parenchymal cell-derived cytoplasmic blebs[9] can be minimized by omitting the incubation step and by immediate separation of parenchymal and nonparenchymal cells using centrifugal elutriation.[10]

Isolation of Nonparenchymal Cells from Rat Liver Using Pronase and Collagenase

Alternative approaches for the isolation of nonparenchymal cells specifically include the selective destruction of parenchymal cells using pronase or enterotoxin. Conditions for the isolation of nonparenchymal cells are generally less critical than those described for parenchymal cells. The smaller nonparenchymal cells are less vulnerable to mechanical stress and low oxygen tension. The medium used in our procedures is Gey's balanced salt solution (GBSS) (68.5 mM NaCl, 2.5 mM KCl, 0.5 mM $CaCl_2 \cdot 2H_2O$, 0.45 mM $MgCl_2 \cdot 6H_2O$, 0.15 mM $MgSO_4 \cdot 7H_2O$, 0.85 mM $Na_2HPO_4 \cdot 2H_2O$, 0.1 mM KH_2PO_4, 1.35 mM $NaHCO_3$, and 5.5 mM glucose). Solutions are also perfused through the portal vein at 10 ml/min.

The nonparenchymal cell suspension is prepared by an initial *in situ* linear perfusion with 0.225% pronase E (Merck, Darmstadt, FRG) for 6 min. During this perfusion, the canula is fixed in the portal vein using two ligatures, and the liver is transferred to a sieve to enable the two following recirculation perfusions. Once the liver is placed on the sieve, dehydration is prevented by covering the liver with a tissue, temperature is maintained by placing the liver under a heat lamp. The liver is then perfused using 0.055% pronase and 0.055% collagenase in 60 ml buffer and recirculated for 30 min. After this perfusion the Glisson's capsule is disrupted. The liver cells are resuspended in 0.055% pronase and 0.025% collagenase and incubated for 30 min. The pH is constantly monitored and kept at 7.4 during the incubation using 1 N NaOH. The cell suspension is filtered over nylon gauze to remove any remaining Glisson's capsule. The cells are harvested by centrifugation at 450 g for 10 min. This incubation yields, on average, 7×10^6 Kupffer cells, 19×10^6 endothelial cells, and $5-10 \times 10^6$ fat-storing cells per gram liver.

The specific advantage of the pronase/collagenase digestion method is

[9] J. F. Nagelkerke, K. P. Barto, and T. J. C. van Berkel, *Exp. Cell Res.* **138,** 183 (1982).
[10] J. F. Nagelkerke, K. P. Barto, and T. J. C. van Berkel, *J. Biol. Chem.* **258,** 30 (1983).

the high yield of fat-storing cells (Table I). The major disadvantage of the isolation method using pronase is that freshly isolated cells will not retain intact membrane receptors. Specific disadvantages of the enterotoxin method center on the availability of the specific type of enterotoxin used in the method described and the lack of evidence for the absence of blebs in the isolated cell fractions.[11]

Isolation of Rat Liver Cells at Low Temperatures

Procedures for the isolation of parenchymal and nonparenchymal cells at low temperatures have been developed (Table I). Basically, parenchymal cells are isolated at 8° using perfusion medium without Ca^{2+} and without dissociating enzymes. The liver is first perfused *in situ* with perfusion medium containing 1 mM $CaCl_2$ for 10 min at 8°, followed by a recirculation perfusion with perfusion medium without $CaCl_2$ but containing 0.32% BSA for 20 min. The liver is cut and then agitated for 5 min in perfusion medium containing 0.32% BSA. Remaining pieces of liver are gently forced through nylon gauze. The gauze is washed with ice-cold perfusion medium containing 0.32% BSA. Agitating and filtering are repeated twice. Clumps of cells in the resulting suspension are left to sediment at 1 *g* for 10 min, the supernatant is removed, and the cells are harvested by centrifugation at 50 *g* for 4 min. This procedure results in an average yield of only 10×10^6 parenchymal cells per gram of liver.

The isolation procedure for the isolation of nonparenchymal cells was developed using a procedure at 10°.[22] The liver is perfused *in situ* with GBSS at 37° for 3 min and subsequently at 10° for 10 min with GBSS containing 0.2% pronase, followed by GBSS containing 1.3% BSA at 10° for 2 min. The liver is disrupted, the cells are suspended in ice-cold BSA-containing GBSS, the pH is adjusted to 7.4 using 1 N NaOH, and the cell suspension is filtered through nylon gauze. The cells are harvested by centrifugation at 450 *g* for 10 min. This incubation yields, on average, 7×10^6 Kupffer cells and 19×10^6 endothelial cells per gram liver.

The advantages of these cell isolation methods are (1) prevention of metabolism of substances ingested *in vivo;* (2) prevention of endocytosis of substances and cell debris during isolation; and (3) substantial retention of membrane receptors involved in endocytosis. Disadvantages of these methods, however, are that cell yield and cell viability may be low.

[11] R. Blomhoff, B. Smedrsod, W. Eskild, P. E. Granum, and T. Berg, *Exp. Cell Res.* **150,** 194 (1984).

Purification Procedures

Purification of Parenchymal Cell Fractions from Liver Cell Isolates

The parenchymal cell isolation methods described above provide cell fractions consisting of individual parenchymal cells containing all other cell types present in the liver. Furthermore, some of the parenchymal cells may still be attached to sinusoidal cells, specifically to fat-storing cells. Therefore, parenchymal cells purified using only differential centrifugation may contain considerable numbers of fat-storing cells. Such contaminations may be the reason for a temporary overestimation of the retinoid content of parenchymal cells.[3]

Parenchymal cell fractions may also be contaminated with parenchymal cell-derived structures, called blebs.[9] Blebs consist mainly of cytosol. These structures are specifically enriched in the top of density gradients (see below). Both the cellular complexes consisting of parenchymal cells with fat-storing cells as well as the blebs are lost during centrifugal elutriation procedures. The best purification of parenchymal cell fractions is obtained using centrifugal elutriation in a JE-6 elutriator rotor preferentially using a Sanderson chamber (Beckman Instruments, Palo Alto, CA). Centrifugal elutriation is based on the balance between a centrifugal force and a centripetal flow of liquid within a separation chamber. The Sanderson chamber can contain up to 100×10^6 parenchymal cells. At a rotor speed of 1350 rpm, parenchymal cells from BN/BiRij rats are elutriated between 32 and 41 ml/min. Theoretical aspects of this method of separation are described in great detail elsewhere.[12] On average, 80% of the cells are recovered after this centrifugal elutriation method. The cell fractions recovered have a purity up to 99%.

Purification of Fat-Storing Cell Fractions from Nonparenchymal Cell Isolates

The nonparenchymal cell suspension is further separated using isotonic density gradients. Several different materials for density gradients are known such as Metrizamide, Nycodenz, and Stractan. Nycodenz has the advantage over metrizamide in that it is autoclavable and less toxic for cells. Stractan is also nontoxic but is more elaborate to prepare. Fat-storing cells are separated by a gradient and not by centrifugal elutriation because the yield is higher.

The isolated nonparenchymal cell suspension is diluted in the gradient-

[12] A. Brouwer, R. J. Barelds, and D. L. Knook, *in* "Centrifugation, a Practical Approach" (D. Rickwood, ed.), p. 183. IRL Press, Oxford and Washington, 1984.

generating material up to a specific percentage. The percentage depends on the density of the fat-storing cells and is therefore directly dependent on the vitamin A status of the animal. In our hands, fat-storing cells from a liver containing 1000–2000 μg retinol equivalents/g are well separated on a gradient of 11.5% Nycodenz. This vitamin A status can be achieved by using rats of a certain minimum age (6–12 months, depending on the diet[3]) or animals that received supplementary vitamin A. In our procedures the total of $200-300 \times 10^6$ nonparenchymal cells are separated on two gradients of a total volume of 15 ml. The osmolality of the gradient is 308 mOsm.

The discontinuous gradient is constructed as follows. The 11.5% density layer of about 8 ml contains the cells to be separated. Under this part of the gradient 5 ml of 17.2% is layered using a Pasteur pipette. This layer is more dense than Kupffer and endothelial cells and less dense than erythrocytes and parenchymal cells. These two solutions are overlayed with 1 ml of GBSS. This gradient is spun at 1400 *g* for 17 min in a stable centrifuge at 8°. Acceleration and deceleration of the gradient during centrifugation has to be slow both to prevent swirling of the tube contents and to smooth centrifugal forces. After the spin the very top of the gradient, containing lipid droplets, is discarded. The top of the second layer (11.5%) contains purified fat-storing cells. The top of the 17.2% layer contains Kupffer and endothelial cells.

The two cell preparations are spun once (450 *g* for 10 min) to remove Nycodenz. An average of 80% of the fat-storing cells is recovered. The cell fraction recovered has a purity up to 80%. This cell fraction is mainly contaminated with endothelial cells and some lymphocytes and void of contaminating parenchymal cell blebs.[13,14]

Separation of Kupffer Cell Fractions and Endothelial Cell Fractions

The combined Kupffer and endothelial cell fraction is further separated into a Kupffer cell fraction and an endothelial cell fraction by centrifugal elutriation. The conditions for centrifugal elutriation of Kupffer and endothelial cells using the same elutriator rotor are a rotor speed of 3250 rpm. At a pump speed of 18 ml/min the lymphocytes and remaining fat-storing cells are elutriated and collected in a 100-ml volume, the endothelial cells are eluted at 33 ml/min in a volume of 150 ml, and the Kupffer cells are eluted at 58 ml/min in a total volume of 150 ml/min. An average of 80% of

[13] W. S. Blaner, H. F. J. Hendriks, A. Brouwer, A. M. de Leeuw, D. L. Knook, and D. S. Goodman, *J. Lipid Res.* **26,** 1241 (1985).

[14] H. F. J. Hendriks, W. S. Blaner, H. M. Wennekers, R. Piantedosi, A. Brouwer, A. M. de Leeuw, D. S. Goodman, and D. L. Knook, *Eur J.Biochem.* **171,** 237 (1988).

the Kupffer cells is recovered, and the cell fraction recovered has a purity up to 80%. An average of 80% of the endothelial cell fraction is recovered, and the cell fraction recovered has a purity up to 85%. Contaminations include mainly endothelial cells and lymphocytes, respectively.[13,14]

Characterization of Isolated Cell Fractions

Studies on isolated liver cells are meaningful only when based on reliable methods for assessment of purity, viability, and composition of the cell isolates. Cellular composition includes not only data on the purity of the cell type of interest, but also identification of the contaminating cell types. The choice of the type of analysis depends on the contamination to be expected and the parameters under study. Sometimes a combination of different markers including both morphological and biochemical techniques are required.

Viability of isolated fractions is crucial and can be easily estimated. Viability based on the integrity of the cell membrane is determined by the trypan blue [0.25% (w/v) in GBSS] exclusion test, but this is only a crude measure.

Light microscopic analysis is performed on the basis of cell size, ratio of nucleus size to cell size, and the presence of large intracellular structures. Furthermore, specific markers have been described for the different liver cell types. Markers which are applicable for light microscopical detection include specific enzyme activities and antigens recognizable by monoclonal or polyclonal antibodies. Characteristics applicable for light microscopical cellular identification include the following: cell size for parenchymal cells; lobulated nucleus, differently sized vesicles with varying intensities, and peroxidase staining for Kupffer cells[12]; round nucleus with a high nucleus to cytoplasm ratio and a staining for nonspecific esterase activity, which should be supported by establishing the absence of peroxidase activity, for endothelial cells[12]; lipid droplets with a highly specific, rapidly fading, pale blue fluorescence when excited with 325 nm light for fat-storing cells.[15] The intermediate filament protein vimentin and possibly also desmin may be used. These markers are not highly specific, however, because they are also present in fibroblasts and myofibroblasts.

The definite identification of liver cells can, however, only be achieved on the basis of ultrastructure, possible in combination with biochemical markers. Cell samples are fixed for electron microscopy in 2% glutaraldehyde in 0.15 *M* sodium cacodylate buffer (pH 7.4) for 15 min, followed by 1% OsO_4 in the same buffer for 20 min. The cell pellets are dehydrated in a

[15] A. M. de Leeuw, S. P. McCarthy, A. Geerts, and D. L. Knook, *Hepatology* **4**, 392 (1984).

series of graded ethanol solutions and embedded in LX 112 (Ladd Research Industry Inc., Burlington, VT). Sections are contrasted with uranyl acetate and lead citrate and examined under an electron microscope. Presently, the only way to unequivocally assess the quality and purity of fat-storing cell preparations, under experimental conditions that alter the cellular composition and phenotype, is by applying morphological criteria using electron microscopy. Markers requiring intact membrane receptors or antigens[16,17] can be applied only after an isolation method without pronase. Such markers include factor VIII-related antigen, Fc and mannose receptors, and the capacity to take up fluorescent acetylated low-density lipoproteins (LDL) for endothelial cells;[12,16] and phagocytosis of particulate material as well as Fc, C_3, mannose, and galactose receptors for Kupffer cells.[12,17,18] The presence of blebs in cell isolates can be tested biochemically[9] and by electron microscopy.

Concluding Remarks

Isolated liver cells have been widely used to study the cellular distribution of retinoids, retinoid-binding proteins, and enzyme activities important in retinoid metabolism.[3,6,13,14] The results obtained are generally consistent with *in vivo* data. Cell isolation procedures are currently being used to define further the respective roles of the different liver cells types in retinoid metabolism.

[16] A. Brouwer, E. Wisse, and D. L. Knook, *in* "The Liver: Biology and Pathobiology" (I. M. Arias, W. B. Jakoby, H. Popper, D. Schachter, and D. A. Shafritz, eds.), 2nd Ed., p. 665. Raven, New York, 1988.

[17] E. A. Jones and J. A. Summerfield, *in* "The Liver: Biology and Pathobiology" (I. M. Arias, W. B. Jakoby, H. Popper, D. Schachter, and D. A. Shafritz, eds.), 2nd Ed., p. 683. Raven, New York, 1988.

[18] H. Pertoft and B. Smedsrod, *in* "Cell Separation: Methods and Selected Applications," Vol. 4, p. 1. Academic Press, Orlando, 1987.

[6] Isolation and Cultivation of Rat Liver Stellate Cells

By RUNE BLOMHOFF and TROND BERG

Introduction

The liver consists of parenchymal cells (hepatocytes) and three main types of nonparenchymal cells: endothelial cells, Kupffer cells, and stellate cells. The parenchymal cells are considerably larger than the nonparenchy-

TABLE I
COMPOSITION OF RAT LIVER[a]

Cell type	Number of cells (% of total)	Mass (% of total)
Parenchymal	65	94
Endothelial	18	2
Kupffer	10	3
Stellate	7	1

[a] Data are taken from A. C. Munthe-Kaas, T. Berg, and R. Seljelid, *Exp. Cell Res.* **99,** 146 (1976); R. Daoust, *in* "Liver Functions" (R. W. Brauer, ed.), Publ. 4, p. 3. American Institute of Biological Sciences, Washington, D.C., 1958; D. L. Knook, A. M. Seffelaar, and A. M. de Leeuw, *Exp. Cell Res.* **139,** 468 (1982); and D. L. Knook, N. Blansjaar, and E. C. Sleyster, *Exp. Cell Res.* **109,** 317 (1977).

mal cells and account for about 94% of the total liver mass (Table I). The nonparenchymal cells represent about 35% of the total number of cells. It has become recognized that the nonparenchymal cells in the liver play an important role in hepatic metabolism of various compounds. The nonparenchymal cells may also cooperate with the parenchymal cells in the handling of nutrients.

Endothelial Cells

Endothelial cells form the continuous lining of the liver sinusoids and are a barrier between the blood and the parenchymal cells (Fig. 1). The endothelial cells in liver differ from most endothelial cells in the body with regard to both structure and function. The liver endothelial cells have fenestrations or pores which are about 0.1 μm in diameter. These fenestrations, which are grouped together in so-called sieve plates, influence the filtration of particles from the blood to the parenchymal cells and vice versa.[1] The pore sizes, and therefore control of filtration, can be influenced by certain hormones.[2] Endothelial cells in liver have an important function

[1] E. Wisse and D. L. Knook, *in* "Progress in Liver Diseases" (H. Popper and F. Schaffner, eds.), Vol. 6, p. 153. Grune & Stratton, New York, 1979.

[2] J. H. Van Dierendonck, J. van der Meulen, R. B. De Zanger, and E. Wisse, *Ultramicroscopy* **4,** 149 (1979).

in the clearance and degradation of various plasma components by receptor-mediated endocytosis.[3-6]

Kupffer Cells

The Kupffer cells are the largest group of tissue macrophages in most mammals. A large part of the cell surface is exposed to the blood flow through the liver and to the space of Disse (Fig. 1). The main function of Kupffer cells is to phagocytose foreign particles, such as bacteria or colloids.[1] They comprise about 25% of the nonparenchymal cells (Table I).

Perisinusoidal Stellate Cells

Stellate cells (also called fat-storing cells, lipocytes, and Ito cells) are localized within the space of Disse parallel to the endothelial lining[7] (Fig. 1). The stellate cells have a role in the metabolism of vitamin A,[8] which is stored in the many fat droplets found in the cytoplasm of these cells.[7] Stellate cells also have a role in the synthesis of connective tissue components and may be involved in the pathological changes observed during the development of liver fibrosis.[7] Stellate cells represent about 7% of the cells in liver (Table I).

Role of Various Liver Cells in Retinoid Metabolism

Most of the newly absorbed retinyl esters accompany chylomicron remnants when they are taken up by the liver parenchymal cells via receptor-mediated endocytosis.[9] The retinol taken up by parenchymal cells may subsequently be transferred to perisinusoidal stellate cells in liver for storage.[10]

Under normal circumstances about 50–80% of the total retinoids in the body are stored in the liver. Mass analysis of retinoids in isolated liver cells has revealed that 80–90% of the total retinol in liver is located in the perisinusoidal stellate cells as retinyl esters, with the rest in liver parenchy-

[3] A. L. Hubbard, G. Wilson, G. Ashwell, and H. Stukenbrok, *J. Cell Biol.* **83,** 47 (1979).

[4] R. Blomhoff, C.A. Drevon, W. Eskild, P. Helgerud, K. R. Norum, and T. Berg, *J.Biol. Chem.* **259,** 8898 (1984).

[5] J. F. Nagelkerke, K. P. Barto, and T. J. C. van Berkel, *J. Biol. Chem.* **258,** 12221 (1983).

[6] B. Smedsrød, H. Pertoft, S. Eriksson, J. R. E. Fraser, and T. C. Laurent, *Biochem. J.* **223,** 617 (1984).

[7] K. Wake, *Int. Rev. Cytol.* **66,** 303 (1980).

[8] R. Blomhoff, *Nutr. Rev.* **45,** 257 (1987).

[9] R. Blomhoff, P. Helgerud, M. Rasmussen, T. Berg, and K. R. Norum, *Proc. Natl. Acad. Sci. U.S.A.* **79,** 7326 (1982).

[10] R. Blomhoff, K. Holte, L. Naess, and T. Berg, *Exp. Cell Res.* **150,** 186 (1984).

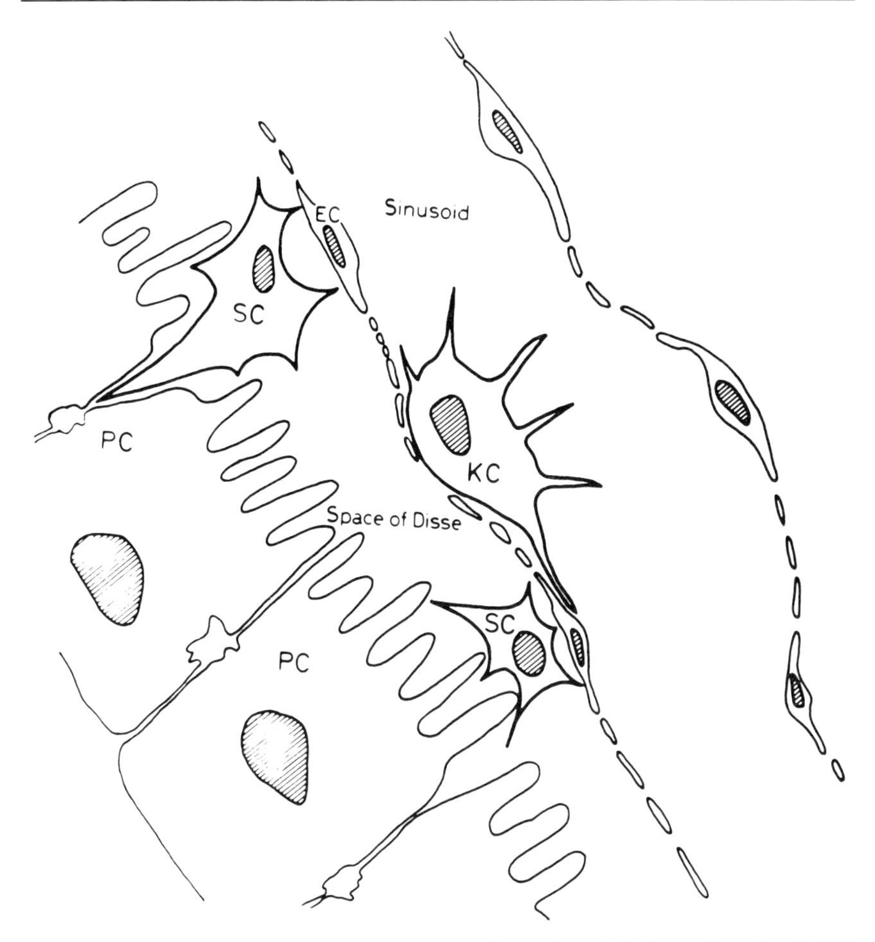

FIG. 1. Schematic diagram of the structure of the liver. PC, Parenchymal cell; EC, endothelial cell; KC, Kupffer cell; SC, stellate cell.

mal cells.[11-12] The liver is a main organ for uptake of the retinol–retinol-binding protein (RBP) complex from plasma.[13,14] Parenchymal as well as

[11] R. Blomhoff, M. Rasmussen, A. Nilsson, K. R. Norum, T. Berg, W. S. Blaner, M. Kato, J. R. Mertz, D. S. Goodman, U. Eriksson, and P. A. Peterson, *J. Biol. Chem.* **260,** 13560 (1985).

[12] W. S. Blaner, H. F. J. Hendriks, A. Brouwer, A. M. de Leeuw, D. L. Knook, and D. S. Goodman, *J. Lipid Res.* **26,** 1241 (1985).

[13] R. Blomhoff, K.R. Norum, and T. Berg, *J. Biol. Chem.* **260,** 13571 (1985).

[14] T. Gjøen, T. Bjerkelund, H. K. Blomhoff, K. R. Norum, T. Berg, and R. Blomhoff, *J. Biol. Chem.* **262,** 10926 (1987).

stellate cells take up approximately the same amount of retinol–RBP complex when expressed as uptake per gram of liver.

We have recently used several approaches to elucidate the mechanism of transfer of retinol between liver parenchymal and stellate cells. Because parenchymal cells secrete retinol bound to RBP, and we had shown that stellate cells may take up retinol–RBP, we tested the possibility that RBP mediates the transfer of retinol from parenchymal to stellate cells in the liver. In an *in situ* liver perfusion system we were able to show that antibodies against RBP completely blocked the transfer of retinol to stellate cells.[15] These findings suggest that retinol secreted from parenchymal cells on RBP is taken up by stellate cells via RBP receptors.

Although rat liver parenchymal cells may be easily recovered in high yield, until recently few simple methods had been developed for the preparation of the individual nonparenchymal cell types. The role of stellate cells in the hepatic metabolism of retinol has been determined owing to development of isolation and cultivation techniques for the nonparenchymal liver cells.

Enzymatic Perfusion of Liver

Most of the newly developed methods include either collagenase perfusion[16] or a combined collagenase/pronase perfusion[17] of intact liver. Collagenase perfusion provides a suspension of all types of liver cells. Nonparenchymal cells can then be prepared after removal of the larger parenchymal cells by low-speed differential centrifugation. Alternatively, protease or enterotoxin treatment of the total cell suspension may be used to selectively destroy the parenchymal cells.[18]

Based on our experience, collagenase perfusion is the best method for enzymatic digestion of the liver. First, in contrast to the collagenase/protease perfusion method, it produces a suspension of all the liver cells, including the parenchymal cells, with high yield and viability. The cells can subsequently be separated into the individual cell types. This is especially important when studying the interplay of the various cells on the hepatic metabolism of components. Second, although protease treatment destroys most cell surface proteins, very little proteolytic degradation of cell surface proteins takes place during the collagenase perfusion period. This is im-

[15] R. Blomhoff, T. Berg, and K. R. Norum, *Proc. Natl. Acad. Sci. U.S.A.* **85,** 3455 (1988).

[16] P. Seglen, *Methods Cell Biol.* **13,** 29 (1976).

[17] D. L. Knook, A. M. Seffelaar, and A. M. de Leeuw, *Exp. Cell Res.* **139,** 468 (1982).

[18] R. Blomhoff, B. Smedsrød, W. Eskild, P. E. Granum, and T. Berg, *Exp. Cell Res.* **150,** 194 (1984).

portant when studying receptor-mediated endocytosis or when using antibodies to characterize cell surface proteins on the isolated cells.

Collagenase Perfusion of Liver

Buffers

Preperfusion buffer: 8.0 g NaCl, 0.4 g KCl, 60 mg $Na_2HPO_4 \cdot 2H_2O$, 47 mg KH_2PO_4, 0.20 g $MgSO_4 \cdot 7H_2O$, and 2.05 g $NaHCO_3$ dissolved in water to 1000 ml final volume

Perfusion buffer: 25–50 mg collagenase (130–150 units/mg solid) and 100 μl of 1.0 *M* $CaCl_2$ made up to 50 ml with the preperfusion buffer

Both the preperfusion and perfusion buffers are gassed with 95% O_2/5% CO_2 to pH 7.5 10 min before and during perfusion.

Incubation buffer: 8.5 g NaCl, 0.4 g KCl, 60 mg $Na_2HPO_4 \cdot 2H_2O$, 47 mg KH_2PO_4, 0.20 g $MgSO_4 \cdot 7H_2O$, 4.76 g HEPES, and 0.29 g $CaCl_2 \cdot 2H_2O$ dissolved in about 800 ml water and adjusted to pH 7.5 with NaOH; make up to 1000 ml with water (the osmolality of this solution is approximately 298 mOsm)

Perfusion Apparatus. A simple perfusion apparatus is shown in Fig. 2. With this apparatus it is possible to perfuse the tissue with solutions preequilibrated at 37°. It also allows switching from the preperfusion buffer to the perfusion buffer without any air bubbles entering the system. Air bubbles cause air blocks and prevent efficient perfusion of the liver. The pump should be capable of a variable flow rate, from 0 to 50 ml/min.

Perfusion. Expose the entire abdomen of a deeply anesthetized rat by a transverse cut across the lower abdomen and a longitudinal cut up to the sternum. Gently move the gut contents to the left side of the rat to expose the inferior vena cava and the vena porta. Pass a ligature around the vena porta about 1 cm from its junction with the liver capsule and tie loosely. With a very fine-pointed scissors, cut through the vena porta, about half its circumference, 0.5 cm on the caudal side of the ligature. Insert the cannula into the vena porta to a point 2–5 mm past the ligature. Do not enter the liver capsule as this will reduce the yield of cells considerably. During this operation the flow rate of the buffer should be reduced to about 10 ml/min. When the cannula is properly in place, tighten the ligature and tie securely. Immediately sever the inferior vena cava to permit a free flow of the preperfusion buffer and then increase the flow rate to 50 ml/min. This treatment disrupts the desmosomal cell junctions which are calcium-dependent. The rat should now be sacrificed by puncturing the diaphragm.

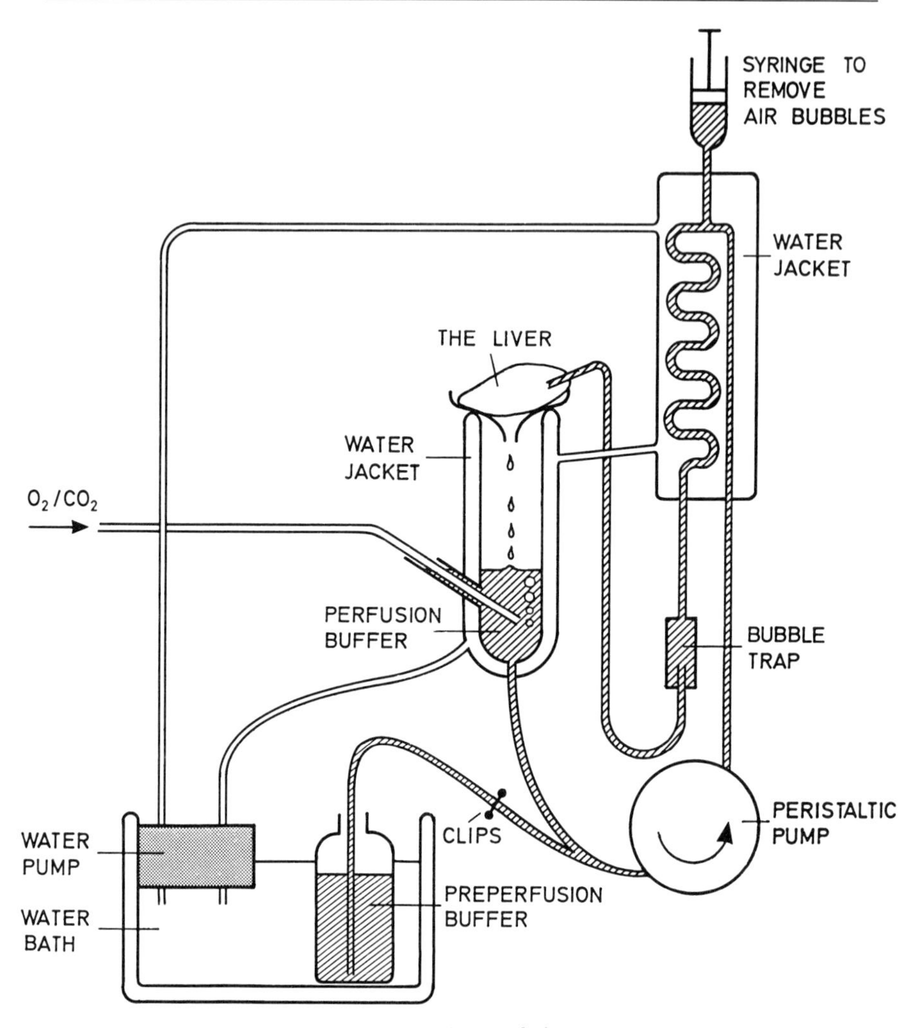

FIG. 2. Recirculating liver perfusion apparatus.

Free the liver by carefully cutting it away from all attachments, taking care not to damage the liver capsule.

Continue the preperfusion for 10 min with the liver *in situ,* the lobes lying in their natural position, thus ensuring a flow of buffer to all parts of the liver. Then switch to the perfusion buffer and allow 30 sec to elapse before transferring the liver to the perfusion dish so that recirculation of the perfusion buffer can take place. Continue the perfusion for 10 min, by which time it should be possible to see signs of the liver disintegrating within the capsule. Disconnect the cannula and place the liver into a petri dish containing ice-cold incubation buffer. Cut away all extraneous tissue from the liver, taking care not to damage the capsule. Transfer the liver to another dish of incubation medium and disrupt the capsule, using forceps, and free the cells into the medium by fairly vigorous agitation of the forceps in the medium. Transfer the capsule remnants to fresh medium and free any remaining cells in the same way. Discard the capsule remnants and combine the contents of the two dishes. The total volume of buffer into which the cells are isolated should be about 50 ml. Subsequent treatment of the cell suspension will depend on how the cells will be used.

Differential Centrifugation of Total Liver Cell Suspension

The parenchymal cells are easily separated from the nonparenchymal cells by differential centrifugation.[19] The average volume of the parenchymal cell is more than 10 times that of the average nonparenchymal cell, and centrifugation of a liver cell suspension at low speed (about 50 *g*) for 20–25 sec at 4° gives a fairly good separation of the parenchymal and nonparenchymal cells. The pellet, containing the bulk of the parenchymal cells, can be washed repeatedly to give a relatively pure suspension of parenchymal cells. (To get rid of all contaminating stellate cells, centrifugal elutriation is required.[14]) The supernatant is enriched in nonparenchymal cells but still contaminated with some parenchymal cells. Dead parenchymal cells do not sediment during low-speed centrifugation and will therefore be selectively retained in the supernatant. Nevertheless, by repeated centrifugation at 600 *g* for 4 min at 4°, a fairly pure preparation of nonparenchymal cells may be obtained from the initial supernatant after sedimentation of the parenchymal cells.

Cell viability can be evaluated by the exclusion of trypan blue (0.04% in incubation buffer): viable cells exclude trypan blue, whereas dead cells accumulate the blue dye. Viability of parenchymal as well as the nonparenchymal cells should always be more than 95%.

The nonparenchymal cells collected using the above methods may be separated into their individual cell types by further steps which involve density-gradient centrifugation, centrifugal elutriation, or selective attachment to culture dishes. The method chosen depends on the purpose of the experiment. Very pure stellate cells may be isolated from the nonparenchymal cell suspension by centrifugal elutriation[17] or Percoll density-gradient centrifugation.[10] These methods will, however, lead to a low yield of stellate cells. About $1-3 \times 10^6$ cells per gram liver may be isolated,

[19] M. Nilsson and T. Berg, *Biochim. Biophys. Acta* **497**, 171 (1977).

representing a recovery of about 8–25%. The yield of cells may increase with the age of the rat. No method is available at present to isolate pure stellate cells in higher yield. As centrifugal elutriation needs a special and expensive centrifugation rotor, and Percoll density-gradient centrifugation is equally efficient, the latter is the method of choice for most purposes.

Density-Gradient Centrifugation in Percoll Gradients

For density-gradient centrifugation in Percoll gradients[10] (Fig. 3), non-parenchymal cells are first separated from parenchymal cells by differential centrifugation of a total liver cell suspension as described above. Percoll self-forming gradients are generated by mixing 7.5 ml Percoll solution [9 parts Percoll and 1 part 9% (w/v) NaCl] with 5 ml cell suspension and centrifuging of 40,000 *g* for 60 min at 4°. In gradient fractions with densities of 1.025–1.035 g/ml almost all cells are stellate cells as judged by vitamin A autofluorescence. The cells contain considerable amounts of vitamin A, making identification by fluorescent microscopy easy. Alternatively, the cells can be identified by transmission electron microscopy (Fig. 4A).

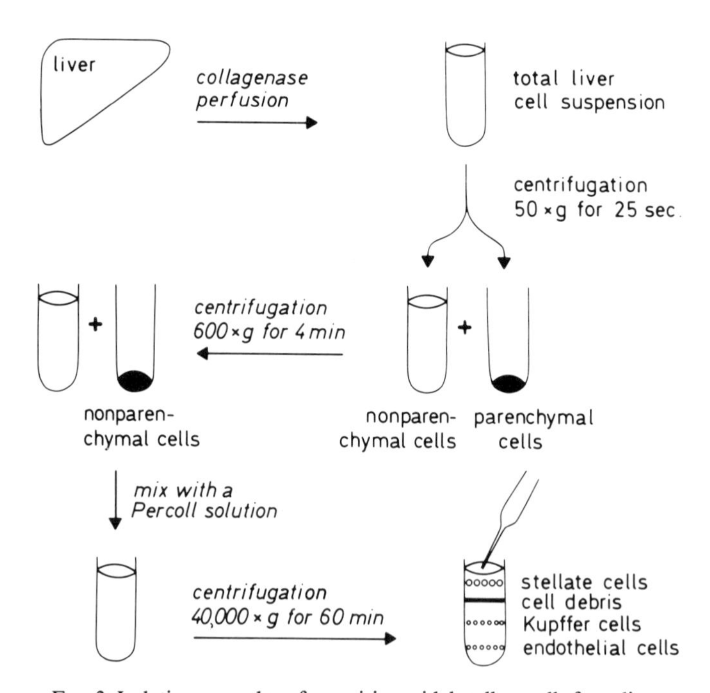

FIG. 3. Isolation procedure for perisinusoidal stellate cells from liver.

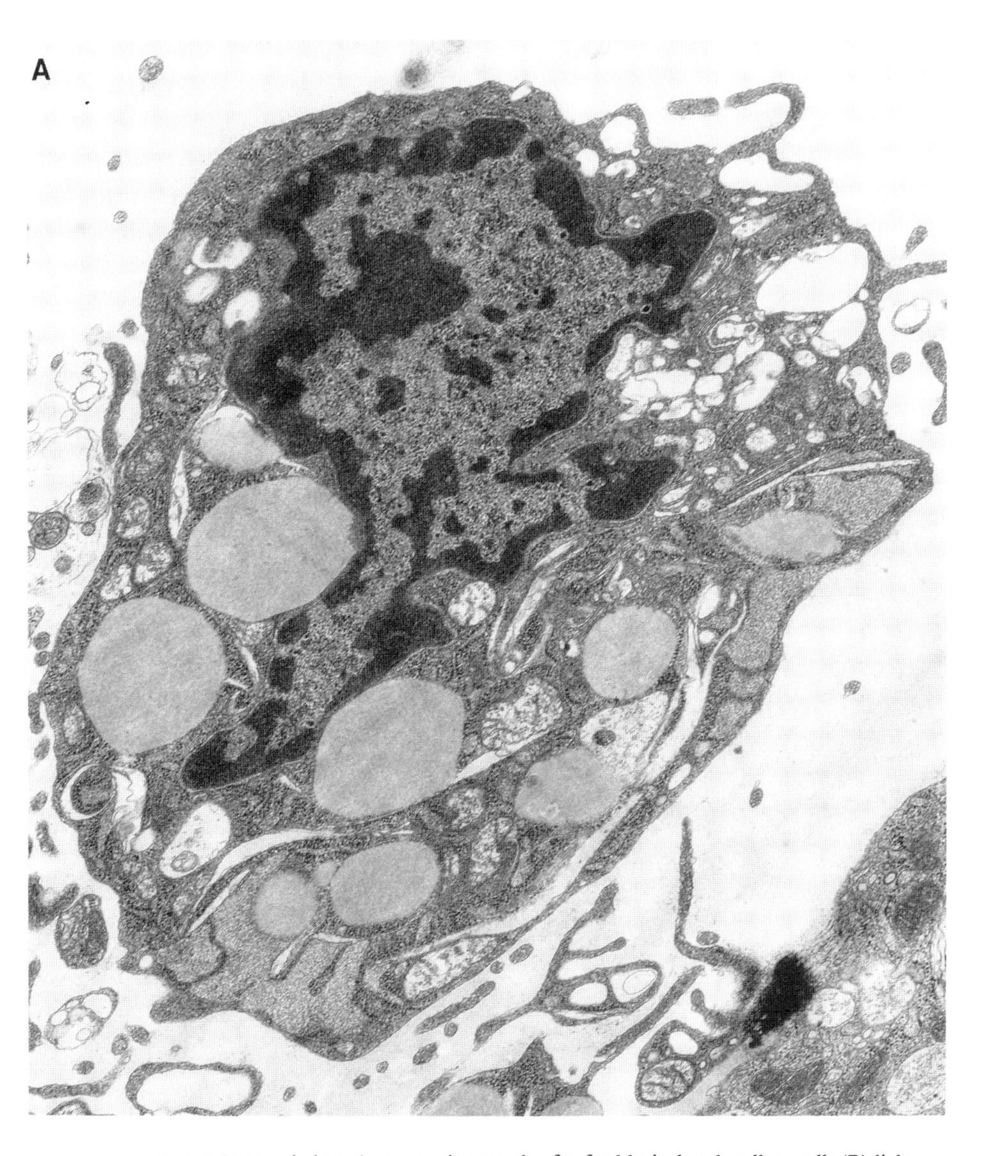

FIG. 4. (A) Transmission electron micrograph of a freshly isolated stellate cell; (B) light micrograph of stellate cells cultivated for 2 weeks; (C) fluorescence micrograph of a cultivated stellate cell.

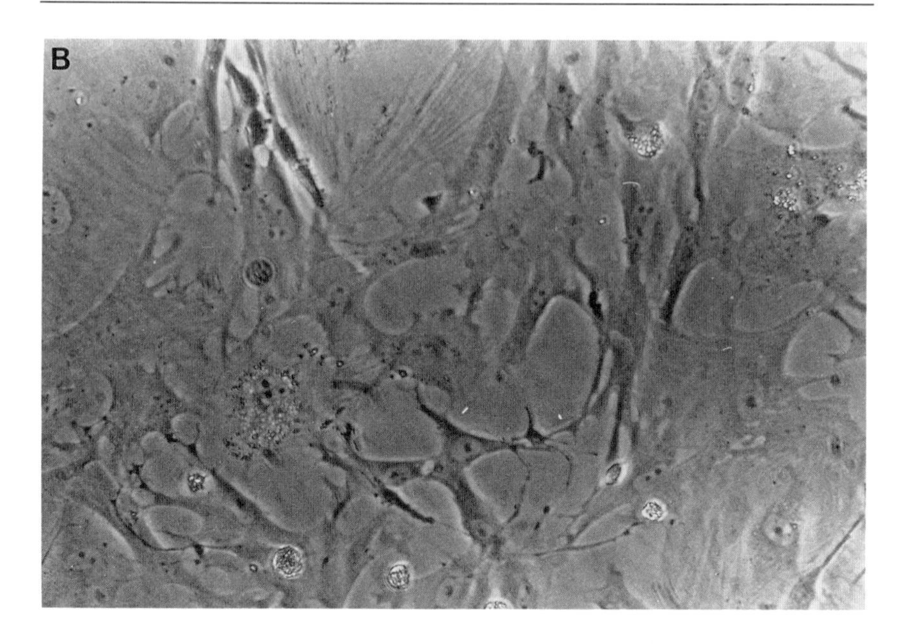

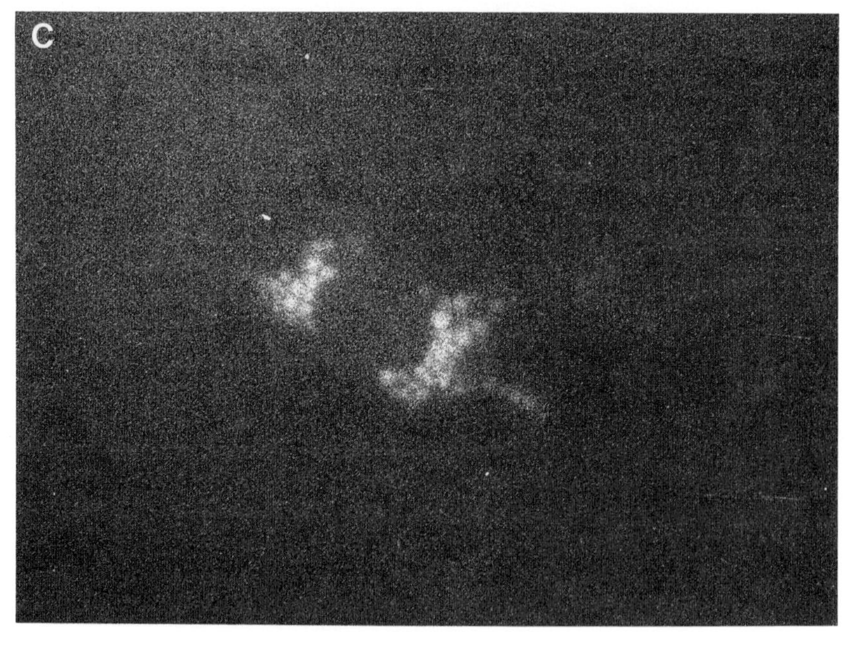

FIG. 4. *(continued)*

Cultivation of Stellate Cells

When isolating cells for cultivation, all solutions should be sterile-filtered. The perfusion apparatus and tubings should be washed thoroughly in 70% ethanol and then in sterile water and buffer before use. All surgical equipment should be stored in 70% ethanol, and the abdomen of the rat shaved and washed with ethanol before surgery.

Primary cultures of stellate cells may be obtained by several methods. The cells isolated by Percoll density-gradient centrifugation or centrifugal elutriation may be washed once or twice in cultivation medium and seeded. As noted earlier, these methods do have the drawback of yielding only a few stellate cells. We have recently used selective detachment during cultivation to obtain a much higher yield of cultivated stellate cells. This procedure is much shorter, and it is easier to obtain sterility. We are now regularly using this method for the cultivation of stellate cells.

Nonparenchymal cells isolated by differential centrifugation of a total liver cell suspension should be washed twice in sterile incubation buffer. No extensive washing to remove contaminating parenchymal cells is needed. Cells are then suspended in Dulbecco's modified Eagle's medium (DMEM) containing L-glutamine (0.6 mg/ml), penicillin (100 U/ml), streptomycin (100 U/ml), and 20% fetal calf serum (FCS) and seeded at a density of about 10^6 cells/cm^2. The cells are cultivated at 37° in a humidified atmosphere containing 5% CO_2 in air. Most of the cells seeded will attach, and the 1-day-old cell culture contains mostly endothelial cells and some Kupffer and stellate cells. The medium is changed after overnight culture and then changed every second day. Under such conditions, endothelial and Kupffer cells do not proliferate, and they detach after 2–4 days in culture. The stellate cells, however, start to proliferate, and after 1 week the culture contains stellate cells exclusively, as judged by desmin immunostaining.[20] After 1–2 weeks the cultures are confluent. The cultivated stellate cells show a characteristic stellate shape with long cytoplasmic processes resembling those present *in situ* (Fig. 4B).

Stellate cells can be subcultivated by trypsination through several passages. The cultures are washed in medium without FCS and trypsinated with DMEM containing trypsin (2.5 mg/ml) and EDTA (0.25 mg/ml) and no FCS. Detaching of the cells can be inspected by microscopy. After 5–10 min FCS is added to a final concentration of 20%, and the cells are suspended carefully by pipetting up and down and washed for 10 min at 450 *g*. After dilution (1:4) in cultivation medium and replating, most cells reattach within 30 min, and the respreading process takes about 6 hr.

[20] Y. Yokoi, T. Namihisa, H. Kuroda, I. Komatsu, A. Miyazaki, S. Watanabe, and K. Usui, *Hepatology (Baltimore)* **4,** 709 (1984).

Thereafter cells start to proliferate, and a confluent culture is obtained after about 1 week.

The characteristic lipid droplets containing vitamin A are present in the primary culture but become smaller with increasing culture time, as judged by transmission electron microscopy. Vitamin A fluorescence (Fig. 4C) can be detected in cells cultured up to the second passage; however, the intensity of fluorescence decreases with time. The decrease in intensity is mainly due to the dilution of vitamin A concentration during cell division and the mobilization of retinol to the medium.

When stellate cells are cultivated on uncoated plastic *in vitro,* they undergo a transition to proliferating myofibroblast-like cells, which is paralleled by a drop in retinyl ester content and a more than 10-fold increase in secretion of collagen (mainly type I).[21] Friedman *et al.*[22] demonstrated that cultivation of stellate cells on plastic coated with a basement membrane-like matrix inhibits such transformation. These results show that the extracellular matrix can modulate matrix production and dedifferentation of stellate cells *in vitro.*

Identification and Characterization of Stellate Cells

Freshly isolated stellate cells may be distinguished from endothelial and Kupffer cells by light microscopy. The lipid droplets of the cells can easily be detected, and the cells resemble a bunch of grapes.[23] Stellate cells may also be identified in a fluorescence microscope with a proper filter block. Retinol and its esters fluoresce after excitation at 330 nm. The lipid droplets of the stellate cells show intense, but rapidly fading vitamin A autofluorescence.[7] When preparing cells for fluorescence microscopy, care should be taken to use dim light and not to expose the cells to sunlight as this will destroy the autofluorescence of the cells.

The lipid droplets in stellate cells can also be detected by transmission electron microscopy.[7] The number and size of the lipid droplets vary considerably. Typically, the nucleus is indented by the lipid droplets. In culture, the characteristic lipid droplets remain present for 1–2 weeks, although the larger ones seemed to be split into droplets of smaller size.

The stellate cells produce several types of cytoskeletal and connective tissue proteins, such as vimentin, actin, tubulin, various collagens, fibronectin, and laminin.[23] Most of these are also present in other cell types

[21] B. H. Davis, B. M. Pratt, and J. A. Madri, *J. Biol. Chem.* **262,** 10280 (1987).

[22] S. L. Friedman, F. J. Roll, J. Boyles, and D. M. Bissell, *Proc. Natl. Acad. Sci. U.S.A.* **82,** 8681 (1985).

[23] A. M. de Leeuw, S. P. McCarthy, A. Geerts, and D. L. Knook, *Hepatology (Baltimore)* **4,** 392 (1984).

such as fibroblasts and may therefore not be used as a specific marker. However, staining with desmin antibodies represent a specific method for identification of the cells.[20] Neither endothelial cells, Kupffer cells, nor fibroblasts are stained with desmin antibodies.

Acknowledgments

Our research has been supported in part by grants from the Norwegian Cancer Society and The Norwegian Research Council for Science and the Humanities.

[7] Sertoli Cells of the Testis: Preparation of Cell Cultures and Effects of Retinoids

By ALICE F. KARL and MICHAEL D. GRISWOLD

Introduction

Vitamin A in the form of retinol is essential for mammalian spermatogenesis.[1] All cell types in the testis including Leydig cells, germinal cells, peritubular myoid cells, and the Sertoli cells are potential candidates for the sites of vitamin A action. A major focus of scientific investigation has been on the Sertoli cells because they are the cells which provide physical and biochemical support to the differentiating germinal cells.[2] The tight junctions between adjacent Sertoli cells separate the seminiferous tubules into a basal and an adluminal compartment so that the biochemical environment of the adluminal compartment in which the meiotic stages of the germinal cells are sequestered is to a large extent determined by the Sertoli cells.[2] A number of the secreted protein products of Sertoli cells have been isolated and characterized, and both antibodies and cDNA probes are available. One such protein, transferrin, has been utilized by a number of investigators as an index of Sertoli cell function.[2]

Actions of Retinoids on Cultured Sertoli Cells

Sertoli cells contain relatively high levels of cellular retinol-binding protein (CRBP) but very little, if any, cellular retinoic acid-binding protein

[1] J. N. Thomson, J. McC. Howell, and G. A. Pitt, *Proc. R. Soc. London B* **159,** 510 (1964).
[2] M. D. Griswold, C. Morales, and S. R. Sylvester, *Oxford Rev. Reprod. Biol.* **10,** 124 (1988).

(CRABP), whereas the situation is reversed in advanced germinal cells.[3-5] Levels of transferrin and transferrin mRNA in cultured Sertoli cells and in vitamin A-deficient rats are responsive to added retinoids. Both retinol and retinoic acid will stimulate the secretion of transferrin from cultured Sertoli cells. The K_D for this response is approximately 0.1 μM, and transferrin secretion is increased approximately 2-fold over controls after 3 days in culture.[6] In the presence of exogenous retinol, transferrin mRNA levels are increased by 50% in cultured Sertoli cells from normal 20-day-old rats and approximately 3-fold in cultured Sertoli cells from vitamin A-deficient rats.[7,8]

Studies of retinoid metabolism in cultured Sertoli cells show that retinol is rapidly taken up and immediately esterified.[9] Sertoli cells contain large pools of retinyl esters but no detectable retinoic acid. Isolated germinal cells also take up retinol and convert it to esters.[10] A model for the action of retinoids has been proposed whereby retinol is taken up by specific receptors on the basal surface of Sertoli cells and esterfied. In this model, part of the ester pool is then transported to the germinal cells. Thus, it is proposed that retinoic acid is unable to support spermatogenesis because it cannot form retinyl esters.[10]

Preparation of Primary Cultures of Sertoli Cells

Techniques for serum-free culture of rat Sertoli cells were initially developed in 1975.[11-13] The procedure has been modified over time by a number of other investigators. We have cultured Sertoli cells from bulls, sheep, dogs, rabbits, mice, and rats. The original techniques were developed for use on 20-day-old rats because the entire array of the germinal cells is not yet present in rats of this age and a relatively pure monolayer of nondividing Sertoli cells can be obtained. It is important to emphasize that

[3] M. K. Skinner and M. D. Griswold, *Biol. Reprod.* **27,** 211 (1982).
[4] J. I. Huggenvik, R. L. Idzerda, L. Haywood, D. C. Lee, G. S. McKnight, and M. D. Griswold, *Endocrinology* **120,** 332 (1987).
[5] S. Hugly and M. D. Griswold, *Dev. Biol.* **121,** 316 (1987).
[6] P. D. Bishop and M. D. Griswold, *Biochemistry* **26,** 7511 (1987).
[7] M. D. Griswold, P. D. Bishop, K. Kim, R. Ping, J. S. Siiteri, and C. Morales, *Ann. N.Y. Acad. Sci.* (in press).
[8] J. Huggenvik and M. D. Griswold, *J. Reprod. Fertil.* **61,** 403 (1981).
[9] S. B. Porter, D. E. Ong, F. Chytil, and M. C. Orgebin-Crist, *J. Androl.* **6,** 197 (1985).
[10] W. S. Blaner, M. Galdieri, and D. S. Goodman, *Biol. Reprod.* **36,** 130 (1987).
[11] J. H. Dorrington and I. B. Fritz, *Endocrinology* **96,** 879 (1975).
[12] A. Steinberger, J. J. Heindel, J. N. Lindsey, J. S. H. Elkington, B. M. Sanborn, and E. Steinberger, *Endocr. Res. Commun.* **2,** 261 (1975).
[13] M. J. Welsh and J. P. Weibe, *Endocrinology* **96,** 618 (1975).

if the cells are obtained from a 15-day-old or older rat, the Sertoli cells will have essentially ceased mitotic activity. Sertoli cells from this age of animal appear to have many biochemical properties similar to those in the mature animal.[2]

Materials Required

Hank's basic salt solution lacking Mg^{2+} and Ca^{2+} (HBS); just before use, the pH is adjusted to 7.4 with sterile 420 m*M* sodium bicarbonate solution

Deoxyribonuclease I, 6.64 mg/ml in HBS (584 U activity/mg; Sigma, St. Louis, MO)

Trypsin 2.5%, 10× solution as supplied by Gibco (Grand Island, NY) and reconstituted with HBS just prior to use

Soybean trypsin inhibitor 0.34 g per 100 ml in HBS (10,000 U activity/mg; Gibco) filter-sterilize (2 μm)

Collagenase type II (70 mg per 100 ml HBS, 420 U/mg; Sigma); filter-sterilize (0.2 μm)

Two sets of surgical scissors with one blunt tip, and forceps

Two small plastic petri dishes

Two sterile, foil-covered Erlenmeyer flasks

Table-top centrifuge and water bath

Procedure. We have found that 10 rats which are 20 days of age make a convenient sized preparation. The yield is approximately 20 tissue culture dishes of 60 × 15 mm which will contain about 10^7 cells per dish. Rats are sacrificed by cervical dislocation, the abdomen is wiped with 70% ethanol, and testes are removed after making an abdominal incision. The tissue close to the testis (fat and epididymis) is cut away, and the testes are immersed in HBS in a petri dish. Rupturing the gut during this procedure must be avoided. With forceps, the testes are then placed on a piece of Parafilm for removal of the tunica. Each testis is grasped at one end, and a small incision is made in the tunica at the opposite end. The tubules are then pushed out of the tunica using the side of a closed scissors blade. The tubule bundle is immersed in HBS in a second petri dish. After dissection of all of the testes the tissue is weighed after removal of HBS and then covered to keep from drying. All procedures from this point are kept sterile, and containers are opened only within a sterile tissue culture hood.

The first Erlenmeyer flask is prepared by addition of trypsin (5 ml of Gibco 10× solution), HBS (45 ml), and DNase (200 μl). These values are for 2 g of wet tissue, the amount which would be obtained from 10 rats at 20 days of age, and should be adjusted for the actual weight. The tissue is placed onto an ethanol-cleaned surface suitable for chopping and cut with razor blade into pieces approximately 10 mm in length. The tissue is added to the Erlenmeyer flask, which is then covered with foil and incubated in a water bath at 32° for 25 min with occasional swirling. During the last 10 min of the digest, the tissue should come apart rapidly until all tubules are separated. The Erlenmeyer flask is then moved to a sterile tissue culture hood, and the tubule fragments are allowed to settle for up to 5 min. Most of the overlying liquid is removed and discarded. Trypsin inhibitor (5 ml) is added, and the mixture is swirled and allowed to stand for at least 1 min. More HBS is added (50 ml), and the tissue is swirled and allowed to settle again. The supernatant is removed, and more HBS is added as before (40 ml). This settling step is repeated 2 × in order to remove single cells, which include some germinal cells and interstitial cells.

After the final wash following trypsin incubation, the supernatant is removed and HBS is added. The volume of the added HBS should be equal to the volume of the collagenase solution that is to be used (in this example, 10 ml). The second Erlenmeyer flask is prepared by adding HBS (5 ml), collagenase (5 ml), and DNase (50 μl). Again, note that these volumes are for every 2 g of starting tissue and should be adjusted accordingly. The tissue suspension is then added to the collagenase solution, the container is covered with foil, and the mixture is swirled and placed in a water bath (32°) and incubated for 15 min. During the last 5 min of incubation, aliquots are checked microscopically to monitor the release of peritubular cells. A drop of the suspension is placed on a glass slide, and the tissue is viewed under a coverslip with an inverted microscope using a 10× objective. The coverslip is moved slightly, and as the tissue moves, the peritubular cells can be seen to peel off. The tissue is very fragile at this stage and must not be allowed to overdigest. Digestion is complete when the surface of the tubular fragments no longer appears smooth owing to removal of the collagen layer and associated cells. At the end of the incubation, the mixture is poured into a 12-ml graduated conical centrifuge tube(s) and centrifuged at 200 *g* for 2 min. This is a gentle centrifugation designed to pellet tubule fragments and leave single cells, which at this stage are primarily peritubular myoid cells in suspension. This wash step is usually done at least twice. A preparation which starts with the tissue from 10 rats at 20 days of age should yield, on average, about 1 ml of lightly packed tubule fragments (measured in a conical 12-ml centrifuge tube).

After the final centrifugation, 1.25 ml HBS per 0.1 ml packed tubule fragments is added, and the tissue is suspended and plated into plastic tissue culture dishes at 0.5 ml per 60 × 15 mm tissue culture dish. Cultures can be maintained in 5 ml Ham's F12 medium supplemented with 0.215 g glutamine per liter, 1.5 m*M* HEPES, and antibiotics. Cells are incubated at 32° in 5% CO_2, in a humidified chamber, and the medium is changed every 48 hr. The tubule fragments will attach in a few hours, and Sertoli cells will begin to spread over the surface of the dish. After 48 hr, the spreading of the Sertoli cells from the tubule fragments should be evident, and the germinal cells will be released into the culture medium or can be seen attached to the upper surface of the Sertoli cells. The cultures can be gently swirled during the first medium change in order to assist in the detachment of germinal cells.

This procedure is a dynamic one that needs frequent monitoring to optimize each step for a satisfactory product. Some absolutes for making the procedure work well are as follows: (1) HBS must be free of Mg^{2+} and Ca^{2+}, and the pH of the HBS solution must be monitored carefully throughout the procedure. (2) DNase activity is unstable, and, therefore, the enzyme should be weighed out and put into solution shortly before use and not used after more than one freeze–thaw cycle. Amounts of DNase used in the procedure are starting points, and if during the digestions the tissue mixture is viscous, more DNase should be added. The most common reason for problems in this procedure is the failure to adequately digest DNA from the broken cells.

Sertoli cell preparations are considered to be enriched but not pure cell preparations. A good preparation should be greater than 95% Sertoli cells, but there are always a small number of myoid cells remaining. Culture conditions requiring addition of serum will allow myoid cells to proliferate, and the cultures will be taken over by them after a few days.

Either retinol or retinoic acid (0.35 μM) is added separately or in conjunction with other hormones at the time of plating of the cells. Retinoid stock solutions are made up in ethanol, stored in light-protected containers, and diluted into medium before addition to the cultures. Optimal conditions require the addition of follicle-stimulating hormone (FSH) (25 ng/ml, NIH S13) and insulin (5 μg/ml). The additives are replaced every 2 days when the medium is changed.[3] Cultures remain viable and continue to secrete transferrin for at least 2 weeks. We utilize most cultures for experiments 4 to 6 days after plating because maximum secretion of transferrin occurs during this period.

[8] Measurement of Retinoid Effects on Epidermal Renewal

By LEONARD M. MILSTONE

Introduction

Retinoids have become important therapeutic tools. They are used to treat a wide array of skin diseases such as acne, ichthyosis, and photoaging. Changes in epidermal turnover always accompany the therapeutic use of retinoids, irrespective of the pathogenesis of the disease being treated or the biochemical mechanism(s) by which the retinoids produce their therapeutic effect. It can be reasonably argued that retinoid-induced amelioration of many of the clinical manifestations of acne, ichthyosis, and photoaging is a direct result of retinoid-induced changes in epidermal turnover. Moreover, the fine scaliness that is a constant side effect of retinoid use is certainly a manifestation of retinoid effects on turnover. The purpose of this chapter is to describe methods for measuring epidermal turnover.

Epidermal turnover is continuous and spatially oriented. To the best of our knowledge, the epidermal keratinocyte is the only cell in the epidermis that participates in the renewal process; turnover of melanocytes, Langerhans cells, or Merkel cells has not been identified. Renewal or turnover occurs to replace the most highly differented keratinocytes (squames) that separate (desquamate) from the outermost layer of the epidermis (stratum corneum). Desquamated cells are replaced from below by cells whose progressive differentiation terminates in the stratum corneum (Fig. 1). The cell replication that fuels the regenerative process occurs almost exclusively in the innermost layer of the epidermis, the basal cell layer. The general principles of renewal as well as the measurements made to quantitate renewal are equally applicable to human epidermis, the epidermis of commonly used experimental animals, and cultivated human keratinocytes.[1] Certain quantitative details, for example, cell cycle time, growth fraction, and number of cell layers, may vary according to the experimental system, the anatomic location, or the age of the subject. Therefore, special caveats of interpretation pertain to each system, and appropriate controls are especially important in renewal experiments.

Three parameters are most often measured to assess epidermal renewal: proliferation, transit time, and horn production or desquamation. Retinoids affect each of them. Two techniques are common to several of the

[1] L. M. Milstone, *J. Invest Dermatol.* **81,** 69s (1983).

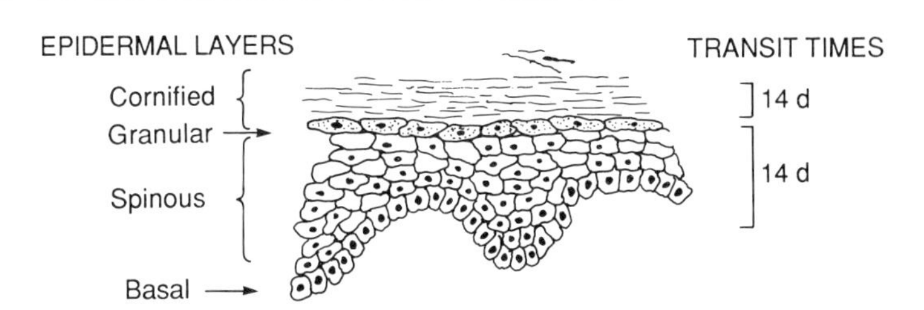

FIG. 1. Diagrammatic representation of the keratinocyte layers in the epidermis. Cells or squames from the cornified layer desquamate approximately 28 days after leaving the basal layer.

methods described: preparation of tissue sections for light microscopy and autoradiography using a photographic emulsion.[2,3]

Proliferation

Proliferation can be measured by the mitotic index, labeling index, or by radioactive precursor incorporation into DNA followed by scintillation counting. The [^{3}H]thymidine labeling index has been the method most commonly used to assess retinoid effects on proliferation.

Method: In Vivo Proliferation. [^{3}H]Thymidine is injected subepidermally, 5 μCi in 0.1 ml (New England Nuclear, Boston, MA; specific activity 2 Ci/mmol). Specimens are obtained 2 hr later using a 4-mm-diameter cylindrical punch, immediately following local anesthesia with 2% lidocaine. The tissue is fixed, embedded, sectioned at 5 μm, deparaffinized, and overlaid with photographic emulsion. After exposure for 2 weeks, the emulsion is developed and the sections counterstained with hematoxylin and eosin. The number of cells with more than five silver grains over their nuclei is counted and expressed as an index (see below).[2]

Method: In Vitro Proliferation. Alternatively, 3-mm cylindrical skin biopsies are trimmed of fat and as much dermis as possible. With a scalpel, 0.5 to 1.0-mm sections, perpendicular to the surface, are cut. Sections are incubated in culture medium containing 50 μCi/ml [^{3}H]thymidine (New England Nuclear; specific activity 2 Ci/mmol) for 2 hr at 37° with agitation by bubbling with 5% CO_2 95% air. A good correlation between this *in vitro* labeling procedure and *in vivo* labeling has been established.[3] Human, but not rodent, epidermis has a great capacity to degrade thymidine enzymatically to products that will not be incorporated into DNA.[4] Therefore,

[2] W. L. Epstein and H. I. Maibach, *Arch. Dermatol.* **92,** 462 (1965).

[3] J. M. Lachapelle and T. Gillman, *Br. J. Dermatol.* **81,** 603 (1969).

[4] P. M. Schwartz, L. C. Kugelman, Y. Coifman, L. M. Hough, and L. M. Milstone, *J. Invest Dermatol.* **90,** 8 (1988).

when dealing with human skin it is essential to keep the length of the radioactive pulse short and the concentration of thymidine high. Cutting the tissue into thin sections (<1 mm) permits rapid diffusion of thymidine to the DNA-synthesizing basal cells and reduces local variations in labeling.

Findings. all-*trans*-Retinoic acid increases keratinocyte proliferation, as shown by an increased labeling index *in vivo*[5] and in confluent cultures of human keratinocytes.[6]

Uses and Limitations. Measurement of [^{3}H]thymidine incorporation is the most common and direct way of assessing epidermal proliferation. Advantages are low cost, technical simplicity, and wide applicability to human, animal, or *in vitro* studies. Disadvantages are interpretive: calculating an index requires choice of a denominator. Total basal cell nuclei, total nuclei in the two lowermost layers of epidermis, or unit surface length are denominators advocated by different investigators. Obviously, experimental situations or diseases which alter the number of cell layers, the cell cycle time, the growth fraction, or the amount of papillomatosis could influence interpretation of these measurements.

Transit Time

Subcorneal Transit in Vivo

[^{3}H]Thymidine is injected intradermally into 10 well-marked sites. Small (3 mm) biopsies are obtained at regular intervals (usually every other day), and prepared as above.[7]

Findings. In vivo, the transit time for normal epidermis is variously reported as 12–16 days.[7] The transit time was decreased to 3 days by 3 weeks of twice daily application of 0.1% *trans*-retinoic acid to the back.[8]

Uses and Limitations. This method measures the minimum time it takes for a cell to move from the basal layer to the layer just beneath the stratum corneum (the granular layer). This measurement, known as the transit time, has been effectively used to assess renewal kinetics in normal skin and in diseases of the epidermis.[7] It assumes that thymidine incorporation occurs only in the basal layer (a reasonable assumption except in certain disease states, such as psoriasis) and that transit "time" has mean-

[5] J. Pohl and E. Christophers, *in* "Retinoids" (C. E. Orfanos, O. Braun-Falco, E. M. Farber, C. Grupper, M. K. Polano, and R. Schuppli, eds.), p. 133. Springer-Verlag, Berlin and New York, 1981.

[6] L. M. Milstone, J. McGuire, and J. F. LaVigne, *J. Invest Dermatol.* **79,** 253 (1982).

[7] G. D. Weinstein and E. J. Van Scott, *J. Invest Dermatol.* **45,** 257 (1965).

[8] G. Plewig and O. Braun-Falco, *Acta Dermatovener, Suppl.* **74,** 87 (1975).

ing in a tissue in which the number of cell layers between basal layer and stratum corneum varies because of rete ridges, prior treatment, and regional variation. One practical limitation on the use of this method in humans is the need to inject a radioactive tracer and to obtain multiple specimens.

Epithelial Transit in Vitro

In stratified, confluent cultures of human keratinocytes, there is no stratum corneum, and cells which are desquamated into the liquid medium retain their nuclei. Moreover, thymidine incorporation into DNA occurs almost exclusively in the basal layer. Transit time through the cultivated epithelium can be measured after a pulse of radioactive thymidine by periodically collecting shed cells from the medium, precipitating the DNA with 10% (w/v) trichloroacetic acid, hydrolyzing the radioactive thymine from the sugar phosphate backbone with 5% (v/v) perchloric acid at 90°, and counting the supernatant in a scintillation counter.[1]

Findings. Figure 2 shows that the minimum transit time, the time at which radioactive DNA first appears in cells shed into the culture medium, is reduced by 0.1 μM *trans*-retinoic acid.

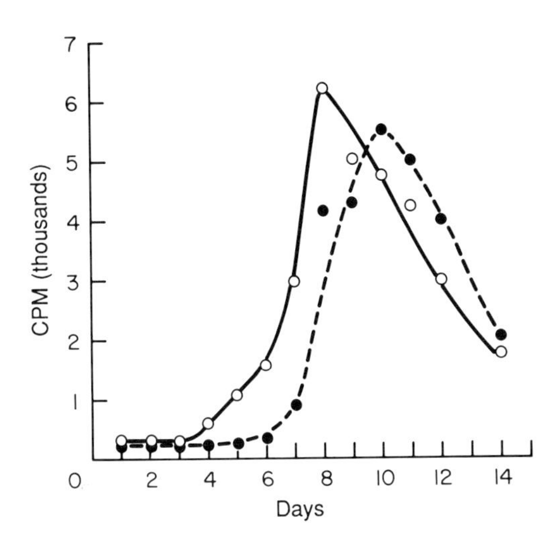

FIG. 2. Retinoic acid speeds transit of cells through an epithelium of human keratinocytes *in vitro.* Confluent cultures of neonatal foreskin keratinocytes were pulse-labeled with [^{3}H]thymidine, collected, and analyzed as described in Epithelial Transit *in Vitro.* In cultures treated with 0.1 mM retinoic acid (○), the earliest significant increase in radioactivity detectable in the desquamated cells occurs 3 days earlier than in control (●) cells, and the radioactivity peaks 2–3 days earlier.

Stratum Corneum Transit

[^{3}H]Glycine Appearance. The [^{3}H]glycine appearance method[9] relies on the fact that glycine is incorporated into protein in all layers of the epidermis up to, but not beyond, the innermost layer of the stratum corneum. Radioactively labeled protein is carried to the surface by transit of cells through the stratum corneum. [^{14}C]Glycine (0.5 μCi in 0.1 ml saline) is injected intradermally into two sites, 1.5 inches apart. Periodically (usually every 3–4 days) a 1-square inch piece of clear cellophane tape is applied to the skin over the injection site, removed, and counted in a scintillation counter. The first detection of a significant increase in radioactivity over background is taken as the transit time through the stratum corneum.

Dansyl Chloride Disappearance. Topically applied dansyl chloride produces fluorescent staining of the stratum corneum but not the epidermal layers beneath it. Therefore, disappearance of cutaneous fluorescence can be used to measure transit through the stratum corneum. Dansyl chloride (5%) in white petrolatum is applied to the skin on a disk of cotton felt held in place by an impermeable tape (Blen-Derm, 3M, St. Paul, MN). A second application is made 8 hr later, and the next day the area is washed thoroughly with soap and water. Daily examination of the region with a Wood's lamp in a darkened room reveals a fluorescent spot on the skin until the full thickness of the stratum has been replaced. The time required for the spot to disappear is the transit time through the stratum corneum.[10]

Findings. Transit time through the normal stratum corneum is 14–16 days by the [^{14}C]glycine method[9] and 20 days by the dansyl chloride method.[10] The glycine method has not been applied to retinoid research. Topical application of all-*trans*-retinoic acid reduced the transit time through the stratum corneum as judged by the dansyl chloride method.[11]

Uses and Limitations. The advantage of these two methods, especially compared to methods for measuring transit through the lower layers of the epidermis, is that they are noninvasive, that is, do not require repeated biopsies. One should be aware of differences in stratum corneum transit time related to anatomic regions being sampled, age of the subject, and method chosen (i.e., dansyl chloride versus [^{3}H]glycine). Since retinoids reduce the thickness of the stratum corneum, the meaning of a reduced transit time is not entirely clear; are cells moving more rapidly through the stratum corneum, or do they simply have less distance to traverse?

[9] D. Porter and S. Shuster, *J. Invest Dermatol.* **49,** 251 (1967).

[10] L. H. Jansen, M. T. Hojyo-Tomoko, and A. M. Kligman, *Br. J. Dermatol.* **90,** 9 (1972).

[11] A. M. Kligman, G. L. Grove, R. Hirose, and J. J. Leyden, *J. Am. Acad. Dermatol.* **15,** 836 (1986).

Desquamation or Soft Horn Production

Method: In Vitro. Measurement of soft horn production *in vitro* requires only counting of desquamated cells at the time of daily medium changes *in vitro.*[6]

Method: In Vivo. An adaptation of the dansyl chloride method measures horn production rather than transit time.[12] Stratum corneum is saturated with dansyl chloride as above, and skin biopsies are obtained daily. The stratum corneum is allowed to swell *in vitro* in NaOH and then examined by fluorescence microscopy. The rate at which nonfluorescent layers appear at the bottom of the statum corneum is a measure of the rate of horn production.

Findings. 13-*cis*- and all-*trans*-retinoic acid increased desquamation from stratified human keratinocyte epithelia *in vitro.*[1] This increase in desquamation can be taken as a direct measure of increased epidermopoiesis, since epithelial mass and cell number were unchanged and several indices of proliferation were simultaneously increased. The dansyl chloride method has not been used to measure retinoid effects on skin.

Uses and Limitations. In the final analysis, epidermal renewal occurs to compensate for desquamation of keratinocytes from the skin surface. Therefore, during times when the epidermis is in a steady state, the rate of renewal equals the rate of desquamation equals the rate of proliferation. Collecting and counting desquamated cells is a direct measure of the renewal process. Although this procedure is readily performed *in vitro,* collecting cells is cumbersome *in vivo.*

[12] A. Johannesson and H. Hammar, *Acta Dermatovener (Stockholm)* **58,** 76 (1978).

[9] Retinoids and Cell Adhesion

By LUIGI M. DE LUCA, SERGIO ADAMO, and SHIGEMI KATO

Introduction

Work from several laboratories[1-5] has confirmed the initial observation[6,7] that retinoids alter the adhesive properties of cultured cells. Initially, studies were conducted in cultured cell systems in the presence or absence

[1] J. S. Bertram, *Cancer Res.* **40,** 3141 (1980).
[2] S. S. Shapiro and J. P. Poon, *Exp. Cell Res.* **119,** 349 (1979).
[3] R. C. Klann and A. C. Marchok, *Cell Tissue Kinet.* **15,** 473 (1982).

of retinoic acid (RA), and adhesive strength was measured in EDTA- and/or protease-mediated detachment assays.[7] More recently we have demonstrated that RA increases the ability of fibroblast cells to attach to laminin or type IV collagen-coated microwell plates.[8] The purpose of this chapter is to summarize the salient methodologies and pertinent data of the adhesive effects of retinoic acid and other retinoids on fibroblastic cell lines.

Studies on Adhesion to Plastic

Cell Culture and Detachability Assay

BALB/3T12-3 mouse fibroblasts (spontaneously transformed derivatives of BALB/3T3) are obtained from the American Type Culture Collection (Rockville, MD) and cultured in T25 flasks (Falcon Plastics, Oxnard, CA) in 3 ml medium. Dulbecco's modified Eagle's minimum essential medium (Grand Island Biological Co., Grand Island, NY) is supplemented with 10% calf serum (Flow Laboratories, Inc., Rockville, MD), 25 m*M* HEPES (pH 7.3), and 50 μg gentamicin/ml (Microbiological Associates, Inc., Walkersville, MD). The medium is changed daily. The initial inoculum is 10,000 cells/cm^2 of growth area.

Trypsinizations (both for cell counting and passaging) are performed as follows: The trypsinization solution contains 2.5 g trypsin/liter (ICN Nutritional Biochemicals, Irvine, CA) and 0.2 g EDTA/liter in Hanks' balanced salt solution (Ca^{2+}-free and Mg^{2+}-free, pH 7.6) with 10 m*M* HEPES, pH 7.3. The trypsin–EDTA solution is divided into 10-ml aliquots and frozen. After removal of the culture medium, the dishes are rinsed twice with 3 ml Dulbecco's phosphate-buffered saline (PBS) (Ca^{2+}- and Mg^{2+}-free). The cells are rinsed quickly with 1.5 ml trypsin–EDTA solution. This solution is discarded, and the cells are incubated for 15 min at 37° and then suspended by pipetting in a suitable volume of either complete medium (for passage) or 10% serum in PBS (for cell counting). Samples of the cell suspensions obtained by trypsinization are routinely screened microscopically: More than 95% of the cells are present as single cells. Moreover, the culture vessels are examined after trypsinization to verify that no

[4] A. C. Marchok, J. N. Clark, and A. Klein-Szanto, *JNCI, J. Natl. Cancer Inst.* **66,** 1165 (1981).

[5] R. Tchao and J. Leighton, *Invest. Urol.* **16,** 476 (1979).

[6] S. Adamo, I. Akalovsky, and L. M. De Luca, *Proc. Am. Assoc. Cancer Res.* **19,** 107 (abstr.) (1978).

[7] S. Adamo, L. M. De Luca, I. Akalovsky, and P. V. Bhat, *JNCI, J. Natl. Cancer Inst.* **62,** 1473 (1979).

[8] S. Kato and L. M. De Luca, *Exp. Cell Res.* **173,** 450 (1987).

significant amount of cells has been left attached after trypsinization. Cells are enumerated with the use of a Model B Coulter counter (Coulter Electronics, Inc., Hialeah, FL). The trypan blue dye exclusion test is performed by incubation of the cells with 1 volume of 0.4% (w/v) trypan blue in PBS and 1 volume of complete medium.

all-*trans*-β-Retinoic acid (Eastman Kodak Co., Rochester, NY) is dissolved in dimethyl sulfoxide (DMSO) (Pierce Chemical Co., Rockford, ILL) at a concentration of 2 mg/ml or less, and the solution (freshly made every week) is stored in the dark at room temperature. Retinoic acid and its derivatives were a gift from Drs. Beverly Pawson and Peter Sorter of the Hoffman-La Roche Company (Nutley, NJ). Retinoic acid in DMSO, DMSO alone, or nothing is added to the culture medium 24 hr after the cells are plated, unless stated otherwise. The DMSO concentration in the culture medium is 0.5% (v/v).

The EDTA-based detachment assay is a modification of that described by Yamada *et al.*[9] and is performed as follows: Cells cultured in T25 flasks are rinsed with 3 ml PBS (Ca^{2+}- and Mg^{2+}-free). After the addition of 1.5 ml EDTA (0.2 g/liter in PBS; Ca^{2+}- and Mg^{2+}-free), the cells are shaken at 110 rpm on a Clay Adams (Parsippany, NJ) variable-speed rotator for 4 min at room temperature. At the end of the incubation, the detached cells are removed, and these as well as the cells still adherent to the culture surface are trypsinized [1.5 ml of 0.25% (w/v) trypsin] for 2 min in the dish; the cells are then transferred to a test tube and incubated for 10 min at 37°. Serum (100 μl) is added to each tube, and the volume is taken to 5 ml with PBS. Cells are dispersed uniformly by pipetting, and aliquots are taken for counting.

Effect of Retinoic Acid on Detachability of BALB/3T12-3 Cells. The effect of different concentrations of retinoic acid on the detachment of BALB/3T12 cells is shown in Fig. 1. Cells are found to adhere more strongly when cultured in the presence of RA for 3 days, and the effect is detected at as low as 1.6×10^{-7} *M* RA concentration. The maximum effect is detected at 3.3×10^{-6} *M* in the same assay which utilized a solution of 0.02% (w/v) EDTA as the detaching agent. Note that the time required for cell detachment varies with the type of plastic and its preparation for tissue culture purposes. More recent preparations of tissue culture flasks have been optimized for cell attachment and require longer treatment time with EDTA and/or trypsin to remove the cells.

Figure 2 shows the relatively early response of spontaneously transformed mouse fibroblasts to RA. Adhesiveness of 3T12 cells increases within the first 48 hr of culture in the presence of 3.3×10^{-5} *M* RA (Fig. 2A). This effect of RA is accompanied by a strong effect on cell morphol-

[9] K. M. Yamada, S. S. Yamada, and I. Pastan, *Proc. Natl. Acad. Sci. U.S.A.* **73,** 1217 (1976).

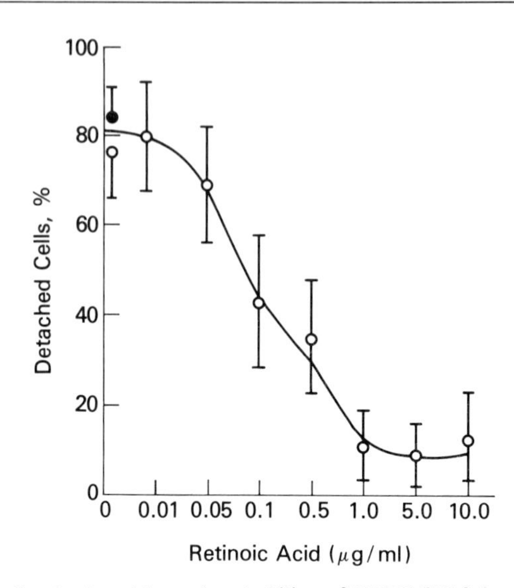

FIG. 1. Effect of retinoic acid on detachability of BALB/3T12-3 cells. The assay was performed at day 4 of culture (day 3 of exposure to the retinoid). The filled black dot refers to cells cultured in the absence of DMSO. Each point represents the mean ± S.D. of three dishes.

ogy. RA does not influence viability or plating efficiency of 3T12 cells. Moreover, the effect on adhesiveness is totally reversible on removal of the retinoid from the culture medium (Fig. 2B).

Structure–Activity Relationships

A study of structure–activity relationships is shown in Table I. Retinoids are tested at concentrations of 0.1, 1, and 10 μg/ml at 4 days of culture (3 days in the presence of the retinoid). Retinol, retinoic acid, and their 5,6-epoxy derivatives are the most active compounds, with activity at 1 μg/ml. 13-*cis*-Retinoic acid, trimethylmethoxyphenylretinoic acid, and dimethylacetylcyclopentenylretinoic acid are active at 10 μg/ml (Table I). Derivatives of retinol devoid of *in vivo* growth-promoting activity, such as anhydroretinol and perhydromonoeneretinol, fail to increase adhesive properties. In addition, derivatives of retinoic acid without vitamin A activity, such as the phenyl analog of RA, also fail to induce cell adhesiveness. Abscisic acid, juvenile hormones I, II, and III (not shown), and β-ionone are also inactive. Esterified or amidated derivatives of active retinoids are inactive. The same phenomenon of inactivation is observed by lactonization of retinoic acid as in the 11-hydroxyretinoic acid lactone.[10]

[10] L. M. De Luca, *J. Am. Acad. Dermatol.* **6,** 611 (1982).

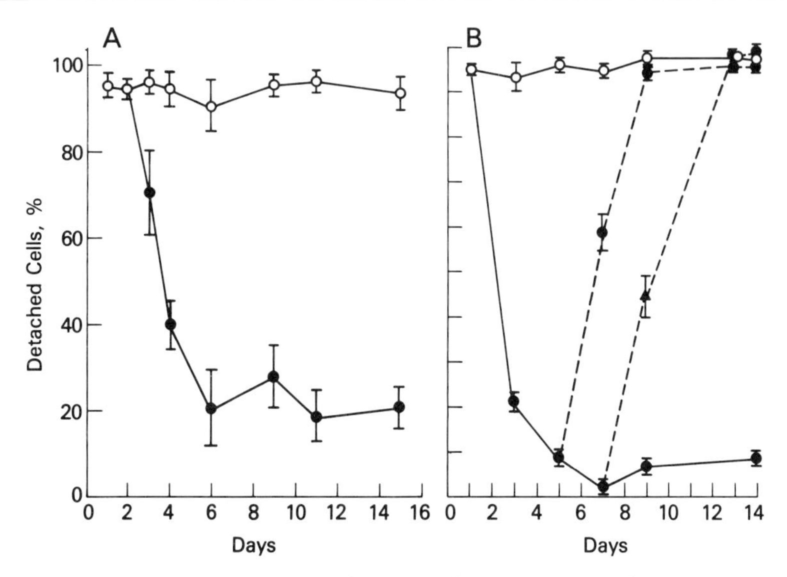

FIG. 2. (A) Time course of the effect of retinoic acid on the detachability of BALB/3T12-3 cells. Retinoic acid (10 μg/ml or 3.3×10^{-5} *M*) in DMSO (●) or DMSO alone (○) was added 24 hr after plating. The adhesion assay was performed on the days indicated. (B) Reversibility of the effect of retinoic acid on the adhesion of BALB/3T12-3 cells. Cells were cultured as described with DMSO or retinoic acid-supplemented medium. At days 5 (●) or 7 (▲), samples treated with retinoic acid were switched to DMSO medium. The detachment assay was performed as described on the days indicated. Each point represents the mean ± S.D. of three dishes.

The work described above shows that retinoic acid has a profound and reversible enhancing effect on the adhesiveness of BALB/3T12-3 cells to plastic substrates. In this work cells are detached from the dish by EDTA treatment. Different batches of tissue culture flasks and/or of FCS may elicit higher adhesion of the cells to the substrate so that only a minimal amount of cells can be detached in the assay described. In such cases, we suggest the use of the following alternative method.

Trypsin-Based Detachment Assay. Cells cultured in 25-cm^2 flasks with or without retinoic acid are rinsed twice with 3 ml of Ca^{2+} and Mg^{2+}-free Dulbecco's phosphate-buffered saline (PBS). Immediately after adding 2 ml of 0.01% (w/v) trypsin in Ca^{2+}- and Mg^{2+}-free PBS, the flasks are placed on a rotary shaker at 110 rpm for 4 min at room temperature. The detached cells are collected, dispersed with 100 μl of bovine calf serum, and counted electronically. The cells remaining attached to the flask are treated briefly with 2 ml of 0.1% (w/v) trypsin in Ca^{2+}- and Mg^{2+}-free PBS, dispersed with 100 μl of bovine calf serum, and counted electronically.

TABLE I

STRUCTURE–ACTIVITY RELATIONSHIPS OF NATURAL AND SYNTHETIC RETINOIDS

Name	Structure	Activity[a] at		
		0.1 μg/ml	1.0 μg/ml	10.0 μg/ml
Retinol	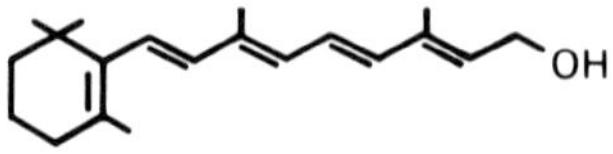	–	++	++
5,6-Epoxyretinol	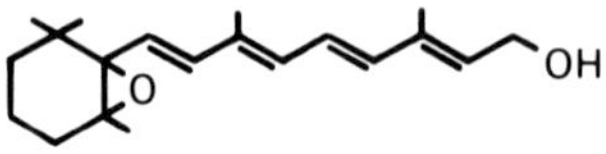	–	++	++
Anhydroretinol	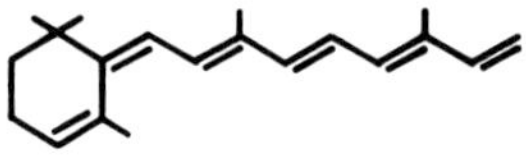	–	–	–
Perhydromonoene retinol	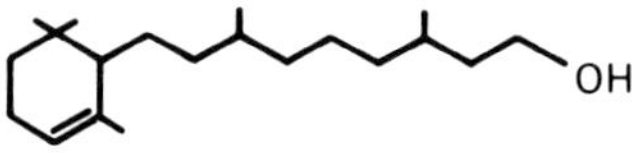	–	–	–
Retinoic acid	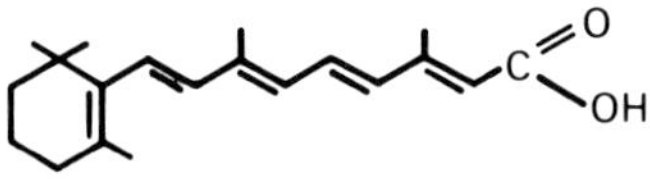	–	+	++
5,6-Epoxyretinoic acid	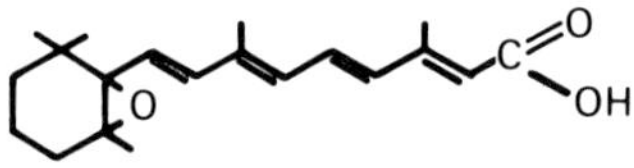	–	+	++
13-*cis*-Retinoic acid	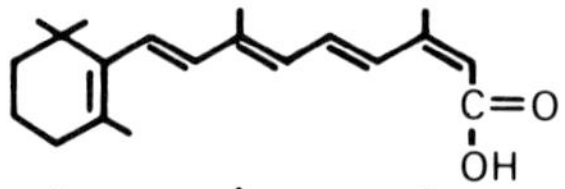	–	–	++
TMMP-Retinoic acid[b]	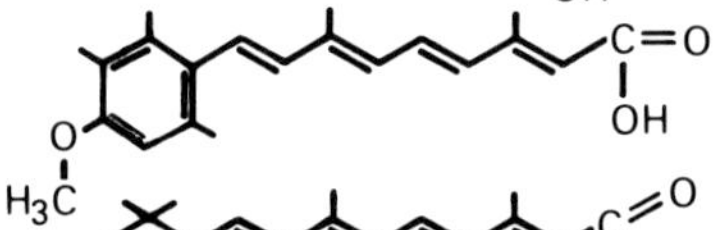	–	–	++
DACP-Retinoic acid	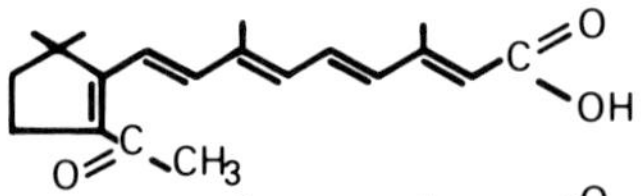	–	–	++
Phenyl analog of retinoic acid	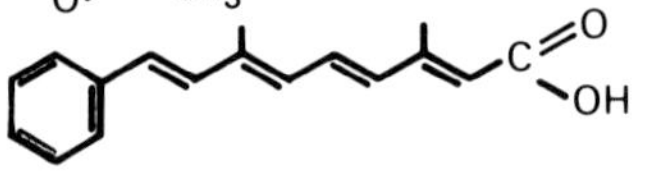	–	–	–
Abscisic acid	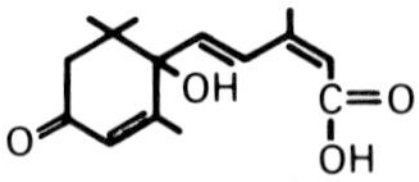	–	–	–
β-Ionone	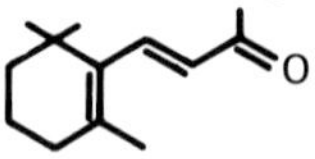	–	–	–
Retinoic acid ethylamide	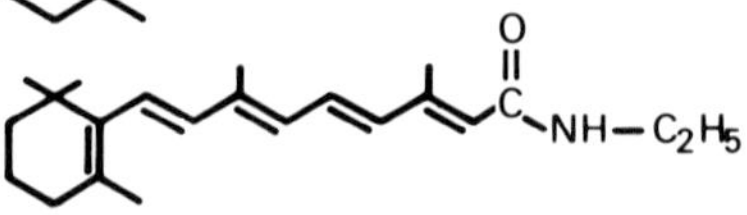	–	–	–

(Continued)

TABLE I
STRUCTURE–ACTIVITY RELATIONSHIPS OF NATURAL AND SYNTHETIC RETINOIDS *(Continued)*

Name	Structure	Activity[a] at 0.1 μg/ml	Activity[a] at 1.0 μg/ml	Activity[a] at 10.0 μg/ml
Retinoic acid 2-hydroxyethylamide	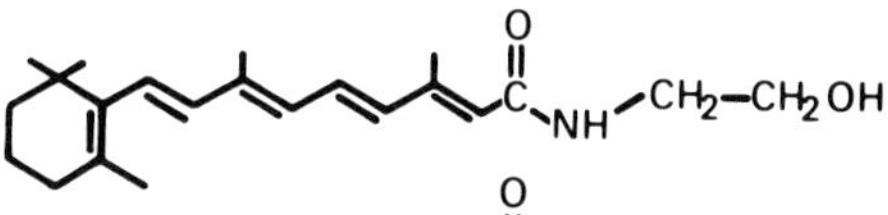 C(=O)–NH–CH₂–CH₂OH	–	–	–
TMMP-Retinoid ethyl ester	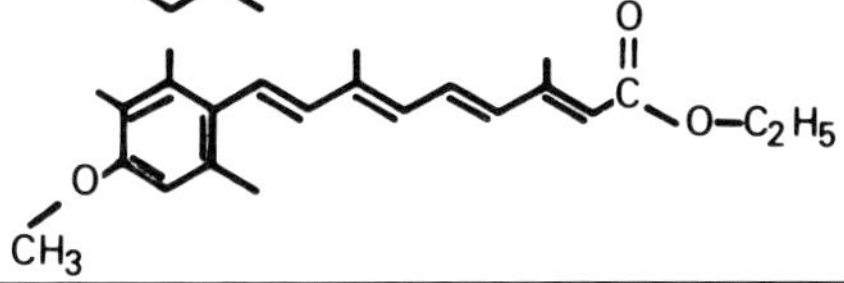 C(=O)–O–C₂H₅; O–CH₃	–	–	–

[a] –, No effect (70–100% of cells detached); +, moderate effect (30–70% of cells detached); ++, large effect (0–30% of cells detached).
[b] TMMP, Triphenylmethoxyphenol.

Data obtained with this assay (not shown) confirm the previous work that RA enhances cell–substratum adhesion.[7,8]

Retinoic Acid Enhances Fibroblast Cell Attachment to Laminin and Type IV Collagen

In addition to increasing the adhesiveness of fibroblasts to plastic, retinoids also enhance the ability of fibroblastic cells to attach to specific extracellular matrix components. In this investigation, the NIH 3T3 fibroblast cell system is employed for most observations, but the observed effect applies to a variety of fibroblastic cells.

Description of Materials and Procedures for Attachment Assays

NIH 3T3 (Dr. Stuart Aaronson, National Cancer Institute, NIH) and MoMSV 3T3 (Moloney mouse sarcoma virus-transformed BALB/3T3) cells are maintained in Dulbecco's modified Eagle's medium (DMEM) (4.5 g/liter glucose) containing 10% bovine calf serum (Hazelton Research Products). 3T3-Swiss, 3T6-Swiss, BALB/3T3, BALB/3T12-3, and SV-T2 (SV40-transformed BALB/3T3) cells are maintained in DMEM (1 g/liter glucose) containing 10% bovine calf serum. All the cell lines except NIH 3T3 are obtained from the American Type Culture Collection.

Cells (1 or 2.5×10^5) are seeded in 25-cm^2 plastic tissue culture flasks (Falcon). Culture media are changed to those containing retinoic acid [10^{-6} *M* with 0.1% (v/v) DMSO, unless otherwise indicated] or DMSO (0.1%, v/v) 1 day after seeding. Cells are used for the assays at 3 (2.5×10^5 cells) or 4 (1×10^5 cells) days of culture.

Preparation of Substrates

Fibronectin from bovine plasma (Calbiochem-Behring, San Diego, CA, 0.6 mg/ml in 0.5 *M* NaCl, 25 m*M* Tris-HCl, pH 7.0, 10% glycerol, 0.1 m*M* phenylmethylsulfonyl fluoride), laminin from mouse EHS sarcoma (Bethesda Research Laboratories, Gaithersburg, MD, 1.1 mg/ml in 50 m*M* Tris-HCl, pH 7.2, 0.15 *M* NaCl), and type IV collagen from mouse EHS sarcoma (Bethesda Research Laboratories, 1 mg/ml in 0.5 *M* acetic acid) are diluted with PBS with Ca^{2+} and Mg^{2+} to yield appropriate concentrations as indicated in Figs. 1 and 2. Multiwell plates (16 mm) (Costar, Cambridge, MA) *not* treated for tissue cultures are incubated with 0.3 ml per well of the above solutions for 1 hr at room temperature. After washing 3 times with 0.5 ml of PBS with Ca^{2+} and Mg^{2+}, plates are incubated with 0.3 ml per well of 10 mg/ml heat-denatured bovine serum albumin for 30 min at room temperature to saturate all remaining protein adsorption sites.[11] Plates are washed 3 times with 0.5 ml of PBS, receive 0.3 ml of DMEM, and are used for the cell attachment assay. Gelatin from pig skin (Sigma, St. Louis, MO, type I) is dissolved in 0.1 *M* sodium carbonate buffer, pH 9.5. Gelatin solution (0.3 ml) is added to each well of 16-mm multiwell plates, and the plates are incubated at room temperature overnight.[12] After coating, the plates are treated as above and are used for the assay.

Cell Attachment Assays

Cells cultured in the presence or absence of retinoic acid are rinsed twice with 3 ml of Ca^{2+}- and Mg^{2+}-free Hanks' balanced salt solution, then 0.4 ml of 0.1% (w/v) trypsin in the same solution is added, and the cells are incubated at room temperature until most of them are detached. After adding 2 ml of 0.5 mg/ml soybean trypsin inhibitor in Ca^{2+}- and Mg^{2+}-free Hanks' solution, the cells are collected and washed with 3 ml of trypsin inhibitor solution. The cells are suspended in DMEM at a concentration of 10^6 cells/ml and are incubated at 37° in a CO_2 incubator for 15 min. The cells are dispersed by gentle pipetting, and 0.1 ml of the cell suspension is added to each well of multiwell plates. An aliquot of cell suspension is used for the determination of viability of trypan blue exclusion. The viability is 80–90% and does not change depending on the retinoic acid treatment. The plates are incubated at 37° in a CO_2 incubator for 1 hr, unless otherwise indicated.

The unattached cells are removed, and each well is rinsed gently with

[11] K. M. Yamada and D. W. Kennedy, *J. Cell Biol.* **99,** 29 (1984).

[12] E. Ruoslahti, E. G. Hayman, M. Pierschbacher, and E. Engvall, this series, Vol. 82, p. 803.

0.5 ml of PBS with Ca^{2+} and Mg^{2+}, twice. The rinsing solutions are combined with the unattached cells, pipetted gently, and then counted electronically. Because none of the cell lines used for the assay attach in control wells (treated with PBS, then with bovine serum albumin), as verified under the microscope, the number of unattached cells in control wells is used for the calculation of percent attachment in the following equation:

$$\% \text{ attachment} = 1 - \frac{[\text{unattached cells in experimental wells}]}{[\text{unattached cells in control wells}]} \times 100$$

When the attached cells are counted after trypsinization, the results are basically the same.

Effects of Retinoic Acid on Cell Attachment

Retinoic acid inhibits the growth of NIH 3T3 cells, and it increases their adhesiveness to the tissue culture surface. This effect is found at concentrations as low as 3×10^{-8} *M* and was greatest at 10^{-6} *M*. NIH 3T3 cells are found to attach efficiently (within 1 hr), and half-maximal attachment is obtained at 1 μg/ml to fibronectin; pretreatment of cells with RA does not seem to influence this attachment (Fig. 3A).

Under physiological conditions fibroblasts abut on matrices containing fibronectin as well as type I and III collagen. NIH 3T3 cells also attach to gelatin, but their attachment is not as efficient as to fibronectin. Therefore, 1.5 hr is used for the studies on attachment to gelatin. As for the attachment to fibronectin, attachment to gelatin is similar for RA-treated and untreated cells. Half-maximal attachment also occurs at about 1 μg/ml (Fig. 3B). Studies on cell attachment to laminin clearly show a marked increase by RA (Fig. 3C). About 5 μg/ml laminin gives half-maximal attachment of RA-treated cells. Cell spreading is not as extensive as that on fibronectin.

RA also enhances the attachment of NIH 3T3 cells to type IV collagen substrates (Fig. 3D). Again, as for gelatin, cells are incubated for 1.5 hr, and half-maximal attachment to collagen is obtained at 10 μg/ml. The increase in cell attachment to laminin could be observed as early as 6 hr after treatment with 10^{-6} *M* RA and reaches its maximum at 2 to 3 days of treatment.

Comparison between Neoplastically Transformed and Nontransformed Mouse Fibroblasts

Table II shows the response of various fibroblasts to RA. The retinoid causes an increased attachment to laminin in all the nontransformed cells

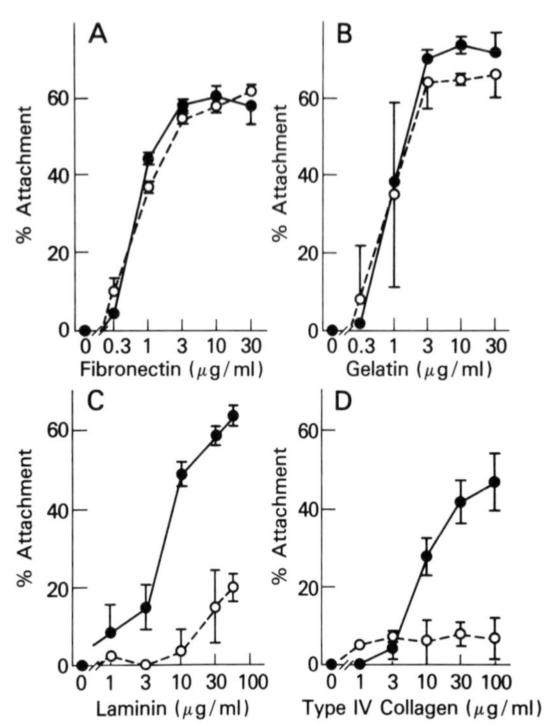

FIG. 3. Effect of retinoic acid treatment on attachment of NIH 3T3 cells to fibronectin (A), gelatin (B), laminin (C), and type IV collagen (D) substrates. The cell attachment assay was done in duplicate as indicated using multiwell plates precoated with proteins at the concentrations indicated. Incubation times were 1 hr for fibronectin and laminin substrates and 1.5 hr for gelatin and type IV collagen substrates. ●, Cells treated with 10^{-6} *M* retinoic acid for 3 days; ○, control cells.

TABLE II
EFFECT OF RETINOIC ACID ON ATTACHMENT OF MOUSE FIBROBLASTS TO LAMININ SUBSTRATES[a]

Cells	Percentage attachment	
	Control cells	Retinoic acid-treated cells
NIH 3T3	7	46
3T3-Swiss	0	23
3T6-Swiss	31	52
BALB/3T3	19	58
BALB/3T12-3	31	52
SV-T2	77	78
MoMSV	60	51

[a] Each fibroblast line was treated with or without 10^{-6} *M* retinoic acid for 3 days. Plates coated with 10 μg/ml laminin were used for the 1-hr attachment assay. Each value represents an average of two experiments done in triplicate.

tested, including NIH 3T3, 3T3-Swiss, 3T6-Swiss, and BALB/3T3. In contrast, transformed fibroblasts show a higher attachment to laminin substrates as compared to their untransformed counterparts. Varied responses to RA are observed. Attachment to laminin is increased in BALB/3T12-3, not affected in SV-T2, and slightly decreased in MoMSV cells.

[10] Retinoids and Control of Epithelial Differentiation and Keratin Biosynthesis in Hamster Trachea

By LUIGI M. DE LUCA, FREESIA L. HUANG, and DENNIS R. ROOP

Introduction

The most evident morphological changes in vitamin A-deficient animals are at the level of epithelial tissues. In particular, mucociliary epithelia such as the tracheal lining are replaced by a stratified squamoid and eventually keratinized type of epithelium. This change in morphology is accompanied by a switch in the products being synthesized from mucin-type to keratin-type proteins. The cell biological interpretation of the sequence of events eventually leading from a mucociliary to a keratinized tissue is still unclear. It is the purpose of this chapter to describe methodologies utilized in the assessment of the morphological and biochemical alterations which occur during vitamin A deficiency in the hamster tracheal epithelium.

Morphological Alterations

General Considerations

The tracheal epithelium is of the pseudostratified type, that is, all its cells are in contact with the basement membrane. Basal cells do not extend to the lumen, whereas ciliated and goblet cells do. The epithelial tracheal surface fulfills its function of clearing foreign particles by virtue of the movement of the cilia. These, in turn, are aided in their motion by the presence of the mucus secreted by the goblet cells (Fig. 1). Mechanical or chemical injury causes focal replacement of the mucocilary epithelium by a squamous metaplastic phenotype. Eventual reestablishment of the mucociliary epithelium after injury takes place gradually. In animals lacking vitamin A, a similar process of loss of the mucociliary epithelium and

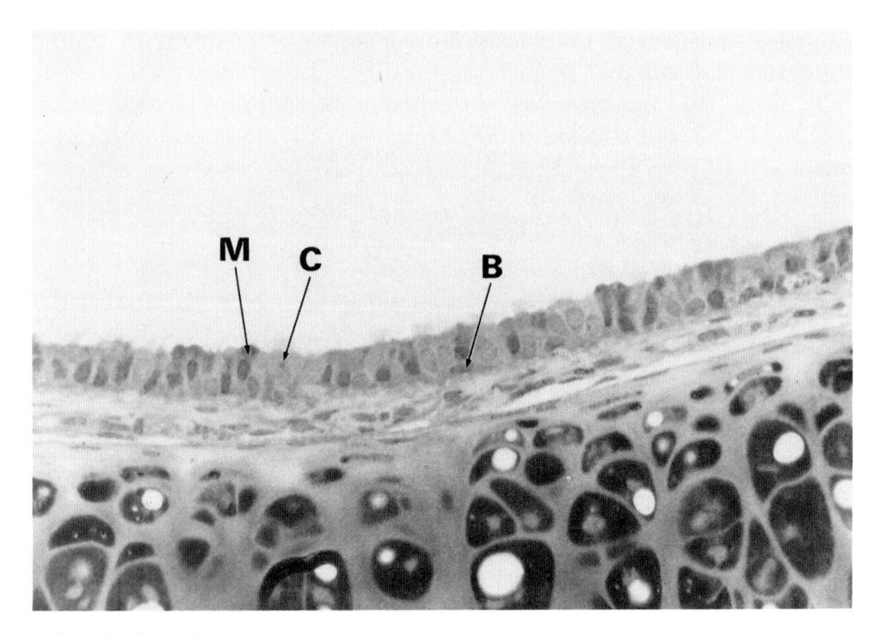

FIG. 1. Normal hamster trachea. The pseudostratified epithelium consists of short basal cells (B) and columnar mucous (M) and ciliated (C) cells. The epithelium lies over the tracheal cartilage. Magnification: ×612.

squamoid metaplasia is also observed. However, these nutritionally deficient animals are unable to regenerate a new mucociliary epithelium; instead, the surface of the trachea remains squamous metaplastic with flat cells resembling epidermal keratinocytes eventually forming a heavily keratinized stratum corneum type of tissue (Fig. 2C,D). This phenomenon has been shown to occur *in vivo*[1] and has been reproduced in tracheas cultured[2,3] in serum-free media and in cell culture.[4] It has been consistently observed that retinoids at very low concentrations permit the regeneration of the normal mucociliary phenotype *in vivo* as well as in organ and cell culture.

Although most epithelia respond to vitamin A deficiency in the manner just described, there are two notable exceptions, namely, the intestinal

[1] S. B. Wolbach and P. R. Howe, *J. Exp. Med.* **2,** 753 (1925).

[2] M. B. Sporn, N. M. Dunlop, D. L. Newton, and J. M. Smith, *Fed. Proc., Fed. Am. Soc. Exp. Biol.* **35,** 1332 (1976).

[3] F. L. Huang, D. R. Roop, and L. M. De Luca, *In Vitro* **22,** 223 (1986).

[4] E. M. McDowell, T. Ben, B. Coleman, S. Chang, C. Newkirk, and L. M. De Luca, *Virchow's Arch. B:* **54,** 38 (1987).

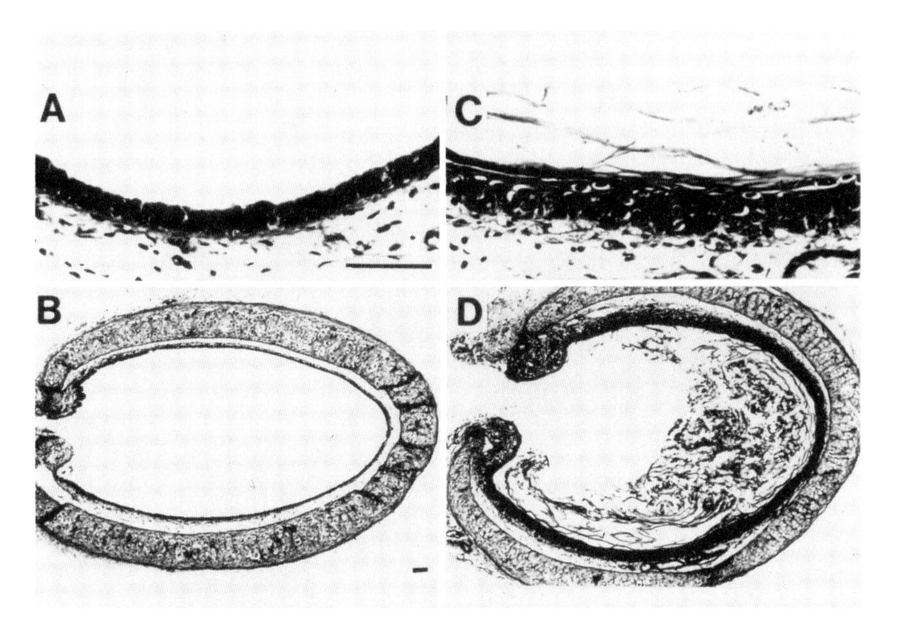

FIG. 2. Morphology of *in vivo* hamster tracheas. (A and B) Normal tracheas; (C and D) tracheas from hamsters kept on a vitamin A-deficient diet for 7 weeks.

epithelium and the epidermis and other tissues which normally display the epidermoid phenotype, such as part of the cornea and the esophagus. The intestinal mucosa fails to become squamoid metaplastic, even under conditions of extreme vitamin A deficiency.[5,6] It does, however, respond to the condition of vitamin A deficiency by an apparent reduction in the number of mucus-secreting goblet cells.[5,6] The epidermis, on the other hand, which normally already displays a keratinizing phenotype, during vitamin A deficiency displays an increased thickness,[7] but the morphological changes are not so profound as in mucociliary epithelia, as also reflected in the pattern of keratin expression, which remains similar in this tissue. Additional effects of retinoic acid, when applied topically *in vivo* on the skin or to cultures of epidermal cells, are very profound and entail a process of inhibition of keratinization and the consequent reduction in certain keratins.[8,9]

[5] L. De Luca, E. P. Little, and G. Wolf, *J. Biol. Chem.* **244,** 701 (1969).
[6] L. De Luca, M. Schumacher, and G. Wolf, *J. Biol. Chem.* **245,** 4551 (1970).
[7] G. S. Kishore and R. K. Boutwell, *J. Nutr. Growth Cancer* **2,** 91 (1985).
[8] E. Fuchs and H. Green, *Cell (Cambridge, Mass.)* **25,** 617 (1981).
[9] B. M. Gilfix and R. L. Eckert, *J. Biol. Chem.* **260,** 14026 (1985).

Preparation of Hamster Tracheas

Syrian golden hamsters are used throughout the studies. Pregnant mothers are housed individually with a 50/50 mixture of laboratory chow and vitamin A-free diet for the deficient group. After birth of the experimental animals, the mothers are fed only the vitamin A-free diet for the deficient group. The hamsters are weaned at 21 days after birth and are kept on control or vitamin A-free diets until use (4 weeks $\pm$ 2 days). At 4 weeks these hamsters have not yet developed severe external signs of vitamin A deficiency, though their liver vitamin A level is already very low (<14 μg/g liver in deficient group versus 65–120 μg/g liver in normal group).

Tracheas are excised aseptically from anesthetized hamsters and are kept in Leibovitz's L15 medium and supplemented with 100 U/ml penicillin and 100 μg/ml streptomycin (GIBCO, Grand Island, NY). Each trachea is cleaned to remove any adherent tissue, cut longitudinally along the dorsal wall, and cultured individually in 60-mm culture dishes. Each dish contains 3 ml of CMRL 1066 serum-free medium (GIBCO) supplemented with 1.0 μg/ml bovine pancreas insulin (Sigma, St. Louis, MO), 0.1 μg/ml hydrocortisone hemisuccinate (Sigma), 2 m*M* glutamine, 100 U/ml penicillin, and 100 μg/ml streptomycin. Retinyl acetate when included is at a concentration of 1×10^{-7} *M*. The cultures are gassed with 5% CO_2, 50% O_2, and 45% N_2 (Matheson Gas Products, Dorsey, MD) in a rocker chamber (Bellco Glass, Vineland, NJ) and are rocked 8 times/min in a 37° incubator.

Morphological Assessment of Tracheal Epithelial Changes

The morphological changes arising from deficiency of vitamin A are usually assessed at the midpoint of the tracheas. This is in part because the larynx is normally squamoid and keratinized. Tracheas are fixed in 10% buffered formalin and embedded in paraffin. Serial horizontal sections of 5 μm are usually cut from the midportions of the organ; they are stained with hematoxylin and eosin or periodic acid–Schiff reagent (PAS) and examined by light microscopy for the development of stratification and squamous metaplasia.[2] Figure 1 shows a typical section obtained from the trachea of a hamster fed a diet containing vitamin A. The three major cell types represented in a normal hamster epithelium are the mucous cell (M of Fig. 1) (~59%), the ciliated cells (C of Fig. 1) (~11%), and the basal cell (B of Fig. 1) (~28%).[10,11] These proportions change markedly during vitamin A deficiency, even in hamsters kept on the deficient diet for only 5 weeks (see Table I).

[10] E. M. McDowell, K. P. Keenan, and M. Huang, *Virchows Arch. B:* **45,** 197 (1984).
[11] E. M. McDowell, K. P. Keenan, and M. Huang, *Virchows Arch. B:* **45,** 221 (1984).

TABLE I
PROPORTIONS OF MAJOR CELL TYPES IN HAMSTER TRACHEAS[a]

Cell	Proportion of total (%)		p value
	Control	A−	
Basal	28.7 ± 2.5	39.7 ± 2.7	.0001
Mucous	59.3 ± 2.6	53.1 ± 1.9	.001
Ciliated	11.0 ± 3.1	6.8 ± 1.8	.05

[a] Data for vitamin A-deprived (A−) cells were derived from epithelia showing minimal changes (5 weeks on diet). Foci of stratification and/or epidermoid metaplasia (about 5% of all epithelial cells) were excluded from the analysis. Reproduced by permission from McDowell *et al.*[10]

The morphology of the pseudostratified respiratory epithelium changes during vitamin A deficiency. The following are observed at the light microscope level: (1) loss of cilia; (2) formation of a contiguous layer of basal cells which now become the only cells to be in contact with the basal lamina (this clearly marks the transition from a pseudostratified to a stratified type of epithelium); and (3) formation of a keratinized epithelium. These events are preceded by a marked reduction in the ability of the mucous cells to divide.[12] Typically, localized lesions (squamous metaplasia) are observed focally (Fig. 3E); these progress by expansion and fusion with other lesions eventually to form a uniform epidermoid epithelium. Keratinization is evident, and eventual occlusion of the lumen of the trachea may be observed. It should be emphasized that the epidermoid phenotype appears more prone to infections by a variety of agents, which eventually may cause pneumonia and death of the animal. *In vitro,* however, the epidermoid epithelium can be maintained for several weeks, but it reverses to mucociliary in the presence of small concentrations of vitamin A or one of its active analogs (Fig. 3).

Keratin Gene Expression during Vitamin A Deficiency of Tracheal Epithelium

Squamous metaplasia of the respiratory epithelium is followed by frank keratinization with keratohyaline granules and shedding of keratinized

[12] L. M. De Luca and E. M. McDowell, *J. Nutr.* **116,** 2064 (1986).

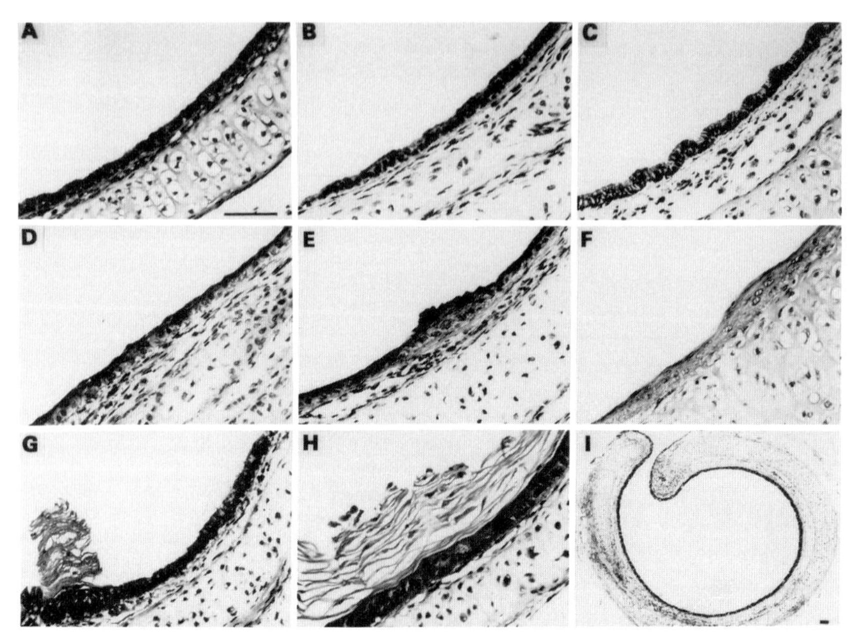

FIG. 3. Micrographs of hamster trachea in organ culture. Sections were stained with hematoxylin and eosin. (A) Trachea of 4-week-old vitamin A-deficient hamster before culturing; (B) trachea of 4-week-old normal hamster before culturing; (C) trachea of deficient hamster cultured 9 days in retinyl acetate-containing medium; (D–H) tracheas of deficient hamsters cultured in vitamin A-free medium for (D) 1 day (E) 3 days, (F and G) 5 days, and (H) 9 days; (I) trachea of deficient hamster cultured in vitamin A-free medium for 9 days and then in retinyl acetate-containing medium for 7 days. Bar: 60 μm. Magnification: A–H, $\times 178$; I, $\times 29$.

fibers into the lumen of the trachea. Investigation of the keratin types synthesized in vitamin A-deficient trachea is usually done by extraction of the fibrous protein, denaturing electrophoretic analysis, and Western blotting. Tracheas from 4-week-old vitamin A-depleted hamsters are cultured in vitamin A-free or retinyl acetate-supplemented (10^{-7} M) medium for various times. Tissues are then washed several times with PBS and are cut in small pieces and homogenized in a tightly fitted Teflon homogenizer with 20 m*M* Tris-HCl buffer, pH 7.6, containing 1 m*M* EDTA and 0.2 m*M* phenylmethylsulfonyl fluoride (PMSF) (buffer A). After centrifugation, the pellet is washed 2 more times with buffer A and 2 more times with buffer A containing 2% (w/v) Nonidet P-40 (NP-40). During each washing the pellet is homogenized in the appropriate buffer and is vigorously vortexed before centrifugation. The final pellet is taken up in a small volume of buffer A containing 2% (w/v) sodium dodecyl sulfate (SDS) and 10 m*M* dithiothreitol (DTT). The mixture is vigorously vortexed while incubating at 37° and is boiled for 3 min and centrifuged to obtain the final cytoskeletal protein extract. Protein concentration in the cytoskeletal extract is determined by the procedure of Bramhall *et al.*[13]

Antikeratin Antibodies

Various antibodies are used in the immunological characterization of keratins. Monoclonal antibodies against human epidermal keratins AE1 and AE3, a gift of Dr. T. T. Sun[14,15] are group specific; AE1 reacts with type I (acidic) keratins and AE3 with type II (basic) keratins.[16] Anti-67, -60, -59, and -55 kilodalton (kDa) keratin sera used in the immunoblot analyses are monospecific.[17,18] Following the nomenclature of the keratins,[19] the 67,000 Da protein is K1 (type II, basic), the 60,000 Da is the K6 (type II, basic), and the 55,000 Da is K14 (type I, acidic); the anti-55 kDa antibody also recognizes a 50,000-Da protein which may be a product of proteolysis of the 55,000 Da form (K14). The antibodies are raised against synthetic peptides corresponding to the carboxy-terminal amino acid sequence of each mouse epidermal keratin deduced from the nucleotide sequence of individual cDNA clones.[17,18]

Immunoblot Analysis of Keratins

Cytoskeletal proteins (50 μg) extracted from the cultures as specified above are separated by SDS–polyacrylamide gel electrophoresis and then electrophoretically transferred to a nitrocellulose membrane (Bio-Rad, Richmond, CA) overnight at 8° using Hoefer's transfer system with the power setting at 60%. Blotted proteins are reacted with monoclonal antibodies AE1 and AE3[14–16] or with monospecific antibodies anti-67, -60, -59, and -55 kDa keratins[17,18] and are detected with a kit using the peroxidase reaction (Bio-Rad). In every specific reaction, analysis with control antibody (P3) for AE1 and AE3 or preimmune serum corresponding to each specific immune serum is carried out simultaneously.

Keratins in cytoskeletal proteins extracted from cultured tracheas are characterized immunologically with AE1 and AE3 after SDS–

[13] S. Bramhall, N. Noack, M. Wu, *et al., Anal. Biochem.* **31,** 146 (1969).
[14] S. C. G. Tseng, M. J. Jarvinen, W. G. Nelson, *et al., Cell (Cambridge, Mass.)* **30,** 361 (1982).
[15] J. Woodcock-Mitchell, R. Eichner, and W. G. Nelson, *J. Cell Biol.* **95,** 580 (1982).
[16] R. Eichner, P. Bonitz, and T. T. Sun, *J. Cell Biol.* **98,** 1388 (1984).
[17] D. Roop, C. K. Cheng, L. Titterington, *et al., J. Biol. Chem.* **259,** 8037 (1984).
[18] D. R. Roop, C. K. Cheng, R. Toftgard, *et al., Ann. N.Y. Acad. Sci.* **455,** 426 (1985).
[19] W. W. Franke, D. L. Schiller, R. Moll, *et al., J. Mol. Biol.* **153,** 933 (1981).

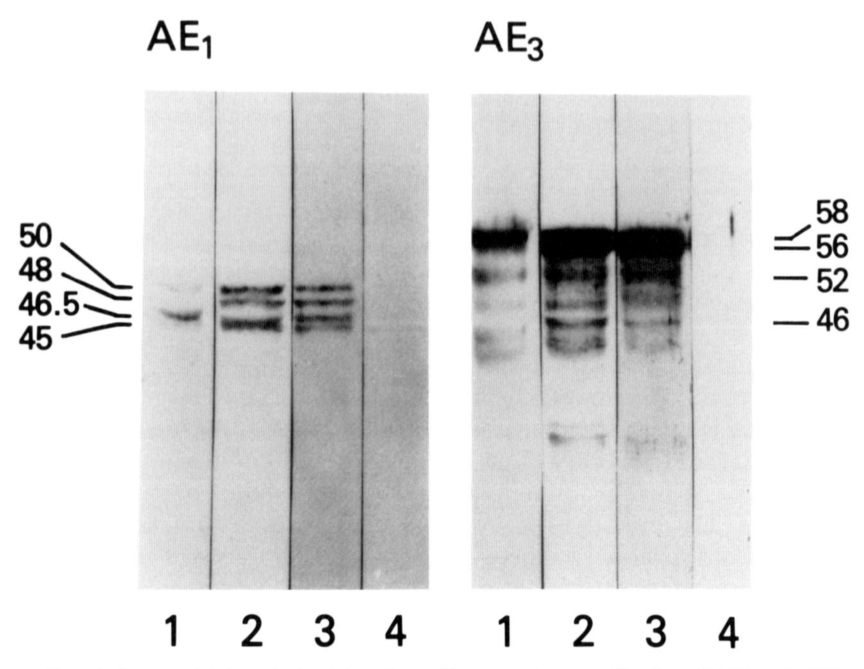

FIG. 4. Immunoblot analysis of keratins with monoclonal antibodies of AE1 and AE3. Values denote the molecular mass $\times 10^{-3}$ of keratins. Extracts were prepared from tracheas of deficient hamsters cultured in vitamin A-free medium for 1 day (lane 1), 8 days (lane 2), and 15 days (lane 3) and in retinyl acetate-containing medium for 15 days (lane 4). Fifty micrograms of cytoskeletal protein was analyzed in each lane. The first antibody was used at 1:2 dilution and the second antibody at 1:2000 dilution.

polyacrylamide gel electrophoresis. AE1 detects four major species of acidic keratins in extracts of squamous metaplastic tracheas (Fig. 4, AE1, lanes 2 and 3) of 50, 48, 46.5, and 45 kDa. Extracts of deficient tracheas cultured 1 day in vitamin A-free medium, though actively synthesizing keratins as evidenced from the results of immunoprecipitation of [^{14}C]methionine-labeled protein (not shown), exhibit only one major band of 46.5 kDa and two faint bands of 50 and 45 kDa when stained with AE1 antibody (Fig. 4, lane 1). AE3 recognized basic keratins of 58 and 56 kDa from extracts of severely squamoid-deficient tracheas (Fig. 4, AE3, lanes 2 and 3). Extracts of 1 day vitamin A-free cultured tracheas show the major keratin band of 58 kDa when stained with AE3, whereas the band of 56 kDa appears only in minor quantity (lane 1). A few other less prominent bands such as 52 and 46 kDa are also consistently detected by AE3 in vitamin A-free cultured tracheas. Mucociliary tracheas (cultured with retinyl acetate), being low in keratin-synthesizing activity, show only a very

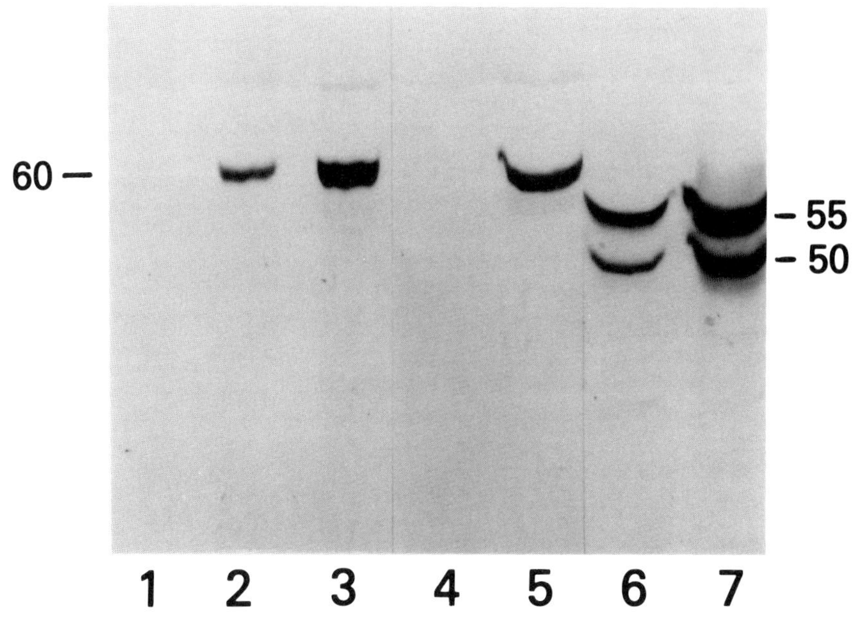

FIG. 5. Immunoblot analysis of keratins with monospecific antisera against the 55- and 60-kDa keratins. Lanes 1 to 5 were reacted with anti-60 kDa (K6) keratin serum; lanes 6 and 7 were reacted with anti-55 kDa (K14) keratin serum. Lanes 1 and 4 show extracts from tracheas of deficient hamsters cultured 8 and 15 days, respectively, in retinyl acetate-containing medium. Lanes 2, 3, and 5 show extracts from tracheas of deficient hamsters cultured 1, 8, and 15 days in vitamin A-free medium. For lane 6 the same extract as for lane 4 was used, and for lane 7 the same extract as for lane 5 was used. Seventy-five micrograms of cytoskeletal protein was analyzed in each lane. The first antibody was used at 1:250 dilution and the second antibody at 1:2000 dilution.

faint band of 45 kDa reactive with AE1 and of 58 kDa reactive with AE3 at all time points, although equivalent amounts of cytoskeletal protein were analyzed (Fig. 4, lane 4) for tracheas cultured with retinyl acetate for 15 days. Analyses of tracheas cultured with retinyl acetate for 1 or 8 days give essentially the same pattern.

Cytoskeletal protein extracts of cultured tracheas are also subjected to immunoblot analysis with polyclonal but monospecific antibodies against mouse epidermal keratins of 67 (K1), 60 (K6), 59 (K5), and 55 kDa (K14). The anti-55 kDa keratin serum recognizes two protein bands at 55 and 50 kDa in extracts of cultured metaplastic tracheas (Fig. 5, lane 7). These two bands are also detected in extracts of mucociliary tracheas but in lower quantity (lane 6). The component of lower mass may be a proteolytic product of the 55-kDa protein. The monospecific anti-60 kDa keratin antibody reacts with a protein of approximately 60 kDa in extracts from metaplastic tracheas (lanes 2, 3, and 5); however, this protein is not detected in extracts from mucociliary tracheas (Fig. 5, lanes 1 and 4). The monospecific anti-67 and -59 kDa keratin sera do not recognize any protein in extracts of these cultured trachea (data not shown). Negative results are obtained with all the preimmune sera used at the same dilution as their corresponding immune sera (data not shown).

[11] Inhibition of Tumor Cell Growth by Retinoids

By REUBEN LOTAN, DAFNA LOTAN, and PETER G. SACKS

Introduction

The growth and differentiation of various normal and malignant cells in culture are modulated (stimulated or inhibited) by retinoids.[1-5] Most of the cultured malignant cells that have been analyzed for responsiveness to retinoids exhibit inhibition of anchorage-dependent growth and anchorage-independent growth.[1-4,6-16] This chapter describes various methods used to analyze the growth inhibitory effects that retinoids exert on cultured cells.

[1] B. Amos and R. Lotan, this volume [23].
[2] R. Lotan, *Biochim. Biophys. Acta* **605,** 33 (1980).
[3] A. M. Jetten, *in* "Growth and Maturation Factors" (G. Guroff, ed.), Vol. 3, p. 251. Wiley, New York, 1985.
[4] E. W. Schroder, E. Rapaport, and P. H. Black, *Cancer Surv.* **2,** 223 (1983).
[5] A. Hiragun, M. Sato, and H. Mitsui, *Exp. Cell Res.* **145,** 71 (1983).
[6] L. D. Dion, J. E. Blalock, and G. E. Gifford, *J. Natl. Cancer Inst.* **58,** 795 (1977).
[7] R. Lotan and G. L. Nicolson, *J. Natl. Cancer Inst.* **59,** 1717 (1977).
[8] L. Patt, K. Itaya, and S.-I. Hakomori, *Nature (London)* **273,** 379 (1978).
[9] D. Douer and H. P. Koeffler, *J. Clin Invest.* **69,** 277 (1982).
[10] M. K. Haddox and D. H. Russell, *Cancer Res.* **39,** 2476 (1979).
[11] L. D. Dion and G. E. Gifford, *Proc. Soc. Exp. Biol. Med.* **163,** 510 (1980).
[12] R. Lotan, G. Neumann, and D. Lotan, *Ann. N.Y. Acad. Sci.* **359,** 150 (1981).
[13] N. Sidell, *J. Natl. Cancer Inst.* **68,** 589 (1981).
[14] F. Traganos, P. J. Higgins, C. Bueti, Z. Darzynkiewicz, and M. R. Melamed, *J. Natl. Cancer Inst.* **73,** 205 (1984).
[15] R. Lotan, P. G. Sacks, D. Lotan, and W. K. Hong, *Int. J. Cancer* **40,** 224 (1987).
[16] R. Lotan, G. Giotta, E. Nork, and G. L. Nicolson, *J. Natl. Cancer Inst.* **60,** 1035 (1978).

Inhibition of Anchorage-Dependent Growth

The exposure of adherent cells that grow as monolayers on plastic tissue culture dishes to medium containing retinoids often result in a decrease in growth rate (Fig. 1A) and a lowered saturation density.[2,3,8] These effects occur in the concentration range of 1 nM to 10 μM of retinoids and are time-dependent and dose-dependent. DNA-synthesis is suppressed,[10-13] and the cells are either arrested in or accumulate in the G_1 phase[10-12,14] or in the S phase of the cell cycle,[3] depending on the cell type. Changes in the cell cycle occur within 12-24 hr after addition of retinoids to the culture medium, and growth inhibition can be detected 24-72 hr after treatment initiation, depending on the doubling time of the cells. Cells that form discrete colonies on plastic tissue culture dishes show a decrease in the number of colonies as well as in colony size after retinoid treatment (Fig. 1B and insets in Fig. 2A,B).[15] The removal of retinoids from the culture medium of pretreated cells often results in a reversal of the growth inhibitory effects within 24-72 hr.[2,16] The growth inhibitory effects of retinoids on anchorage-dependent growth can be determined by the following methods.

Treatment of Cells with Retinoids. Retinoids are stored as a crystalline powder in an N_2 atmosphere in sealed dark containers at $-70°$. All procedures involving retinoids are performed under subdued light. Before an experiment, the retinoid is dissolved in either ethanol or dimethyl sulfoxide (DMSO) at a concentration of 10 mM, and this stock solution can be stored at $-70°$ under N_2 for up to 3 weeks. The retinoid is diluted from the stock solution in a series of 1 to 10 dilutions in ethanol or in DMSO to obtain solutions ranging in concentration from 0.1 μM to 10 mM. These solutions are further diluted 1 to 1000 directly in the culture medium to obtain final concentrations in the range 0.1 nM to 10 μM.

A variety of growth media including modified Eagle's medium, Dulbecco's modified Eagle's medium (DMEM), Ham's F10 and F12, RPMI 1640, Waymouth's MB 752/1, and McCoy's 5A medium have been used for treatment of cells with retinoids and are usually those that best support the growth of the untreated cells. The media are often supplemented with 10% fetal calf serum, as most cells require serum for efficient growth. However, retinoids have been shown to affect cell growth in serum-free medium as well.[5] Usually, the toxicity of high doses of retinoids[17,18] is enhanced in serum-free conditions,[5,18] possibly because of the absence of serum albumin, which is known to bind retinoids. Some studies have

[17] M. Audette and M. Page, *Cancer Detect. Prev.* **6,** 497 (1983).
[18] R. Lotan and D. Lotan, unpublished observation (1987).

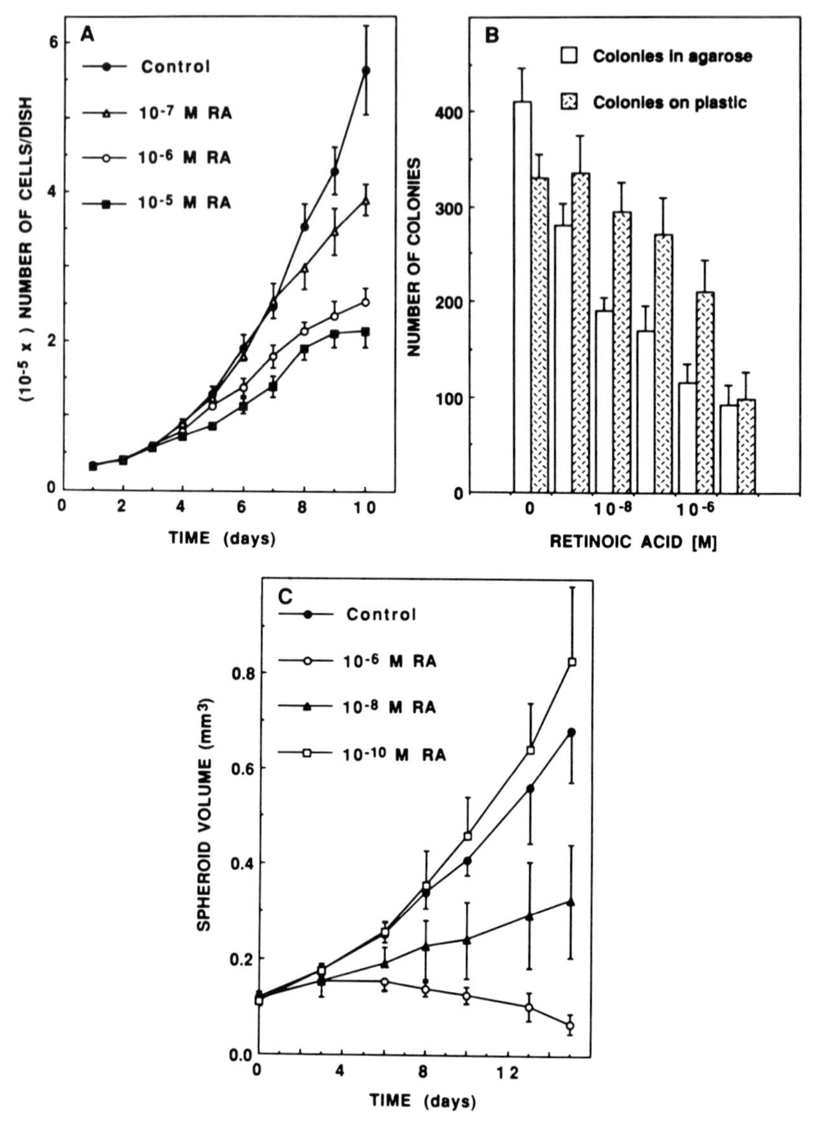

FIG. 1. β-all-*trans*-Retinoic acid dose-dependent and time-dependent inhibition of human head and neck squamous carcinoma cell (HNSCC) proliferation, multicellular spheroid growth, and colony formation. (A) Cells were seeded in a series of 35-mm-diameter tissue culture dishes at 10^4 cells per dish. After 24 hr, and at 72-hr intervals thereafter, the spent medium was replaced with fresh control medium (DMEM/F12, 1:1; 10% fetal calf serum; 0.01% dimethyl sulfoxide) or medium containing the indicated concentrations of retinoic acid. The cells were cultured at 37° in a humidified atmosphere composed of 94% air and 6% CO_2. Dishes were removed from the incubator at each of the indicated times, the cells were detached after a brief exposure to a solution containing 0.1% trypsin and 2 m*M* EDTA in phosphate-buffered saline, pH 7.2, and suspended repeatedly to give a single-cell suspension, and the cells were counted using an electronic particle counter (Coulter model ZBI, Hialeah, FL). The results are presented as the means ± S.E. of values obtained in triplicate dishes.

shown that the serum can alter the effect of retinoids on cells. For example, the growth of murine sarcoma virus-transformed 3T3 cells is not altered by retinoic acid in serum-containing medium but is stimulated by retinoic acid in serum-free medium.[5] However, murine S91-C2 melanoma cells are about 100 times more sensitive to retinoic acid in serum-free medium (DMEM/F12) supplemented with insulin, transferrin, and selenious acid than in the presence of 10% fetal calf serum.[18]

It is advisable to allow cells to adhere and spread on the plastic dishes for 24 hr prior to the addition of retinoid-containing medium because the adhesive properties of cells can be altered by retinoids,[2] thus affecting plating efficiency. The control cultures should receive medium supplemented with no more than $<0.1\%$ ethanol or DMSO. The cells are cultured at 37° in a humidified atmosphere consisting of 94% air and 6% CO_2 (v/v). After 72 hr, and at 72-hr intervals thereafter, the spent medium is replaced with fresh medium with or without retinoids to replenish the retinoid, which may be depleted to 30–50% of the original concentration during a 72-hr incubation with cells.

Analysis of Growth Inhibition by Cell Counting. Cells are cultured in the absence or presence of retinoids for different times, and the cells are then detached after a brief incubation in either 2 m*M* EDTA or 0.1% trypsin/2 m*M* EDTA in calcium-free and magnesium-free phosphate-buf-

(B) Retinoic acid dose-dependent inhibition of colony formation on plastic or in agarose by 1483 HNSCC cells. To analyze the effect of retinoic acid on colony formation on plastic the cells were seeded at 3×10^3 cells per dish in a series of 100-mm-diameter dishes, and after 24 hr, and at 72-hr intervals thereafter, the cells were treated with retinoic acid as described for (A). After 14 days the cells were fixed, stained with crystal violet, and the number of colonies containing more than 16 cells determined using an inverted microscope. To analyze the effect of retinoic acid on colony formation in semisolid agarose, the cells were suspended in 0.5% agarose at 4×10^4 cells/ml in control medium (containing 0.1% DMSO) or in medium containing retinoic acid in the concentration range between 10^{-9} and 10^{-5} *M*, 1-ml samples of the suspension were placed in 35-mm dishes on top of an agarose layer, and the dishes were placed in a 37° incubator. One milliliter of fresh medium, with or without retinoic acid, was placed on top of the cell-containing agarose gel after 72 hr, and at 72-hr intervals thereafter. After 14 days the number of colonies per dish was determined using an inverted microscope. The results are presented as the means ± S.E. of values obtained in triplicate dishes. (C) Retinoic acid dose-dependent and time-dependent inhibition of the growth of human MDA 886Ln HNSCC multicellular spheroids. Spheroids of similar diameter were placed individually in each well of a 24-multiwell cluster plate that had been precoated with 1.25% agarose. The growth medium consisted of DMEM/F12 (1:1) containing 15% fetal calf serum and 0.01% DMSO (control) or of medium supplemented with the indicated concentrations of retinoic acid. After 72 hr, and at 72-hr intervals thereafter, the spent medium was replaced with fresh medium. At the indicated times two perpendicular diameters of the spheroids were measured using an inverted microscope with a calibrated reticle and the cell volume calculated. The values represent the means ± S.E. of six spheroids.

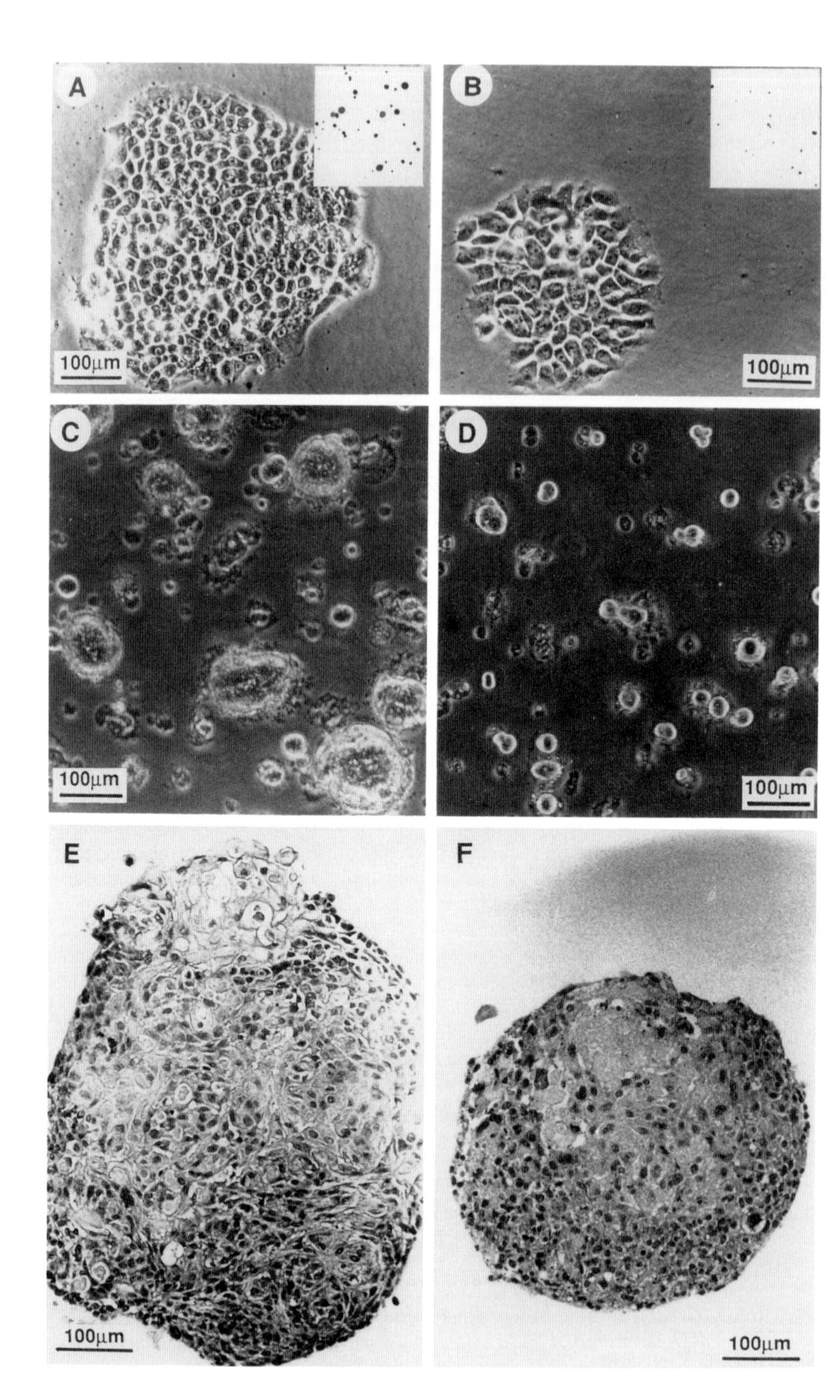
A
100μm
B
100μm
C
100μm
D
100μm
E
100μm
F
100μm

fered saline, pH 7.2 (CMF–PBS). The cells are suspended in CMF–PBS by repeated pipetting to give a single-cell suspension, and the number of cells in samples of the suspensions are counted using an electronic particle counter.[16] Triplicate cultures are used for each time point and retinoid dose. The number of cells per dish is used to construct growth curves (Fig. 1A) and to calculate growth inhibition according to Eq. (1)

$$\% \text{ growth inhibition} = 100\,(1 - R/C) \qquad (1)$$

where R and C represent the number of cells in treated and control cultures, respectively. Cell viability is determined by incubating cells in 0.1% (w/v) trypan blue in PBS and counting the total number of cells and the number of those excluding the stain (the viable ones), using a hemacytometer. Retinoic acid is toxic to most cells at concentrations higher than 50 μM.[17,18] If cell viability in retinoid-treated cultures is lower than 80% then it is not valid to refer to the effect of the retinoid as growth inhibition but rather as a cytotoxic effect.

Cells that tend to aggregate even after trypsin treatment cannot be counted accurately by either the hemacytometer or a particle counter. In such instances it is possible to use a colorimetric cell quantitation assay. After retinoid treatment, the cell monolayers are washed 5 times with PBS and fixed and stained by incubation for 30 min in 0.5% crystal violet dissolved in 70% ethanol. The cell monolayers are then washed 3 times with PBS, and the stained cells are solubilized in 30% (v/v) acetic acid. The absorbance of the cell lysates at 590 nm is then measured on individual samples using a spectrophotometer or, after placing the samples in a 96-well plate, using a Titertech Multiskan reader (Flow Laboratories, McLean, VA).[18] A standard curve can be obtained by staining a predetermined number of cells. The absorbances of treated cultures are then compared to those of control cultures. The colorimetric assay is also useful for microassays on small numbers of cells by plating the cells in 96-well plates and treating them with retinoids in the microwells.[18]

FIG. 2. Morphology of human head and neck squamous carcinoma cells grown in the absence (A, C, E) or presence (B, D, F) of 1 μM retinoic acid. (A, B) 1483 HNSCC cells were grown as monolayers on plastic tissue culture dishes for 12 days, and a representative colony from untreated (A) and treated (B) cultures was photographed. The insets in (A) and (B) show a 1.75-cm^2 area of dishes on which the colonies were stained with crystal violet. (C, D) Colonies of 1483 cells formed during a 14-day growth period in 0.5% agarose. (E, F) Four-micron sections of multicellular tumor spheroids of MDA 886Ln HNSCC grown for 10 days, then fixed in 10% formalin, processed for histology, and stained with hematoxylin and eosin. Bars: 100 μm. Magnification of insets in (A) and (B): ×1. (A–D) Phase-contrast micrographs; (E, F) bright-field micrographs. [Figure 2E,F reproduced from P. G. Sacks, V. Oke, T. Vasey, and R. Lotan, *Head Neck* **11**, 219 (1989).]

Analysis of Growth Inhibition by Counting Colony Number. Growth inhibition of cells that form discrete colonies on the plastic tissue culture dishes (Fig. 2A,B) can be evaluated by comparing the number of colonies in untreated and treated cultures. The cells are seeded at clonal density (i.e., 10–100 cells per cm^2) in plastic tissue culture dishes and incubated in control medium or in medium containing retinoids for about 2 weeks. The cells are then fixed with methanol/acetic acid (3:1, v/v), stained with 0.05% crystal violet (dissolved in 20% methanol in water), and counted under a dissecting microscope (Fig. 1B).[15]

Cell Cycle Analysis. Cells are cultured in the absence or presence of retinoids in a series of tissue culture dishes, and at different times (e.g., 6, 12, 18, 24, 48, 72, 96, 120 hr) cells are detached, suspended, fixed in ethanol, and stained with a DNA staining reagent (e.g., mithramycin, propidium iodide).[10,12] Alternative methods involve dissolution of the cell membrane by a nonionic detergent (1% Triton X-100) followed by staining the nuclei with one of the above-mentioned reagents.[14] The stained cells or nuclei are subjected to analysis by a flow microfluorimeter.[3,10,12,14] This analysis provides the distribution of fluorescence intensity per cell in the form of two peaks connected by a shoulder. The first peak represents cells in the G_1 phase of the cell cycle and the second peak those in the G_2 plus M phases. The shoulder represents the cells that are in the S phase. Computer analysis of the data provides information on the percentage of cells in each phase of the cell cycle in untreated and retinoid-treated cultures. Accumulation in the G_1 phase of the cell cycle was reported for Chinese hamster ovary cells treated with retinol,[10] as well as for human HeLa carcinoma cells,[11] murine S91-C2 melanoma cells,[12] and murine Friend erythroleukemia cells[14] treated with retinoic acid. Retinoid-treated cells may also accumulate in the S phase of the cell cycle as was found for murine Swiss 3T3 cells.[3]

Analysis of DNA Synthesis. Cell growth inhibition is almost invariably associated with a decrease in DNA synthesis. To determine the effect of retinoid treatment on DNA synthesis, the cells are treated for different periods of time, and [*methyl*-^{3}H]thymidine is added to the culture medium at a concentration of 1 to 5 μCi/ml. The cells are incubated for 1 to 18 hr, and then they are detached after trypsinization and suspended, with samples of the suspension being analyzed for cell number or for DNA content and for the incorporation of [^{3}H]thymidine into DNA. The latter parameter is determined by collecting a predetermined number of cells on glass fiber disks or filter paper disks and washing the filters successively with ice-cold 10% trichloroacetic acid (TCA), 5% cold TCA, and 95% (v/v) ethanol. The filters are then dried, and the radioactivity into the TCA-insoluble material is measured by scintillation spectrometry. The amount of radioactivity incorporated per cell or per microgram DNA in untreated and treated cultures is then compared. The TCA-soluble material is also collected during the washes, and the radioactivity in this fraction is counted and used to determine whether the treatment alters thymidine uptake or thymidine incorporation into DNA, or both. Variations of this procedure have been used in several studies with retinoids.[10-13,19]

Inhibition of Anchorage-Independent Growth

Several types of cells, such as lymphoid cells, grow normally in suspension. Untransformed cells derived from solid tissues require attachment to and spreading on a substrate to proliferate in culture, and they fail to proliferate if cell attachment and flattening are prevented by suspending the cells in a round configuration in semisolid media or above a nonadhesive substrate.[20,21] In contrast, many tumor cells are able to grow even if denied attachment because they have lost their dependence on anchorage during the malignant transformation.[20,21] Treatment of various transformed and tumor cells with retinoids restores anchorage dependence and inhibits the ability of the cells to grow in suspension in liquid medium or in semisolid medium including methyl cellulose,[22] agar,[23] or agarose[12,15,24] (Figs. 1B and 2C,2D). Some tumor cells can grow as tightly packed multicellular spheroids suspended in liquid medium. The growth of such spheroids is also suppressed by retinoids.[25]

Assay of Inhibition of Lymphoid Cell Growth. The inhibition by retinoids of the proliferation of cells growing in suspension is analyzed by withdrawing samples of the suspension at different times after initiation of treatment and counting the number of cells in untreated and treated cultures using an electronic particle counter as described above.

Restoration of Shape-Dependent Growth Control. To obtain substrates on which cells exhibit different degrees of cell flattening, culture dishes are coated with a nonadhesive polymer, poly(2-hydroxyethyl methacrylate) [poly(HEMA), Interferon Sciences, New Brunswick, NJ]. The poly(HEMA) is suspended in 95% ethanol to make a 12% (w/v) suspension. After a vigorous mixing the suspension is centrifuged at 10,000 *g*

[19] R. Lotan, T. Stolarsky, D. Lotan, and A. Ben-Ze'ev, *Int. J. Cancer* **33,** 115 (1984).
[20] I. MacPherson and L. Montagnier, *Virology* **23,** 291 (1964).
[21] J. Folkman and A. Moscona, *Nature (London)* **273,** 345 (1978).
[22] L. D. Dion, J. E. Blalock, and G. E. Gifford, *Exp. Cell Res.* **117,** 15 (1978).
[23] F. L. Meyskens and S. E. Salman, *Cancer Res.* **39,** 4055 (1979).
[24] R. Lotan, D. Lotan, and A. Kadouri, *Exp. Cell Res.* **141,** 79 (1982).
[25] P. G. Sacks, V. Oke, T. Vasey, and R. Lotan, *Head Neck* **11,** 219 (1989).

(23°, 15 min), and the supernatant is collected and used as a stock solution. This solution is diluted in absolute ethanol to obtain dilutions of 1:100, 1:200, 1:500, 1:1000, and 1:2000. To each well of a 24-well cluster plate 200 μl of each poly(HEMA) dilution is added, and the ethanol is evaporated by incubating the dishes at 37° for 36–38 hr. This procedure provides surfaces coated with polymer films of decreasing thickness and correspondingly increasing adhesiveness.[21]

Cells grown as monolayers on plastic tissue culture dishes are pretreated with retinoids, and then they are detached and placed in unmodified wells or in poly(HEMA)-coated wells at a density of 10^5 cells/ml in 0.5 ml of medium with or without retinoid. After 20 hr of incubation at 37°, the cells are pulse-labeled for 60 min with 5 μCi per well of [*methyl*-^{3}H]thymidine. The cells are then suspended and analyzed for thymidine incorporation into TCA-insoluble material as described above. The radioactivity incorporated into DNA of cells incubated on unmodified plastic wells serves as control (100%). Retinoic acid treatment of various tumor cell lines suppressed their ability to synthesize DNA when cell attachment and flattening was restricted by the poly(HEMA)-coated substratum.[19] Thus, retinoic acid treatment caused the cells to behave like untransformed cells. Cells placed on poly(HEMA)-coated substratum tend to aggregate;[19] therefore, this assay does not provide information on the response of individual clones within the population.

Inhibition of Colony Formation in Semisolid Medium. There are several methods for suspending cells and maintaining them at clonal density using semisolid media.[12,22–24] One such medium is methy cellulose. Cells are suspended at 10^3 to 10^5 cells/ml, depending on their colony-forming efficiency, in medium containing 1.17% (w/v) methyl cellulose, 10% serum, and 0.1% ethanol or 0.1% DMSO or different concentrations of a retinoid. Samples containing 2 ml cell suspension are placed in each well of a multiwell plate or into 35-mm-diameter tissue culture dishes, in which 2 ml of 0.5% agar (w/v) has been previously placed, and allowed to gel. The plates or dishes are then incubated at 37° in a humidified atmosphere consisting of 94% air and 6% CO_2 for 14 days, and the number of colonies is determined using an inverted microscope.[22] Agar has also been used as a semisolid medium to analyze the effect of retinoids on tumor cells and primary cells derived from fresh tumor biopsies. The cells are suspended at 5×10^5 cells/ml in 0.3% agar in medium containing serum and either solvent or retinoid (at twice the desired final concentration), and 1-ml samples of the suspension are placed in 35-mm plastic dishes on top of a precast underlayer of 1 ml of 0.5% agar in serum-containing medium. The dishes are then incubated for 14 days, and the number of colonies in treated cultures is compared to the number in control cultures.[9,23] In this procedure the retinoid is added only at the beginning of the culture and is not replenished, although it is known that the retinoid is degraded within a few days.

Another method utilizes low-temperature-gelling agarose, which provides improved colon-forming efficiency for certain cell types. Cells are suspended at 10^3 to 10^5 cells/ml, depending on their colony-forming efficiency, in 0.5% agarose (w/v) in medium containing serum and solvent or retinoid, and 1-ml samples are placed in 35-mm-diameter plastic dishes on top of a layer of 1% agarose (in the same medium), which has been allowed to gel previously. The dishes are kept at 4° for 10 min to gel the cell-containing agarose, and then the dishes are incubated at 37° as described above. During this incubation the cultures are reexposed to fresh medium with or without retinoids every 72 hr by initially placing and subsequently replacing 1 ml of medium on top of the cell-containing agarose layer. After 14 days, the number colonies with over 50 cells per colony is determined under an inverted microscope.[12,24] The observation of colonies can be facilitated by staining the viable cells. To stain the colonies in agarose the medium is removed, and 1 ml of a solution containing 1 mg of the tetrazolium salt 2-(*p*-iodophenyl)-3-(nitrophenyl)-5-phenyltetrazolium chloride per milliliter of 0.9% sodium chloride solution is placed over the agarose. After an additional 24-hr incubation, the colonies stained bright red with the precipitated formazan are counted. An example for the retinoic acid dose-dependent suppression of colony formation by human squamous carcinoma cells is presented in Fig. 1B, and the morphology of such colonies (unstained) is shown in Fig. 2C,D.

Suppression of Growth of Multicellular Spheroids. There are some tumor cells that can grow in suspension as tightly-packed three-dimensional multicellular spheroids. The volume of such spheroids increases over time, and they are used as models for preclinical study of chemotherapeutic agents, including retinoids.[25] Spheroids are formed by suspending cells from monolayer cultures at the logarithmic growth phase, placing them on plastic petri dishes which had been precoated with 1.25% agarose in normal serum-containing medium, and incubating them for 3 days at 37°. Small spheroidal aggregates formed during this time are transferred to spinner flasks, and the flasks are gassed with 5% CO_2 and incubated at 37°. For growth inhibition experiments, spheroids are sized using a calibrated reticle under a dissecting microscope, and spheroids of a similar initial diameter are picked and placed in individual wells of a 24-well plate, which had been coated with agarose as above, in control medium or in medium containing different concentrations of retinoids. On the first day and at 72-hr intervals, when the medium is changed, spheroids are sized by measuring two perpendicular diameters (a and b) of the spheroids using the microscope and reticle. The volume of the spheroids is calculated according to Eq. (2)

$$V = \frac{4}{3}\pi(\sqrt{ab}/2)^3 \qquad (2)$$

as described elsewhere.[26] Using this procedure it is possible to obtain information on the time and dose dependence of growth inhibition of the spheroids by retinoids as demonstrated in Fig. 1C. The spheroids can also be fixed and processed for histological analysis as shown in Fig. 2E,F.

Concluding Remarks

Retinoids inhibit the growth of numerous tumor cells and suppress the expression of the transformed phenotype as represented by anchorage-independent growth. The effective concentrations of retinoids for inhibition of tumor cell growth in culture are often pharmacological and not physiological. However, quite a large number of tumor cells are inhibited considerably even at physiological dose (1 μM for retinol and 10 nM for retinoic acid). Often cells that are only marginally inhibited in monolayer culture show marked inhibition of anchorage-independent growth in agarose, suggesting that the inhibition of anchorage-independent growth is a more sensitive assay for the suppression of the growth of tumor cells by retinoids.

[26] T. Nederman and P. Twentyman, *Rec. Results Cancer Res.* **95,** 84 (1984).

[12] Maintenance and Use of F9 Teratocarcinoma Cells

By ANDREW L. DARROW, RICHARD J. RICKLES, and SIDNEY STRICKLAND

Introduction

F9 cells are an established cell line of mouse embryonal carcinoma cells that differentiate spontaneously at very low frequency. Although F9 cells are routinely maintained and passaged as stem cells, treatment with retinoic acid (RA) induces differentiation in culture, producing cells that resemble primitive extraembryonic endoderm of the mouse embryo.[1] RA-treated F9 cells are bipotential: aggregation results in the appearance of visceral endoderm,[2] whereas further treatment with cyclic AMP (RA/

[1] S. Strickland and V. Mahdavi, *Cell (Cambridge, Mass.)* **15,** 393 (1978).
[2] B. L. M. Hogan, A. Taylor, and E. Adamson, *Nature (London)* **291,** 235 (1981).

cAMP) leads to differentiation of the cells into parietal endoderm.[3] Interestingly, treatment with dibutyryl-cAMP alone does not cause F9 cells to differentiate; the cells must first be exposed to RA to be permissive for the effects of cAMP. In addition to their use for studying certain aspects of mammalian differentiation, F9 cells are also a useful system for the study of the mechanisms by which retinoids and cAMP affect gene expression.

Maintenance of F9 Cells

F9 cells can be maintained in a conventional cell culture facility. We propagate the cells in Dulbecco's modified Eagle's medium containing 15% fetal bovine serum (DME/15% FBS). The cells must be passaged frequently to avoid overcrowding, which may induce spontaneous differentiation. They should therefore be subcultured every other day, and plating intervals should not exceed 3 days.

For subculturing F9 cells, we use 80-cm^2 tissue culture flasks treated with a sterile 0.1% gelatin solution. This solution should be autoclaved, allowed to cool to room temperature, and autoclaved again in order to destroy possible fungal spores. Gelatin is added to culture dishes or flasks, which are then refrigerated for at least 20 min or kept at room temperature for a few hours. The solution is removed by aspiration prior to seeding of cells. Dishes treated in this manner may be used immediately or stored sterilely indefinitely.

Subconfluent monolayers of F9 cells may be detached using versene. Versene contains 0.2 g KCl, 0.2 g KH_2PO_4, 8.0 g NaCl, 2.17 g $Na_2HPO_4 \cdot 7H_2O$, and 0.2 g Na_2 EDTA per liter, pH 7.4, and is sterilized by autoclaving. Conditioned medium is first removed by aspiration, and about one-half that volume of versene is added to rinse the monolayer. The rinse is removed by aspiration and replaced with an equal volume of fresh versene. Place the flask at 37° for 5–10 min. The monolayer is then removed by pipette and added to an equal volume of DME/15% FBS in a plastic conical centrifuge tube. Cells are pelleted at 2000 *g* and resuspended in approximately 2 ml of fresh DME/15% FBS. It is important to obtain a single cell suspension, and this can be achieved by vigorously pipetting up and down using a Pasteur pipette. Count the cells using a hemacytometer as this will allow an assessment of the degree of dispersion. If the cells remain clumped, disruption is repeated with a Pasteur pipette. Seed the single cell suspension in 10 ml of media at 10^6 and 5×10^5 for a backup culture.

[3] S. Strickland, K. K. Smith, and K. R. Marotti, *Cell (Cambridge, Mass.)* **21**, 347 (1980).

TABLE I
RECOMMENDED PLATING DENSITIES OF VARIOUS F9 CELL DERIVATIVES

Cell type	Density (cells/cm²)	Inducer
96 hr induction without refeeding		
F9 stem	5.7×10^2	None
F9-cAMP	5.7×10^2	10^{-3} *M* DBC
F9-RA	1.7×10^3	10^{-7} *M* RA
F9-RA/cAMP	3.4×10^3	10^{-7} *M* RA + 10^{-3} *M* DBC
96 hr induction with refeeding at 80 hr		
F9 stem	2.8×10^3	None
F9-cAMP	2.8×10^3	10^{-3} *M* DBC
F9-RA	5.7×10^3	10^{-7} *M* RA
F9-RA/cAMP	1.1×10^4	10^{-7} *M* RA + 10^{-3} *M* DBC

[a] Plating densities yield approximately equal numbers of cells at the end of the 96-hr induction period. For 10-cm dishes (79 cm²) we use 10 ml of media and 25 ml for 15-cm plates (177 cm²).

Differentiation of F9 Cells

F9 stem cells have a doubling time of 12–16 hr when cultured under normal growth conditions. The doubling time increases on differentiation in response to RA and increases still further when the cells are fully differentiated using RA and cAMP. In order to compensate for the reduction in cell division following differentiation, it is useful to seed the cells at a higher density in the presence of inducers. The number of cells to be plated should be determined empirically since growth rates may vary with different populations of F9 cells or batches of fetal bovine serum. We routinely differentiate cells for a minimum of 96 hr. Refeeding during this period is usually necessary if the cells are plated at a high density.[4] Two protocols we have followed are outlined in Table I.

For best results we recommend preparing the inducers just before use. all-*trans*-retinoic acid (RA) (Eastman Kodak, Rochester, NY) is stored as a powder at −20° in a lightproof container owing to its photosensitivity. It should be dispensed in the dark and dissolved in 100% ethanol at a concentration of 3.0 mg/ml. As some preparations of RA may not dissolve in ethanol immediately, the tube should be wrapped in aluminum foil and occasionally vortexed until the RA is entirely dissolved. We dilute the 3.0

[4] R. J. Rickles, A. L. Darrow, and S. Strickland, *J. Biol. Chem.* **263**, 1563 (1988).

mg/ml stock 1:10 in 100% ethanol and finally 1:100 in DME/15% FBS to prepare a 10^{-5} *M* stock solution. This stock need not be filtered if all manipulations after the ethanol dilution are carried out sterilely. The 10^{-5} *M* stock in DME/15% FBS can be kept for a few days at 4° in the dark. RA in ethanol is not stable and should be discarded after use. We have used RA at concentrations ranging between 1 and 5×10^{-7} *M* to induce differentiation. However, the concentration of RA optimal for differentiation may vary; when working with F9 cells for the first time, when working with a new batch of F9 cells, or when using new stocks of RA, cells should be characterized to determine the extent of differentiation. At concentrations at or above 10^{-6} *M*, RA is toxic to F9 cells.[1,5]

Dibutyryl-cAMP (DBC) (Sigma, St. Louis, MO) can be dissolved as either a 10^{-2} *M* (10×) or a 2×10^{-2} *M* (20×) stock directly in DME/15% FBS. It is filtered using a syringe fitted with a 0.2-μm filter and then used immediately. An alternative means of increasing the intracellular concentration of cAMP is to use a combination of cholera toxin (Schwarz-Mann, Cleveland, OH) at a final concentration of 10^{-10} *M* (8.4 ng/ml) and the cyclic phosphodiesterase inhibitor 3-isobutyl-1-methylxanthine (MIX) (Sigma) at a final concentration of 10^{-4} *M*. These two inducers are prepared as 100× filtered stock solutions in DME/15% FBS and are stable for at least several days at 4°. The MIX is first dissolved in dimethyl sulfoxide (DMSO) at 0.1 *M* then diluted 1:10 in DME/15% FBS and filtered to prepare the 10^{-2} *M* (100×) stock.

When F9 cells differentiate, there are associated morphological changes that are apparent by 48–72 hr of induction. With RA treatment the cells become larger, more spread out, and accumulate intracellular granules. Treatment with both RA and DBC causes the cells to become more rounded in appearance. Figure 1 shows the morphology of F9 stem cells or cells after exposure to RA or both RA and DBC for 96 hr. Assessing the extent of differentiation has relied on monitoring the levels of differentiation-specific gene products. It has been demonstrated that as F9 cells differentiate they synthesize several secreted proteins such as laminin, type IV collagen, and plasminogen activator (PA).[3] Two endodermal cytoskeletal proteins, endo A and B, also accumulate upon differentiation.[6] Another reliable marker for the differentiated state is the loss of the stage-specific embryonic antigen SSEA-1.[7] A convenient quantitative measure we employ is the examination of conditioned media for the presence of PA

[5] S.-Y. Wang and L. J. Gudas, *J. Biol. Chem.* **259**, 5899 (1984).
[6] R. G. Oshima, *J. Biol. Chem.* **256**, 8124 (1981).
[7] D. Solter and B. Knowles, *Proc. Natl. Acad. Sci. U.S.A.* **75**, 5565 (1978).

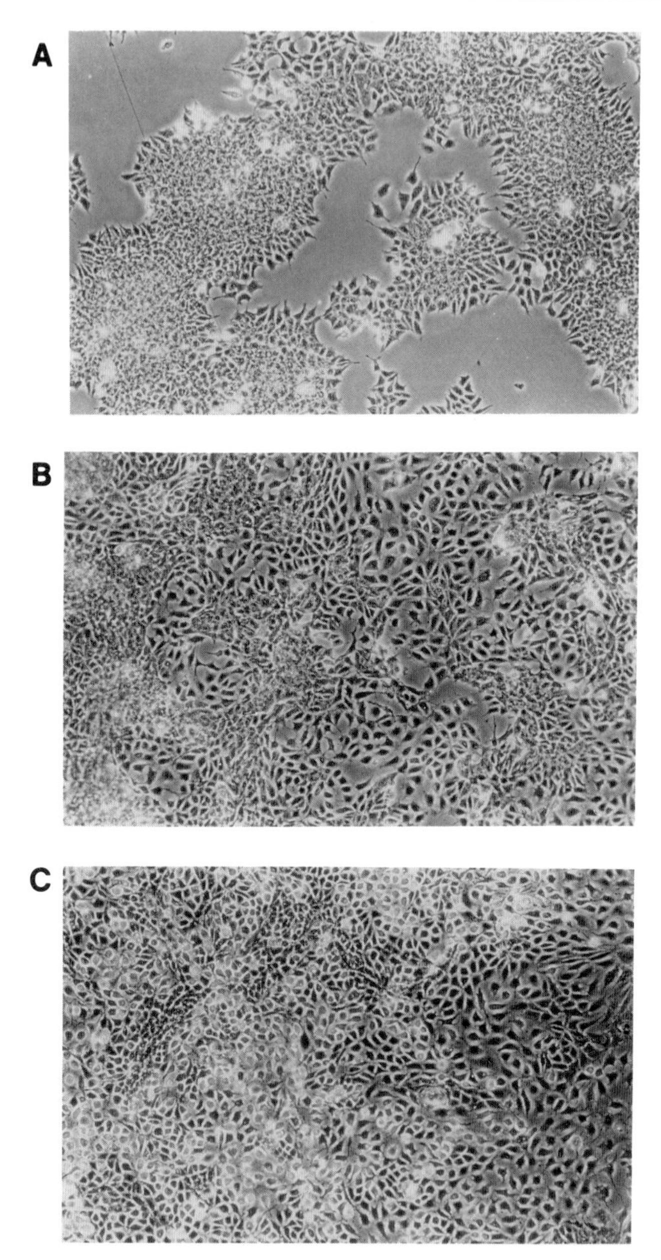

FIG. 1. Morphological changes observed on differentiation of F9 cells. F9 cells were seeded according to the low density protocol and cultured for 96 hr. (A) Cells in the absence of inducers (F9 stem); (B) cells treated with 10^{-7} *M* RA (F9-RA); (C) cells in the presence of 10^{-7} *M* RA and 10^{-3} *M* DBC (F9-RA/cAMP).

activity using a chromogenic amidolytic assay.[8] In addition, cDNA clones[4,9] or antibodies[10] to differentiation specific markers are available.

Transient Transfection of F9 Cells and Differentiated Derivatives

The F9 cell system provides a means to identify mechanisms that control gene expression upon differentiation. Gene transfer experiments are allowing investigators to examine this regulation at the molecular level. To introduce DNA into F9 cells, we have optimized a transfection protocol which is a modification of the calcium phosphate DNA precipitation technique first described by Graham and van der Eb.[11] This procedure can be used to compare the strength of promoter or enhancer elements linked to a reporter gene upon introduction into either F9 stem or differentiated cells. The transfection procedure can also be used to determine the importance of sequences present in an mRNA which confer transcript stability or translational control.

F9 cells are seeded according to the low density protocol in 15-cm dishes and differentiated without refeeding for 96 hr. At 96 hr they are collected sterilely using Versene and replated at approximately 1.4×10^4 cells/cm^2 in DME/15% FBS on 0.1% gelatin-coated dishes. RA/cAMP differentiated cells (which divide slowly) should be seeded at twice this density. Maintaining the presence of inducers from this step onward does not influence the level of expression of the transfected differentiation-responsive tissue plasminogen activator (t-PA) promoter.[12] However, other promoters may not behave in a similar manner and should be analyzed for their hormone requirements posttransfection.

Transfections are carried out 20–24 hr later as follows.[12] Each 10-cm dish receives 1.0 ml of a 1× DNA–calcium phosphate (DNA/CaP) slurry containing 68 m*M* NaCl, 2.5 m*M* KCl, 5.6 m*M* glucose, 104 m*M* HEPES (*N*-2-hydroxyethylpiperazine-*N'*-2-ethanesulfonic acid), 0.7 m*M* Na_2HPO_4, 125 m*M* $CaCl_2$, and 20–40 μg supercoiled DNA. It was found that with a concentration of 104 m*M* HEPES in the DNA/CaP slurry, elevated reporter gene activity was observed in F9 stem cells, which was attributed to increased DNA uptake. When comparing reporter gene plasmids of different size, a constant molar quantity of the constructs should be used, and differences in the overall amount of DNA transfected is

[8] P. Andrade-Gordon and S. Strickland, *Biochemistry* **25,** 4033 (1986).

[9] S.-Y. Wang, G. J. LaRosa, and L. J. Gudas, *Dev. Biol.* **107,** 75 (1985).

[10] L. M. Silver, G. R. Martin, and S. Strickland, "Teratocarcinoma Stem Cells." Cold Spring Harbor Laboratory, Cold Spring Harbor, New York, 1983.

[11] F. L. Graham and A. J. van der Eb, *Virology* **52,** 456 (1973).

[12] R. J. Rickles, A. L. Darrow, and S. Strickland, *Mol. Cell. Biol.* **9,** 1691 (1989).

compensated for by the addition of a supercoiled plasmid DNA, such as pUC9. Plasmid DNA can be prepared using the alkaline lysis procedure[13] and further purified by equilibrium density-gradient centrifugation. Refeeding the cells 4 hr prior to transfection has no significant effect on transfection efficiency of any of the F9 cell types.[12]

Careful preparation of the DNA/CaP slurry is important to control variability in the transfection efficiency. Multiple slurries can be prepared by standard methods just prior to transfection.[14] A 2X Ca–DNA solution (250 mM $CaCl_2$, 40–80 μg/ml plasmid DNA) is added to an equal volume of a 2X HEPES buffered phosphate solution (136 mM NaCl, 5.0 mM KCl, 11.2 mM glucose, 208 mM HEPES, 1.4 mM Na_2HPO_4, pH 7.1). Consistent results are obtained by dropwise addition with simultaneous gentle agitation using a vortex mixer followed by introducing bubbles with an air-driven pipettor. After a 30-min incubation at room temperature, slurries are added to the plates at 1.0 ml per 10 ml DME/15% FBS. The cells are cultured overnight and harvested 20–24 hr later in order to determine the relative levels of reporter gene product.

Relative Transfection Efficiency of F9 Stem Cells and Differentiated Derivatives

A question that arises in comparing the reporter gene activities in transient assays of distinct cell types is that of transfection efficiency. For the transfection protocol outlined above, we have examined the relative transfection efficiencies of the different F9 cell types (Fig. 2). F9 stem, F9-cAMP, F9-RA, and F9-RA/cAMP cells are transfected with supercoiled plasmid DNA and 24 hr later harvested for DNA extraction. The monolayers are first washed 3 times with ice-cold phosphate-buffered saline containing 3 mM EGTA [ethylene glycol bis(β-aminoethyl ether)-N,N,N',N'-tetracetic acid]. Cells are collected in versene, pelleted, and then resuspended and pelleted three times in the PBS/EGTA solution to extensively wash the cells. Total genomic DNA is isolated and subjected to Southern blot analysis.[13] Two ^{32}P-labeled random-primed probes are used. (1) An 1800-base pair (bp) fragment of the mouse t-PA cDNA[4] is a normalization probe used to detect the single-copy endogenous t-PA gene. (2) An 800-bp fragment of the chloramphenicol acetyltransferase (CAT)

[13] T. Maniatis, E. F. Fritsch, and J. Sambrook, "Molecular Cloning: A Laboratory Manual." Cold Spring Harbor Laboratory, Cold Spring Harbor, New York, 1982.

[14] F. L. Graham, S. Bacchetti, R. McKinnon, C. Stanners, B. Cordell, and H. M. Goodman, *in* "Transformation of Mammalian Cells with DNA Using the Calcium Phosphate Technique" (R. Baserga, C. Croce, and G. Rovera, eds.), p. 3. Alan R. Liss, New York, 1980.

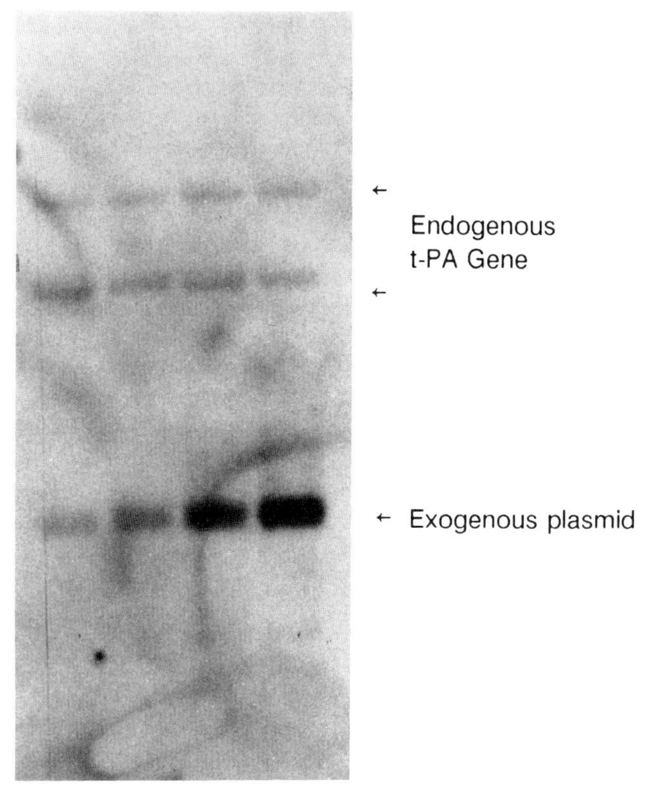

FIG. 2. Comparison of F9 cell transfection efficiencies. Southern analysis of total DNA isolated from transfected cells. The cell types examined are indicated above the autoradiogram. Also indicated are the positions of bands corresponding to the endogenous t-PA gene and internalized exogenous plasmid DNA.

gene is used to examine the relative amount of transfected DNA within each cell type.

Densitometric scanning of the autoradiogram in Fig. 2 indicates that the relative transfection efficiencies are as follows: F9 stem, 1; F9-cAMP, 1; F9-RA, 2.5; F9-RA/cAMP, 5. Thus, using the above protocol, there exists a difference in the efficiency with which derivatives of F9 cells internalize DNA. It is important to note, however, that the efficiency of DNA uptake is likely to depend on the methods used for introducing DNA.

[13] Growth and Differentiation of Human Myeloid Leukemia Cell Line HL60

By THEODORE R. BREITMAN

Introduction

In recent years the development of human myelomonocytic cell lines has provided useful models for studying regulation of both cell proliferation and differentiation. This can be very important for studies on the treatment of leukemia because the finding that some of these cell lines are induced by a wide variety of compounds, including retinoic acid (RA), to terminally differentiate has suggested an alternative approach to the therapy of certain types of leukemias. These studies have had an impact on the approach to the treatment of some leukemias, with a recent clinical study showing complete remissions in patients with acute promyelocytic leukemia treated with RA.[1]

The growth advantage that myeloid leukemia cells have *in vivo* over normal cells is not that of a more rapid growth rate but rather an apparent inability to mature to functional, terminally differentiated end cells. It is possible that some leukemia cells do not mature either because they have a decreased ability to respond to exogenous differentiative factors or because the production of specific gene products obligatory for differentiation is altered.

Human Myeloid Leukemia Cell Line HL60

HL60 is the most widely used of the human myeloid leukemia cell lines for studying differentiation. It was the first human cell line with distinct myeloid features to be developed.[2] The HL60 cell line was isolated from the blood of a patient with what has recently been rediagnosed as acute myeloid leukemia with maturation.[3] It proliferates continuously in suspension culture and consists predominantly of promyelocytes. HL60 cells are induced by exposure to a wide variety of compounds, including dimethyl sulfoxide (DMSO), hypoxanthine, dimethylformamide (DMF), actinomy-

[1] M.-E. Huang, Y.-C. Ye, S.-R. Chen, J.-R. Chai, J.-X. Lu, L. Zhoa, L.-J. Gu, and Z.-Y. Wang, *Blood* **72,** 567 (1988).

[2] S. J. Collins, R. C. Gallo, and R. E. Gallagher, *Nature (London)* **270,** 347 (1977).

[3] W. T. Dalton, Jr., M. J. Ahearn, K. B. McCredie, E. J. Freireich, S. A. Stass, and J. M. Trujillo, *Blood* **71,** 242 (1988).

cin D, and RA,[4-6] to terminally differentiate into cells having many of the morphological features of mature granulocytes. HL60 cells are induced by compounds such as 12-*O*-tetradecanoylphorbol 13-acetate (TPA), 1,25-dihydroxyvitamin D_3, and sodium butyrate to terminally differentiate into cells having morphological features of monocytes/macrophages.[7-9] Many of these induced HL60 cells have functional characteristics of normal human peripheral blood granulocytes and monocytes/macrophages including phagocytosis, lysosomal enzyme release, complement receptors, chemotaxis, hexose monophosphate shunt activity, superoxide anion generation, and the ability to reduce nitro blue tetrazolium (NBT).[4,10,11]

Growth of HL60 Cells

HL60 cells grow in stationary cultures in many nutrient media containing fetal bovine serum. HL60 cells can also be grown in a serum-free nutrient medium supplemented only with 5 μg of insulin/ml and 5 μg of transferrin/ml.[12] The nutrient medium can be either RPMI 1640 containing 10 m*M* *N*-2-hydroxyethylpiperazine-*N*′-2-ethanesulfonic acid (HEPES), pH 7.3, or a 1:1 mixture of Dulbecco's modified Eagle's medium and Ham's F12 medium containing 14.3 m*M* $NaHCO_3$ and 15 m*M* HEPES at pH 7.3.[13] In this insulin- and transferrin-supplemented medium, referred to as defined medium, long-term growth of HL60 continues at a rate approximately 80% of that occurring in medium supplemented with serum. The saturation density is lower in defined medium (1.5×10^6 cell/ml) than in serum-supplemented medium (3×10^6 cells/ml). Growth of HL60 cells has continued in defined medium for over 30 passages, including at least 60 population doublings. This translates into a 10^{18}-fold

[4] S. J. Collins, F. W. Ruscetti, R. E. Gallagher, and R. C. Gallo, *Proc. Natl. Acad. Sci. U.S.A.* **75,** 2458 (1978).

[5] T. R. Breitman, S. E. Selonick, and S. J. Collins, *Proc. Natl. Acad. Sci. U.S.A.* **77,** 2936 (1980).

[6] Y. Honma, K. Takenaga, T. Kasukabe, and M. Hozumi, *Biochem. Biophys. Res. Commun.* **95,** 507 (1980).

[7] G. Rovera, D. Santoli, and C. Damsky, *Proc. Natl. Acad. Sci. U.S.A.* **76,** 2779 (1979).

[8] D. M. McCarthy, J. F. San Miguel, H. C. Freake, P. M. Green, H. Zola, D. Catovsky, and J. M. Goldman, *Leuk. Res.* **7,** 51 (1983).

[9] A. W. Boyd and D. Metcalf, *Leuk. Res.* **8,** 27 (1984).

[10] S. J. Collins, R. E. Ruscetti, R. E. Gallagher, and R. C. Gallo, *J. Exp. Med.* **149,** 969 (1979).

[11] P. E. Newburger, M. E. Chovaniec, J. S. Greenberger, and H. J. Cohen, *J. Cell Biol.* **82,** 315 (1979).

[12] T. R. Breitman, S. J. Collins, and B. R. Keene, *Exp. Cell Res.* **126,** 494 (1980).

[13] J. P. Mather and G. H. Sato, *Exp. Cell Res.* **120,** 191 (1979).

increase in cell numbers without serum, making it unlikely that residual serum components are contributing to the growth of the cells.

Defined medium has been very useful in investigating differentiation of HL60 where studies have been carried out without undefined serum inhibitors or enhancers of differentiation, thus providing a useful means of assessing the HL60 response to physiological differentiation-including compounds that have relevance to the control mechanisms of normal granulopoiesis.

Retinoic Acid Induced Differentiation of HL60 Cells

all-*trans*-β-Retinoic Acid is a potent inducer of granulocytic differentiation of HL60.[5] This compound induces relatively extensive morphological differentiation[5] and, probably more importantly, induces at concentrations that are physiological.[14]

The possibility that RA acts on HL60 after its conversion to 4-hydroxy- and 4-keto-RA[15] was investigated indirectly by testing the ability of these two metabolites to induce differentiation.[16,17] Both compounds are about one-tenth as potent as RA. In these as well as other studies with retinoidal benzoic acid derivatives[18] and with heteroarotinoids,[19] the most effective retinoid inducers of HL60 differentiation possess a carboxylic acid function at a position corresponding to the C-15 terminal carbon of RA. This activity is retained, although somewhat diminished, in spite of alterations in the ring as in the 4-hydroxy- and 4-keto-substituted derivatives and α-RA. Substitutions at the C-15 position result in essentially a complete loss of activity (retinal, retinol, retinyl acetate). These activity–structure relationships emphasize the specificity of the RA effect on HL60 and make more likely the possibility that this phenomenon, observed *in vitro,* is an expression of a true physiological process.

[14] J. L. Napoli, B. C. Pramanik, J. B. Williams, M. I. Dawson, and P. D. Hobbs, *J. Lipid Res.* **26,** 387 (1985).

[15] A. B. Roberts and C. A. Frolik, *Fed. Proc., Fed. Am. Soc. Exp. Biol.* **38,** 2524 (1979).

[16] T. R. Breitman, *in* "Expression of Differentiated Functions in Cancer Cells" (R. P. Revoltella, G. M. Pontieri, C. Basilico, G. Rovera, R. C. Gallo, and J. H. Subak-Sharpe, eds.), p. 257. Raven, New York, 1982.

[17] H. Hemmi and T. R. Breitman, *in* "Retinoids: New Trends in Research and Therapy" (J. H. Saurat, ed.), p. 48. Karger, Basel, 1984.

[18] S. Strickland, T. R. Breitman, F. Frickel, A Nurrenbach, E. Hadicke, and M. B. Sporn, *Cancer Res.* **43,** 5268 (1983).

[19] L. W. Spruce, S. N. Rajadhyaksha, K. D. Berlin, J. B. Gale, E. T. Miranda, W. T. Ford, E. C. Blossey, A. K. Verma, M. B. Hossain, D. van der Helm, and T. R. Breitman, *J. Med. Chem.* **30,** 1474 (1987).

Procedures for Studying Growth and Differentiation

Maintenance and Growth of HL60 Cells

General Equipment

Laminar flow hood
Incubator with a humidified atmosphere of 5% CO_2 in air
Inverted microscope
Light microscope
Centrifuge
Electronic particle counter (Coulter Electronics, Hialeah, FL)
Cytospin centrifuge (Shandon Elliott, Sewickley, PA)
Water bath
Liquid nitrogen refrigerator

General Supplies

Dulbecco's modified Eagle's medium/Nutrient Mixture F12 (Ham's), 1:1 (GIBCO, Grand Island, NY, Cat. No. 320-1330)
RPMI medium 1640 (GIBCO Cat. No. 320-1875 or equivalent)
Dulbecco's phosphate-buffered saline without Ca^{2+} and Mg^{2+} (DPBS)
1 *M* HEPES buffer, pH 7.3 (GIBCO Cat. No. 380-5630)
Bovine insulin, zinc-free (Collaborative Research, Lexington, MA, or Upstate Biotechnology, Lake Placid, NY)
Human transferrin (Collaborative Research or Upstate Biotechnology)
Tissue culture flasks
0.4% Trypan blue solution (GIBCO Cat. No. 630-5250)
Polypropylene freezing tubes
0.45-μm (500 ml) and 0.2-μm (115 ml) sterilization filter units (Nalge, Rochester, NY, or Corning, Corning, NY)
Fetal bovine serum (FBS)
Dimethyl sulfoxide
Cell freezer (Biotech Research Laboratories, Rockville, MD)

Procedure. The nutrient medium is either Dulbecco's modified Eagle's medium/Nutrient Mixture F12 (Ham's), 1:1, or RPMI medium 1640 supplemented with 10 m*M* HEPES, pH 7.3. Stock cultures are grown in nutrient medium supplemented with 10% FBS. Cells are fed 4 or 5 days after splitting and are subcultured (passaged) once a week. All HL60 experiments are performed on passages 12–40. HL60 cells are obtained from the American Type Culture Collection (Rockville, MD, Cat. No. CCL 241). Antibiotics should not be added to stock cultures. Antibiotics can mask problems with equipment or technique and probably will not prevent mycoplasmal infection, which often accompanies bacterial infection. Serum-free medium is composed of nutrient medium supplemented with 5 μg of insulin/ml and 5 μg of transferrin/ml.

HL60 cells are harvested by centrifugation at 250 *g* for 7 min and resuspended in growth medium at 2.5×10^5 cells/ml or higher. Incubations are at 37° in a humidified atmosphere of 5% CO_2 in air. Cell numbers are determined with a Coulter counter, and viability, by trypan blue dye exclusion.

Freezing and Reconstitution of HL60 Cells

HL60 cells are stored frozen in a freezing medium containing 10% dimethyl sulfoxide and 10–20% FBS in nutrient medium. Cells are harvested by centrifugation and resuspended in the freezing medium at a density of at least 3×10^6 cells/ml, and portions (1 ml) are added to polypropylene freezing tubes. The tubes are placed in the Biotech cell freezer unit, and after 24 hr at −70° the tubes are transferred to a liquid nitrogen refrigerator at −170°.

Frozen stock is reconstituted by immersing the freezing vial in a 37° water bath to thaw the cells as quickly as possible. Shake gently until the last piece of ice melts. Do not allow the cells to become warm. The cells are transferred to a centrifuge tube containing an equal volume of medium at room temperature (about 25°); after 5 min add an equal volume of medium; wait 5 min; repeat once more. Centrifuge at 130 *g* for 5 min. Resuspend the cells in nutrient medium containing 20% FBS at 5×10^5 viable cells/ml. Transfer culture to CO_2 incubator. Refeed every 2 days. The time for recovery from freezing is highly variable and may take longer than 2 weeks.

Nitro Blue Tetrazolium Reduction Assay for Differentiation of HL60 Cells

Principle. Differentiated HL60 cells as well as normal monocytes (macrophages) and mature granulocytes produce superoxide anions (O_2^-) when stimulated with TPA.[11] Superoxide reduces the water-soluble NBT to produce blue-black cell-associated nitro blue diformazan (NBD) deposits. Morphological assessment of Wright–Giemsa-stained cytospin slides shows good correlation between NBT reduction and the number of cells that are monocytes/macrophages or at or past the myelocyte stage of granulocytic differentiation. Tests are done routinely on induced or uninduced HL60 cells grown for about 4 days.

Two NBT tests are described. The qualitative NBT test determines which cells reduce NBT (% NBD^+ cells), and the quantitative NBT test determines how much NBD is produced (nmol NBD/10^6 cells).

Materials

NBT (Sigma, St. Louis, MO)
TPA (Pharmacia P-L Biochemicals, Inc., Milwaukee, WI)
Nutrient medium
FBS
Wright–Giemsa stain (LeukoStat Stain Kit, Fisher Scientific, Pittsburgh, PA)
Safranin stain solution (1 g safranin O dissolved in 100 ml of 95% ethanol)
Slides
Cytospin 2 centrifuge (Shandon)
Minifold microsample manifold system (Schleicher & Schuell, Keene, NH)
37° water bath
Centrifuge tubes (15 ml)
Dulbecco's phosphate-buffered saline without Ca^{2+} and Mg^{2+} (DPBS)
Hemacytometer
Trypan blue stain, 0.4% (GIBCO)
Spectrophotometer

Procedure. Prepare a 324 μM stock solution of TPA in dimethyl sulfoxide (200 μg TPA/ml). Store at $-20°$. Prepare a 1.223 mM stock solution of NBT in DPBS (1 mg of NBT/ml). This can be stored at 4° for at least 1 week. Also prepare nutrient medium with 20% FBS. Warm all reagents to 37°.

Cells are counted with a Coulter counter, and viability is assessed by trypan blue exclusion. Harvest 2×10^6 viable cells/condition by centrifugation at 250 *g* for 7 min. Resuspend the cells in 1 ml of nutrient medium containing 20% FBS. Remove 0.5 ml and transfer to another tube containing 0.5 ml nutrient medium without serum. These cells will be used for the slide for morphological assessment.

Make a 1:1000 (v/v) dilution of the 324 μM TPA stock solution into a portion of the 1.233 mM NBT stock solution (add 1 μl of TPA stock solution for every milliliter of NBT stock solution) and add 0.5 ml of the NBT–TPA mixture to the remaining 0.5 ml of cell suspension. The final concentration of reagents in the NBT reaction tube is as follows: NBT, 0.61 mM; TPA, 162 nM; FBS, 10%; 1×10^6 cells/ml. Vortex the NBT tubes gently; remove the caps, and incubate for 25 min at 37° in a 5% CO_2 incubator. Put NBT tubes on ice after the reaction is complete.

Make cytospin slides of both NBT-treated and untreated cells for each condition. Use about 0.1 ml of cell suspension per slide and centrifuge 4 min at 400 rpm in the Cytospin 2 centrifuge. Check each slide for proper cell density and dispersement before air-drying and staining. The slides from the NBT-treated suspensions are placed into the safranin solution for 5 min, rinsed in water, and then air-dried. NBD^+ cells will have blue-black granules associated with the cells. The cytospin slides of cells without NBT are stained with Wright–Giemsa. A minimum of 100 cells is counted under light microscopy at a magnification of about $\times 1000$.

An alternative method for analysis is the quantitative NBT test.[20] A 0.5-ml portion from the NBT reaction tube is transferred to a well of a Minifold microsample manifold system assembled with a glass fiber paper sheet or individual 7-mm-diameter glass fiber paper disks (#30 Glass, Schleicher & Schuell). The samples are collected, washed consecutively with DPBS and 0.01 *N* HCl, and then air dried. If the glass fiber sheet is used, the areas containing NBD are cut out and transferred to 4-ml glass screw-cap vials. If the disks are used, they are transferred to the glass vials. The NBD in each vial is dissolved by adding 1 ml of DMF and then 1/100 volume of freshly prepared 1 *N* NaOH (10 μl/ml DMF). The DMF and NaOH can be mixed first and then added to the vial. However, CO_2 is absorbed rapidly, causing the formation of $NaCO_3$ crystals. The air in each vial is removed by a gentle stream of N_2 gas, and the vials are sealed immediately with aluminum foil-lined caps. The rate of dissolution of the NBD is increased by heating at 50° with shaking. The efficiency of extraction can be estimated visually and is usually complete within 60 min. NBD has absorbance maxima of about 710 nm in DMF–10 m*M* NaOH and about 560 nm in DMF. The molar absorption coefficients are 99,700 M^{-1} cm^{-1} at 710 nm in DMF–NaOH and 23,580 M^{-1} cm^{-1} at 560 nm in DMF.

Although the increased sensitivity in DMF–NaOH justifies the use of alkaline conditions, another reason is that NBD is extracted more efficiently from cells by alkaline DMF than by DMF alone. Pure NBD dissolves readily in DMF, but the NBD associated with the cells is apparently trapped and is released quantitatively only after the NaOH solubilizes cell constituents. Thus, the NaOH serves two purposes: releasing the trapped NBD and increasing the absorbance.

It is convenient to transfer samples from each vial into a flow cell cuvette with an automatic filling device. If there is difficulty in maintaining the high pH necessary to read the samples at 710 nm the samples are

[20] M. Imaizumi and T. R. Breitman, *Blood* **67,** 1273 (1986).

exposed to air until the color changes from green to blue. The absorption is then measured at 560 nm.

The light microscopy assessment of the percentage of NBT-reactive cells can be troublesome for two reasons. One problem is the definition of a positive cell. The other problem is a correction for dead cells. However, in this assay, cells completely covered with NBD deposits will be counted as equals with cells that have very few NBD granules. Direct comparison between control cells and treated cells will usually self-teach an observer for this determination. We judge a cell to be NBD^+ if it has any NBD deposit associated with it.

The final value of $\%NBD^+$ cells should be corrected for viability. Many inductions of differentiation of HL60 cells are accompanied by cytotoxicity.The NBT test is a viability determination. Dead cells do not reduce NBT. Thus, cells that differentiated before dying will also be NBD^-. The most direct way to deal with viability is to count each cell on the NBT slide and correct the $\%NBD^+$ value for total cells (live plus dead) with the percent viability value determined with trypan blue exclusion. Thus, if the NBT test is performed on a cell population that is 50% viable and the initial $\%NDB^+$ value is 50%, the corrected $\%NBD^+$ value is 100%. With experience, dead cells can be recognized on the NBT slide as misshapened, poorly staining cells and just not counted. In our laboratory, we have obtained the same values for $\%NBD^+$ cells with either method.

The measurement of percentage of differentiated cells is most commonly used in the HL60 system for assessing the relative activities of various compounds. For reasons discussed above, this measurement is most meaningful when the cytotoxicity is low. However, under cytotoxic conditions a portion of an apparent increase in the percentage of differentiated cells can be the result of an enrichment of the preexisting population of differentiated cells, which in control cultures of HL60 can have values of 5 to 15%. This can be a major problem if the cytotoxicity is directed particularly against growing, undifferentiated cells. This is because the percentage of differentiated cells is determined by dividing the number of viable differentiated cells by the total number of viable cells. There is no simple method to correct quantitatively for this possible enrichment. However, induction of differentiation in culture is indicated if there is an increase in the viable cell concentration as well as a net increase in the concentration of mature cells that is greater than what could have occurred in the control culture at the same density. The higher the percent viability the better the NBT test value will be as a true measure of differentiation. The lower the percent viability, the greater will be the possibility that enrichment will influence the results. Therefore, viability determinations are critical for an accurate assessment of inducers of differentiation that are also cytotoxic.

Cytochrome c Reduction Assay for Differentiation of HL60 Cells

Principle. The production of superoxide anion (O_2^-) can be assayed by measuring the TPA-activated reduction of ferricytochrome *c* to ferrocytochrome *c* by intact cells.[21]

Materials

Ferricytochrome *c* (Sigma)
Superoxide dismutase (SOD) (Sigma)
Dulbecco's phosphate-buffered saline without Ca^{2+} and Mg^{2+} (DPBS)
TPA (Pharmacia P-L Biochemicals)
Centrifuge (Microfuge 12) (Beckman, Palo Alto, CA)
Microcentrifuge tubes (1.5 ml) (Sarstedt, Princeton, NJ)
Spectrophotometer

Procedure. Prepare a 160 μM solution of cytochrome *c* in DPBS (1.98 mg/ml) and a solution of SOD at 1.5 mg/ml in DPBS and store at 4°. Prepare a 324 μM TPA stock solution as described for the NBT procedure. A 1/20 dilution of this stock (16.2 μM) in DPBS is made fresh daily. All reagents are kept in an ice-water bath.

Label two sets of tubes per condition and keep in an ice bath. Add 20 μl SOD solution to one set and 500 μl cytochrome *c* solution to all tubes. Count cells and centrifuge 3.1×10^6 viable cells per condition at 1000 *g* for 10 min. Resuspend in 3.1 ml cold DPBS. Add 56.4 μl of 16.2 μM TPA solution and mix well. Add 0.5 ml of this cell suspension to each tube (5×10^5 cells per tube).

Mix by inversion. Incubate at 37° for 20 min. The tubes are iced and then centrifuged at about 12,000 *g* for 1.5 min. The absorbance of the supernatants is measured at 550 nm, and the results with duplicate or triplicate tubes are averaged. Results are converted to nanomoles of cytochrome *c* reduced (nmol O_2^- produced) using the molar extinction coefficient ($E_{550\,nm}$) of $2.1 \times 10^4\ M^{-1}\ cm^{-1}$. The tubes containing SOD serve as blanks. It has been found repeatedly that O_2^- reduction of cytochrome *c* is completely inhibited in HL60 cells by SOD. Therefore a blank composed of 80 μM ferricytochrome *c* in DPBS is used for the negative SOD control.

[21] R. B. Johnston, Jr., B. B. Keele, Jr., H. P. Misra, J. E. Lehmeyer, L. S. Webb, R. L. Baehner, and K. V. Rajagopalan, *J. Clin. Invest.* **55**, 1357 (1975).

Hexose Monophosphate Shunt Assay for Differentiation of HL60 Cells

Principle. Differentiating HL60 cells as well as normal monocytes/macrophages and granulocytes exhibit a TPA-activated increase in glucose utilization by the hexose monophosphate shunt (HMPS), also called the phosphogluconate pathway. The amount of $^{14}CO_2$ released from [1-^{14}C]glucose is used as a measure of this pathway.[11]

Materials

PBS (Dulbecco's phosphate-buffered saline with Ca^{2+} and Mg^{2+})
2 m*M* glucose in PBS
[1-^{14}C]Glucose (55.8 Ci/mol) (Du Pont-New England Nuclear, Boston, MA)
TPA (Pharmacia P-L Biochemicals)
Polypropylene tubes (17 × 100 mm)
Tape
Glass fiber disks (2.4 cm diameter)
Needles (16 and 19 gauge)
Syringes (1 ml)
1.8 *M* sulfuric acid
Center wells (Cat. No. 882320, Kontes, Vineland, NJ)
Rubber stoppers (Cat. No. 882310, Kontes)
Sterilization filter units (0.2 μm, 500 ml) (Nalge Co., Rochester, NY)
1 *N* NaOH

Procedure. Make a stock solution of 2 m*M* glucose by dissolving 180.16 mg of glucose in 500 ml PBS, filter-sterilize, and store at 4°. Add 0.4 μCi [1-^{14}C]glucose/ml of 2 m*M* glucose solution for experimental reaction mixture and put in an ice bath. Prepare the TPA stock as described for the NBT procedure. A 1/20 dilution (16.2 μ*M*) in PBS of the 324 μ*M* stock is made fresh daily and kept cold.

Label two sets of duplicate 17 × 100 mm polypropylene tubes per experimental condition. Add 10 μl of 16.2 μ*M* TPA to one set of tubes. Count cells and determine the viability with trypan blue. Centrifuge 4.2 × 10^6 cells per condition at 1000 g for 10 min. Resuspend cells in 4.2 ml reaction mixture (2 m*M* glucose containing 0.4 μCi [1-^{14}C]glucose/ml). Carefully pipette 1 ml of cell suspension into the bottom of the tube. Take care not to touch the top half of the tube with the radioactive mixture.

Insert center wells into the rubber closure and put one-quarter of a 2.4-cm glass fiber disk into each center well. Add 100 μl of 1 *N* NaOH into the center well and carefully close tubes with stoppers and center wells. Incubate 1 hr in a 37° water bath. Punch a hole in each tube with a 16-gauge needle below the center well. With a 1-ml syringe and a 19-gauge needle, inject 0.1 ml of 1.8 *M* H_2SO_4 into each tube. Close hole with tape. Put all tubes in a 60° water bath for 1 hr to release the dissolved CO_2 from the acidified reaction mixture. Cut off the center wells into scintillation vials, add 1 ml of water and 10 ml of a water-miscible scintillation fluid, and count in a scintillation counter. For calculation of total counts, add 50 μl of cell reaction mixture to a scintillation vial. Add 1 ml of water and 10 ml of scintillation fluid and count.

Calculate the nanomoles of CO_2 released per 10^6 cells per hour by the following formula:

$$\text{Nanomoles CO}_2\text{ released} = \frac{\text{dpm CO}_2\text{ released}}{\text{total dpm/ml reaction mixture}} \times 2{,}000\text{ nmol glucose} \times \frac{100}{\text{viability (\%)}}$$

Subtract the nanomoles CO_2 released without TPA from the value with TPA to obtain a value for TPA-activated HMPS activity.

Fc Receptors and Immune Phagocytosis Assay for Differentiation of HL60 Cells

Principle. Immune phagocytosis requires the initial attachment of a particle-bound ligand to a specific receptor on a macrophage or granulocyte plasma membrane.[22-24] When the ligand is an immunoglobulin (IgG or IgM) or complement (C3), the phagocytic process is called immune phagocytosis. Mature granulocytes and monocytes/macrophages have specific receptors (Fc receptors) on the plasma membrane for the Fc portion of immunoglobulin. These receptors are absent or at low levels in immature cells. An increase in the percentage of cells having Fc receptors is therefore a measure of maturation. By using erythrocytes (E) coated with antierythrocyte antibodies (A), one can measure in one assay the percentage of cells that have Fc receptors (EA rosettes) and that can phagocytose erythrocytes (immunoerythrophagocytosis).

Materials

Sheep erythrocytes (SRBC), obtained from various commercial sources or from a farm as follows: sheep blood is collected into equal volumes of Alsever's solution [0.42% NaCl; 0.8% sodium citrate, dihydrate; 2.05% glucose; pH adjusted to 6.1 with 10% citric acid

[22] F. M. Griffin, Jr., J. A. Griffin, J. E. Leider, and S. C. Silverstein, *J. Exp. Med.* **142,** 1263 (1975).
[23] W. S. Walker and A. Demus, *J. Immunol.* **114,** 765 (1975).
[24] R. Snyderman, M. C. Pike, D. G. Fischer, and H. S. Koren, *J. Immunol.* **119,** 2060 (1977).

(GIBCO, Cat. No. 670-5190)] and is used after storage at 5° for 10–16 days

Dulbecco's phosphate-buffered saline without Ca^{2+} and Mg^{2+} (DPBS)

Anti-SRBC IgG, obtained commercially (Cordis Laboratories, Miami, FL, Cat. No. 768-970); the optimal dilution is determined experimentally but will probably be 1:250–1:500 in DPBS

Defined medium: nutrient medium with 5 μg of insulin/ml and 5 μg of transferrin/ml

Lysing buffer: 140 mM NH_4Cl–10 mM Tris-HCl prepared by mixing 1 part of 1.4 M NH_4Cl with 9 parts of 11 mM Tris-HCl, pH 7.2

Procedure: Preparation of IgG-Coated SRBC. SRBC ($4-5 \times 10^9$/ml in Alsever's solution) are washed 3 times with DPBS by centrifugation (500 g for 10 min) and then adjusted to 1×10^9 ml in DPBS. Cell number is determined by direct counting (hemacytometer or Coulter counter) or indirectly by reading the hemoglobin absorption. A 1:25 dilution of 1×10^9 erythrocytes/ml in lysing buffer has an $A_{541\text{ nm}}$ of 0.420.

Add an equal volume of anti-SRBC IgG solution to the SRBC suspension dropwise with constant shaking. Incubate in a shaking water bath for 30 min at 30°. After incubation, wash the SRBC 2 times with DPBS. Resuspend the SRBC in DPBS to a concentration of 1.5×10^8/ml. A 1:25 dilution of this concentration in lysing buffer gives an $A_{412\text{ nm}}$ of 0.560.

Procedure: Assay for EA Rosettes and Phagocytosis. Mix 1.2×10^6 test cells in 0.3 ml of defined medium with 0.3 ml of the IgG-coated SRBC suspension prepared above. Remove 0.1 ml and transfer to a separate tube. Incubate both tubes at 37° in a 5% CO_2 incubator. After 30 min, chill the tubes on ice. Add 10 μl of 0.4% trypan blue solution to the tube containing 0.1 ml and transfer the suspension to a hemacytometer. Observe cells under a magnification of $\times 1000$. Viable cells (those excluding the dye) binding three or more SRBC are scored as positive for the EA rosette test. Count at least 100–200 viable cells.

Add 3 ml of lysing buffer to the tube containing 0.5 ml, mix well, and then centrifuge at 400 *g* for 5 min. In this step, nonphagocytosed SRBC are lysed. Resuspend the cell pellet in 1 ml of defined medium and make Cytospin slides. Stain slide with Wright–Giemsa stain and count 200–300 cells under the microscope to determine the percentage of cells that have ingested at least one SRBC.

5′-Nucleotidase Activity Assay for Differentiation of HL60 Cells

Principles. 5′-Nucleotidase (5′-ribonucleotide phosphohydrolase, EC 3.1.3.5) is a plasma membrane-associated enzyme that because of its high activity in monocytes/macrophages can be used as a marker for these cell types.[25] The enzyme hydrolyzes ribonucleoside 5′-monophosphates to their corresponding nucleosides with the release of inorganic phosphate. The assay measures the capacity of intact cells to release inorganic phosphate (P_i) from AMP. Inorganic phosphate is measured by the procedure of Chen *et al.*[26] An alternate test, specific for monocytes/macrophages, is the qualitative nonspecific esterase assay.[27] Reagents and detailed instructions for this test are available as a kit (Kit No. 91-A, Sigma).

Materials

Buffer A: 50 m*M* Tris-HCl, pH 7.4, containing 145 m*M* NaCl and 10 m*M* $MgCl_2$
5 m*M* AMP in buffer A
Reagent A: ascorbic acid, 10%; make up fresh monthly and store at 5°
Reagent B: 0.42% ammonium molybdate tetrahydrate in 1 *N* H_2SO_4 (solution is stable on the shelf)
Reagent C: 1 part reagent A to 6 parts reagent B; prepare fresh each day

Procedure. Cells are harvested by centrifugation, washed 2 times with buffer A, and then suspended in buffer A at a cell density of 10^7/ml. The reaction mixture, in a 1.5-ml microcentrifuge tube, contains 22 μl of 5 m*M* AMP, 88 μl of buffer A, and 110 μl of washed cell suspension (1.1×10^6 cells) in a total volume of 220 μl. The reaction mixture is incubated at 37° for 30–60 min in a shaking water bath. The reaction is stopped by placing the tubes in an ice-water bath. The mixture is centrifuged at 12,000 *g* for 2 min. A portion of the supernatant fraction (200 μl) is added to a tube containing 800 μl of reagent C. This mixture is incubated at 45° for 20 min, and the $A_{820\,nm}$ is then measured. This reading is compared to a standard P_i curve.

[25] Z. Werb and Z. A. Cohn, *J. Biol. Chem.* **247,** 2439 (1972).
[26] P. S. Chen, Jr., T. Y. Toribara, and H. Warner, *Anal. Chem.* **28,** 1756 (1956).
[27] L. T. Yam, C. Y. Li, and W. H. Crosby, *Am. J. Clin. Pathol.* **55,** 283 (1971).

[14] Assays for Expression of Genes Regulated by Retinoic Acid in Murine Teratocarcinoma Cell Lines

By LORRAINE J. GUDAS

Introduction

Teratocarcinoma stem cell lines, and in particular the F9 line, have been extensively employed as model systems to study certain aspects of embryonic differentiation. Several different genes that exhibit increases in expression at the mRNA level during the retinoic acid-induced differentiation of murine teratocarcinoma stem cell lines have already been identified and cloned.[1] A number of these genes were identified by differential screening of cDNA libraries,[1] whereas one gene, ERA-1, was identified by screening a cDNA library with a "subtracted" probe.[2] As F9 teratocarcinoma stem cells differentiate into parietal endoderm in response to retinoic acid (RA) and dibutyryl-cAMP treatment of the cells for 48–72 hr, the steady-state levels of mRNAs encoded by genes such as laminin B1, laminin B2, collagen IV(α1), J6, and SPARC (J31) increase by 30- to 50-fold (e.g., Table I).

An understanding of the retinoic acid-associated differentiation of teratocarcinoma stem cells requires knowledge of which cellular genes are activated by RA and what functions these genes have. We have identified only a small proportion of the genes that are regulated by RA (e.g., Table I), and many more will certainly be discovered in the future. For all of the genes, one important question that should be asked is whether the increase in gene expression results from a direct or primary action of the RA–receptor complex at the transcriptional level. Evidence for such a primary action is that the complete induction of the mRNA takes place in the absence of protein synthesis. Another characteristic of a primary response is the rapidity of the induction of the mRNA following RA addition. Finally, a correlation exists often between the half-maximal induction of the mRNA and the concentration of hormone that half-saturates receptors.

[1] S. Y. Wang, G. J. LaRosa, and L. J. Gudas, *Dev. Biol.* **107,** 75 (1985); S. Y. Wang and L. J. Gudas, *Proc. Natl. Acad. Sci. U.S.A.* **80,** 5880 (1983); D. Barlow, N. Green, M. Kurkinen, and B. Hogan, *EMBO J.* **3,** 2355 (1984); I. Mason, A. Taylor, J. Williams, H. Sage, and B. Hogan, *EMBO J.* **5,** 1465 (1986); G. J. LaRosa and L. J. Gudas, *Mol. Cell. Biol.* **8,** 3906 (1988); S. P. Murphy, J. Garbern, W. Odenwald, R. Lazzarini, and E. Linney, *Proc. Natl. Acad. Sci. U.S.A.* **85,** 5587 (1988); R. J. Rickles, A. L. Darrow, and S. Strickland, *Mol. Cell. Biol.* **9,** 1691 (1989).

[2] G. J. LaRosa and L. J. Gudas, *Proc. Natl. Acad. Sci. U.S.A.* **85,** 329 (1988).

TABLE I

ABUNDANCE OF RETINOIC ACID-INDUCIBLE GENES[a]

Gene	Amount of DNA immobilized on filter paper	Stem cells[b]			Differentiated cells[b]			
		Total hybridization (cpm)[c]	Percentage of total poly(A)$^+$ RNA[d]	mRNA copies per stem cell[e]	Total hybridization (cpm)[f]	Percentage of total poly(A)$^+$ RNA[d]	mRNA copies per differentiated cell[e]	Induction (-fold) of genes in differentiated cells[b]
pBR322	1[g]	33	—	—	52	—	—	—
	2[g]	27	—	—	62	—	—	—
I5	1	128	0.028	98	5282	1.7351	6073	54
[collagen IV(α1)]	2	141	0.032	112	4578	1.5016	5256	—
I56	1	81	0.014	49	2272	0.7368	2579	54
(laminin B1)	2	83	0.014	49	2424	0.7873	2756	—
J6	1	74	0.0126	44	1248	0.3973	1391	30
	2	88	0.0166	58	1494	0.4822	1688	—
J31	1	96	0.0186	65	2264	0.7342	2570	31
(SPARC)	2	113	0.0236	83	1826	0.5890	2062	—
F117	1	52	0.006	21	2234	0.7243	2535	113
	2	54	0.0066	23	2166	0.7017	2456	—
ERA-1 (Hox 1.6)[h]	—	—	~0.001–0.002[h]	~5	—	~0.03–0.06[h]	~158	25–30

[a] Reproduced in part from S. Y. Wang, G. J. LaRosa, and L. J. Gudas, *Dev. Biol.* **107,** 75 (1985).

[b] Stem cells: F9 teratocarcinoma. Differentiated cells: F9 cells treated with 5×10^{-7} *M* retinoic acid, 500 μM dibutyryl-cAMP, and 500 μM theophylline for 72 hr.

[c] Input is [^{32}P]cDNA made to total RNA from F9 stem cells; 7.00×10^5 trichloroacetic acid-precipitable counts per minute (cpm).

[d] This value was calculated by using a hybridization efficiency of 50%. This percentage was determined by hybridizing ^{32}P-labeled I5-specific cDNA to nitrocellulose containing immobilized pcI5. The efficiency was constant in the range of 1000 to 5000 cpm bound, indicating that hybridization had gone to completion and that plasmid DNA was bound to the filter in excess of the input cDNA.

[e] This number was calculated by assuming approximately 350,000 mRNAs per cell.

[f] Input is [^{32}P]cDNA made to total RNA from F9 differentiated cells; 6.03×10^5 trichloroacetic acid-precipitable counts per minute.

[g] The numbers indicate the relative amount of recombinant plasmid DNA immobilized on the nitrocellulose filters; 2 indicates twice as much DNA as 1 (corresponding to 5 and 10 μg, respectively).

[h] G. J. LaRosa and L. J. Gudas, *Proc. Natl. Acad. Sci. U.S.A.* **85,** 329 (1988); ERA-1 mRNA percentage is based on the number of ERA-1 clones isolated from an unamplified cDNA RACT-8-hr λgt10 library divided by the total number of recombinant plaques screened.

Of course, definitive proof of the direct interaction of an RA–receptor complex with a promoter of a particular gene would require some type of footprinting assay to detect the binding of the purified RA–receptor complex to specific promoter sequences. How do these ideas, which were developed in the steroid hormone field, relate to the action of RA in the teratocarcinoma stem cell differentiation pathway?

Genes such as laminin B1 and collagen IV(α1) exhibit increases in their mRNAs at relatively late times (24–48 hr) after RA addition,[1] whereas genes such as ERA-1 are more rapidly induced by RA in a protein synthesis-independent fashion.[2] These kinetics of induction after RA addition suggest that genes such as the ERA-1 gene and the Hox 1.3 gene are examples of primary responses to RA and that genes such as laminin B1, collagen IV(α1), H-2, and β_2-microglobulin are secondary responses, their transcriptional activation resulting from the interaction of a protein product of a primary response gene with their promoters. However, experimental evidence from detailed promoter analysis in support of these ideas has not yet been obtained.

In fact, the situation with respect to RA–receptor gene activation in teratocarcinoma cells is certainly more complicated than that of steroid–receptor gene activation for the following reasons. First, at least one type of low molecular weight, high-affinity RA-binding protein, the CRABP, exists, and its mRNA level is regulated in some manner by RA addition.[3] Whether the level of CRABP expression regulates the internal [RA] is unclear, but variations in expression of this protein may influence the RA dose–response relationship of RA-inducible genes. Second, three nuclear RA receptors have been identified to date.[4] Messages for two of these, RARα and RARγ, are present in undifferentiated F9 teratocarcinoma stem cells,[5,6] whereas the third, the RARβ gene, exhibits characteristics of a primary RA response gene in F9 cells.[5] An increase in the steady-state level of RARβ mRNA commences within 6–12 hr after RA addition, and this RA-associated increase in message occurs in the presence of protein synthesis inhibitors.[5] Thus, a gene could exhibit an mRNA increase at a late time after RA addition (24–48 hr) and still be considered a "primary response" gene if its transcriptional activation resulted from interaction of the RARβ–RA complex with its promoter.

[3] C. M. Stoner and L. J. Gudas, *Cancer Res.* **49,** 1497 (1989).

[4] V. Giguere, E. Ong, P. Segui, and R. M. Evans, *Nature (London)* **330,** 624 (1987); M. Petkovich, N. J. Brand, A. Krust, and P. Chambon, *Nature (London)* **330,** 444 (1987); N. Brand, M. Petkovich, A. Krust, P. Chambon, H. de The, A. Marchio, P. Tiollais, and A. Dejean, *Nature (London)* **332,** 850 (1988).

[5] L. Hu and L. J. Gudas, *Mol. Cell. Biol.* **10,** 391 (1990).

[6] A. Krust, P. Kastner, M. Petkovich, A. Zelent, and P. Chambon, *Proc. Natl. Acad. Sci. U.S.A.* **86,** 5310 (1989).

Challenging questions for future experiments include the following. What is the role of each of the three known retinoic acid receptors (and possibly more RA receptors that will be cloned in the future) in the regulation of specific genes? Which cell types express which receptors? Does a particular gene interact with more than one endogenously expressed receptor? Which genes are "RA primary response" genes, as defined above, and which genes are activated as the result of "secondary" interactions with primary response gene products? Although this aspect is not discussed at length in this chapter, cAMP analogs often enhance the transcriptional activation of genes by RA,[7] and they also influence the levels of RARα and RARβ mRNAs.[5] How do cyclic AMP analogs exert these effects? Such questions are not limited to the RA-associated teratocarcinoma differentiation pathway, but rather these and related questions are relevant to all cell types that exhibit a differentiation response after treatment with retinol, retinoic acid, and/or other retinoid analogs.

A complete analysis of the expression of genes regulated in retinoic acid (RA)-inducible differentiation systems first requires that one determine the time course and magnitude of the increase in steady-state mRNA level of a particular gene by Northern analysis or RNase protection experiments. Then, whether the increase in steady-state mRNA results from an increase in the rate of transcription can be determined by *in vitro* nuclear runoff experiments. Whether the steady-state mRNA increase results from an RA-associated change in the stability of the particular mRNA can also be measured. Methods for these assays are described in the following sections. Finally, if RA can be shown to influence the gene of interest at the transcriptional level, then analysis of the promoter should be performed; this usually requires the introduction of an expression vector containing the promoter of interest positioned immediately 5′ to a reporter gene (see Transfection of Teratocarcinoma Cells).

Northern Analysis

RNA Isolation by the Guanidine Hydrochloride Method

The guanidine hydrochloride method for RNA isolation, adapted from those of Cox[8] and Strohman *et al.*,[8] is useful for the preparation of multiple samples. Gloves should be worn. All glassware should be baked. The steps are given below.

[7] L. J. Gudas, J. F. Grippo, K. W. Kim, G. J. LaRosa, and C. M. Stoner, *Ann. N.Y. Acad. Sci.* **580,** 245 (1990).

[8] R. A. Cox, this series, Vol. 12B, p. 120; R. C. Strohman, P. S. Moss, J. M. Eastwood, D. Spector, A. Przybyla, and B. Paterson, *Cell (Cambridge, Mass.)* **10,** 265 (1977).

1. Teratocarcinoma cells are plated prior to RNA isolation (see this volume, [12] and [16]) on teratocarcinoma/embryonal carcinoma lines). Approximately one dense T-150 plate ($3-5 \times 10^7$ cells) should yield 100–200 μg of total RNA.

2. Tissue culture plates are washed 3 times with cold phosphate-buffered saline (PBS) in the cold room. Remove all PBS prior to the next step.

3. One milliliter of guanidine mix is added per 10^6 cells, on ice in the cold room. Transfer the guanidine mix from plate to plate if isolating RNA from many sparse plates, and scrape each plate with a sterile rubber policeman. The ratio of guanidine mix to cells is crucial for obtaining good RNA yields.

Ten milliliters of guanidine mix contains 10 ml of 7 *M* guanidine hydrochloride (33.4 g in 50 ml sterile water), 0.1 ml of 2 *M* potassium acetate (pH 7.0), 0.1 ml of 500 m*M* EDTA (pH 7.4), and 1.5 mg dithiothreitol (DTT). The guanidine-HCl is not sterilized. Using a *sterile* spatula, the guanidine-HCl is weighed into a 50-ml sterile plastic centrifuge tube, and sterile water is added to the guanidine-HCl. The potassium acetate and EDTA solutions are autoclaved, followed by filtering (0.2 μm Millex-GS sterile filter, Millipore, Bedford, MA) into a new sterile bottle to remove all protein. The guanidine mix must be made *fresh* just prior to RNA isolation.

4. The guanidine mix is scraped into a snap-cap sterile polypropylene tube and passed through a sterile 18-gauge needle 10 times.

5. A half-volume of 95% ethanol is added (95% (v/v) ethanol is kept at −20°).

6. The RNA solution is precipitated at −20° overnight. The solution is then centrifuged at 7000 rpm at −10° for 25 min.

7. The supernatant is drained off completely, and the pellet is resuspended in 5 ml of a solution containing 4.75 ml of guanidine hydrochloride (7 *M*) plus 0.25 ml EDTA (500 m*M* stock), *or* use roughly one-half the volume used in Step 3, that is, 0.5 ml/10^6 lysed cells.

8. Two molar potassium acetate (pH 7.0) is added to a final concentration of 20 m*M*, and one-half volume of 95% ethanol is added.

9. The RNA is precipitated at −20° overnight and then centrifuged at 5000 rpm, −10°, for 25 min.

10. The supernatant is poured off completely, and the pellet is resuspended in extraction buffer (EB) at room temperature by vortexing. Use a volume of EB that is one-half that of guanidine/EDTA in Step 7. Transfer the RNA in EB solution into a sterile polypropylene centrifuge tube before the phenol extraction step.

Extraction buffer (EB) consists of (at a final concentration) 50 m*M*

Tris-HCl, pH 9.0, 5 m*M* EDTA, 100 m*M* NaCl, and 0.5% sodium dodecyl sulfate (SDS). The first three solutions are autoclaved. The SDS solution is not autoclaved. *After* the EB mix is made up, it is filtered through a Millipore GS filter (0.2 μm) into a 50-ml sterile plastic centrifuge tube at room temperature. The stock SDS solution (10%) (w/v) is kept at room temperature.

11. At room temperature, an equal volume of phenol solution is added to the EB buffer solution. The phenol solution consists of 5 ml distilled phenol (must *not* look yellowish) 5 ml chloroform, 1 ml extraction buffer, and 100 μl isoamyl alcohol. Use a sterile plastic pipette or a sterile baked glass pipette to add phenol. After shaking, the solution is centrifuged at 2000 rpm for 10 min at room temperature. The upper aqueous layer is removed and put into a 15-ml sterile polypropylene tube. This phenol extraction is then repeated.

12. Sodium chloride (5 M) (autoclaved and filtered) is added to a final concentration of 0.2 *M*, and 2.5 volumes of 95% cold ethanol is added. The solution is left overnight at $-20°$.

13. The RNA is centrifuged at 9500 rpm for 25 min at $-10°$.

14. The pellet is washed with cold 70% ethanol (1–2 ml) (do not vortex the pellet), and centrifuged at 10,000 rpm, 25 min, $-10°$. The ethanol is drained off carefully (the pellet is loose), the RNA is lyophilized for 10 min, and 1 ml cold TE buffer (10 m*M* Tris-HCl, pH 7.4; 1 m*M* EDTA) is added. The RNA is vortexed at room temperature.

15. The solution is brought to 2.5 *M* ammonium acetate, and 2.5 volumes ethanol (95%) is added. The solution is then left overnight at $-20°$, or for at least 45 min at $-70°$ in an alcohol bath.

16. Steps 13 and 14 above are repeated.

17. At this point, the RNA is transferred to an autoclaved microcentrifuge tube. Keep the RNA on ice while in TE buffer. A 5- or 20-μl aliquot is removed from the tube with a sterile pipette and put into 0.3 ml distilled water to obtain the optical density (OD) at both 260 and 280 nm (1 OD unit equals ~48 μg RNA per ml) A 260/280 nm ratio of 1.8 to 2.0 is good. If the RNA concentrations in different samples vary greatly, it is important to adjust the concentration of the RNA in each sample so that all of the samples are within a 2- to 3-fold range of each other, and then to obtain the optical densities again prior to loading RNA samples on a Northern gel. The RNA can be stored in the TE buffer at $-70°$.

Electrophoresis, Blotting, and Hybridization

Because of space limitations, detailed descriptions of electrophoresis, blotting, and hybridization techniques cannot be included. Briefly, to detect relatively abundant (0.1–2% of total mRNA) mRNAs, 5 μg total RNA/lane is loaded on a 1% agarose/2.2 *M* formaldehyde gel,[9] electrophoresed,[9] and transferred to nitrocellulose[10] using 20× SSC buffer without prior NaOH treatment to fragment the RNA. cDNA probes are isolated and labeled with [^{32}P] dCTP by the low-melting agarose/random primer method.[11]

The nitrocellulose filters are prehybridized for at least 4 hr, then hybridized at 42° in 50% (v/v) formamide, 5× SSC (1 × SSC is 0.15 *M* NaCl/15 mM sodium citrate), 50 m*M* NaH_2PO_4/Na_2HPO_4 (pH 7.4), 5m*M* EDTA, 0.08% each Ficoll and poly(vinylpyrrolidone), 0.1% SDS, 100 μg/ml denatured salmon sperm DNA, 50 μg/ml poly(A), and 10% (w/v) dextran sulfate. After hybridization for 16 hr, the filters are usually washed with 0.2× SSC and 0.1% SDS sulfate at 60°. These hybridization conditions are generally useful for hybridizations involving murine RNA and murine cDNA probes.

In Vitro Transcription/Nuclear Runoff Assays

Nuclear runoff assays are used to determine if an increase in the steady-state level of a particular mRNA results from an increase in the rate of transcription of the gene. In this laboratory, the method of Greenberg and Ziff[12] for nuclei isolation and preparation of ^{32}P-labeled RNA is used, with the following modification for preparation of ^{32}P-labeled RNA. After the transcription reactions are terminated by the addition of deoxyribonuclease I, a deproteinization step is performed, followed by a phenol–chloroform extraction, based on the procedure of Groudine *et al.*[13] RNA is then precipitated in the presence of carrier yeast tRNA at 4° with 3 ml of 10% trichloroacetic acid/60 m*M* sodium pyrophosphate, centrifuged at 10,000 rpm at 4° for 15 min, washed with 100 times the nuclei volume of 5% trichloroacetic acid/30 m*M* sodium pyrophosphate, and centrifuged again. This step is slightly different from that of Groudine *et al.*[13] The RNA pellet is then dissolved and DNase I is added, followed by another proteinase K digestion and phenol–chloroform extraction. NaOH is added as described,[12] but then this reaction is neutralized with HEPES free acid, and RNA is precipitated by addition of 5.3 times the nuclei volume of 3 *M* sodium acetate and 145 times the nuclei volume of 95% ethanol, followed

[9] T. Maniatis, E. F. Fritsch, and J. Sambrook, "Molecular Cloning: A Laboratory Manual," pp. 202–203 Cold Spring Harbor Laboratory, Cold Spring Harbor, New York, 1982.

[10] B. Seed, *Nucleic Acids Res.* **10,** 1799 (1982).

[11] A. P. Feinberg and B. Vogelstein, *Anal. Biochem.* **137,** 266 (1984).

[12] M. Greenberg and E. B. Ziff, *Nature (London)* **311,** 433 (1984).

[13] M. Groudine, M. Peretz, and H. Weintraub, *Mol. Cell. Biol.* **1,** 281 (1981).

by incubation overnight at $-20°$.[7] The RNA is then centrifuged, dissolved in water, and quantitated by scintillation counting.[7] ^{32}P-Labeled RNA is then ready to use in the hybridizations.

To prepare DNA slot blots, plasmid DNA is purified by CsCl gradient centrifugation, linearized, and extracted once with phenol, once with phenol/chloroform/isoamyl alcohol (10:9.6:0.4, v/v), and once with chloroform/isoamyl alcohol (24:1, v/v). Ten micrograms of DNA per slot is used. The DNA is dissolved in 22 μl TE buffer (10 m*M* Tris, pH 8.0; 1 m*M* EDTA), mixed with 1 μl of 5 *N* NaOH, and incubated at 25° for 30 min. The sample is then neutralized by addition of 37 μl of 7.5 *M* ammonium acetate and 0.19 ml water on ice. This 250 μl DNA sample is then applied to a slot and washed with 375 μl of 1 *M* ammonium acetate. The nitrocellulose blot is then baked for 120 min at 80° *in vacuo*. The hybridization and filter washing conditions are as previously described.[12]

Measurement of mRNA Stability

Although a number of methods have been designed for measuring the half-life of a particular mRNA species, some of these are useful only when studying very abundant messages. We have used the following method to analyze the relative stabilities of both the ERA-1 mRNA[14] (not very abundant, see Table I) and the RARβ mRNA[5] before and after retinoic acid (RA) treatment. This method employs RNA synthesis inhibitors such as actinomycin D; once such an inhibitor is added to cells, no new RNA synthesis occurs. With time after addition of actinomycin D the cellular mRNA levels decline, and from the rate of decrease of a particular mRNA with time in the presence of actinomycin D, the half-life of the mRNA can be determined.

Multiple plates of F9 cells are grown for 6, 24, or 48 hr in the absence or presence of RA. At each of these times, RNA is immediately isolated from one plate, while cells in five other plates are incubated in the presence of 2 μg/ml of the RNA synthesis inhibitor actinomycin D, added directly to the tissue culture medium. RNA is isolated at 30, 60, 120, 180, and 240 min after actinomycin D addition from each of the five plates. These times for RNA isolation can be varied for further experiments depending on whether the particular message has a long or short half-life. However, long exposure of the cells (>10 hr) to actinomycin D should be avoided. Another inhibitor that can be used as an alternative to actinomycin D is DRB (5,6-dichloro-1-β-*D*-ribofuranosylbenzimidazole) at a final concentration of 25 μg/ml.[14] The total RNA from all of the plates of cells is then fractionated, blotted, and hybridized to a specific cDNA probe. The RNA

[14] G. J. LaRosa and L. J. Gudas, *Mol. Cell. Biol.* **8,** 3906 (1988).

signals on the resulting autoradiogram are quantitated by densitometry, graphed, and the half-life is determined.

Using this procedure, the half-lives of ERA-1 mRNA and RARβ mRNA have been determined to be 90–100 min[14] and 60–70 min,[5] respectively, in both untreated F9 stem cells and RA-trated F9 cells. In contrast, actin mRNA[14] and laminin B1 mRNA[14] are much more stable messages in both F9 stem and RA-treated F9 cells.

Transfection of Teratocarcinoma Cells

If a gene has been shown by nuclear runoff experiments to be transcriptionally activated by RA, this gene should be subjected to further promoter analysis. Once genomic clones for a particular mRNA have been characterized and the 5′-flanking region containing the promoter has been isolated and sequenced, further analysis of the promoter region requires the introduction of a vector containing the promoter region linked to a reporter gene [i.e., chloramphenicol acetyltransferase (CAT), luciferase, β-galactosidase] into cells that have or have not been treated with RA. Studies of various promoter deletions should then allow the delineation of sequences important in the RA-associated transcriptional increase. (It is important to note that although most genes contain enhancer sequences in their promoters, some genes may possess enhancer sequences in other positions, e.g., an intron.)

The introduction of expression plasmids into cells can be accomplished by a number of different techniques, but the most commonly used procedures for teratocarcinoma cells are calcium phosphate transfection and electroporation. As parameters of the calcium phosphate transfection procedure must be altered for different cell types, the procedure used in this laboratory for a number of different RA-responsive teratocarcinoma cell lines (e.g., P19, F9, PSA) is described below.

As a control for variations in the transfection efficiencies in different plates of cells, this laboratory uses a murine β-actin/β-galactosidase expression vector constructed by Vasios *et al.*[15] The β-actin promoter is a strong promoter in these cells, and the transcription of the actin gene changes only slightly after RA addition.[16] We use the chloramphenicol acetyltransferase (CAT) gene as a reporter in the experimental expression vector.

Reagents. The solutions for the calcium phosphate coprecipitation are as follows:

[15] G. W. Vasios, J. D. Gold, M. Petkovich, P. Chambon, and L. J. Gudas, *Proc. Natl. Acad. Sci. U.S.A.* **86,** 9099 (1989).

[16] S. Y. Wang, G. J. LaRosa, and L. J. Gudas, *Dev. Biol.* **107,** 75 (1985).

1 m*M* Tris-HCl, pH 7.4
2 *M* $CaCl_2 \cdot 2H_2O$
2× HBS, pH 7.05±0.05: 280 m*M* NaCl, 50 m*M* HEPES, 1.5 m*M* Na_2HPO_4
TBS, pH 7.4: 25 m*M* Tris-HCl, pH 7.4, 137 m*M* NaCl, 5 m*M* KCl, 0.7 m*M* $CaCL_2 \cdot 2H_2O$,
0.5 m*M* $MgCl_2 \cdot 6H_2O$,
0.6 m*M* Na_2HPO_4; sterilize by passage through 0.22-μm Millipore filter

Procedure. F9 stem cells are plated at 1×10^6 cells/100-mm culture dish 24 hr prior to transfection. For the transfection, 5–10 μg of test promoter/CAT plasmid (sterile, CsCl purified) is added to 420 μl of 1 m*M* Tris-HCl, pH 7.4, in a 15-ml sterile plastic tube. Five micrograms of pβ-actin/β-gal plasmid are added, and the total concentration of DNA is adjusted to 20 μg with pUC9 plasmid DNA. Then, 60 μl of 2 *M* $CaCl_2$ is added dropwise with gentle agitation. The solution is mixed by hand.

Then, using one pipette, the DNA–$CaCl_2$ solution is added dropwise to a clear sterile 15-ml plastic tube containing 480 μl of 2× HBS, while simultaneously, with a second pipette, a constant bubbling of air within the 2× HBS solution is created. The rate of bubble formation should be slow to moderate. This solution is mixed by hand, then remains at 22° for 30 min. The precipitate is slowly added in a dropwise fashion directly to the medium and cells in the culture dish, and the culture dish is gently swirled. The dishes are placed in the cell culture incubator for 6 to 12 hr. The media and the precipitate are then pipetted off, the cells are washed 2 times with 5 ml TBS, and fresh medium, with or without RA, is added. The cells are grown for 24 hr and then harvested for the β-galactosidase[17] and CAT[18] assays.

Acknowledgments

This work was supported by R01CA43796, R0139036, and R01HD24319 from the National Institutes of Health. L.J.G. is an Established Investigator of the American Heart Association.

[17] P. Herbomel, B. Bourachot, and M. Yaniv, *Cell (Cambridge, Mass.)* **39,** 653 (1984).
[18] C. M. Gorman, L. F. Moffatt, and B. H. Howard, *Mol. Cell. Biol.* **2,** 1044 (1982).

[15] Retinoid-Binding Proteins in Retinoblastoma Cells

By SHAO-LING FONG and C. D. B. BRIDGES

Introduction

The role of retinoids in cancer has not been elucidated, but there have been many studies on retinoid-binding proteins in tumor cells.[1] This chapter deals with retinoblastoma, the most common intraocular malignancy in children.[2] Retinoblastoma cells have features in common with cells of the embryonic retina,[3,4] and many cultured cell lines derived from different retinoblastoma tumors have been investigated for their ability to express four retinoid-binding proteins, namely cellular retinol-, retinoic acid-, and retinal-binding proteins (CRBP, CRABP, and CRAlBP) and interstitial (or interphotoreceptor) retinol-binding protein (IRBP).[5-8] The expression of retinoid-binding proteins is found to be variable. With the notable exception of the Y-79 line, the only retinoid-binding protein that is consistently expressed by retinoblastoma tumors and cultured retinoblastoma cells is IRBP,[8] an extracellular glycoprotein normally associated with the maturing retina.[9,10]

Retinoblastoma Tumor Cell Culture

Two well-established cell lines that have been available for over 10 years are Y-79[11] and WERI-Rb1.[12] In the authors' laboratory, Y-79 cells

[1] F. Chytil and D. E. Ong, *in* "The Retinoids" (M. B. Sporn, A. B. Roberts, and D. S. Goodman, eds.), p. 89. Academic Press, New York, 1984.
[2] A. B. Reese, *Arch. Ophthalmol.* **42,** 119 (1949).
[3] L. E. Becker and D. Hinton, *Hum. Pathol.* **14,** 538 (1983).
[4] B. L. Gallie and R. A. Phillips, *Ophthalmology* **91,** 666 (1984).
[5] J. C. Saari, S. Futterman, G. W. Stubbs, J. T. Heffernan, L. Bredberg, K. Y. Chan, and D. M. Albert, *Invest. Ophthalmol. Visual Sci.* **17,** 988 (1978).
[6] G. J. Chader, B. Wiggert, P. Russell, and M. Tanaka, *Ann. N.Y. Acad. Sci.* **359,** 115 (1981).
[7] A. P. Kyritsis, B. Wiggert, L. Lee, and G. J. Chader, *J. Cell. Physiol.* **124,** 233 (1985).
[8] S. L. Fong, H. Balakier, M. Canton, C. D. B. Bridges, and B. Gallie, *Cancer Res.* **48,** 1124 (1988).
[9] S. L. Fong, G. I. Liou, R. A. Landers, R. A. Alvarez, F. Gonzales-Fernandez, P. A. Glazebrook, D. M. K. Lam, and C. D. B. Bridges, *J. Neurochem.* **42,** 1667 (1984).
[10] A. T. Johnson, F. L. Kretzer, H. M. Hittner, P. A. Glazebrook, C. D. B. Bridges, and D. M. K. Lam, *J. Comp. Neurol.* **233,** (1985).
[11] T. W. Reid, D. M. Albert, A. S. Rabson, P. Russell, J. Craft, E. W. Chu, T. S. Tralka, and J. L. Wilcox, *J. Natl. Cancer Inst.* **53,** 347 (1974).
[12] R. C. McFall, T. W. Sery, and M. Makadon, *Cancer Res.* **37,** 1003 (1977).

were obained from the American Type Culture Collection (Rockville, MD) and WERI-Rb1 from Dr. Larry Donoso (Wills Eye Hospital, Philadelphia, PA).

Other cell lines referred to in this chapter were developed by Dr. Brenda Gallie (Hospital for Sick Children, Toronto, Canada), originally from retinoblastoma surgical specimens. The tumors were grown initially as xenografts in the anterior eye chambers of athymic nude mice[13] and/or in tissue culture.[14] Some tumors were started on Dexter[15] mouse bone marrow stromal feeder layers. Retinoblastoma cells were then harvested and grown with weekly feeding in suspension cell cultures in Dulbecco's modified Eagle's medium (DME), 15% (v/v) fetal calf serum, 10 μg/ml bovine insulin, and 5×10^{-5} *M* 2-mercaptoethanol.

Once established, the cells are grown in suspension in 25- to 150-cm^2 plastic tissue culture flasks. With the exception of the WERI-Rb1 cells, all cells are cultured in DME supplemented with 15% (v/v) heat-inactivated calf serum, 10 μg/ml bovine insulin, 5×10^{-5} *M* 2-mercaptoethanol, and 1% penicillin–streptomycin. WERI-RB1 cells are grown in RPMI 1640, 10% (v/v) heat-inactivated fetal calf serum, 1% L-glutamine, and 1% penicillin–streptomycin. All cells are grown at 37° under 5% CO_2. The cells are fed 2 or 3 times weekly.

Retinoblastoma cells may be stored for indefinite periods of time in liquid nitrogen. They are frozen in their respective media containing 10% dimethyl sulfoxide (DMSO). The reestablishment of a healthy culture from frozen cells may take between 1 and 3 months.

Retinoid-Binding Proteins Determined by Radiolabeled Ligand Binding

Cells are harvested, washed free of medium with phosphate-buffered saline (PBS; 150 m*M* NaCl, 5 m*M* sodium phosphate, pH 7.5), homogenized in PBS, and centrifuged at 100,000 *g* for 1 hr at 4° to yield the cytosolic supernatant. The protein content is determined,[16] and the volume is adjusted to give 500 μg protein/ml. Two hundred-microliter volumes (100 μg protein) are incubated in replicate with 10^6–10^7 disintegrations per minute (dpm) all-*trans*-[11,12-^{3}H]retinoic acid (36 Ci/mmol,

[13] B. L. Gallie, D. M. Albert, J. J. Y. Wong, N. Buyukmichi, and C. A. Puliafito, *Invest. Ophthalmol. Visual Sci.* **16,** 256 (1977).

[14] B. L. Gallie, W. Holmes, and R. A. Phillips, *Cancer Res.* **42,** 301 (1982).

[15] T. M. Dexter, T. D. Allen, and L. G. Lajtha, *J. Cell. Physiol.* **91,** 335 (1977).

[16] O. H. Lowry, N. J. Rosebrough, A. L. Farr, and R. J. Randall, *J. Biol. Chem.* **193,** 265 (1951).

Hoffman-La Roche, Nutley, NJ) or all-*trans*-[11,12-^{3}H]retinol (40–60 Ci/mmol, Amersham, Arlington Heights, IL) dissolved in 2- to 10-μl volumes of ethanol. The mixtures are incubated for 12–15 hr in the dark at 4°, then mixed vigorously (vortex mixer) with 80 μl of a dextran-coated charcoal suspension (for preparation, see below), left for 10 min, and centrifuged for 4 min in a Fisher (Fairlawn, NJ) Model 235 centrifuge. A 200-μl volume is then injected onto a calibrated 7.5 × 300 mm Bio-Sil TSK-250 column (Bio-Rad, Richmond, CA) and eluted at 1 ml/min with 0.1 *M* Na_2SO_4, 20 m*M* sodium phosphate (pH 6.8). Fractions are collected at 0.5-min intervals and counted.

The peaks of radioactivity arising from CRBP and CRABP elute close to the position of myoglobin (17,000 daltons). A typical result obtained by incubating RB344 cell cytosol with [^{3}H]retinol is shown in Fig. 1. A large peak of radioactivity (fractions 12–14) is evident at the elution position of authentic CRBP (16,000 daltons[1]). The peaks indicated by the arrows coincide with void (thyroglobulin) and included (vitamin B_{12}) volumes respectively. These peaks are presumably due to free retinol (that at the void volume being in the form of a colloidal dispersion): their size is minimized by using small volumes of ethanol to deliver the ligand and by treating the mixture with dextran-coated charcoal. Under these conditions, the amount of radioactivity associated with binding protein is found to increase linearly up to as least 500 μg protein per incubation.

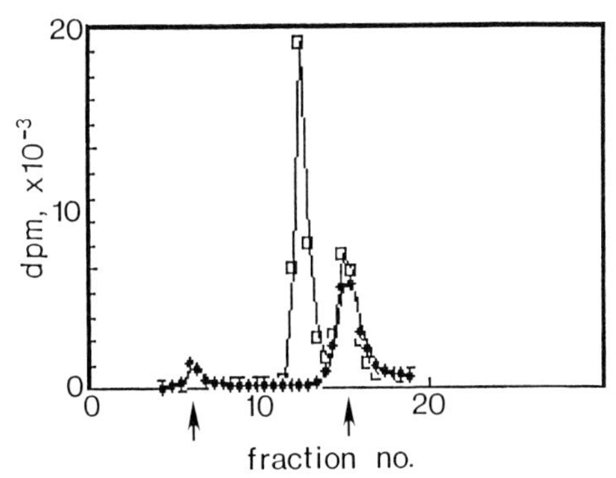

FIG. 1. High-performance size-exclusion chromatographic analysis of CRBP in the cytosol of RB344 cells. Aliquots of cytosol (100 μg protein) were incubated with [^{3}H]retinol, and approximately two-thirds of the incubation mixture was chromatographed on a calibrated Bio-Rad TSK size-exclusion column. Fractions were collected at 0.5-min intervals. Arrows indicate the elution positions of thyroglobulin (void volume) and vitamin B_{12} (included volume). Replicate samples were incubated with [^{3}H]retinol in the absence (□) and presence (+) of 4 m*M* *p*-chloromercuribenzene sulfonate.

Dextran-coated charcoal is prepared by adding 1.5 g washed Norit A charcoal (Fisher) to 0.15 g dextran 80 (Fluka, Ronkonkoma, NY) in 299 ml of 10 m*M* Tris-HCl (pH 7.5) containing 1 m*M* EDTA. The suspension can be stored for up to 2 months at 4°.

Specific binding of retinoids to CRBP and CRABP can be abolished by adding the organomercurial *p*-chloromercuri-benzene sulfonate (4 m*M*) before incubating with radiolabeled ligand.[17] The crosses in Fig. 1 show the results of such a control experiment. The peaks of presumptive free retinol at the void and included volumes are virtually unaffected by this treatment, but the radioactivity in the large peak due to radiolabeled retinol bound to CRBP is reduced to baseline levels.

Western Blot Analysis

Western blot analysis has been used mainly for IRBP. To date, there has been no evidence from Western blotting for the expression of CRAlBP in any retinoblastoma cell line. IRBP is secreted by cultured retinoblastoma cells into the medium, and only traces are found in the cytosol.[8] The large amount of protein arising from the presence of serum in the medium interferes with the determination of IRBP by Western blotting; consequently, the cells should be transferred to serum-free medium for about 3 days. Alternatively, the cells can be transferred to a serum-free medium at an initial concentration of 3×10^5 cells/ml and the medium collected over a 2- to 3-week period.

That amount of medium or cytosol corresponding to 10^4-10^5 cells is mixed with 5 volumes of acetone and kept on ice for 60 min, and the precipitated proteins are collected by centrifuging. The pellet is suspended in water containing 0.1 m*M* phenylmethylsulfonyl fluoride (PMSF); aliquots are loaded and electrophoresed on sodium dodecyl sulfate (SDS)–polyacrylamide slab gels, then electrophoretically transferred to nitrocellulose paper.[18,19] Immunochemical visualization of IRBP is carried out by incubating with rabbit anti-bovine IRBP IgG followed by peroxidase- or alkaline phosphatase-conjugated goat anti-rabbit IgG. Samples containing 0.5 to 10.0 ng of purified bovine IRBP as standards should also be electrophoresed in the same gel and transblotted to the same membrane (Fig. 2;

[17] E. A. Allegretto, M. A. Kelly, C. A. Donaldson, N. Levine, J. W. Pike, and M. R. Haussler, *Biochem. Biophys. Res. Commun.* **116,** 75 (1983).

[18] F. Gonzalez-Fernandez, R. A. Landers, P. A. Glazebrook, S.-L. Fong, G. I. Liou, D. M. K. Lam, and C. D. B. Bridges, *J. Cell Biol.* **99,** 2092 (1984).

[19] F. Gonzalez-Fernandez, R. A. Landers, P. A. Glazebrook, S.-L. Fong, G. I. Liou, D. M. K. Lam, and C. D. B. Bridges, *Neurochem. Int.* **7,** 533 (1985).

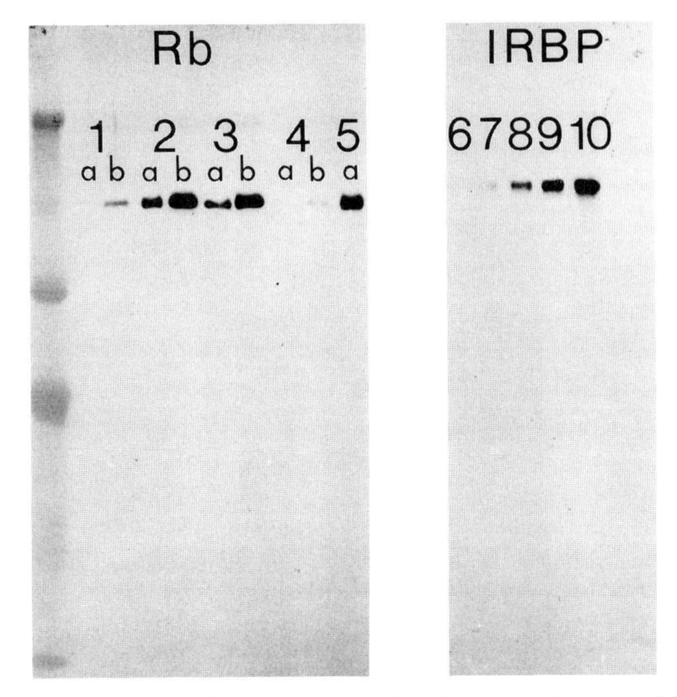

FIG. 2. Western blot analysis of serum-free medium from retinoblastoma cell lines and bovine IRBP standards. Rabbit antibodies were directed against bovine IRBP. In each case, medium was collected from $10^6 - 10^7$ cells. Lanes 1, RB412; lanes 2, RB247; lanes 3, RB383; lanes 4, RB344; lane 5, RB429; lanes 6–10, 0.5, 1.0, 2.5, 5.0, and 10.0 ng bovine IRBP. Retinoblastoma samples were loaded at two levels: those in lanes a contained 0.5–5.0% of the medium whereas those in lanes b contained 2.0–20% of the medium. The unnumbered lane at the left-hand side contains prestained M_r markers; from the top, they are 200K, 97K, and 67K.

see also Ref. 8). Using a suitable scanner (e.g., an LKB Ultroscan XL), the intensities of the bovine bands are then matched against the samples.

The sensitivity of the method can correspond to about 0.5 ng of bovine IRBP loaded on the gel, although in Fig. 2 the sensitivity is only 1.0 ng. The use of anti-bovine IRBP antiserum and of bovine IRBP as a standard is convenient because bovine eyes are available in quantity and bovine IRBP can be prepared from them in comparatively large amounts. The disadvantage is that bovine IRBP is not serologically identical with the corresponding human protein.[9] Therefore, the amounts of IRBP calculated by this method can only be used to compare different cell lines subjected to different treatments: if it is necessary to determine absolute amounts, then anti-human IRBP must be prepared or the system calibrated against purified human IRBP.

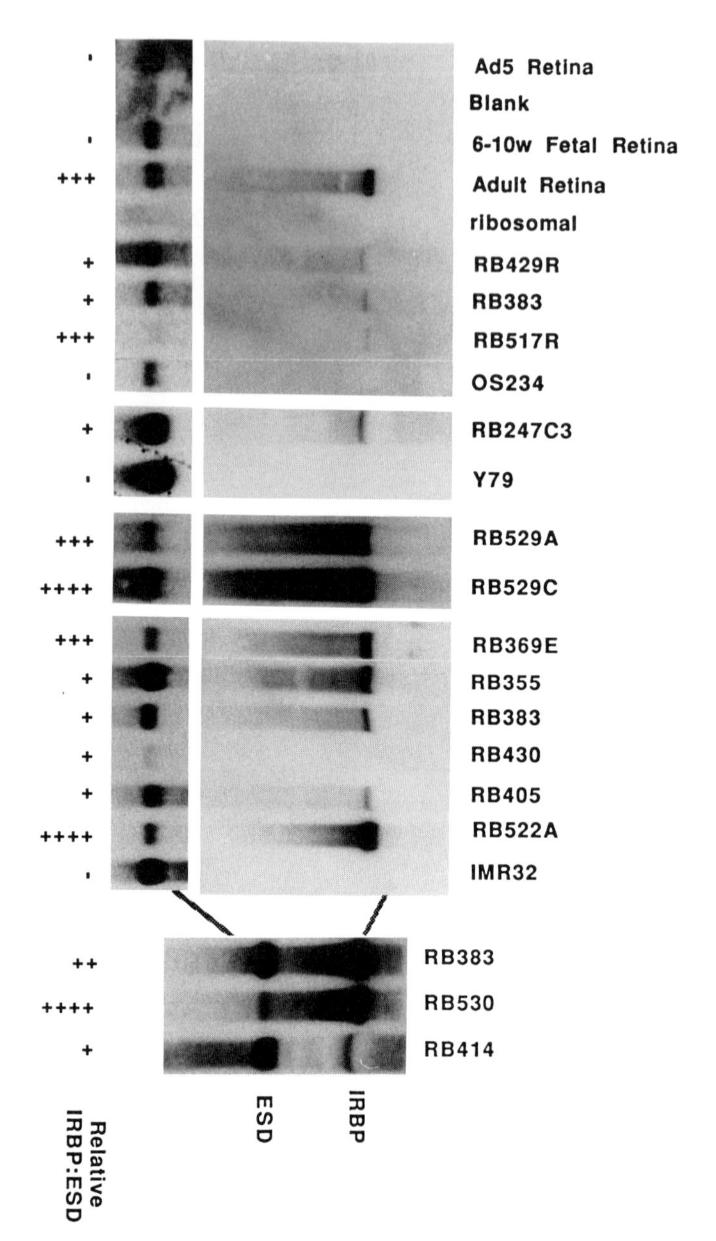

FIG. 3. Northern blot analyses of RNA from 14 retinoblastoma tumor cell lines, normal human fetal and adult retina, an adenovirus 5-transformed human fetal retina cell line (Ad5 retina), an osteosarcoma cell line (OS234), and a neuroblastoma cell line (IMR32). Polyadenylated RNA was used in all lanes except for RB414, where total RNA was present. Blots were probed with the 2.1-kilobase human H.4 IRBP cDNA clone and with the ED1 subclone of esterase D (ESD), either simultaneously or sequentially. The relative levels of IRBP compared with ESD are indicated at the bottom of the blots and range from not detectable (–) to abundant (++++). [From S. L. Fong, H. Balakier, M. Canton, C. D. B. Bridges, and B. Gallie, *Cancer Res.* **48,** 1124 (1988).]

Northern Blot Analysis of Interstitial Retinol-Binding Protein mRNA from Retinoblastoma Cells

Total retinoblastoma cellular RNA may be prepared by the guanidinium isothiocyanate–cesium chloride gradient method.[20,21] The cells are rinsed free of medium then lysed in 4 *M* guanidium solution, the viscosity of the lysate is reduced by shearing through a 20-gauge needle, and the RNA is pelleted by centrifuging on a cesium chloride step gradient. The yield of total RNA from 10^8–10^9 cells is variable but usually ranges from 100 to 300 μg. Polyadenylated RNA can be prepared, if so desired, by oligodeoxythymidylate–cellulose[22] or poly(U)-Sephadex affinity chromatography. The advantage of preparing polyadenylated RNA is that the IRBP mRNA has a sizes of 4.3 kilobases (kb) close to the abundant 28 S rRNA (5.1 kb). This can lead to occasional distortion and loss of clarity of the IRBP signal (see RB414 in Fig. 3).

Electrophoresis in 1% agarose (glyoxal or formamide) and Northern blotting of the RNA to GeneScreen or nitrocellulose may be carried out according to Ausubel *et al.*[21] The human IRBP cDNA clones H.4 or H.12 IRBP[23] labeled with [^{32}P]dCTP by random priming (Boehringer-Mannheim, Indianapolis, IN) have been used by the authors to probe Northern blots of retinoblastoma RNA, as exemplified in Fig. 3 (Ref. 8). Because esterase D is ubiquitously expressed, the ED1 fragment of esterase D[24] may be used as a reference probe, either simultaneously with the IRBP probe, or sequentially.[8]

[20] J. M. Chirgwin, A. E. Przybyla, R. J. MacDonald, and W. I. Rutter, *Biochemistry* **18,** 5294 (1979).

[21] F. M. Ausubel, R. Brent, R. E. Kingston, D. D. Moore, J. G. Seidman, J. A. Smith, and K. Struhl, "Current Protocols in Molecular Biology." Green Publ. and Wiley (Interscience), New York, 1989.

[22] H. Aviv and P. Leder, *Proc. Natl. Acad. Sci. U.S.A.* **69,** 1408 (1972).

[23] S.-L. Fong and C. D. B. Bridges, *J. Biol. Chem.* **263,** 15330 (1988).

[24] J. A. Squire, T. P. Dryja, J. Dunn, A. D. Goddard, T. Hofmann, M. Musarella, H. F. Willard, A. J. Becker, B. L. Gallie, and R. A. Phillips, *Proc. Natl. Acad. Sci. U.S.A.* **83,** 6573 (1986).

[16] Retinoid-Binding Proteins in Embryonal Carcinoma Cells

By JOSEPH F. GRIPPO and MICHAEL I. SHERMAN

Introduction

The mechanism by which retinoids promote cell differentiation has yet to be elucidated, but current evidence suggests that retinoids alter the transcription of regulatory genes to initiate a cascade of events leading to the differentiated phenotype. By analogy with the steroid hormones, it has been proposed that retinoids interact with a receptor which can bind both its ligand and specific regulatory regions of the genome to modulate gene expression at the level of transcription.[1]

Considerable effort has gone into characterizing the proteins or receptors which presumably mediate retinoid action. Most research has centered on two retinoid-binding proteins, namely, the cellular retinoic acid-binding protein (CRABP) and the cellular retinol-binding protein (CRBP). Both CRABP and CRBP belong to a family of low molecular weight (15,000) proteins with eight known members, including CRBP II, three fatty acid-binding proteins, an adipocyte protein (aP2), and the P2 protein of peripheral nerve myelin.[2,3] CRABP and CRBP exhibit high affinity, as well as selectivity, for their respective ligands, with apparent dissociation constants in the range of 10–20 n*M*.[4–7] Whereas specific nuclear binding sites for CRABP and CRBP appear to exist (see Ref. 8), data from Chytil's laboratory suggest that both CRABP and CRBP deliver their ligands to nuclear acceptor sites, but do not appear to be retained in the nuclear compartment.[9,10] In keeping with this observation, primary sequence data

[1] F. Chytil and D. Ong, *Fed. Proc., Fed. Am. Soc. Exp. Biol.* **38,** 2510 (1979).

[2] L. A. Demmer, E. H. Birkenmeier, D. A. Sweetser, M. S. Levin, S. Zollman, R. S. Sparkes, T. Mohandas, A. J. Lusis, and J. I. Gordon, *J. Biol. Chem.* **262,** 2458 (1987).

[3] J. Sundelin, S. R. Das, U. Eriksson, L. Rask, and P. A. Peterson, *J. Biol. Chem.* **260,** 6494 (1985).

[4] D. E. Ong and F. Chytil, *J. Biol. Chem.* **250,** 6113 (1975).

[5] D. E. Ong and F. Chytil, *J. Biol. Chem.* **253,** 828 (1978).

[6] A. M. Jetten and M. E. R. Jetten, *Nature (London)* **278,** 180 (1979).

[7] J. F. Grippo and L. J. Gudas, *J. Biol. Chem.* **262,** 4492 (1987).

[8] M. I. Sherman, *in* "Retinoids and Cell Differentiation" (M. I. Sherman, ed.), p. 161. CRC Press, Boca Raton, Florida, 1986.

[9] S. Takase, D. E. Ong, and F. Chytil, *Proc. Natl. Acad. Sci. U.S.A.* **76,** 2204 (1979).

[10] S. Takase, D. E. Ong, and F. Chytil, *Arch. Biochem. Biophys.* **247,** 328 (1986).

for both bovine adrenal CRABP[3] and rat liver CRBP[11] do not suggest the presence of DNA-binding domains.

The recent discovery of nuclear retinoic acid receptors (RAR), which resemble the steroid receptor superfamily of transcriptional regulatory proteins,[12-15] suggests that these are the proteins which directly modulate the effects of retinoids (or at least retinoic acids) on gene expression. This in turn provides a plausible role for the cellular retinoid-binding proteins, namely, to shuttle retinoids from the cytoplasm to the nucleus, where they interact with the RAR. Members of this class of proteins include RARα,[12,13] RARβ,[14] and the newly described RARγ.[15a] They are 50-kDa proteins containing both ligand-binding and DNA-binding domains. As yet, no nuclear receptor specific for retinol binding has been described. Methods to characterize RAR are presented elsewhere.[16]

This chapter focuses on procedures that can be employed in the characterization of the role of CRABP and CRBP in retinoid-related cellular behavior. Embryonal carcinoma (EC) cells are described here as prototypes for such studies because they, like many other cell types, possess CRABP, CRBP, and RAR. Embryonal carcinoma cells differentiate readily in response to retinoids, an event which can be monitored in a number of convenient ways.[17] Furthermore, EC lines can be induced to metabolize retinoids,[18] a phenomenon which might involve the retinoid-binding proteins.[19] We examine below methods to characterize CRABP and CRBP protein and RNA levels, and we also outline strategies for producing mutant cell lines that lack detectable CRABP activity.

[11] J. Sundelin, H. Anundi, L. Tragardh, U. Eriksson, P. Lind, H. Ronne, P. A. Peterson, and L. Rask, *J. Biol. Chem.* **260,** 6488 (1985).

[12] M. Petkovich, N. J. Brand, A. Krust, and P. Chambon, *Nature (London)* **330,** 444 (1987).

[13] V. Giguere, E. S. Ong, P. Segui, and R. M. Evans, *Nature (London)* **330,** 624 (1987).

[14] N. Brand, M. Petkovich, A. Krust, P. Chambon, H. de The, A. Marchio, P. Tiollais, and A. Dejean, *Nature (London)* **332,** 850 (1988).

[15] D. Benbrook, E. Lernhardt, and M. Pfahl, *Nature (London)* **333,** 669 (1988).

[15a] A. Zelent, A. Krust, M. Petkovich, P. Kastner, and P. Chambon, *Nature (London)* **339,** 714 (1989).

[16] A. M. Jetten, J. F. Grippo, and C. Nervi, this series Vol. 189 [25].

[17] P. Abarzua and M. I. Sherman, this series, Vol. 189 [35].

[18] M. L. Gubler and M. I. Sherman, this series, Vol. 189 [60].

[19] M. L. Gubler and M. I. Sherman, *J. Biol. Chem.* **260,** 9552 (1985).

Cell Lines

Numerous murine EC cell lines are available.[20] Two EC cell lines commonly used are F9[21,22] and PCC4.aza1R.[23,24] Both F9 and PCC4.aza1R cells are cultured in 5% CO_2 at 37° in Dulbecco's modified Eagle's medium (DMEM), supplemented with 2 m*M* glutamine containing 10% heat-inactivated fetal calf serum. Under these conditions, the generation time of the cells is very short (10 hr or less). F9 cells can be grown in the same medium supplemented with 10% calf serum[7] (generation time 15–18 hr). In either case, the cells fare best when grown on tissue culture plates coated with 0.15–0.3% gelatin (see Ref. 17). Cell passage number should be recorded as cells should be discarded after 30–50 passages in order to avoid phenotypic drift. Typically, cells are plated at a density of 0.7×10^6/100-mm dish or 1.5×10^6 cells/150-mm dish and allowed to attach overnight.

Cytosol Preparation

Approximately 5×10^8 EC cells growing in monolayer (6–15 150-mm dishes of confluent cells, depending on the cell line and the culture conditions) are washed with cold (Ca^{2+}- and Mg^{2+}-free phosphate-buffered saline (PBS) and removed from the plates in PBS with a rubber policeman or by trypsin treatment.[17] The cells are collected by centrifugation at 600 *g* for 5 min at 4° and disrupted by Dounce homogenization (40 strokes) in 1 packed cell volume of homogenization buffer (10 m*M*) Tris buffer, pH 7.4, 2 m*M* $MgCl_2$, and 7 m*M* 2-mercaptoethanol.[6] Similar results are obtained when the $MgCl_2$ is omitted.[7] The cell homogenates are centrifuged at 100,000 *g* for 60 min at 4° and stored in small aliquots at −70° until assayed. Freezing and thawing do not seem to affect retinoid-binding activity.[25]

[20] G. R. Martin, *in* "Teratocarcinoma Stem Cells" (L. M. Silver, G. R. Martin, and S. Strickland, eds.), p. 690. Cold Spring Harbor Laboratory, Cold Spring Harbor, New York, 1983.

[21] E. G. Bernstine, M. L. Hooper, S. Grandchamp, and B. Ephrussi, *Proc. Natl. Acad. Sci. U.S.A.* **70,** 3899 (1973).

[22] S. Strickland and V. Mahdavi, *Cell (Cambridge, Mass.)* **15,** 393 (1978).

[23] H. Jakob, T. Boon, J. Gaillard, J. F. Nicolas, and F. Jacob, *Ann. Microbiol. Inst. Pasteur* **124B,** 269 (1973).

[24] A. M. Jetten, M. E. R. Jetten, and M. I. Sherman, *Exp. Cell Res.* **124,** 381 (1979).

[25] J. Schindler, K. I. Matthaei, and M. I. Sherman, *Proc. Natl. Acad. Sci. U.S.A.* **78,** 1077 (1981).

Cellular Retinoid-Binding Assays

Several methods are available for detecting CRABP and CRBP activity. These include sucrose gradient analyses,[7,25] dextran-treated charcol,[7] gel filtration and ion-exchange chromatography,[26–28] high-performance size-exclusion chromatography (HPSEC),[29] radioimmunoassay, and immunohistochemistry.[30–32] We describe below the sucrose gradient, dextran-charcoal, and Sephadex assays that we have used. In these procedures, retinoid-binding proteins are measured in high-speed cytosols. Attempts to measure this activity in particulate fractions are complicated by considerable levels of nonspecific binding. Even when cytosolic fractions are used, there should be a side-by-side control for each sample containing a 100- to 200-fold molar excess of unlabeled retinoic acid retinol in order to control for nonspecific binding. This applies for all of the procedures described below.

Sucrose Gradient Analysis

Sucrose gradient analysis is the procedure used originally to characterize retinoid-binding protein activities.[4,5] It is accurate and effective but requires more time and cell extract protein than other techniques, and the availability of rotors and centrifuges limits the number of samples that can be analyzed simultaneously. At least 600 μg cytosol protein is mixed with 50 n*M* all-*trans*-[^{3}H]retinoic acid (RA) for CRABP or [^{3}H]retinol for CRBP activity determinations (both at 10–40 Ci/mmol) and adequate homogenization buffer to bring the total volume to 200 μl. Ethanol is used to dissolve the retinoids in a volume not exceeding 4% of the total incubation volume. It should be stressed that inconsistent results are obtained with protein concentrations less than 2–3 mg/ml in this assay.[7] Samples are incubated for 3–5 hr at 4°.

After incubation, unbound ligand is removed by mixing with dextran-coated charcoal suspension (130 μl) followed by centrifugation. [Dextran-

[26] J. C. Saari, S. Futterman, G. W. Stubbs, J. T. Heffernan, L. Bredberg, K. Y. Chan, and D. W. Albert, *Invest. Ophthalmol. Visual Sci.* **17,** 988 (1978).

[27] M. I. Sherman, M. L. Paternoster, and M. Taketo, *Cancer Res.* **43,** 4283 (1983).

[28] U. Barkai and M. I. Sherman, *J. Cell Biol.* **104,** 671 (1987).

[29] H. E. Shubeita, M. D. Patel, and A. M. McCormick, *Arch. Biochem. Biophys.* **247,** 280 (1986).

[30] M. Kato, K. Kato, W. S. Blaner, B. S. Chertow, and D. S. Goodman, *Proc. Natl. Acad. Sci. U.S.A.* **82,** 2488 (1985).

[31] U. Eriksson, E. Hansson, M. Nilsson, K.-H. Jonsson, J. Sundelin, and P. A. Peterson, *Cancer Res.* **46,** 717 (1986).

[32] M. Maden, D. E. Ong, D. Summerbell, and F. Chytil, *Nature (London)* **335,** 733 (1988).

coated charcoal is prepared by mixing 2.5% activated charcoal and 0.2% dextran sulfate (w/v) overnight at 4°. Fines are removed by low-speed centrifugation, and the suspension is stored at 4°.] Two hundred fifty microliters of supernate is applied to a 5–20% sucrose gradient (5 ml) and spun at 220,000 *g* for 18 hr at 4° using a swinging-bucket rotor (SW50.1). After centrifugation, 280-μl aliquots (7–8 drops), withdrawn from the bottom of each tube, are collected into scintillation vials (5 ml cocktail) and counted in a liquid scintillation spectrometer. Myoglobin (2 S) and bovine serum albumin (4.6 S) can be used as sedimentation markers. CRABP and CRBP sediment with the 2 S marker.[7,25]

Dextran-Coated Charcoal Assay

In the dextran-coated charcoal assay, 450 μg (125 μl) of cytosol protein is incubated with [^{3}H]retinoid in the assay mix described in the previous section. Unbound ligand is removed by mixing with 100 μl of dextran-coated charcoal suspension (prepared as described above), followed by centrifugation (3000 *g*, 20 min) to clarify the supernatant. As noted above, levels of nonspecific [^{3}H]retinoid binding are determined by incubation of sample in the presence of a 100-fold molar excess of unlabeled retinoid. One hundred microliters of the clear supernatant is counted by liquid scintillation spectrometry, and specific [^{3}H]retinoid binding, determined by subtracting nonspecific from total binding, is generally expressed as picomoles [^{3}H]retinoid bound per milligram supernatant protein.

Sephadex Assays

We have used two procedures involving Sephadex chromatography to measure retinoid-binding protein activities.[27,28] Both are rapid and convenient and require only small amounts of cell extract for analysis of activity. The first assay[27] combines the use of dextran-coated charcoal with Sephadex chromatography, which results in low background levels of radioactivity, thereby allowing the use of reduced amounts of cellular protein compared with the above procedures. The assay mixture is prepared and incubation carried out as described above except that only 100–250 μg cytosolic protein is added and the total volume is 150 μl. To the assay mix is added 100 μl dextran-coated charcoal solution. After centrifugation (3000 *g*) for 20 min at 4°, 100 μl of supernatant is passed through a 7-cm Sephadex G-25–150 column in a Kimble Pasteur pipette (No. 72050). The bound counts (in the void volume) are eluted with 1 ml homogenization buffer and are collected directly into scintillation vials for radioactivity determinations.

In an alternative procedure,[28] the binding reaction is carried out on a disk from a Tetralute kit purchased from Ames-Yissum (Jerusalem, Israel);

the disk is in turn placed on a 1×2 cm Sephadex G-25 column, which is used to separate bound from unbound [^{3}H]retinoid after the incubation period has ended. Carrying out the incubation on the disk appears to facilitate the binding reaction, thus permitting shorter incubation times and the use of smaller amounts of radioactivity. This, in turn, obviates the need for the dextran-coated charcoal extraction step. Cytosol protein (50–200 μg) is pipetted onto the disk followed by [^{3}H]retinoic acid or [^{3}H]retinol ($3-4 \times 10^5$ cpm in 10 μl ethanol). Incubation buffer (25 m*M* Tris, pH 7.4, 25 m*M* NaCl) is added to give a total reaction volume of 300 μl. Control samples to determine nonspecific binding should contain no more than a 100-fold excess of the appropriate unlabeled retinoid. When [^{3}H]retinol is used, 0.3 *M* sucrose is added to the incubation mix (in the absence of sucrose, spuriously high values for retinol binding are obtained for reasons that are unclear). After 90 min at 4°, the reaction mix is allowed to flow through the column followed by 500 μl incubation buffer. Bound radioactive retinoid is then eluted into scintillation vials with 900 μl incubation buffer.

All of these methods have been used successfully to determine levels of CRABP or CRBP. In each case, we have validated the assays carried out with F9, PCC4.aza1R, and/or Nulli-SCC1 EC cells by comparing the results with those obtained simultaneously by sucrose gradient centrifugation; data obtained are in agreement at the level of within 10%. It must, however, be noted that neither the charcoal–dextran nor the Sephadex assay confirms that the retinoids are indeed bound to their respective binding proteins. Therefore, when these assays are being used for the first time, it would be prudent to validate them against the sucrose density gradient or HPSEC[29] assay. None of the techniques described above will physically separate CRABP from CRBP. These binding proteins can be distinguished pharmacologically, as a 200-fold molar excess of retinol will not displace [^{3}H]retinoic acid binding from CRABP and a 200-fold molar excess of retinoic acid will not displace [^{3}H]retinol binding from CRBP. Physical separation of the two proteins can be achieved by ion-exchange chromatography.[26]

RNA Isolation and Northern Analysis

cDNA clones for CRABP[33,34] and for CRBP[35] have been obtained and have been used to characterize mRNA expression for these retinoid-bind-

[33] H. E. Shubeita, J. F. Sambrook, and A. M. McCormick, *Proc. Natl. Acad. Sci. U.S.A.* **84,** 5645 (1987).

[34] C. Stoner and L. J. Gudas, *Cancer Res.* **49,** 1497 (1989).

[35] V. Colantuoni, R. Cortese, M. Nilsson, J. Lundvall, C. Bavik, U. Eriksson, P. A. Peterson, and J. Sundelin, *Biochem. Biophys. Res. Commun.* **130,** 431 (1985).

ing proteins in F9 EC cells.[36] We have routinely used a guanidine-HCl method to isolate RNA from monolayer cultures of EC cells. Confluent dishes of EC cells are rinsed with PBS and scraped into ice-cold 7*M* guanidine-HCl, 20 m*M* potassium acetate, pH 7.0, 5 m*M* EDTA, and 1 m*M* dithiothreitol and precipitated twice at −20° from this solution by addition of one-half volume of ethanol. Following centrifugation the pellet is redissolved in 50 m*M* Tris-HCl, pH 9.0, 5 m*M* EDTA, 100 m*M* NaCl, and 0.5% sodium dodecyl sulfate and extracted twice with an equal volume of phenol/chloroform/isoamyl alcohol (10:9.6:0.4, v/v). RNA is then precipitated with 0.2 *M* NaCl and 2.5 volumes ethanol.

RNA blot analysis is performed by standard procedures. RNA is fractionated by electrophoresis in 1% agarose/2.2 *M* formaldehyde slab gels, stained with ethidium bromide, transferred to nitrocellulose or nylon membranes, and attached to the filters by baking at 80° *in vacuo.* Filters are prehybridized in a suitable hybridization mixture[33] and hybridized with at least 10^6 cpm/ml labeled cDNA probe.

When harvesting more than 3×10^7 cells, we use approximately 0.5 ml guanidine-HCl/10^6 cells. Poly(A)$^+$ RNA can be selected, but 15–30 μg total F9 EC RNA can be used to measure RNA levels for CRABP and CRBP. Both mouse CRABP and human CRBP cDNA probes hybridize with a low abundance, single mRNA species in F9 cells of about 800 nucleotides (CRABP mRNA in human Y79 retinoblastoma cells is slightly larger, 1000 nucleotides).[37]

Isolation of Embryonal Carcinoma Cell Lines Lacking Cellular Retinoic Acid-Binding Protein Activity

The methods described above can be used to characterize the specific retinoid-binding capacity and expression of retinoid-binding protein mRNA species in EC cells. As one means of investigating the role of CRABP in mediating retinoid effects, mutant EC cell lines that lack functional CRABP activity can be isolated (the role, if any, of CRBP in retinoid-induced EC cell differentiation has yet to be elucidated,[8] and the isolation of mutant cells devoid of CRBP has not been reported). Resistance to retinoic acid-induced differentiation can be used as a convenient selection procedure for obtaining these lines because differentiated cells grow poorly, if at all, at clonal density.[17]

Two independent efforts have resulted in generation of CRABP-deficient EC cell lines.[35,37] In the first, PCC4.aza1R cells were mutagenized by

[36] S.-Y. Wang, G. J. LaRosa, and L. J. Gudas, *Dev. Biol.* **107,** 75 (1985).

[37] S.-Y. Wang and L. J. Gudas, *J. Biol. Chem.* **259,** 5899 (1984).

exposure to *N*-methyl-*N*′-nitronitrosoguanidine (MNNG) at 3 μg/ml for 18 hr followed by plating at clonal density. Retinoic acid-refractory cells were selected for clonal growth in 10 μ*M* retinoic acid. Approximately 1 in 10^5 cells formed clones with an undifferentiated, EC-like appearance. Cells from these clones were passaged several times in the presence of 10 μ*M* retinoic acid and were finally subcloned in the same retinoic acid-containing medium. Two differentiation-defective clonal EC cell lines, PCC4(RA)$^-$ 1 and PCC4(RA)$^-$ 2, which contain low or undetectable levels of CRABP activity, have been isolated in this way.

The F9 RA3-10 line derived from F9 cells was obtained[36] by treating F9 wild-type cells ($1-2 \times 10^7$ cells per T150 flask) with 5 μg/ml MNNG for 5 hr. Surviving cells were grown under nonselective conditions for 5 days and were then treated with 1 μ*M* retinoic acid for 24 hr. The cells were removed from the flask with trypsin and cloned on plates containing 0.34% agarose over a mouse embryo fibroblast feeder layer in the presence of retinoic acid. This selection procedure is based on the observation that wild-type F9 cells will form colonies in soft agar whereas retinoic acid-treated differentiated derivatives will not. Colonies picked from these plates were expanded for further analysis. The F9 RA3-10 line isolated in this manner was found to lack mRNA and binding capacity for CRABP and also failed to differentiate in response to retinoic acid.

It should be noted that these procedures can also result in the generation of differentiation-defective EC cells which nevertheless possess CRABP.[36,38] Therefore, any differentiation-defective lines which are obtained should be analyzed to identify those lacking CRABP mRNA and/or binding activity.

[38] P. A. McCue, K. I. Matthaei, M. Taketo, and M. I. Sherman, *Dev. Biol.* **96,** 416 (1983).

[17] Acyl-CoA:Retinol Acyltransferase and Lecithin:Retinol Acyltransferase Activities of Bovine Retinal Pigment Epithelial Microsomes

By JOHN C. SAARI and D. LUCILLE BREDBERG

Introduction

The reactions that accompany the bleaching and subsequent regeneration of visual pigment make up the visual or regeneration cycle.[1] Although visual pigment bleaching takes place in photoreceptor cells, most of the enzymatic reactions of the cycle, including the isomerization regenerating the 11-cis configuration, take place in adjacent retinal pigment epithelial (RPE) cells.[2] Retinyl esters were previously thought to be solely storage forms of vitamin A in RPE; however, recent work suggests that retinyl ester is the substrate for the isomerase.[3] The main fatty acids esterified to retinol are palmitate (57%), stearate (24%), and oleate (10%).[4] Unusual features of retinyl ester synthesis in RPE, noted in other studies, include its occurrence in CoA-dependent and CoA-independent modes[5] (Fig. 1), the presence of an endogenous acyl donor in RPE microsomes,[6-8] and the large amount of activity present.[5]

Recent studies have shed new light on several aspects of retinyl ester synthesis in RPE. Enzymatic transfer of acyl groups from the 1-position of phosphatidylcholine (PC) to retinol has been demonstrated in RPE microsomes, suggesting that lecithin:retinol acyltransferase (LRAT) accounts for the CoA-independent retinyl esterification and that microsomal PC is the endogenous acyl donor.[9,10] Apparent acyl-CoA:retinol acyltransferase (ARAT,EC 2.3.1.76, retinol fatty-acyltransferase) activity has been shown to result, at least in part, from an initial transfer of the fatty acyl group of acyl-CoA to a lipid component of the microsomes and subsequently to

[1] G. Wald, *Science* **162,** 230 (1968).

[2] J. C. Saari, *in* "Progress in Retinal Research" (N. N. Osborne and J. G. Chader, eds.), Vol. 9, p. 363. Pergamon, Oxford and New York, 1990.

[3] P. S. Deigner, W. C. Law, F. J. Canada, and R. R. Rando, *Science* **244,** 968 (1989).

[4] S. Futterman and J. S. Andrews, *J. Biol. Chem.* **239,** 81 (1964).

[5] J. C. Saari and D. L. Bredberg, *J. Biol. Chem.* **263,** 8084 (1988).

[6] N. I. Krinsky, *J. Biol. Chem.* **232,** 881 (1958).

[7] E. R. Berman, J. Horowitz, N. Segal, S. Fischer, and L. Feeney-Burns, *Biochim. Biophys. Acta* **630,** 36 (1980).

[8] J. S. Andrews and S. Futterman, *J. Biol. Chem.* **239,** 4073 (1964).

[9] J. C. Saari and D. L. Bredberg, *J. Biol. Chem.* **264,** 8638 (1989).

[10] R. J. Barry, F. J. Canada, and R. R. Rando, *J. Biol. Chem.* **264,** 9231 (1989).

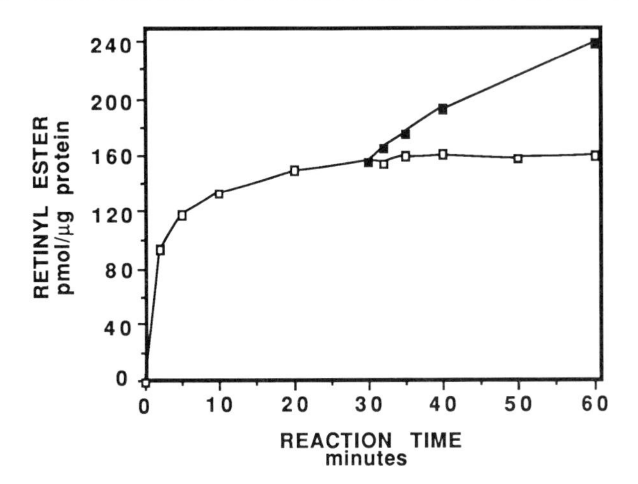

FIG. 1. Progress curve for retinyl esterification. Addition of [^{3}H] retinol to buffered microsomes from bovine retinal pigment epithelium produces an immediate burst of retinyl ester synthesis, demonstrating a CoA-independent component of the reaction (□). After 30 min further addition of [^{3}H]retinol produces little additional retinyl ester synthesis (□). However, additional retinyl ester is synthesized in response to an addition of retinol and palmitoyl-CoA (■). (Modified from Reference 5 with permission of the American Society for Biochemistry and Molecular Biology.)

retinol.[5] The lipid component is likely to be a lysolecithin, although this has not been demonstrated directly. A proposed relationship of CoA-dependent and CoA-independent activities is shown in Fig. 2. LRAT has been described in rat intestinal mucosal cells[11] and is likely to catalyze a general reaction for retinol esterification.

Methods are presented here for the determination of CoA-independent retinyl ester synthesis (lecithin:retinol acyltransferase, LRAT) with an endogenous acyl donor and with *in situ* synthesized PC, and of CoA-dependent retinyl ester synthesis (apparent ARAT).

Assay Methods

Principle. Retinyl ester synthesis is a two-substrate reaction requiring an acyl donor and retinol. An endogenous acyl donor is present in RPE microsomes, giving retinyl ester synthesis on addition of retinol alone.[5-8,12] (LRAT with endogenous acyl donor). In addition, RPE microsomes will catalyze the synthesis of [1-^{14}C]palmitoyl-PC during an incubation with

[11] P. N. MacDonald and D. E. Ong, *J. Biol. Chem.* **263,** 12478 (1988).

[12] C. D. B. Bridges, M. S. Oka, S.-L. Fong, G. I. Liou, and R. A. Alvarez, *Neurochem. Int.* **8,** 527 (1980).

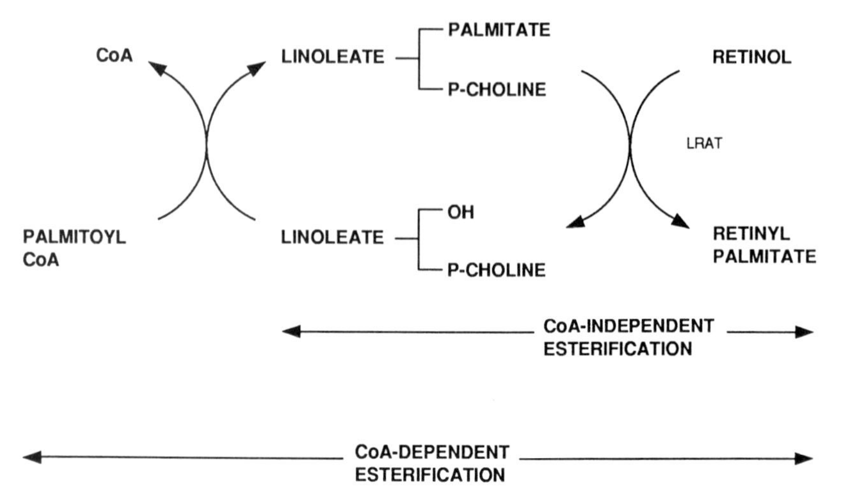

FIG. 2. Proposed relationship between CoA-independent (LRAT) and CoA-dependent (ARAT) retinyl ester synthesis in bovine retinal pigment epithelial microsomes. A contribution arising from direct transfer of palmitate from palmitoyl-CoA to retinol has not been ruled out. [Reproduced with permission from J. C. Saari, *in* "Progress in Retinal Research" (N. N. Osborne and J. G. Chader, eds.), Vol. 9, p. 363. Pergamon, Oxford and New York, 1990.]

[1-^{14}C]palmitoyllyso-PC and a fatty acyl-CoA, producing a ^{14}C-labeled acyl donor for retinyl ester synthesis (LRAT with exogenous acyl donor).[9] Finally, apparent ARAT activity can be measured with [^{14}C]palmitoyl-CoA as an acyl donor for the reaction.[5]

The assay procedure involves extraction of [^{3}H]retinol and [^{3}H]retinyl esters from reaction mixtures and their separation on disposable alumina columns. Quantification is obtained by liquid scintillation counting.

Reagents

Alumina columns: 0.4 g of 6% water-deactivated alumina;[13] dry in a Pasteur pipette loosely plugged with glass wool

Assay buffer: 40 m*M* Tris–acetate, pH 8, containing 60 μM bovine serum albumin (BSA) and 1 m*M* dithiothreitol (DTT)

Bovine serum albumin, fatty acid poor (Sigma, St. Louis, MO)

Buffered sucrose: 0.25 *M* sucrose buffered with 25 m*M* Tris–acetate, pH 7, containing 1 m*M* DTT

Chloroform/methanol, 2:1, v/v[14]

[13] R. Hubbard, P. K. Brown, and D. Bownds, this series, Vol. 18 part C, p. 615.

[14] J. Folch, M. Lees, and G. A. Sloane-Stanley, *J.Biol. Chem.* **226,** 497 (1957).

Fatty acyl-CoAs (Sigma), assayed and purified by high-performance liquid chromatography (HPLC) if necessary[5]
Thin layer chromatography (TLC) sheets, ITLC-SG silica-impregnated glass fiber sheets (Gelman Sciences, Ann Arbor, MI)
TLC solvent: hexane/2-propanol/water, 6:8:1 (v/v/v), for separation of lyso-PC and PC
Liquid scintillation cocktail (Ecolume, ICN Biomedicals, Inc., Costa Mesa, CA)
[1-^{14}C]Palmitoyllyso-PC (Du Pont–New England Nuclear, Boston, MA), purified by TLC if necessary[9] (128,000 dpm/nmol)
[^{14}C]Palmitoyl-CoA (Du Pont–NEN), purified by HPLC;[5] a working solution of 1 m*M* in assay buffer (~32,000 dpm/nmol) is prepared
Petroleum ether (30–60° boiling range)
all-*trans*-Retinol (unlabeled) (Sigma), purified by HPLC[15]
all-*trans*-[^{3}H]Retinol (Du Pont–NEN), purified by HPLC; a working solution of 2 m*M* in ethanol (specific activity ~20,000 dpm/nmol) is prepared
Ruby red bulbs, 40 W (Westinghouse)

General Procedures. All operations involving retinoids are performed in darkness or under dim red illumination. Buffers are bubbled with argon before use, and tubes are flushed with argon and sealed during incubations. Butylated hydroxytoluene (BHT) is added to all solutions containing lipids (1:100, w/w, BHT/lipid). Bovine serum albumin (60 μM) is included in all solutions of retinoids and fatty acyl-CoAs to minimize the known detergent effects of the latter[16] and to enhance the water solubility of the former.

Isolation of Retinal Pigment Epithelial Microsomes. Bovine (calf or adult) eyes are hemisected and the anterior portion and vitreous discarded. Neural retina is removed with a blunt-tipped capsule forceps (Storz Ophthalmic Instruments, St. Louis, MO). Cold buffered sucrose (1 ml/eye) is added to the eye cup, and RPE cells are dislodged by gently stroking from center to periphery with a small brush. The dark suspension of cells is removed with a Pasteur pipette. After repeating the procedure, the cells are homogenized with 8–10 passes of a motor-driven Teflon–glass homogenizer and centrifuged at 20,000 *g* for 20 min at 5°. The supernatant is centrifuged at 150,000 *g* for 1 hr at 5°. The microsomal pellet from this spin is resuspended in 10 m*M* Tris–acetate, pH 7, 1 m*M* DTT (50 or 100 μl/calf or adult eye, respectively) and stored at −80° in 50-μl portions. Protein concentrations of the preparations are obtained using the Folin phenol procedure.[17]

[15] J. C. Saari, L. Bredberg, and G. G. Garwin, *J. Biol. Chem.* **257,** 13329 (1982).
[16] P. P. Constantinides and J. M. Steim, *J. Biol. Chem.* **260,** 7573 (1985).
[17] G. L. Peterson, this series, Vol. 91, p. 95.

Quantification of Retinyl Ester. Enzymatic reactions are stopped by the addition of 2 volumes of ice-cold ethanol. Following the addition of 6 volumes of petroleum ether, retinoids are extracted by forcing the lower phase into the upper phase with a Pasteur pipette about 6 times. Three volumes of water are then added and the tube centrifuged to separate the phases. The upper phase is removed, extracted with 3 volumes of water to remove traces of ethanol, and applied to an alumina column, which is washed with 1 ml of petroleum ether (bp 35–60°), followed by 2 ml of 3% diethyl ether in petroleum ether to elute retinyl esters and 2 ml of 50% diethyl ether in petroleum ether to elute retinol. After collection of retinoid-containing fractions in scintillation vials, the solvent is evaporated at 37° with a stream of argon (Reacti-Vap, Pierce, Rockford, IL), and the radioactivity is determined by liquid scintillation.

Assay for LRAT, Endogenous Acyl Donor. all-*trans*-[^{3}H]Retinol is added to assay buffer at 37°, to give a final retinol concentration of 10 μM. After a 30-sec incubation at 37°, the reaction is initiated by the addition of microsomes (10–50 μg protein/ml, final concentration) and the incubation continued. For initial rate studies an assay time of 0.5 min is employed. The reaction is stopped with the addition of ice-cold ethanol, and retinoids are extracted as described.

Assay for LRAT, Exogenous Acyl Donor. The assay using an exogenous acyl donor involves enzymatic generation, during a preincubation, of labeled PC that serves as an acyl donor for subsequent retinyl ester synthesis. Microsomes (10–50 μg protein/ml) are preincubated with 10 μM [1-^{14}C]palmitoyllyso-PC and linoleoyl-CoA (60 μM) in assay buffer containing 150 μM BSA. After 5 min, [^{3}H]retinol is added (10 μM final concentration) and the incubation continued. Following the addition of 2 volumes of cold ethanol, retinol and retinyl esters are extracted and quantified as described above. Retinyl ester synthesis in this assay takes place with acyl groups provided by the endogenous acyl donor (~80%) and the *in situ* synthesized PC (~20%). Determination of the rate of incorporation of [^{14}C]palmitate into retinyl esters is a measure of the latter process, whereas the rate of incorporation of [^{3}H]retinol into retinyl esters measures the overall process. Linoleoyl-CoA generated the most active substrate of the acyl-CoAs tested.[9]

Conversion of lyso-PC to PC can be determined by thin-layer chromatography on silica-impregnated glass fiber sheets following extraction of 1 volume of the reaction mixture with 4 volumes of $CHCl_3$–methanol.[9]

Assay of ARAT Activity. Initial rates of ARAT activity can be observed following the addition of [^{14}C]palmitoyl-CoA and retinol to microsomes. [^{14}C]Palmitoyl-CoA (20 μM) is added to microsomes (10–50 μg protein/ml) and preincubated for 30-sec. The reaction is initiated by the addition of [^{3}H]retinol (10 μM final concentration). After an appropriate time (0.5 – 1.5 min for routine assays), 2 volumes of cold ethanol are added and ^{14}C and ^{3}H incorporation into retinyl esters determined as described above.

An effect of palmitoyl-CoA on the rate of incorporation of [^{3}H]retinol into retinyl ester is observed following a preincubation of microsomes with retinol to allow retinyl ester synthesis to apparently cease (Fig. 1). all-*trans*-[^{3}H]Retinol is added to RPE microsomes in assay buffer to give a concentration of 10 μM. Portions are removed at intervals during incubation at 37° and the extent of incorporation of [^{3}H]retinol into retinyl ester determined. When the reaction has apparently ceased (~20 min, Fig. 1), palmitoyl-CoA (final concentration 20 μM) and [^{3}H]retinol (final concentration 10 μM) are added and portions removed at intervals for approximately 20 min for analysis of retinyl esters.

General Comments

Stability. Both LRAT and ARAT activities are stable for at least 6 months in microsomal fractions stored at −80° in the presence of DTT. Storage and/or assay in the absence of DTT results in a loss of ARAT activity with retention of LRAT activity. It is likely that conflicting accounts[6–8,12] of the effect of added palmitoyl-CoA are attributable to lack of adequate sulfhydryl protection.

At 37° in the absence of substrate, the LRAT half-life ($t_{\frac{1}{2}}$) is 17 min. The $t_{\frac{1}{2}}$ of palmitoyl-CoA at 37° and pH 8, measured in the presence of microsomes (60 μg/ml) but absence of retinol, is 9 min.

Progress Curve of Retinyl Ester Synthesis. Addition of retinol to buffered microsomes from bovine RPE results in a burst of CoA-independent retinyl ester synthesis that appears to stop after approximately 20 min (Fig. 1). Addition of palmitoyl-CoA and retinol at this time allows retinyl ester synthesis to resume, demonstrating a CoA-dependent component of the reaction.

The CoA-independent component of the reaction appears to be catalyzed by lecithin:retinol acyltransferase (LRAT), an enzyme that transfers a fatty acid from the 1-position of phosphatidylcholine to retinol.[9,10] Exogenous dipalmitoyl-PC is poorly integrated into microsomal membranes, and its addition results in little LRAT activity.[5] However, substrate can be generated *in situ* as RPE microsomes catalyze synthesis of [1^{14}C]palmitoyl-2-acyl-PC from 1-[^{14}C]palmitoyllyso-PC and fatty acyl-CoA. Addition of retinol then results in the transfer of [^{14}C]palmitate to retinol, demonstrating LRAT activity. Results obtained employing different fatty acyl-CoAs for PC synthesis suggest that the fatty acid in the 2-position is important in determining the efficiency of the substrate.[9]

TABLE I
CoA-INDEPENDENT RETINYL ESTER SYNTHESIS: SPECIFIC ACTIVITY OF MICROSOMAL PREPARATIONS

Source of microsomes	Retinyl ester synthesis (nmol/min/mg)
Bovine retinal pigment epithelium[a]	102
Rat intestine[b]	0.031
Rat liver[c]	0.145
Rat mammary gland[d]	0.06
Rat testes[e]	0.094

[a] C. Saari and D. L. Bredberg, *J. Biol. Chem.* **263,** 8084 (1988).
[b] D. E. Ong, B. Kakkad, and P. N. MacDonald, *J. Biol. Chem.* **262,** 2729 (1987).
[c] R. W. Yost, E. H. Harrison, and A. C. Ross, *J. Biol. Chem.* **263,** 18693 (1988).
[d] A. C. Ross, *J. Lipid Res.* **23,** 133 (1982).
[e] L. R. Chandhary and E. C. Nelson, *Biochim. Biophys. Acta* **917,** 24 (1987).

The apparent rapid cessation of retinyl ester synthesis (Fig. 1) was initially interpreted as depletion of an endogenous acyl donor.[5] However, addition of [^{3}H]retinol to microsomes that had been preincubated with retinol for 20 min results in rapid appearance of the label in retinyl ester,[18] demonstrating that retinyl ester synthesis is continuing without a net increase in accumulation. Thus, the plateau in the progress curve (Fig. 1) is likely to be due to reversal of the LRAT reaction and/or hydrolase activity balancing the rate of retinyl ester synthesis.

Detailed studies of the ARAT reaction indicate that it can be divided into two steps.[5] Addition of [^{14}C]palmitoyl-CoA to microsomes in the absence of retinol results in an initial acylation of a microsomal lipid. Incorporation of [^{14}C]palmitate into retinyl palmitate, in the absence of palmitoyl-CoA, can then be demonstrated following the addition of retinol, demonstrating the two-step nature of the reaction. Thus, in RPE it appears that ARAT activity can be explained, at least in part, by an initial transfer of an acyl group from acyl-CoA to a microsomal lipid followed by acyl transfer to retinol (LRAT) (Fig. 2). Direct acyl transfer from palmitoyl-CoA to retinol as an additional reaction has not been ruled out.

Kinetic Properties. The specific LRAT activity of RPE microsomes with the endogenous acyl donor is the highest reported for retinyl ester

[18] J. C. Saari and D. L. Bredberg, unpublished.

synthesis, with a value approximately 10^3 times that reported for microsomes from liver, intestinal mucosa, testes, or rat mammary gland (Table I). K_m values of approximately 2 μM for all-*trans*- and 11-*cis*-retinol are observed for LRAT activity;[5] it is not known whether both substrates are processed by the same or different enzymes. The rate of the reaction is optimal between pH 7.5 and 8.

Acknowledgments

Supported in part by U.S. Public Health Service Grants EY02317 and EY01730, and in part by an award from Research to Prevent Blindness, Inc.

[18] High-Performance Liquid Chromatography of Natural and Synthetic Retinoids in Human Skin Samples

By ANDERS VAHLQUIST, HANS TÖRMÄ, OLA ROLLMAN, and EVA ANDERSSON

Introduction

The effects of retinoids on the skin are described in detail elsewhere in this series. Typically, advanced hypovitaminosis A elicits epidermal hyperkeratinization and follicular plugging. Hypervitaminosis A, in contrast, stimulates epidermal cell turnover and retards keratinocyte differentiation. In each of these situations it may be of interest to know not only the serum concentration of vitamin A but also the skin concentration. A small, superficial skin biopsy will usually suffice for screening the endogenous concentrations of retinol and 3,4-didehydroretinol, the two predominating forms of vitamin A in human epidermis,[1] provided that a sensitive analytical technique is available. The skin is also a primary target tissue for retinoid therapy. For several years, synthetic retinoids have been used topically and systemically in skin disorders such as acne, psoriasis, and keratinizing genodermatoses. In these situations, retinoid analysis of a skin biopsy may be of value for monitoring the therapeutic drug concentrations and establishing the blood–tissue concentration relationship.

We describe a technique for routine analysis of picomolar amounts of both neutral (retinol, 3,4-didehydroretinol) and acidic (retinoic acid, iso-

[1] A. Vahlquist, *J. Invest. Dermatol.* **79**, 89 (1982).

tretinoin, etretinate, acitretin) retinoids in human skin biopsies. With minor modifications the technique is also applicable to other tissues which, like human skin, are difficult to homogenize and to extract.

Materials and Methods

Reagents

Spectrographic ethanol
Potassium hydroxide, 14 *M*
HPLC-grade *n*-hexane, methanol, chloroform, and acetonitrile
Deionized water
Butylated hydroxytoluene (BHT, 2mg/ml in methanol)
Authentic retinoids: all-*trans*-retinol, 3,4-didehydroretinol (vitamin A_2, Ro 04-3791), and Ro 12-0586 [all-*trans*-9-(4-methoxy-2,3,6-trimethylphenyl)-3,7-dimethyl-2,4,6,8-nonatetraen-1-ol, i.e., TMMP-retinol]

Additional Reagents for Analysis of Acidic Retinoids

Sodium hydroxide, 1 *M*
Hydrochloric acid, 1 and 5 *M*
Glacial acetic acid
Authentic retinoids: all-*trans*-retinoic acid, 13-*cis*-retinoic acid (isotretinoin, Ro 4-3780), 4-oxoisotretinoin (Ro 22-6595), acitretin (etretin, Ro 10-1670), 13-*cis*-acitretin (Ro 13-7652), and etretinate (Ro 10-9359)

Preparation of Retinoid Solutions

Working solutions of reference retinoids and internal standards at a concentration of 1 – 2 μg/ml are prepared in methanol and stored in 200-μl plastic vials in the dark at $-70°$. all-*trans*-Retinol, retinoic acid, and 13-*cis*-retinoic acid may be purchased from Sigma Chemical Company (Poole, UK). Other retinoids are usually available on request from Hoffmann-La Roche (Basel, Switzerland, or Nutley, NJ). Retinoid concentrations are determined spectrophotometrically using the following absorptivities ($E_{1\%}^{1cm}$, in ethanol): retinol, 1830 at 325 nm; 3,4-didehydroretinol, 1454 at 350 nm; TMMP-retinol, 1596 at 325 nm; retinoic acid, 1455 at 350 nm; isotretinoin, 1290 at 357 nm; acitretin, 1270 at 361 nm; 13-*cis*-acitretin, 1241 at 361 nm; etretinate, 1256 at 362 nm; and 4-oxoisotretinoin, 1242 at 361 nm.

Chromatography

High-performance liquid chromatography (HPLC) with ultraviolet (UV) detection is a convenient, sensitive, and specific method for the identification of a wide range of closely related retinoids in biological samples. A standard HPLC apparatus is generally satisfactory, although the low concentrations of endogenous vitamin A present in human skin tissues require a UV monitor capable of detecting picomole quantities of retinoids.

Our present equipment includes a Waters M-45 solvent pump (Waters Associates Inc., Milford, MA), a Rheodyne Model 7125 syringe-loading injection valve fitted with an external 50-μl sample loop (Rheodyne Inc., Berkeley, CA), and two UV detectors (UV Monitor D, LDC/Milton Roy, Riviera Beach, Fl) connected in series and operating at fixed wavelengths 326 and 360 nm, respectively. Two dual-pen recorders (Hitachi, Ltd., Tokyo, Japan) set between 1 and 10 mV, respectively, allow for a wide range of optimal detection (0.02–0.0005 full-scale absorbency). A Guard-Pak precolumn module (Waters) serves to extend column life.

Skin Samples

Human skin samples are obtained by cutting parallel to the surface of anesthetized skin with a razor blade (shave biopsy) or by the use of a Castroviejo keratotome set between 0.1 and 0.5 mm, depending on the nature of the skin under study. The biopsies, which contain epidermis and a minor proportion of papillary dermis, can be used as such. If desirable, epidermis and dermis can also be studied separately simply by heating the skin sample for 45 sec at 54° and gently peeling the epidermis from the dermis with fine curved forceps. The size of the biopsy should be approximately 4 cm^2 or 20–50 mg wet weight.

Analysis: Step 1 (Neutral Retinoids)

Hydrolysis and Extraction

Skin samples require hydrolysis in order to extract the retinoids quantitatively. Although hydrolysis leads to saponifying the esters and inevitably producing some artifactual cis–trans isomerizations of the retinoids, this technique is recommended when the level of total retinoid is the objective without taking into consideration the conjugated state or the geometric configuration of the molecules. Because several compounds (e.g., retinol and retinyl esters) appear in a common HPLC peak after hydrolysis, the detection limit of the assay is improved.

Procedure. All work should be performed in a hood under dim yellow or red light. A weighed tissue sample (20–50 mg of epidermis or dermis, 5–10 mg subcutis) or serum/plasma (25–100 μl) is added to a Teflon-lined screw-capped, round-bottomed glass test tube (16 × 95 mm) containing 0.5 ml ethanol, 0.1 ml of 14 *M* potassium hydroxide, 10 μl BHT as antioxidant, and 7.5 ng of internal standard, TMMP-retinol. The test tube is flushed with nitrogen, sealed, and heated for 20 min at 80° on a block heater. During hydrolysis the mixture is gently agitated a few times and then cooled in ice water for 5 min. Extraction is performed after adding 0.5 ml water and 4 ml hexane to the hydrolyzate. The tubes are flushed with nitrogen, sealed, and vigorously mixed for 3 min at room temperature. Phase separation is accelerated by centrifugation for 3 min at 1000 *g*. The hexane layer is collected with a Pasteur pipette and the extraction procedure repeated with another 3-ml portion of hexane. The hexane layers are pooled and evaporated under nitrogen at 40°. To collect any residues of retinoids left dry on the glass surfaces, the sides of vials are rinsed with 250 μl of chloroform using a Hamilton microsyringe. After evaporation, redissolution in 60 μl of methanol, ultrasonication for 2 min, and centrifugation for 5 min at 1000 *g*, the sample is subjected to HPLC analysis. The aqueous phase is used for subsequent analysis of acidic retinoids or is retained for determination of the protein content using the biuret technique (see Optional Methods).

Chromatography of Neutral Retinoids

Inject 25–50 μl of a methanolic solution of the retinoids. Separation is effected on a reversed-phase octadecylsilane (ODS) column at room temperature with a mixture of 12–22% water in acetonitrile as the mobile phase. Using a Nucleosil 5-μm C_{18} column (4.6 mm i.d. × 20 cm, Macherey-Nagel & Co., Düren, FRG) eluted with water/acetonitrile (18:82, v/v) at a flow rate of 1.2 ml/min, the retention times of the internal standard (TMMP-retinol), 3,4-didehydroretinol, and all-*trans*-retinol are about 6.9, 9.2, and 12.2 min, respectively (Fig. 1). Peak identifications are based on the retention times and absorbency ratios at 326/360 nm using authentic retinoids as references. To confirm the identity of the retinoids, the peak fractions corresponding to retinol and 3,4-didehydroretinol may be collected for further chromatography using different eluents or a normal-phase HPLC system. Another convenient method for identification of retinol and 3,4-didehydroretinol is by anhydro derivatization and subsequent rechromatography.[2,3]

[2] P. E. Dunagin and J. A. Olson, this series, Vol. 15, p. 289.
[3] H. Törmä and A. Vahlquist, this volume [22].

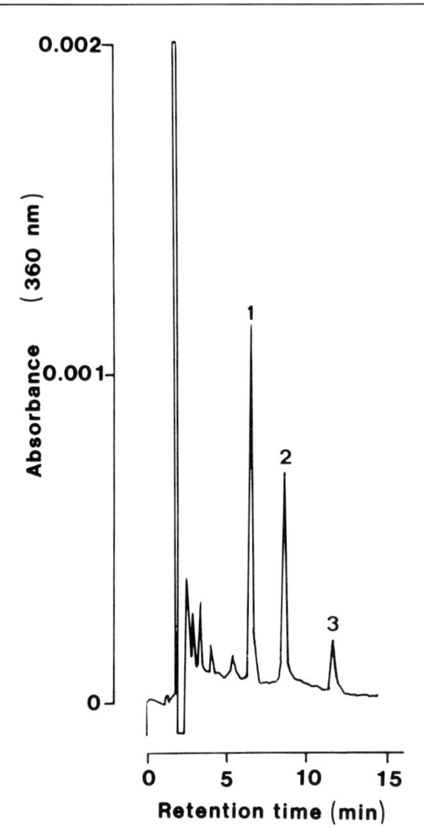

FIG. 1. Chromatogram of neutral (endogenous) retinoids extracted from a hydrolyzed skin sample from a patient suffering from ichthyosis. Column: Nucleosil ODS (5 μm); mobile phase: water/acetonitrile (18:82, v/v); flow rate: 1.2 ml/min; detection: 360 nm. Peak 1, TMMP-retinol; peak 2, 3,4-didehydroretinol; peak 3, all-*trans*-retinol.

Alternatively, an isocratic normal-phase HPLC system can be applied for separation of retinoid alcohols.[4] A Nucleosil silica 5-μm column (4.6 mm i.d. $\times$ 15 cm) eluted with hexane/ethyl acetate/ethanol (84.5:15:0.5, v/v/v) at a flow rate of 1.2 ml/min will retain retinol, 3,4-didehydroretinol, and TMMP-retinol for 7.5, 8.3, and 10.4 min, respectively. A major disadvantage, however, in using a hexane-based system is its tendency to create irreproducible elution times owing to variable hydration of the silica gel.

[4] O. Rollman and A. Vahlquist, *J. Invest. Dermatol.* **86,** 384 (1986).

Calibration and Quantification

Quantification of endogenous 3,4-didehydroretinol and all-*trans*-retinol can be achieved by relating the peak heights (or areas) of 3,4-didehydroretinol and all-*trans*-retinol to that of the internal standard (TMMP-retinol). These three compounds are only minimally isomerized as a result of the alkaline hydrolysis. By adding standard amounts of 3,4-didehydroretinol and all-*trans*-retinol, linear relationships between the peak height and mass ratios can be established (Fig. 2).

The precision of the method is determined by repeated analysis of 3,4-didehydroretinol and all-*trans*-retinol from a single epidermal specimen. In our laboratory the coefficient of variation (within-run) for the analysis of these compounds is 12 and 9%, respectively.[1]

Analysis: Step 2 (Acidic Retinoids)

Extraction

Many of the synthetic retinoids used in current clinical practice (retinoic acid, isotretinoin, and acitretin) are acidic compounds possessing a free carboxylic acid moiety. The pK_a is 6.0 for retinoic acid and 6.4 for acitretin;[1] hence, they are ionized at basic pH and maintained in the aqueous phase following extraction of the alkaline hydrolyzate. This will

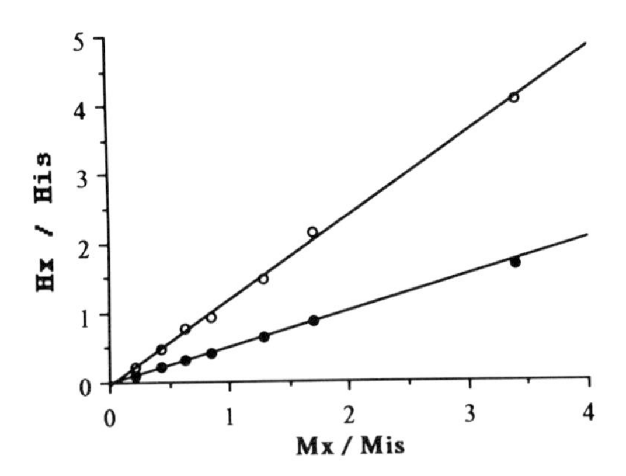

FIG. 2. Calibration curves for retinol (●) and 3,4-didehydroretinol (○). Mx/Mis denotes the mass ratio of the retinoid of interest (x) relative to the internal standard (is). Peak heights (at 360 nm) of each retinoid of interest (Hx) and the internal standard (His).

allow them to be recovered separately after acidification of the hydrolysis mixture.

Internal Standard. Depending on the amount of acidic retinoid expected to be present in the sample, 5 to 25 ng of the internal standard is added prior to the hydrolysis (see under Step 1). An all-purpose internal standard has not yet been established. Retinoic acid is the preferred standard for quantification of acitretin (or its metabolic precursor etretinate) and its main metabolite 13-*cis*-acitretin, whereas acitretin is a useful standard for quantitation of isotretinoin and retinoic acid. Both standard compounds have chemical and chromatographic characteristics similar to those of the retinoids of interest. The extremely low concentrations of endogenous retinoic acid present in human skin are hardly detectable by the assay and hence negligible in pharmacological studies of retinoids.

There are exceptional instances when these internal standards are not applicable. Obviously, when analyzing isotretinoin in a sample from a patient previously treated with etretinate, acitretin should not be selected as internal standard. Similarly, retinoic acid should not be used as standard when determining acitretin or etretinate in a patient who has recently received isotretinoin therapy. In both instances, 13-demethylretinoic acid may be used as an alternative internal standard.

Procedure. Any organic residues on top of the aqueous phase remaining after extraction of neutral retinoids are carefully aspirated and discarded. To extract the acidic retinoids the pH of the aqueous phase is reduced to 4–5 by adding appropriate amounts of 5 *M* HCl. Final adjustment to the desired pH range is obtained by dropwise addition of 1 *M HCl or 1 M* NaOH. Water is added to give a final volume of 500 μl, and then 4 ml of hexane is added. The tubes are flushed with nitrogen, sealed, and thoroughly shaken for 5 min at room temperature. Following centrifugation for 3 min at 1000 *g*, the hexane layer is transferred to a screw-capped conical test tube (16 $\times$ 95 mm) to allow small sample isolation. The hexane is evaporated and the sample prepared for HPLC as described above for neutral retinoids. The aqueous phase is retained for determination of the protein content (see Optional Methods).

Recovery of the acidic retinoids (72–83%) using hexane/ethanol extraction is lower than for the neutral retinoids, but repeated hexane extractions do not significantly increase the recovery. Using an extraction mixture of methyl *tert*-butyl ether/methanol, the extraction efficiency of 4-oxoisotretinoin (the main polar metabolite of isotretinoin in humans) increases by 10%, but the recovery of retinoic acid, isotretinoin, and acitretin actually decreases by 15–25%. Furthermore, the ether will extract UV-absorbing impurities more efficiently than hexane, hence affecting the subsequent HPLC analysis.

Chromatography of Acidic Retinoids

A single isocratic reversed-phase HPLC procedure allows complete resolution of etretinate, acitretin, 13-*cis*-acitretin, retinoic acid, and isotretinoin. The HPLC equipment and stationary phase recommended for separation of the neutral retinoids is also applicable for the acidic ones. Precolumn purification may not be necessary.

Procedure. Inject an 25- to 50-μl aliquot of the clear methanol extract onto a reversed-phase ODS column. A Nucleosil C_{18} column (4.6 mm i.d. $\times$ 20 cm) running isocratically at 1.2 ml/min with water/acetonitrile/acetic acid (18:82:0.05, v/v/v) will elute the retinoids at the following times: 13-*cis*-acitretin, 5.4 min; acitretin, 6.1 min; isotretinoin, 8.2 min; all-*trans*-retinoic acid, 10.0 min; and etretinate, 17.6 min. Peak identification is made by comparing absorbancy ratios at the fixed wavelengths 326 and 360 nm with the values obtained using pure retinoids. A drawback of using isocratic reversed-phased HPLC when analyzing isotretinoin is the long retention time of the parent drug as compared to its more polar metabolite 4-oxoisotretinoin. Their analysis in a single run is best achieved by gradient elution on reversed-phase[5] or normal-phase HPLC.[4]

Figure 3 shows a typical HPLC profile of an acidic skin extract obtained from a patient receiving oral etretinate therapy. The parent drug ester is completely converted to its corresponding free acid acitretin during the saponification. It is apparent from the chromatogram that considerable *in vitro* cis isomerization (usually 15–30%) of the internal standard, retinoic acid, occurs. A similar degree of cis isomerization takes place when acitretin is used as internal standard in hydrolyzed specimens. These artifactual isomerizations are not always proportional to those of the retinoids of interest, and they occur in addition to the metabolic isomerizations *in vivo.* Because of these complex interconversions, quantifications of geometric retinoid isomers should be performed with caution.

Calibrations and Quantification. The total retinoid can be determined from the hydrolyzed extract taking into account the major artifactual isomerizations that occur *in vitro* at C-13 position. Thus, quantification of acitretin or etretinate (subsequently converted to acitretin by the alkaline hydrolysis) from the chromatogram is accomplished by relating the peak heights (or areas) of acitretin (sum of the 13-*cis*- and all-*trans*-acitretin peaks) to those of the internal standard (13-*cis*- plus all-*trans*-retinoic acid). Likewise, the total amount of isotretinoin or retinoic acid present in the sample can be calculated from the peak sizes of 13-*cis*- plus all-*trans*-retinoic acid relative to those of the internal standard (all-*trans*-acitretin plus

[5] F. M. Vane, J. K. Stoltenberg, and C. J. L. Buggé, *J. Chromatogr.* **227,** 471 (1982).

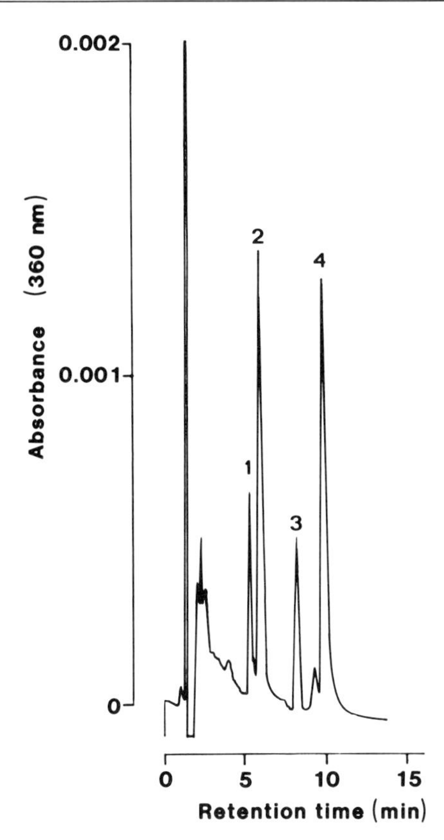

FIG. 3. Chromatogram of acidic retinoids extracted from a hydrolyzed epidermal sample from a patient treated with oral acitretin. Column: Nucleosil ODS (5 μm); mobile phase: water/acetonitrile/acetic acid (18:82:0.05, v/v/v); flow rate: 1.2 ml/min; detection: 360 nm. Peak 1, 13-*cis*-acitretin; peak 2, acitretin; peak 3, 13-*cis*-retinoic acid; peak 4, all-*trans*-retinoic acid (internal standard).

13-*cis*-acitretin). By this means good linearity of the calibration plots are achieved (Fig. 4). Alterations of the hydrolysis conditions with respect to potassium hydroxide concentration (0.2–80%), temperature (4–90°), and time (0–60 min) result in minimal changes in the recovery of retinoids (serum spiked with acitretin or retinoic acid). However, the cis–trans isomerization varies with the hydrolysis conditions used, being highest at low temperature and long incubation time (results not shown).

The proportion of 13-cis and all-trans isomers present in the original sample can be determined from the direct extracting procedure, which produces only minor isomerizations (see below). By comparing the relative

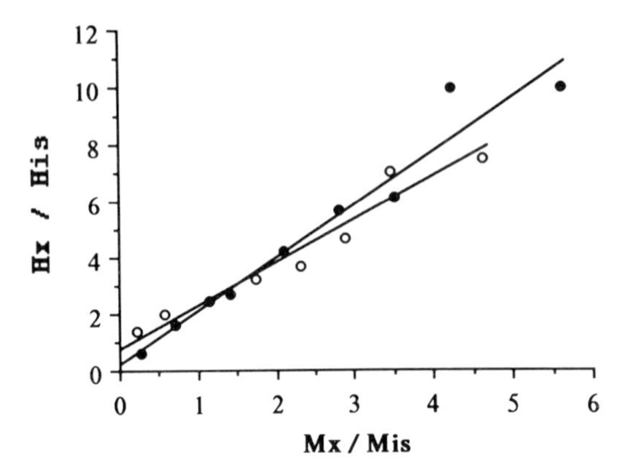

FIG. 4. Calibration curves for acitretin (●) and retinoic acid (○). Various amounts of the retinoids were added to a constant amount of the internal standard and specimens of retinoid-depleted skin. Peak heights (at 360 nm) of all-trans and 13-cis isomers are summed for each retinoid of interest (Hx) and the internal standard (His), respectively.

absorbancies at the actual detection wavelength, the percentages of the geometric isomers can be established from the chromatogram from their peak size ratios. Knowing the total amount of the retinoid from the hydrolyzed sample, the absolute amounts of the all-trans and 13-cis-isomers, respectively, may then easily be calculated. At a signal-to-noise ratio of 3 the lower limit of detection is 10–20 ng retinoid/g wet tissue of skin and 5–10 ng retinoid/ml of serum, using ordinary sample sizes. Coefficients of variation of the analysis of acitretin are 8.6% (within-run) and 10.2% (between-run), whereas the corresponding values for 13-*cis*-retinoic acid are 7.7 and 4.3%, respectively.

Optional Methods

Semiquantitative Analysis of Retinoids by Direct Extraction

For measurements of retinoid conjugates and isomers, a nonhydrolytic sample preparation method should be used. The sample (30–50 mg of minced skin or 100–200 μl of plasma) is homogenized in the presence of 750 μl water and 250 μl ethanol. The pH of the skin homogenate is adjusted to 4–5 by adding appropriate amounts (usually about 50 μl) of diluted acetic acid, pH 2. The appropriate internal standard and 4 ml of hexane are then added. After vigorous shaking and a brief centrifugation, the upper phase (containing neutral *and* acidic retinoids) is transferred to a conical test tube and evaporated to dryness under nitrogen at 40°. Methanol (60 μl) is added, the retinoids are redissolved by ultrasonication, and the sample is centrifuged at 1000 *g* for 5 min. An aliquot of the supernatant is then subjected to HPLC as described below. In order to avoid comigration of all-*trans*-retinoic acid and 13-*cis*-retinol, the proportion of water of the eluent should preferably be in the range of 20–24% by volume.

The use of nonhydrolytic conditions during homogenization is less destructive to the retinoids and causes only minor cis–trans isomerations at the C-13 position of retinoic acid (2.8–4.8%), isotretinoin (0.8–2.9%), and acitretin (0.4–1.0%). Hence, determination of geometric isomers of retinoids is preferably achieved by omitting sample saponification. The nonhydrolytic technique should be used when it is desirable to distinguish etretinate from its main metabolite (and hydrolysis product), acitretin. Similarly, the vitamin A esters can be separated from retinol and 3,4-didehydroretinol. The recovery of these retinoids is, however, below 70% (A. Vahlquist, 1989, unpublished observation).

Protein Analysis

It is recommended that the retinoid concentration be expressed both on a wet weight and on a protein weight basis. Many of the commercially available methods for protein analysis are unreliable when the sample has been exposed to high concentrations of KOH. The biuret technique is less influenced by extreme pH values and, although less sensitive than more modern techniques, is usually sufficient for the amount of protein (0.1–2.0 mg) under study. Using this method, hydrolysis (80°, 20 min) will only reduce the protein value by 15–20% as compared to a nonhydrolyzed sample. The protein content is determined in duplicate. Briefly, the aqueous phase of a hydrolyzed skin sample is centrifuged at 1200 *g* for 5 min to remove debris, then 0.1–0.25 ml of the clear hydrolyzate is mixed with 2 ml biuret reagent. The sample is allowed to stand for 30 min, filtered through a 0.2-μm filter (Dynagard, Microgon Inc., Laguna Hills, CA), and the absorption of the complex measured within 5 min in a spectrophotometer set at 550 nm. The extinction value is read against a standard curve produced with bovine serum albumin.

Comments

Following the introduction in 1979 of an HPLC technique for the analysis of retinoids in the skin, several thousand biopsies have been analyzed in our laboratory. Some of the most intriguing findings have been: (1) low retinol concentrations in certain skin tumors and in normal

[6] O. Rollman and A. Vahlquist, *Br. J. Dermatol.* **109**, 439 (1983).

skin after UV irradiation,[7,8] (2) high retinol values in chronic renal failure and after isotretinoin therapy,[4,9] (3) extremely low 3,4-didehydroretinol levels after isotretinoin therapy,[3] and (4) strikingly high 3,4-didehydroretinol levels in several keratinizing disorders (psoriasis, keratosis follicularis, ichthyosis) and in certain skin tumors (basal cell carcinoma, keratoacanthoma, and squamous cell carcinoma).[7,10-13] In addition, drug monitoring in skin biopsies obtained from patients undergoing chronic etretinate treatment has helped to identify a storage compartment in subcutaneous fat responsible for the exceedingly long biological half-life (about 100 days) of the drug.[6,14]

The technique described may appear somewhat laborious, but the advantages of separate extraction (phase distribution) of neutral and acidic retinoids are numerous: the problem of comigration of all-*trans*-retinoic acid and 13-*cis*-retinol in subsequent HPLC analysis is avoided, the appearance of a retinoid in one or the other of the extracts helps to identify the compound, and the alkaline extract contains very few contaminants that interfere with HPLC detection of neutral retinoids. Generally, more UV-absorbing components appear in acidic extracts. In practice, however, most UV-absorbing contaminants tend to appear shortly after the void volume during reversed-phase HPLC and do not interfere significantly with the peaks of interest.

Acknowledgments

We thank Dr. Ed J. Wood for kindly revising the English text. Supported by the Swedish Medical Research Council (Project No. 03x-07133) and the Edvard Welander Foundation.

[7] O. Rollman and A. Vahlquist, *Arch. Dermatol. Res.* **270,** 193 (1981).
[8] B. Berne, M. Nilsson, and A. Vahlquist, *J. Invest. Dermatol.* **83,** 401 (1984).
[9] A. Vahlquist, B. Berne, and C. Berne, *Eur. J. Clin. Invest.* **12,** 63 (1982).
[10] O. Rollman and A. Vahlquist, *Arch. Dermatol. Res.* **278,** 17 (1985).
[11] A. Vahlquist, J. B. Lee, and G. Michaëlsson, *Arch. Dermatol.* **118,** 389 (1982).
[12] A. Vahlquist, H. Törmä, O. Rollman, and B. Berne, *in* "Retinoids: New Trends in Research and Therapy" (J.H. Saurat, ed.), p. 159, Karger, Basel, 1985.
[13] O. Rollman and A. Vahlquist, *Br. J. Dermatol.* **113,** 405 (1985).
[14] A. Vahlquist, O. Rollman, and I. Pihl-Lundin, *Acta Derm. Venereol.* **66,** 431 (1986).

[19] Retinoids and Rheumatoid Arthritis: Modulation of Extracellular Matrix by Controlling Expression of Collagenase

By CONSTANCE E. BRINCKERHOFF

Introduction

Collagenase is a member of a gene family of metalloproteinases, enzymes that are active at neutral pH, have Zn^{2+} as an integral component, and are secreted in inactive latent forms.[1-3] Collagenase is the only enzyme active at neutral pH that can degrade collagen types I, II, and III, the most abundant proteins in the body. A related metalloproteinase (stromelysin) degrades the noncollagenous components of the extracellular matrix, namely, laminin, fibronectin, proteoglycans, and together the two enzymes can digest nearly all components of the extracellular matrix.[1-3] Despite the fact that expression of these enzymes is often coordinate, this chapter focuses on collagenase because, in contrast to stromelysin, an assay that is singularly specific for collagenase has allowed us to monitor accurately the activity of this enzyme.

Collagenase and stromelysin are produced in excessive amounts by fibroblastlike cells that comprise the membrane (synovium) that lines the joints of patients with rheumatoid arthritis.[3] Rheumatoid arthritis is a chronic disease in which there is progressive destruction of the joints, with subsequent deformity and crippling, owing to the fact that the synovium proliferates into a huge mass of tissue which invades and destroys the extracellular matrix. Because rheumatoid synovial cells produce such an overabundance of collagenase and stromelysin, these cells provide an excellent system for studying mechanisms controlling the expression of metalloproteinases. Furthermore, because of the potential of the enzymes for destroying connective tissue, there is interest in compounds that inhibit either their activity or synthesis. The activity of metalloproteinases can be inhibited by at least two naturally occurring compounds, namely, the serum component α_2-macroglobulin and the tissue inhibitor of metalloproteinases (TIMP),[1,2] as well as by synthetic inhibitors.[4] These inhibitors

[1] D. E. Woolley and J. M. Evanson (eds.), "Collagenase in Normal and Pathological Connective Tissues." Wiley, New York, 1980.
[2] E. D. Harris, Jr., H. G. Welgus, and S. M. Krane, *Collagen Relat. Res.* **4,** 493 (1984).
[3] E. D. Harris, Jr., *in* "Textbook of Rheumatology" (W. N. Kelley, E. D. Harris, Jr., S. Ruddy, and C. B. Sledge, eds.), p. 886. Saunders, Philadelphia, Pennsylvania, 1985.
[4] M. S. Stack and R. D. Gray, *J. Biol. Chem.* **264,** 4277 (1989).

complex with active enzyme, binding in such a way that proteolytic activity is blocked.

This chapter discusses two classes of compounds that inhibit the synthesis of collagenase: the glucocorticoid hormones and the vitamin A analogs, that is, retinoids. Glucocorticoids have been the prototype compounds,[5] and evidence suggests that they decrease collagenase synthesis at the level of transcription.[6,7] We have observed that all-*trans*-retinoic acid and the synthetic retinoids could inhibit collagenase synthesis as successfully as the glucocorticoids.[8-10] This has prompted studies to determine the mechanism(s) by which retinoids act. Research has followed two main lines: (1) *in vitro* studies with cultures of synovial fibroblasts on the basic mechanisms of retinoid action and (2) *in vivo* studies on the effects of administering oral retinoids to animals with experimental arthritis.

Cell Cultures

Many of the *in vitro* studies have used monolayer cultures of rabbit synovial fibroblasts.[6,7,9,10] Fresh cultures are established from the knee synovium of young (4- to 6-week-old) New Zealand White rabbits. The animals are sacrificed by intravenous injection of 1 ml of T-61 (American Hoechst Corp., Somerville, NJ). The knees are bathed in 70% (v/v) alcohol, the skin and subcutaneous layers are reflected back, and the knee joint is entered with sterile scissors and forceps. Tissue (synovium) adhering to the articular cartilage is removed to a tube containing Hanks' balanced salt solution (HBSS), as is the fat pad and the meniscus. Then, in a tissue culture hood, the tissue is dissected away from the meniscus and the fat pad and placed in 4 ml of crude clostridial bacterial collagenase (4 mg/ml in HBSS; Worthington Scientific, Freehold, NJ). The synovial tissue derived from four knees is usually combined in a single tube of bacterial collagenase. The tube is placed at 37° and shaken vigorously every 15–20 min. After 2–3 h of incubation, when the tissue is dissociated thoroughly, the mixture is filtered through sterile gauze, centrifuged, and resuspended

[5] J.-M. Dayer, S. M. Krane, R. G. G. Russell, and D. R. Robinson, *Proc. Natl. Acad. Sci. U.S.A.* **73,** 945 (1976).

[6] C. E. Brinckerhoff, I. M. Plucinska, L. A. Sheldon, and G. T. O'Connor, *Biochemistry* **25,** 6378 (1986).

[7] M. E. Fini, M. J. Karmilowicz, P. L. Ruby, K. A. Borges, and C. E. Brinckerhoff, *Arthritis Rheum.* **30,** 1255 (1987).

[8] C. E. Brinckerhoff, R. M. McMillan, J.-M. Dayer, and E. D. Harris, Jr., *N. Engl. J. Med.* **303,** 432 (1980).

[9] C. E. Brinckerhoff, C. A. Vater, and E. D. Harris, Jr., *in* "Cellular Interactions" (J. T. Dingle and J. L. Gordon, eds.), p. 215. Elsevier/North-Holland, New York, 1981.

[10] C. E. Brinckerhoff and E. D. Harris, Jr., *Biochim. Biophys. Acta* **677,** 424 (1981).

in Dulbecco's modified Eagle's medium (DMEM) with 20% fetal calf serum (FCS) and then plated into four 60-mm-diameter culture dishes. After 2–3 days at 37° and 5% CO_2, the plates are usually confluent. The contents are trypsinized, and the cells from each dish are passed 1:1 (passage 1) into a 100-mm-diameter culture dish containing DMEM–10% FCS. Upon confluence, the cultures are trypsinized and passaged 1:2 in 100-mm-diameter culture dishes. For experiments, rabbit cells are usually used between passages 5 and 8.

For studies with human cells, either rheumatoid synovial fibroblast or skin fibroblasts are used. Synovial cells may be established as primary cultures from rheumatoid tissue obtained at the time of synovectomy.[5] These cultures are obtained by sequential digestion of the minced rheumatoid synovium with bacterial collagenase and trypsin, with subsequent plating of the single cell suspension, essentially as described above for the rabbit fibroblasts. However, these primary cultures of rheumatoid cells contain a mixture of cell types (T and B cells, polymorphonuclear leukocytes, monocytes/macrophages, and the fibroblastlike synovial cells) that allows them to secrete large amounts of metalloproteinases "spontaneously."[5] When these cells are passaged, however, only the fibroblastlike synovial cells are carried forward, and, in the absence of cytokines secreted by the monocytes and T and B cells, the passaged cells resemble cultures of normal fibroblasts. They do not secrete collagenase and stromelysin unless stimulated experimentally. Because these cells are often derived from elderly patients in whom the number of cell doublings remaining is limited, they may grow slowly, and there may not be many cells available for experimental manipulations.

Consequently, we have used human skin fibroblasts as a convenient source of human cells. The collagenase produced by these fibroblasts is identical to that produced by synovial or gingival fibroblasts.[11–13] Cultures of these cells can be derived from foreskins of newborns, using the technique of dissociation with bacterial collagenase. We use these cells between passages 5 and 12. They can probably be used at later passages; as responsiveness may change with passage, however, it is important to document which passage is used for experiments.

Cultures of rabbit synovial fibroblasts (or human skin fibroblasts) have several advantages. They are readily available, grow well, and, perhaps

[11] G. L. Goldberg, S. M. Wilhelm, E. A. Kronberger, E. A. Bauer, G. A. Grant, and A. Z. Eisen, *J. Biol. Chem.* **261,** 6000 (1986).

[12] S. E. Whitham, G. Murphy, P. Angel, H.-J. Ramsdorf, B. J. Smith, A. Lyons, T. J. Harris, J. J. Reynolds, K. P. Herrlich, and A. P. Docherty, *Biochem. J.* **240,** 913 (1986).

[13] C. E. Brinckerhoff, P. L. Ruby, S. D. Austin, M. E. Fini, and H. D. White, *J. Clin. Invest.* **79,** 542 (1987).

most importantly, are amenable to experimental manipulations. Unlike the cultures of primary rheumatoid synovial tissue, these cultures do not spontaneously produce collagenase. However, they can be stimulated with a variety of agents, for example, interleukin 1 (IL-1; 1 – 10 ng/ml), phorbol myristate acetate (PMA; 10 ng/ml; 10^{-8} *M*), or crystals of monosodium urate monohydrate (250 μg/ml), to secrete large quantities of collagenase.[1,2,6,7,10,14] Induction of collagenase is mediated at a transcriptional level: mRNA is detected within the cells by 4 h, and protein is found in the culture medium shortly thereafter.[14] The enzyme is not stored within the cell: on synthesis it is rapidly exported to the extracellular space. Indeed, pulse – chase studies with ^{3}H-labeled collagenase have shown that the time required for synthesis and secretion of the enzyme is approximately 40 min.[15]

In Vitro Studies

In our studies, we often treat cells with glucocorticoids so that we have a basis for comparison with retinoids.[6-10] To measure suppression of collagenase synthesis by retinoids or glucocorticoids, it is important to consider both the concentration of retinoid (or steroid) that is used and the time frame over which the action of these compounds is measured. Concentrations exceeding 10^{-6} *M* of either compound in tissue culture do not represent pharmacologic/physiological doses achieved *in vivo.* Thus, results obtained with high concentrations may well be physiologically irrelevant and, perhaps, due to nonspecific toxicity. Dexamethasone (Sigma Chemical Co., St. Louis, MO) is the glucocorticoid usually used. A stock solution of 10^{-2} *M* prepared in 95% ethanol and stored at 4° is stable for months. Just prior to use, the stock is diluted to a final concentration of 10^{-7} *M* (Fig. 1).[8]

To study retinoids, we have used all-*trans*-retinoic acid as a model. It is inexpensive and commercially available (Sigma). A stock of 10^{-3} *M* is prepared weekly in dimethyl sulfoxide (DMSO) and stored in the dark at 4°. It is diluted 1 : 100 into an aqueous solution (more concentrated dilutions are not soluble) and then into the cultures for a final concentration of 10^{-6} or 10^{-7} *M* (Fig. 1a).[8] These concentrations are effective in serum-free medium, namely, DMEM supplemented with 0.2% (w/v) lactalbumin hydrolyzate (LH) as a source of additional protein. Higher concentrations are cytotoxic. If serum-containing medium is used, then 10^{-5} *M* is not toxic

[14] R. H. Gross, L. A. Sheldon, C. F. Fletcher, and C. E. Brinckerhoff, *Proc. Natl. Acad. Sci. U.S.A.* **81,** 1981 (1984).

[15] H. Nagase, C. E. Brinckerhoff, C. A. Vater, and E. D. Harris, Jr., *Biochem. J.* **214,** 281 (1983).

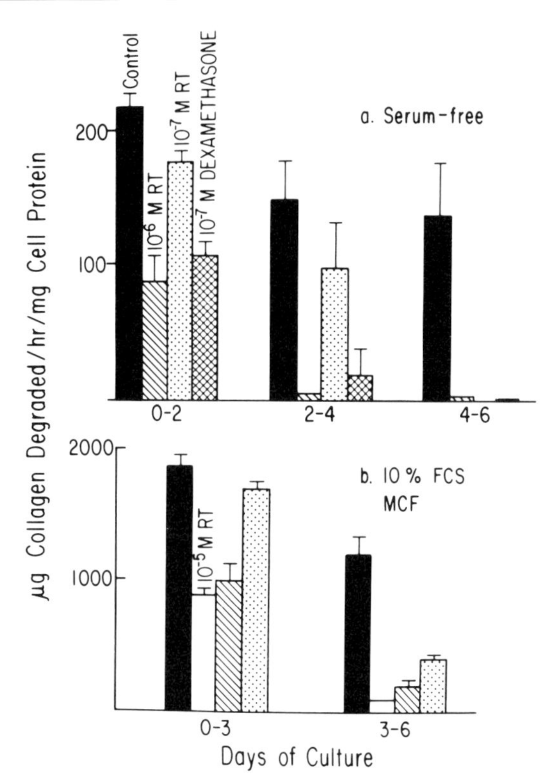

FIG. 1. Effect of all-*trans*-retinoic acid on collagenase production by rheumatoid synovial cells. Primary cultures of human rheumatoid synovial cells were established in 35-mm-diameter culture dishes from tissue received from the operating room of Mary Hitchcock Memorial Hospital (Hanover, NH). After 2–5 days in culture when the cells were confluent, selected cultures were washed 3 times in HBSS to remove traces of serum and then transferred to serum-free medium (DMEM–LH) with various concentrations of all-*trans*-retinoic acid (RT) or dexamethasone. Other cultures received fresh medium containing 10% fetal calf serum with partially purified mononuclear cell factor all-*trans*-retinoic acid, [interleukin 1(IL-1)] or dexamethasone. Serum-free medium and drugs were replaced at Days 2 and 4 (a); serum-containing medium, IL-1, and drugs were replaced on Day 3 (b). The culture medium was assayed for collagenase activity after activation of latent collagenase with trypsin. The total protein content in monolayers was determined [O. H. Lowry, N. J. Rosebrough, A. L. Farr, and R. J. Randall, *J. Biol. Chem.* **193,** 265 (1951)], with the collagenase activity being expressed as micrograms collagen degraded per hour per milligram cell protein at 37°. (From Ref. 8, with permission.)

(Fig. 1b), probably because some of the free retinoic acid is complexed with serum retinoic acid-binding protein, thereby reducing the effective concentration of the retinoid.[16]

[16] M. B. Sporn, A. B. Roberts, and D. S. Goodman (eds.), "The Retinoids," Vols. 1 and 2. Academic Press, New York, 1984.

Experiments to test the effects of glucocorticoids and retinoids on metalloproteinase synthesis have been done in two ways. First, induction of the enzymes may be antagonized by simultaneous addition of an inducer along with the glucocorticoid or retinoid.[8,9] Alternatively, these suppressors can be added to cells that are already induced, in a "turn-off" experiment (Figs. 1 and 2).[6-10] These turnoff experiments can be performed on primary cultures of human rheumatoid cells or on cultures of rabbit or human fibroblasts that have been experimentally stimulated to

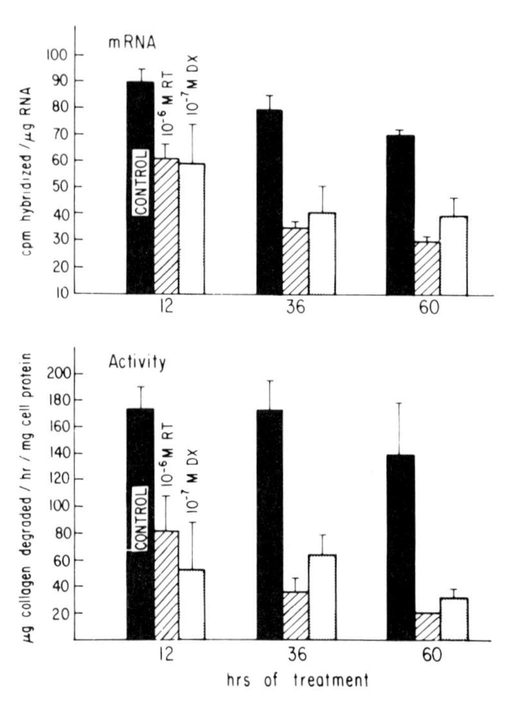

FIG. 2. Effect of all-*trans*-retinoic acid and dexamethasone on collagenase mRNA and protein levels in rabbit synovial fibroblasts. Confluent cultures of rabbit synovial fibroblasts in serum-free medium were stimulated to produce collagenase by 48 hr of treatment with PMA (10^{-8} *M*). PMA was removed from all cultures, and medium was replaced; selected cultures received all-*trans*-retinoic acid (RT, 10^{-6} *M*) or dexamethasone (DX, 10^{-7} *M*). At intervals, cultures were terminated by harvesting the medium and RNA, and medium and drugs were replaced in remaining cultures. Collagenase mRNA was measured by dot-blot analysis, and 2 μg of RNA was spotted onto nitrocellulose filters. The filters were hybridized with a cDNA clone for collagenase, radiolabeled by nick translation with ^{32}P (T. Maniatis, E. F. Fritsch, and J. Sambrook, "Molecular Cloning: A Laboratory Manual." Cold Spring Harbor Laboratory, Cold Spring Harbor, New York, 1982). The amount of hybridization was quantified by excising the dots and counting them in PCS liquid scintillation fluid. Collagenase activity was determined in a standard fibril assay with ^{14}C-labeled collagen. (From Ref. 6 with permission.)

produce large quantities of collagenase. These latter experiments are important because they mimic the primary cultures of rheumatoid cells where enzyme production is spontaneous.

The protocol is as follows. Primary cultures of rheumatoid synovial cells are allowed to become established and are used 2–5 days after plating when they are nearly confluent. The medium is replaced and drugs are added. If cultures of rabbit or human fibroblasts are used, confluent monolayers are treated for 48 hr with an inducer such as PMA (10^{-8} *M*) or IL-1 (10 ng/ml) to induce large quantities of collagenase mRNA and protein. The inducer is then removed, and all cultures are given fresh culture medium together with a retinoid or glucocorticoid. With both primary rheumatoid cells and stimulated fibroblasts, after 24–48 hr the medium and drugs are replaced for an additional 24–48 hr, when the procedure is again repeated.

The medium can then be assayed for collagenase activity with a ^{14}C-labeled collagen fibril assay.[17,18] The latent collagenase must be activated by treatment with *N*-tosyl-L-phenylalanine chloromethyl ketone (TPCK)–trypsin for 30 min at room temperature, followed by addition of a 4-fold excess of soybean trypsin inhibitor. For serum-free culture medium, the final concentration of trypsin needed to activate latent collagenase is 10 μg/ml, whereas for medium containing 10% serum, a concentration of 150–200 μg/ml trypsin is required. In the presence of serum, the higher concentration of trypsin is needed to first saturate the active sites of α_2-macroglobulin, thus permitting trypsin-activated collagenase to remain uncomplexed in serum.[1,2] Alternatively, latent collagenase in serum can be activated by organomercurial compounds, such as aminophenylmercuric acetate (APMA) at a final concentration of 10 m*M*.[19] The APMA is added to culture medium and incubated at 37° for 1 hr to maximally activate the latent enzyme, which is then added to the ^{14}C-labeled collagen fibril. As the APMA solution is alkaline, it should be added to culture medium containing enzyme only in the presence of buffer.

It is important to emphasize that the effects of retinoids on collagenase synthesis are measured in days, not minutes or hours (Figs. 1 and 2). Note also that treatment with a lower concentration for a longer period of time may be equally effective (Fig. 1a). Thus, whereas 10^{-7} *M* all-*trans*-retinoic acid was ineffective after 4 days, a time when 10^{-6} *M* had lowered collagenase levels below detection, by 6 days, the lower concentration had also decreased collagenase to undetectable levels.[8] We have also found that a decrease in collagenase protein, measured in the ^{14}C-labeled collagen fibril

[17] E. D. Harris, D. R. DiBona, and S. M. Krane, *J. Clin. Invest.* **48,** 2104 (1969).
[18] M. T. Gisslow and B. C. McBride, *Anal. Biochem.* **68,** 70 (1975).
[19] C. A. Vater, H. Nagase, and E. D. Harris, Jr., *J. Biol. Chem.* **258,** 9374 (1983).

assay, is paralleled by a decrease in collagenase mRNA, measured by dot-blot (see below and Fig. 2)[6] or Northern hybridization.[14,20] Thus, measurement of either protein or mRNA levels can be used as an end point to assess the effects of retinoids on collagenase production.

The time required for retinoids or steroids to mediate suppression of collagenase synthesis varies directly with the magnitude of the collagenase response; in other words, the greater the level of enzyme induction, the longer it takes for retinoids or steroids to exert their effects, perhaps because retinoids and steroids do not affect the half-life or translation of mRNA already transcribed.[6] This conclusion is based on the results of pulse–chase experiments with [^{3}H]uridine to measure the half-life of collagenase mRNA.[6] In these studies, transcription of the collagenase gene is induced by overnight (18 hr) stimulation of monolayer cultures of rabbit synovial fibroblasts with 10^{-8} *M* PMA in the presence of [^{3}H]uridine (40 μCi/ml, 25 μM; Amersham, Chicago, IL) in DMEM–LH. Following the pulse, the cultures are chased by placing them in nonradioactive serum-free medium containing 5 m*M* uridine and 2.5 m*M* cytidine. The chase consists of two parts, a prechase of 8 hr to deplete the intracellular pool of [^{3}H]UTP and assure that no further radioactivity is incorporated into mRNA during the chase, then a chase during which the $t_{1/2}$ of collagenase mRNA is measured. To measure the decay of [^{3}H]uridine collagenase mRNA, ^{3}H-labeled mRNA is hybridized to 4 μg of collagenase cDNA that has been denatured by boiling in 2 *M* NaCl and 0.2 *M* NH_4OH and then quick-cooled. The cDNA is spotted onto nitrocellulose prewet with 20× SSC and baked for 2 hr at 80°. We have determined that 4 μg of plasmid DNA contains at least a 10-fold excess of collagenase sequences relative to the added mRNA. As a control, filters spotted with pBR322 and blank filters are used. Filters are prewashed for 30 min with 0.3 *M* NaCl, 2 m*M* EDTA, 10 m*M* Tris, pH 7.4, and 0.1% sodium dodecyl sulfate (SDS), and prehybridized for 4 hr at 42° in 50% (w/v) formamide, 0.1 *M* PIPES, pH 6.4, 4× SSPE, 100 μg/ml yeast tRNA, and 5× Denhardt's solution. The mRNA samples (2 μg) are then hybridized in this same solution but with 1× Denhardt's for 48 hr at 42°. The filters are washed 3 times for 1 hr each in 0.3 *M* NaCl, 2 m*M* EDTA, 10 m*M* Tris, pH 7.5, and 0.1% (w/v) SDS. Radioactivity is then eluted with 40 m*M* NaOH, neutralized with acetic acid, and quantitated by counting in PCS II scintillation fluid in a Beckman liquid scintillation counter. For these experiments, it is necessary to use poly(A)$^+$ RNA, obtained by one passage over an oligo(dT)-cellulose column, in order to eliminate nonspecific hybridization.

[20] T. Maniatis, E. F. Fritsch, and J. Sambrook, "Molecular Cloning: A Laboratory Manual." Cold Spring Harbor Laboratory, Cold Spring Harbor, New York, 1982.

The half-life for collagenase mRNA is calculated from the data points from each experiment.[6] The best-fitting exponential curve is obtained by using the method of least squares on log-transformed data.[6,21] The coefficient of determination and the slope of the regression line are calculated by using standard methods, and the half-life is determined from the slope of the regression line.[6,22]

To measure total collagenase mRNA, slot-blot hybridizations are performed.[6] Samples of poly(A)$^+$ RNA at 0.33, 0.67, and 1 μg from each experimental time point are denatured at 65° in 6.15 *M* formaldehyde and 10× SSC, and they are spotted onto nitrocellulose and baked as described above. The filters are then prehybridized for 6 hr at 42° in 50% formamide, 5× Denhardt's, 5× SSPE, and 200 μg denatured salmon sperm DNA. Following prehybridization, a ^{32}P-labeled cDNA clone for collagenase, prepared by nick translation, is added, and hybridization proceeds for 36 hr at 42°. Filters are washed 3 times for 10 min each in 20× SSC and 0.1% SDS at 65°. The blots are then autoradiographed and quantitated by liquid scintillation counting of the filters. Our success with these techniques is due, at least in part, to the fact that collagenase is a major gene product of induced cells, representing about 2% of the mRNA.[6,14] Consequently, we could effectively monitor the decay of the [^{3}H]uridine-labeled collagenase mRNA.

Results obtained with these techniques allowed us to make the following conclusions: (1) PMA increased transcription of the collagenase gene, (2) PMA increased the $t_{1/2}$ of collagenase mRNA, and (3) neither retinoids nor steroids influenced this half-life (Fig. 3).[6] These conclusions are based on two experiments which showed that, at time zero, the level of total collagenase mRNA in Experiment 1 was nearly twice that of Experiment 2 (Fig. 3a). Further, total collagenase mRNA levels in Experiment 1 remained high during the chase period in the presence of PMA, whereas those in Experiment 2 declined rapidly during a chase without PMA. Thus, the half-life of collagenase mRNA in Experiment 1 was approximately 35 hr (Fig. 3b), whereas in Experiment 2 it was about 12 hr (Fig. 3c). In neither case, however, was the $t_{1/2}$ affected by retinoic acid or dexamethasone. These observations are consistent with the hypothesis stated earlier that both classes of compounds interfere with the transcription of new collagenase mRNA but not with the translation of existing mRNA.

So far, the discussion has focused on the use of all-*trans*-retinoic acid. Several synthetic retinoids are also available, including 13-*cis*-retinoic acid

[21] D. Kleinbaum and L. Kupper, *in* "Applied Regression Analysis and Other Multivariable Methods," p. 50. Duxbury Press, Boston, Massachusetts, 1978.

[22] D. A. Franklin and G. B. Newman, "A Guide to Medical Mathematics," p. 110. Blackwell, London, 1973.

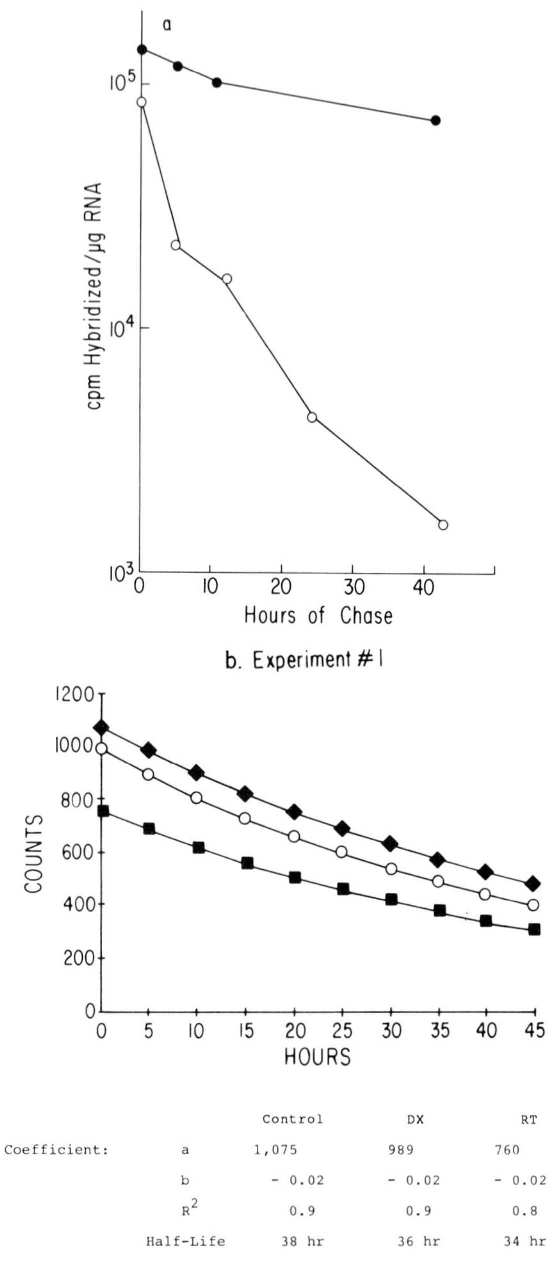

		Control	DX	RT
Coefficient:	a	1,075	989	760
	b	- 0.02	- 0.02	- 0.02
	R^2	0.9	0.9	0.8
	Half-Life	38 hr	36 hr	34 hr

FIG. 3. Effect of PMA, all-*trans*-retinoic acid, and dexamethasone on the half-life of collagenase mRNA. Confluent cultures of rabbit synovial fibroblasts in 150-mm-diameter culture dishes were pulse-labeled with [^{3}H]uridine and PMA for 18 hr. The cells were then placed in chase medium for an 8-hr prechase to deplete the pool of [^{3}H]UTP. Following the prechase, the chase period was begun by replacing the medium and adding all-*trans*-retinoic

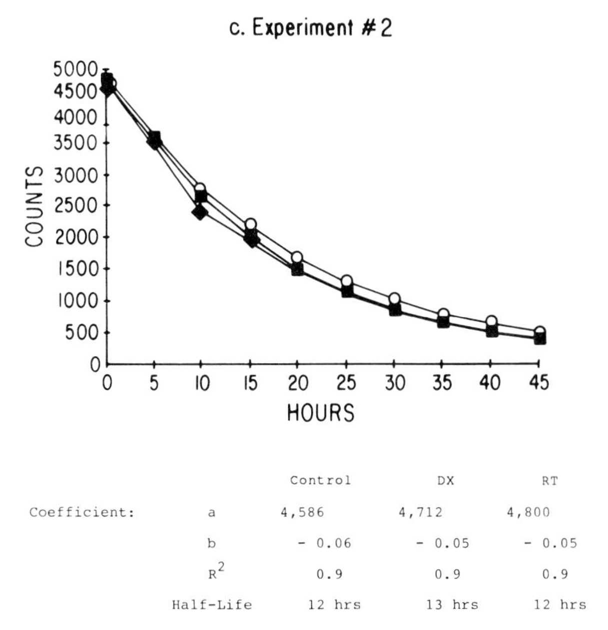

		Control	DX	RT
Coefficient:	a	4,586	4,712	4,800
	b	- 0.06	- 0.05	- 0.05
	R^2	0.9	0.9	0.9
	Half-Life	12 hrs	13 hrs	12 hrs

FIG. 3 *(continued)* acid (RT, 10^{-6} *M*) or dexamethasone (DX, 10^{-7} *M*) to selected cultures. In Experiment 1, PMA was present throughout the pulse, prechase, and chase periods; in Experiment 2, PMA was removed after the 18-hr pulse. At various times during the chase, cultures were terminated, and the amounts of total mRNA and ^{3}H-labeled collagenase mRNA were measured in triplicate. Three dishes were used for each experimental point. (a) Total collagenase mRNA; Experiment 1 (●) and Experiment 2 (○); (b, c) [^{3}H]uridine-labeled collagenase mRNA: control (◆), dexamethasone-treated (○), and retinoic acid-treated (■). (From Ref. 6 with permission.)

and 4-hydroxyphenylretinamide (4-HPR).[9,16] The potency *in vitro* of 13-*cis*-retinoic acid exactly parallels that of all-*trans*-retinoic acid,[8] supporting the theory that an isomerase within the cell may readily interconvert these two retinoids.[16] 4-HPR may be somewhat less potent in reducing collagenase levels: a 10^{-6} *M* concentration may require a slightly longer time period in which to act (Fig. 4).[9] However, this compound is less toxic than either all-*trans*- or 13-*cis*-retinoic acid.[16]

In Vivo Studies

Both 13-*cis*-retinoic acid and 4-HPR have successfully ameliorated two models of experimental arthritis in rats: adjuvant-induced arthritis[23] and

[23] C. E. Brinckerhoff, J. W. Coffey, and A. C. Sullivan, *Science* **221,** 756 (1983).

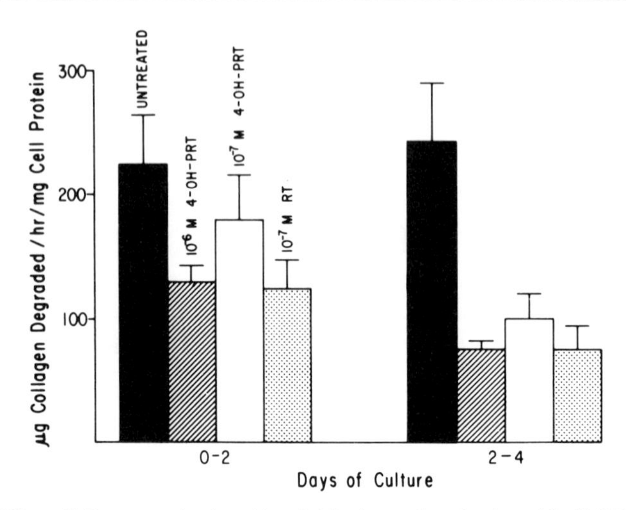

FIG. 4. Effect of all-*trans*-retinoic acid and 4-hydroxyphenylretinamide (4-HPR) on collagenase production by PMA-stimulated rabbit synovial fibroblasts. Confluent monolayers of rabbit synovial fibroblasts in 35-mm-diameter culture dishes were incubated in DMEM–LH and PMA (10^{-8} *M*) for 3 days. The cultures were washed 3 times in HBSS, and fresh medium without PMA but with all-*trans*-retinoic acid (RT, 10^{-7} *M*) or 4-HPR (10^{-6} or 10^{-7} *M*) was added. Medium and retinoids were replaced every 2 days, the medium was assayed for collagenase activity, and the protein content of the monolayers was determined [O. H. Lowry, N. J. Rosebrough, A. L. Farr, and R. J. Randall, *J. Biol. Chem.* **193,** 265 (1951)].

streptococcal cell wall-induced arthritis.[24] Adjuvant arthritis is induced in male Charles River rats weighing around 120 g by subplantar injection of 50 µl of adjuvant [desiccated *Mycobacterium butyricum* suspended in heavy mineral oil, 0.5% (w/v), containing 0.2% digitonin] into the right hind paw. Streptococcal cell wall arthritis is induced in 100-g LEW/N female rats by a single intraperitoneal (i.p.) injection of cell walls in phosphate-buffered saline (PBS), pH 7.4, at a dose equivalent to 15 µg/g body weight. At the time the arthritis is induced, the animals are started on orally administered retinoids. For adjuvant arthritis, the drug is given by gastric intubation at 40 or 160 mg/kg body weight,[23] whereas for streptococcal cell wall-induced arthritis, it is given in the chow at 1–2 mg/kg diet.[24] To administer retinoids by daily intubation, 13-*cis*-retinoic acid is suspended in 0.5% carboxymethyl cellulose, containing 0.9% (w/v) NaCl, 0.39% (v/v) Tween 80, and 0.86% benzyl alcohol. When the retinoid 4-HPR is given in chow, either 1100 or 2200 µg is dissolved in a solution of

[24] B. Haraoui, R. L. Wilder, J. B. Allen, M. B. Sporn, R. K. Helfgott, and C. E. Brinckerhoff, *Int. J. Immunopharmacol.* **7,** 903 (1985).

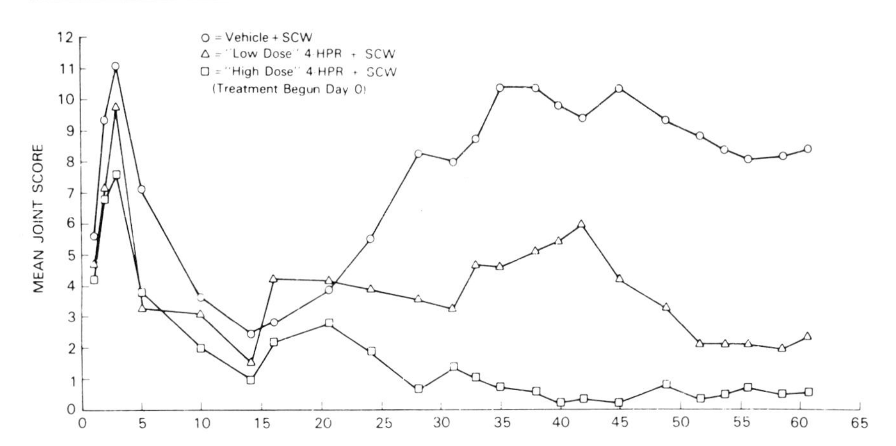

FIG. 5. Effects of 4-hydroxyphenylretinamide (4-HPR) on the development of experimental arthritis in rats. Streptococcal cell wall-induced arthritis was obtained in female LEW/N rats by i.p. injection of purified peptidoglycan-polysaccharide-rich cell wall fragments from group A streptococci. Animals were begun on chow containing either vehicle, 1 mmol 4-HPR/kg diet (low dose), or 2 mmol 4-HPR/kg diet (high dose) on day 0. The mean joint score is a clinical articular index, based on the degree of erythema, swelling, and distortion, and was determined daily for each rat for the first week and 3 times a week thereafter. Each extremity was graded on a scale of 0–4, and scores for each extremity were summed. The maximum score possible was 16. The animals were scored by two observers without knowledge of the treatment group. A mean score was then calculated for each group. (From Ref. 24, with permission.)

ethanol (44 ml) and trioctanoin (120 ml), 1.4 ml of antioxidant Tenox 20, and 1.4 ml of α-DL-tocopherol. These are mixed in the dark with 2660 g of powdered rat chow. These formulations yield 1 or 2 mmol 4-HPR/kg chow. The food is prepared freshly each week and is protected from light. A 2.5-kg batch of chow supplies food for 10 rats for 1 week.

One potential difficulty with oral administration of retinoids is calculating exactly how much drug has been ingested and absorbed. For example, when the drug is mixed in the diet, approximately 20–25 mg/kg body weight may be eaten per rat per day. It is uncertain, however, how much is absorbed. Similarly, drug given by intubation is suspended in an oily vehicle which helps solubilize it. However, this does not necessarily facilitate absorption in the intestine. Thus, the actual blood and tissue levels of the retinoids may be only a fraction of those present in the diet.

The effectiveness of orally administered retinoids on experimental arthritis is shown in Fig. 5.[24] Clinical disease is significantly ablated in a dose-dependent manner. When arthritic animals are sacrificed and synovial tissue is removed from knees and ankles (weight-bearing joints where the disease is most severe), dissociated, and cultured *in vitro* as

TABLE I
COLLAGENASE PRODUCTION BY SYNOVIAL CELLS FROM RATS WITH STREPTOCOCCAL CELL WALL-INDUCED ARTHRITIS[a]

Treatment	Collagen degraded (μg/hr/mg cell protein)
Control	<5
Vehicle	81 ± 15
1 mmol 4-HPR	68 ± 5
2 mmol 4-HPR	16 ± 2

[a] Streptococcal cell wall-induced arthritis was elicited in female LEW/N rats on day 0. The animals received vehicle 1 mmol 4-HPR, or 2 mmol 4-HPR orally for 61 days. The rats were sacrificed, and the synovium was removed and cultured for 48 hr in DMEM with 10% fetal calf serum. Collagenase activity was determined in a ^{14}C-labeled collagen fibril assay after activation of latent collagenase with 200 μg/ml TPCK–trypsin (30 min, room temperature) followed by addition of a 4-fold excess of soybean trypsin inhibitor. Protein content of cultures was determined by the method of Lowry *et al.* [O. H. Lowry, N. J. Rosebrough, and R. J. Randall, *J. Biol. Chem.* **193,** 265 (1951)].

described above, decreased levels of collagenase are seen in tissues taken from retinoid-treated rats (Table I).[24] When culturing tissue from retinoid-treated rats, it is important to work quickly. Extended manipulations and incubations will permit drug present in the tissues to diffuse away, thus lowering the effective concentrations in the tissues that will be cultured *in vitro* and tested.

Acknowledgments

Supported by grants from the U.S. Public Health Service (NIH-AR-26599), The Council for Tobacco Research, and The RGK Foundation (Austin, Texas).

[20] Regenerating Limbs

By DAVID L. STOCUM and MALCOLM MADEN

Introduction

Retinoids have long been known to modify the proliferation and differentiation of normal and neoplastic cells *in vitro*[1] and to be potent teratogens that cause craniofacial and limb abnormalities during embryogenesis.[2] One of their most unique biological effects, however, is to modify the spatial pattern of differentiation of amphibian limb regenerates. The recent demonstration that a nuclear retinoic acid receptor (RAR) is expressed at high levels in regenerating limbs[3] suggests that retinoid–receptor complexes modify regenerate patterns by binding to regulatory response elements of DNA, thereby modulating transcription.[4,5] Retinoids are therefore important tools with which to probe the molecular mechanisms of patterning, including the identification of genes involved in this process. Here we describe biological methods for using retinoids, in particular retinoic acid (RA), to study pattern regulation in regenerating amphibian limbs.

Amputation of Limbs

Animals are anesthetized in a small finger bowl containing 100–200 ml of a 0.02% (w/v) solution of benzocaine (ethyl *p*-aminobenzoate; Sigma, St. Louis MO), prepared by dissolving 0.2 g of benzocaine in 10 ml of absolute ethanol and diluting to 1 liter with water or an appropriately diluted balanced salt solution (usually Holtfreter's solution). The animals are then removed to a petri dish containing water or diluted salt solution, and limbs are amputated at the desired level with iridectomy scissors (larvae) or dissecting scissors (adults). Alternatively, animals can be laid on a wet paper towel, or a platform of dental wax in a petri dish, and the limbs amputated with a scalpel or razor blade. The limb bones will protrude within 1 or 2 min after amputation, owing to retraction of the skin and muscles from the wound surface. It may be useful to trim away the

[1] M. B. Sporn and A. B. Roberts, *Cancer Res.* **43,** 3034 (1983).
[2] D. M. Kochhar, J. D. Penner, and C. I. Tellone, *Teratog. Carcinog. Mutagen.* **4,** 377 (1984).
[3] V. Giguere, E. S. Ong, R. M. Evans, and C. J. Tabin, *Nature (London)* **337,** 566 (1989).
[4] M. Petkovich, N. J. Brand, A. Krust, and P. Chambon, *Nature (London)* **330,** 444 (1987).
[5] V. Giguere, E. S. Ong, P. Segui, and R. M. Evans, *Nature (London)* **330,** 624 (1987).

protruding parts of the bones, as they can sometimes interfere with wound healing and thus regeneration.

Administration of Retinoic Acid

Intraperitoneal Injection Method[6,7]

A fresh 0.166 *M* solution of RA is made by dissolving 50 mg of all-*trans*-RA (Sigma) in 1 ml of spectrophotometric-grade dimethyl sulfoxide (DMSO; Mallinckrodt, St. Louis, MO) under yellow light, or under subdued daylight, to minimize photoisomerization. This solution is used only once. With the room in subdued daylight or under yellow light, batches of 5–10 animals (anesthetized as described above) are placed on their backs under a dissecting microscope on a wet paper towel. A 27-gauge needle is used to make a prepuncture through the skin and abdominal muscles into the intraperitoneal (i.p.) cavity of each animal, a few millimeters anterior to one hindlimb and lateral to the midline. The volume of solution containing the desired dose of RA is then injected into the i.p. cavity with a 50- to 100-μl capacity Hamilton microsyringe. Retinoids are hydrophobic, and the RA precipitates at the injection site, being visible through the skin as a yellow, more or less compact mass. Control animals are injected with an amount of DMSO equal to the amount of RA solution injected. After injection, each animal is returned immediately to water or rearing medium. The RA slowly dissolves into the bloodstream, probably by binding to serum albumin as it does in the bloodstream of mammals.[8] The concentration of RA undoubtedly reaches a steady-state level in the blood under these conditions, but this concentration has not yet been measured.

Immersion Method[9]

The appropriate amount of RA for 1 liter (10–60 mg) is sonicated for several minutes in 200 ml of tap water and added to the remaining 800 ml of water in the plastic container in which the animals are to be kept. Since RA is insoluble in water, the crystals remain as a suspension. Other retinoids such as retinol palmitate or retinyl acetate are available as a complex with corn starch (Sigma), and when added to water these form milky solutions into which the experimental animals are placed.

[6] S. D. Thoms and D. L. Stocum, *Dev. Biol.* **103,** 319 (1984).
[7] W.-S. Kim and D. L. Stocum, *Dev. Biol.* **114,** 170 (1986).
[8] J. E. Smith, P. O. Milch, Y. Muto, and D. S. Goodman, *Biochem. J.* **132,** 821 (1973).
[9] M. Maden, *J. Embryol. Exp. Morphol.* **77,** 273 (1983).

Local Implant Method[10]

One milliliter of Silastic 382 medical elastomer (Dow Corning, Midland, MI) is placed into a small plastic petri dish. The amount of silastin is measured by slowly drawing up the very viscous liquid (improved by warming) with a 1-ml syringe. The appropriate amount (e.g., 100 mg) of RA is added to the silastin and mixed in with a spatula until an even consistency is obtained and the silastin is uniformly yellow. Two drops of the setting compound (stannous octoate) are added and mixed in. After setting, which takes several hours, the silastin can be cut up into pieces of any size and shape under the dissecting microscope, and, assuming an even distribution, the amount of RA in any particular piece can be calculated according to its volume.

Animals are anesthetized as described above and placed under a dissecting microscope. The skin of the limb adjacent to the blastema is lifted with tweezers, and a small scissor cut made. By inserting tweezers into the cut, a tunnel can be made between the skin and soft tissues of the limb, and into this the piece of silastin containing RA is placed. The animals are returned to water. By incorporating radioactive RA into silastin, the rates of release from different sized pieces in different locations have been calculated.[10]

Gastric Intubation Method[11]

The gastric intubation method has been used to administer retinol palmitate to adult newts. The solution is shaken under a nitrogen atmosphere just before use, and the resulting emulsion is squirted into the stomach of the anesthetized animal through a thin plastic tube.

Effects of Retinoic Acid on Different Limb Axes

Proximodistal Axis

The effects of exogenous retinoids on the proximodistal (PD) axial regenerate pattern have been tested on tadpoles of the anuran species *Bufo andersonii*,[12] *Bufo melanosticus*,[13] *Rana breviceps*,[14] *Rana temporania*,[15]

[10] M. Maden, S. Keeble, and R. A. Cox, *Roux's Arch. Dev. Biol.* **194,** 228 (1985).
[11] S. Koussoulakos, K. K. Sharma, and H. J. Anton, *Biol. Struct. Morphog.* **1,** 43 (1988).
[12] I. A. Niazi and S. Saxena, *Folia Biol. (Krakow)* **26,** 3 (1978).
[13] O. P. Jangir and I. A. Niazi, *Indian J. Exp. Biol.* **16,** 438 (1978).
[14] K. K. Sharma and I. A. Niazi, *Experentia* **35,** 1571 (1979).
[15] M. Maden, *Dev. Biol.* **98,** 409 (1983).

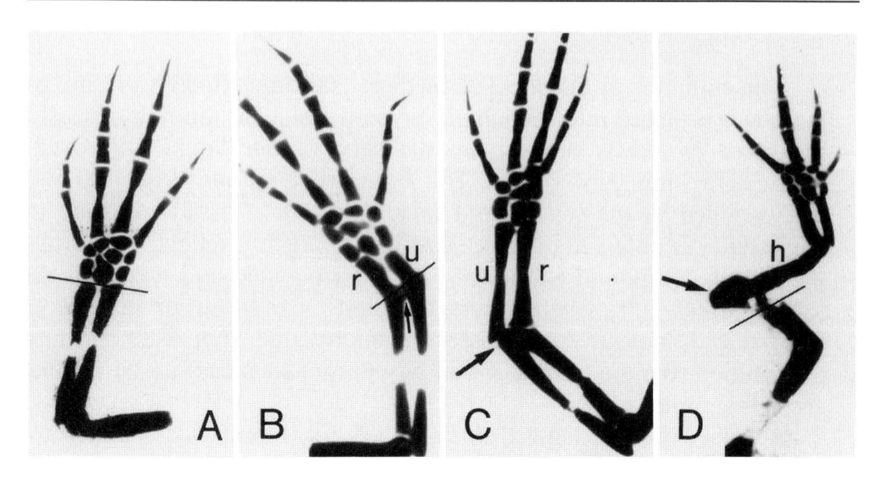

FIG. 1. Dose-dependent effects of RA on the PD axis of axolotl limbs amputated through the distal end of the zeugopodium. (A) Regenerate of a control animal injected with dimethyl sulfoxide; regeneration was normal. (B) Regenerate of an animal injected with 50 μg RA/g body weight. The distal third of the radius (r) and ulna (u) was duplicated. The small arrow points to a cartilage bridge joining the radius and ulna at their distal ends. (C) Regenerate of an animal injected with 75 μg RA/g body weight. A complete radius and ulna was duplicated. (D) Regenerate of an animal injected with 100 μg RA/g body weight. A partial shoulder girdle (arrow), humerus (h), radius, and ulna were duplicated. This is the maximum duplication that is observed. The bars in A, B, and D, and the arrow in C, indicate the regenerate–stump junction. Using a scoring system in which each duplicated half-segment or less is counted as 0.5, the duplication indices of the limbs in B, C, and D are 0.5, 1.0, and 2.5, respectively. [From D. L. Stocum and K. Crawford, *Biochem. Cell Biol.* **65,** 750 (1987), with permission.]

and *Xenopus laevis*[16] as well as the urodele species *Ambystoma mexicanum,*[6,17] *Plurodeles waltl,*[18,19] *Triturus vulgaris,*[20] *Triturus helveticus,*[20] *Triturus alpestris,*[11] *Triturus cristatus,*[21] and *Notopthalmus viridescens.*[6] In all species tested, retinoids proximalize the regenerate pattern in the PD axis; in other words, instead of reproducing just the parts that were lost by amputation, structures proximal to the amputation plane are duplicated in the regenerate (Fig. 1).

To quantitate the degree of proximalization, each half-segment of the limb can be assigned either the same numerical value (0.5) or a progres-

[16] S. R. Scadding and M. Maden, *J. Embryol. Exp. Morphol.* **91,** 35 (1986).

[17] M. Maden, *Nature (London)* **295,** 672 (1982).

[18] I. A. Niazi, M. J. Pescitelli, Jr., and D. L. Stocum, *Roux's Arch. Dev. Biol.* **194,** 355 (1985).

[19] E. Lheureux, S. D. Thoms, and F. Carey, *J. Embryol. Exp. Morphol.* **92,** 165 (1986).

[20] S. Koussoulakos, V. Kiortsis, and H. J. Anton, *IRCS Med. Sci.* **14,** 1093 (1986).

[21] K. K. Sharma and H. J. Anton, *in* "Progress in Developmental Biology (H. Slavkin, ed.), Part A, p. 105. Alan R. Liss, New York, 1986.

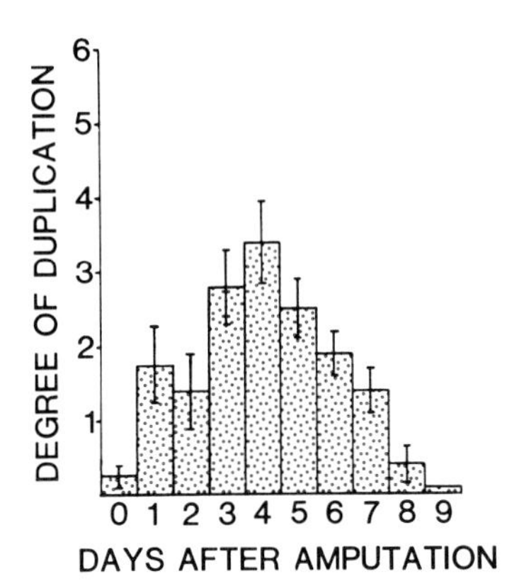

FIG. 2. Stage dependency of the duplication phenomenon in axolotl limbs treated with RA (100 mg/ml) by the implant method. The maximum duplication is achieved by administering RA 4 days postamputation, during the dedifferentiation stage. The same is true when RA is administered by other methods. [From M. Maden, S. Keeble, and R. A. Cox, *Roux's Arch. Dev. Biol.* **194,** 228 (1985), with permission.]

sively higher numerical value: 1 for carpals, 2 for the distal half of the zeugopodium, 3 for the proximal half of the zeugopodium, and so on (Fig. 1). The greater the level to which duplication occurs, the higher the numerical value.

The degree of proximalization is dose- and stage-dependent (Figs. 1 and 2), regardless of the method of administration. Proximalization is maximal when retinoid is administered during the dedifferentiation stage of regeneration (administration at later stages results in inhibition of regeneration). In the injection method, 100–150 μg of RA/g body weight is the dose that evokes maximum duplication (to the level of the shoulder girdle) in axolotls.

Dedifferentiated cells must be exposed for a period of time that is inversely proportional to the concentration of retinoid. This is best illustrated using the immersion or local implant methods of administration, which can control the duration of treatment. Nine natural and synthetic retinoids have been tested by the local implant method for their ability to cause PD duplication,[22] and a hierarchy of potency has been established (Fig. 3). The most active retinoids are those with a carboxyl end group. Retinoids in which the carboxyl group is converted to an ester, alcohol, or aldehyde have little or no potency. Addition of rings enhances the potency if the COOH group is maintained, as exemplified by tetrahydrotetrameth-

[22] S. Keeble and M. Maden, *Dev. Biol.* **132,** 26 (1989).

Retinoid	Structure	Axolotl	
		Potency	Affinity
Retinoic acid	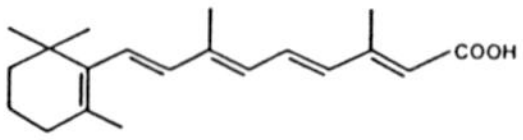 COOH	1	4×10^{-7}
Ro 13-7410	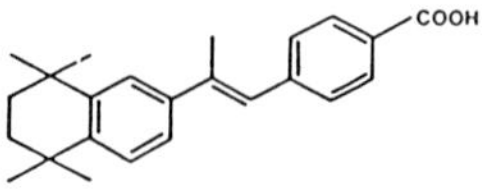 COOH	100+	2×10^{-7}
Arotinoid	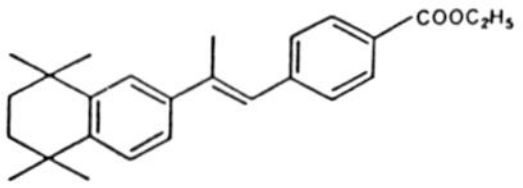 $COOC_2H_5$	10	1.2×10^{-6}
Ro 10-1670	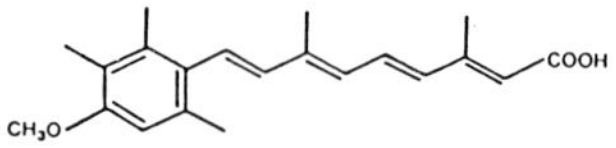 CH_3O COOH	0.1	6.7×10^{-7}
Ro 10-9359	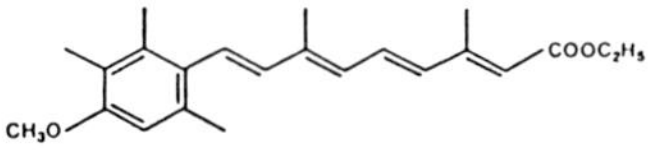 CH_3O $COOC_2H_5$	0	0
Retinal	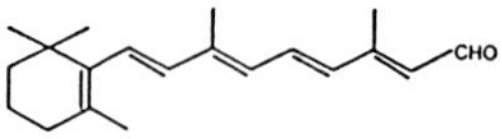 CHO	0	1.4×10^{-5}
Retinol	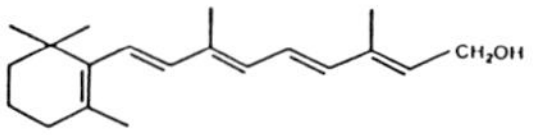 CH_2OH	0	0
Retinyl acetate	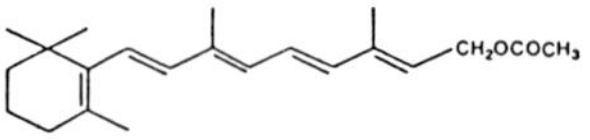 CH_2OCOCH_3	0	0
Retinyl palmitate	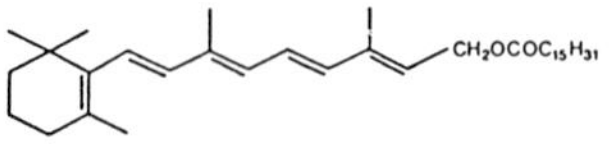 $CH_2OCOC_{15}H_{31}$	0	0

FIG. 3. Relative potencies (ability to proximalize pattern) and binding affinities to cellular retinoic acid-binding protein (CRABP) (measured as IC_{50}) of nine different retinoids tested on regenerating axolotl limbs. Potencies are assessed relative to all-*trans*-retinoic acid. Ro numbers refer to Hofmann-La Roche code numbers.

ylnaphthalenylpropenylbenzoic acid (TTNPB). It is likely that the ester end group of arotinoid, which is more potent than RA, is metabolized to a COOH group.

Anteroposterior and Dorsoventral Axes

The effects of RA on the anterioposterior (AP) and dorsoventral (DV) axes of regenerating urodele limbs are manifested only when structure is removed in the AP or DV planes of the limb prior to amputation. RA-induced pattern modification in the transverse axes is most clearly revealed on surgically constructed double half-zeugopodia (lower arms or legs), as shown in Fig. 4. To make double anterior and posterior zeugopodia, animals are anesthetized in 0.007% (w/v) benzocaine dissolved in 100% Holtfreter's solution. This concentration allows prolonged anesthesia. A cut is made in the dorsoventral plane between the radius and ulna, to the level of the elbow or knee, on both right and left limbs. By means of transverse cuts the anterior half is removed from one limb, and the posterior half from the contralateral limb. The transverse cut should be made so that a small piece of bone remains at the elbow or knee to stabilize the

A. Double Half Anterior and Posterior Zeugopodia

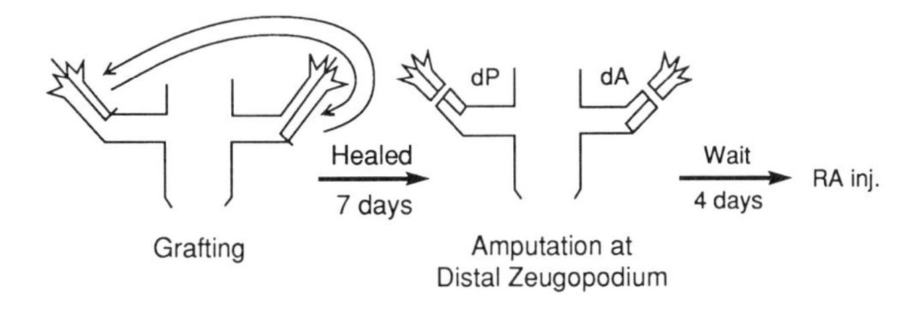

B. Double Half Dorsal and Ventral Zeugopodia

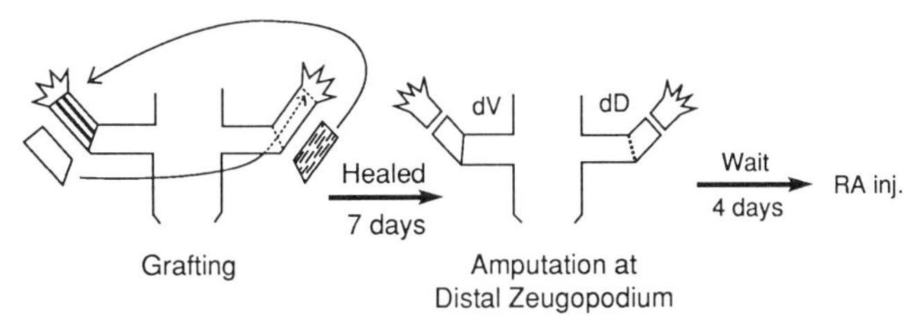

FIG. 4. Method of surgically constructing double half-zeugopodia. Anterior and posterior, or dorsal and ventral, halves are exchanged between left and right fore- or hindlimbs. After healing, the limbs are amputated and injected with RA 4 days postamputation.

grafted tissue. The bone will protrude from the proximal end of each half, and the protrusion should be trimmed flush with the skin and muscle. The halves are then exchanged.

Double dorsal and ventral zeugopodia are made by making a transverse cut at the level of the wrist on the dorsal side of one limb and on the ventral side of the contralateral limb. The skin and extensor muscles are then removed from the dorsal side of one limb, and the skin and flexor muscles from the ventral side of the contralateral limb, by making a longitudinal cut in the AP plane and a transverse cut at the level of the elbow or knee. Only the skin and muscles are used because it is impossible to make a cut that halves the zeugopodial bones in the anteroposterior plane. The dorsal and ventral "halves" are then exchanged, and the grafts allowed to heal for several hours.

The plasma which clots at the wound site serves to anchor the graft initially, and no suturing is necessary. The most difficult part of the double half-limb procedure is in getting the graft and host tissues to adhere closely so that epidermis has to migrate only a minimal distance to join the edges of graft and host. Tight adherence requires pressing the graft and host wound surfaces together for the several hours it takes for epidermal wound healing to occur. Good results are obtained by placing the animal on moist filter paper in a petri dish, with its limbs supported in the air by small balls of modeling clay. The host limbs are oriented with their wound surfaces up, so that the grafts can be pressed onto them and held tightly by gravity during the period of epidermal wound healing. The bodies of the animals are periodically moistened with water to keep them from drying out.

After the grafts have healed in for several hours, the animals are transferred to the rearing medium and healing allowed to continue for 1 week, after which the limbs are amputated through the distal zeugopodium. At 4 days postamputation (stage of dedifferentiation), the animals are injected i.p. with 100–150 μg RA/g body weight, the conditions that give maximum duplication in the PD axis. Other doses and stages have not yet been tested.

The effects of RA treatment on double half-zeugopodia of axolotls and newts are illustrated in Fig. 5. Control double half-zeugopodia produce symmetrical regenerates that are hypomorphic in the AP axis to varying degrees with respect to autopodial (wrist and hand, or ankle and foot) skeletal elements. Double anterior and double ventral regenerates are more hypomorphic than double posterior or double ventral regenerates. The DV muscle pattern of a normal limb is asymmetric, with the dorsal (extensor) and ventral (flexor) muscles having very distinct morphologies. Control double dorsal regenerates have two sets of extensor muscles, and control double ventral regenerates have two sets of flexor muscles.

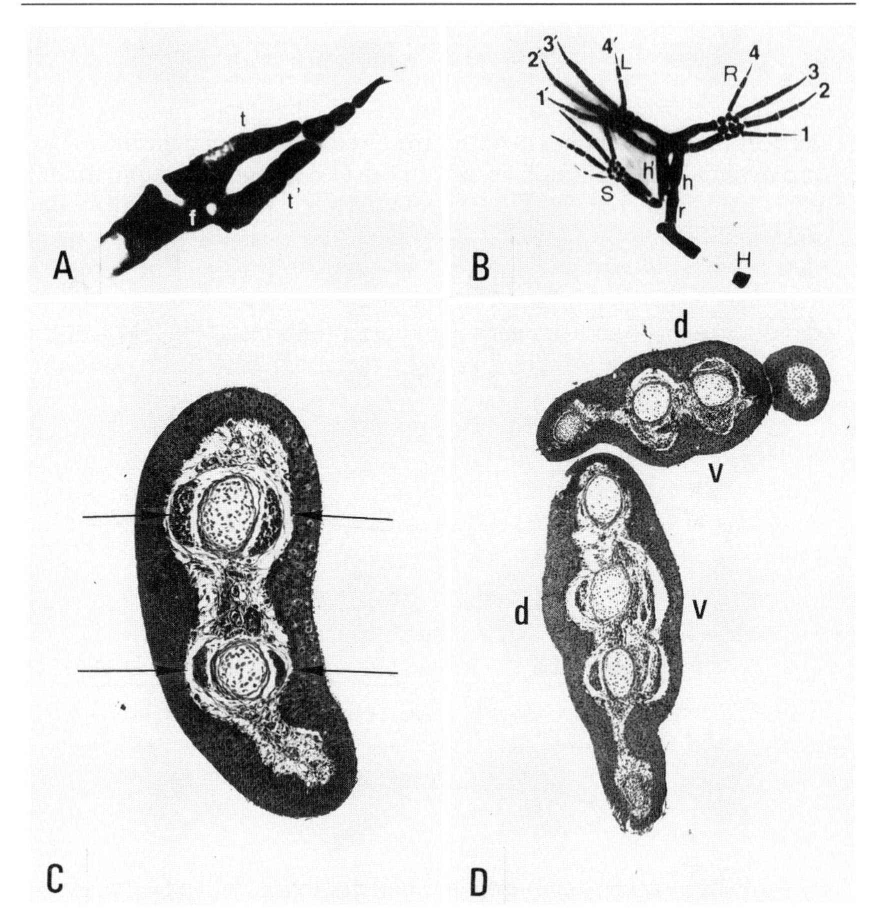

FIG. 5. Completion of the missing half-pattern in RA-treated double anterior and double dorsal zeugopodia. (A) Regenerate of a control double anterior zeugopodium of a hindlimb, amputated through the ends of the tibiae (t, host tibia; t′, graft tibia). (B) Regenerate of RA-treated double anterior forelimb. H, r, Humerus and one of the radii of the double anterior limb stump; h and h′, humeri of the left (L) and right (R) regenerates, which are duplicated in the PD axis. The AP pattern is completely normal, including the digits (numbered from anterior to posterior). S indicates a supernumerary limb that is not a result of RA treatment. (C) Transverse section through a control double dorsal zeugopodium at the level of the metacarpals. The muscle pattern of a normal limb is asymmetric, but note the symmetrical pattern here in which there are two sets of extensor muscles (arrows). (D) Transverse section through the twin regenerates produced by a RA-treated double dorsal zeugopodium. Both have the asymmetric muscle pattern of a normal regenerate. d, Dorsal; v, ventral.

RA treatment restores the missing posterior half-pattern from each half of a double anterior half-zeugopodium[7,23] and restores the missing ventral half-pattern from each half of a double dorsal zeugopodium,[24] producing mirror-image regenerates in both cases. At the same time, the mirror-image regenerates are duplicated in the PD axis because of the proximalizing effect of RA. In contrast, RA treatment inhibits regeneration from double posterior or double ventral zeugopodia. These results indicate that RA posteriorizes the pattern in the AP axis and ventralizes it in the DV axis. This conclusion is reinforced by the results of administering retinoids to developing chick and regenerating anuran limb buds.[15,25,26] Mirror-image limb structures are produced in both cases, but only from tissue on the anterior face of the bud.

A variety of grafting experiments indicate that blastema cells carry a memory of their position in the PD, AP, and DV axes that determines the level from which regeneration takes place. The effects of RA on pattern in each axis of the regenerate have been interpreted to mean that it is this positional memory that is unidirectionally modified by RA.[27]

Bioassays for Position-Dependent Differences in Blastema Cell Affinity

In vitro and *in vivo* assays have been designed to demonstrate the existence of position-related differences in blastema cell adhesivity and affinity along the PD axis of the regenerating axolotl limb.

In Vitro Assay[28]

In the *in vitro* assay, mid to late bud blastemas derived from the wrist, elbow, and mid-upper arm levels are cultured in all of the possible 9 binary combinations in hanging drops. In these combinations, one member of the pair is radiolabeled by injecting the animal twice at 12-hr intervals 1 day prior to culture with 25 μCi of [^{3}H]thymidine dissolved in 50 μl of sterile Steinberg's solution. To enhance the contrast between labeled and unlabeled cell populations, the labeled blastemas are derived from dark-colored animals, whereas the unlabeled ones are derived from white animals.

[23] D. L. Stocum and S. D. Thoms, *J. Exp. Zool.* **232,** 207 (1984).

[24] D. Ludolph, J. A. Cameron, and D. L. Stocum, *Dev. Biol.* **140,** 41 (1990).

[25] D. Summerbell, *J. Embryol. Exp. Morphol.* **78,** 269 (1983).

[26] C. Tickle, J. Lee, and G. Eichele, *Dev. Biol.* **109,** 82 (1985).

[27] D. L. Stocum, *in* "Recent Trends in Regeneration Research" (S. Koussolakos, V. Kiortsis, H. Wallace, and H. J. Anton, eds.), 295 Plenum, New York, 1989.

[28] J. B. Nardi and D. L. Stocum, *Differentiation* **27,** 13 (1983).

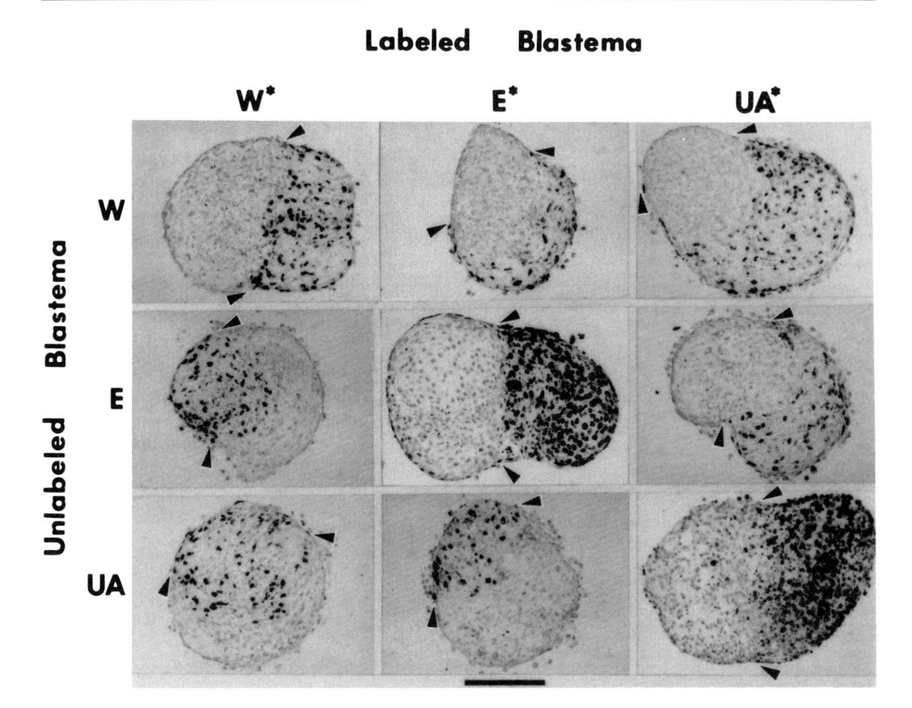

FIG. 6. Representative autoradiographs of the 9 different binary combinations of wrist (W), elbow (E), and upper arm (UA) blastemal mesenchymes. Asterisks indicate the labeled member of each pair. Arrowheads indicate the extent of engulfment by the proximal blastema. Bar: 0.5 mm. [From J. B. Nardi and D. L. Stocum, *Differentiation* **27,** 13 (1983), with permission.]

To prepare the cultures, blastemas plus some stump tissue are placed in dishes containing Earle's balanced salt solution. The stump skin and blastemal epidermis are removed with jeweler's forceps, and the remaining tissue is rinsed 3 times with sterile Earle's solution, followed by 3 more rinses in culture medium. In the last rinse, the blastema mesenchymes are cut away from the stump tissue with iridectomy scissors, and the proximal surfaces of the blastemas in each binary combination are pressed together. After 2–4 hr of healing, each pair is transferred to a hanging drop. The culture medium is a modification of minimal essential medium and has been fully described by Schrag and Cameron.[29] Alternatively, an L-15-based medium that requires no gassing can be used.[30]

The results of this assay are illustrated in Fig. 6. The proximal blastema

[29] J. A. Schrag and J. A. Cameron, *J. Embryol. Exp. Morphol.* **77,** 255 (1983).
[30] D. L. Stocum, *Dev. Biol.* **18,** 441 (1968).

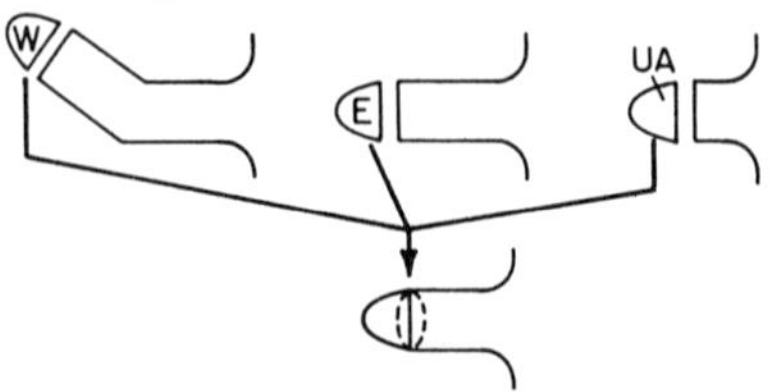

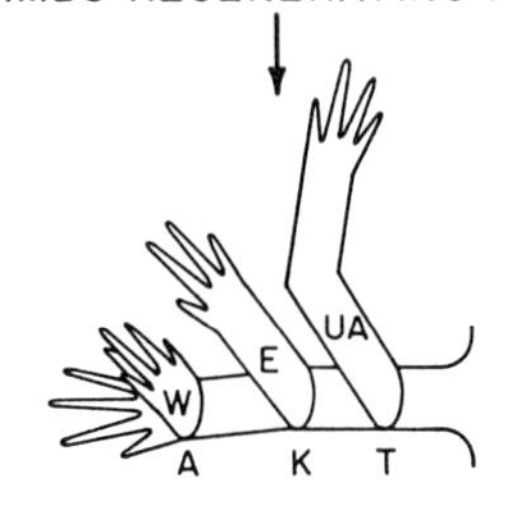

FIG. 7. Diagram of the differential affinity assay *in vivo.* The dashed oval indicates the area where a patch of blastema epidermis and stump skin was removed from the dorsal surface of the host hindlimb, and the line bisecting this oval indicates the blastema–stump junction. The grafted blastemas displace to their corresponding host levels and develop according to their origin, as forelimbs. W, E, UA, Wrist, elbow, and mid-upper arm. A, K, T, Ankle, knee, and midthigh. [From K. Crawford and D. L. Stocum, *Development* **102,** 687 (1988), with permission.]

of each pair surrounds the distal one in the following hierarchy: upper arm engulfs elbow or wrist; elbow engulfs wrist. Blastemas from the same level exhibit no engulfment behavior, but rather fuse in a straight line. These results are consistent with Steinberg's differential adhesion hypothesis,[31] and they suggest the existence of a graded distribution of cell surface adhesive properties along the PD axis of the limb, with adhesivity increasing distally.

In Vivo Assay[32]

In the *in vivo* assay, mid to late bud blastemas from the wrist, elbow, and mid-upper arm levels are grafted to a hindlimb regenerating from the midthigh (Fig. 7). The graft is made by removing a small patch of skin and

[31] M. S. Steinberg, *J. Exp. Zool.* **173,** 395 (1970).

[32] K. Crawford and D. L. Stocum, *Development* **102,** 687 (1988).

epidermis from the dorsal surface of the blastema–stump junction of the host limb with jeweler's forceps and placing the blastema on the wound bed with its AP axis aligned with that of the host limb. The host animal is anesthetized in 0.007% benzocaine dissolved in 100% Holtfreter's solution, and the grafting procedure is done with the animal submerged in this solution in deep petri dishes lined with filter paper. The animals are left in the petri dishes for 5–6 hr at room temperature to facilitate healing. They are then transferred to 100% Holtfreter's solution and left undisturbed in the dark for 2 days, after which they are transferred to normal rearing medium (10–20% Holtfreter's solution). Under these conditions, the grafts heal in within a few hours, and circulation has been reestablished within them by 1–2 days. The grafts subsequently develop according to their origin, that is, as forelimbs.

Position-related differential affinity is manifested in this assay by the distal displacement of the grafted blastemas so that the fully developed grafts articulate with their corresponding axial level of the host regenerate (Fig. 7). This phenomenon of sorting out along an axis to reestablish normal cell neighbors has been termed "affinophoresis." RA-treated (100–150 μg/g body weight at the dedifferentiation stage) wrist and elbow blastemas, which are proximalized to the level of the upper stylopodium or girdle, fail to displace in the affinophoresis assay. This result indicates that positional memory and blastema cell affinity are coordinately proximalized and are therefore directly related.

[21] Targeted Slow-Release of Retinoids into Chick Embryos

By SARAH WEDDEN, CHRISTINA THALLER, and GREGOR EICHELE

Introduction

Local application of retinoids from small ion-exchange beads implanted into embryos is a valuable tool for studies of the mechanism of action of vitamin A compounds in vertebrate development.[1] In contrast to systemic application, local application *targets* the retinoid to a specific region of the embryo. Moreover, the beads release the adsorbed retinoic acid for approximately 1 day in a continuous fashion.[2] The biological

[1] C. Tickle, J. Lee, and G. Eichele, *Dev. Biol.* **109,** 82 (1985).
[2] G. Eichele, C. Tickle, and B. M. Alberts, *Anal. Biochem.* **142,** 542 (1984).

rationale for employing site-specific, sustained targeting is that many signaling agents important in embryonic development (e.g., hormones, growth factors, local chemical mediators) are generated *in situ* in spatially distinct signaling regions ("organizers"). Providing a signaling agent from a specifically positioned exogenous source is a good way to mimic the action of endogenous signaling regions.

Here we describe how retinoids can be locally applied to the developing chick embryo. In addition, we show how applied radioactive retinoids can subsequently be extracted from embryonic chick tissue and be analyzed and identified by high-performance liquid chromatography (HPLC).

Local Application of Retinoic Acid and Congeners to Chick Embryos

An ion-exchange bead is first impregnated with the retinoid of interest, and, after thorough washing, it is implanted at the anterior margin of the right limb bud of a stage 20 chick embryo (Fig. 1b). The embryo is then returned to the incubator and left to develop for 6 to 7 days. At that time it is sacrificed, and the skeletal pattern of the treated wing is inspected. Treatment of wing buds with retinoic acid will invariably change the normal digit pattern to a pattern with one to three additional digits (see Fig. 1d).

Materials Required

Chick embryos (3.5 days old, Hamilton–Hamburger stage 20)[3] are used. AG1-X2 beads in formate form ranging in size from 50 to 200 mesh are from Bio-Rad (Richmond, CA). Other reagents are dimethyl sulfoxide, all-*trans*-retinoic acid, phosphate-buffered saline (PBS) containing 0.1% Phenol Red (PBSP; 100 ml PBS and 500 μl of a 2 mg/ml Phenol Red solution in ethanol), 5% (w/v) trichloroacetic acid solution, Alcian Green or Alcian Blue dye solution (0.5 g of dye in 500 ml 70% (v/v) ethanol containing 1% HCl), 70% ethanol containing 1% HCl, absolute ethanol, and methylsalicylate. Equipment required includes 1.5-ml microcentrifuge tubes, microcentrifuge, microcentrifuge tube shaker, egg incubator, dissecting microscope (e.g., Wild-Leitz, Rockleigh, NJ or Zeiss, Thornwood, NY) equipped with a low-voltage illumination, heat protection filter, green filter, and eyepiece with graticule, scissors, a pair of normal and of watchmaker's forceps (type 5), dissection needles, and sharpened tungsten needles. To generate tungsten needles, mount a 5-cm-long piece of tungsten wire of 0.3 mm diameter into an appropriate handle. One outlet of a microscope transformer is connected to the tungsten wire using an alligator clip. The other outlet is attached to a paper clip that is immersed in a

[3] V. Hamburger and H. Hamilton, *J. Morphol.* **88,** 49 (1951).

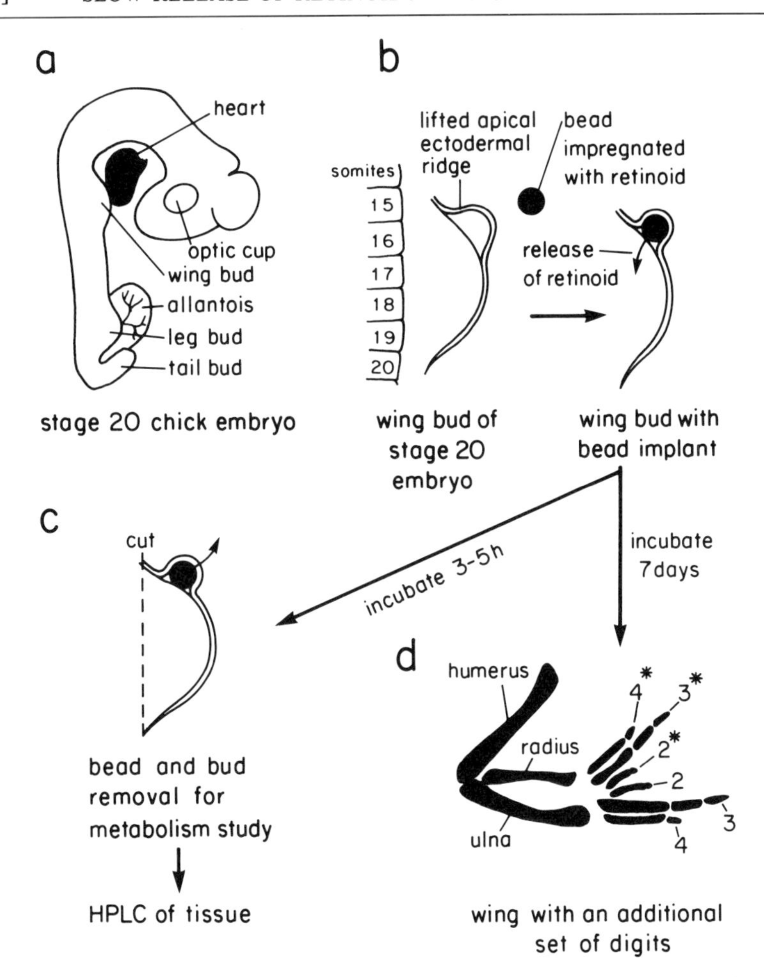

FIG. 1. Schematic representation of a stage 20 chick embryo ($3\frac{1}{4}$ days of incubation) (a) and of the procedure to implant a retinoid-impregnated ion-exchange bead into a wing bud underneath the apical ectodermal ridge (b). Buds treated with radioactive retinoid are cut off after a few hours of incubation, and the metabolites generated are analyzed by HPLC (c). Alternatively, embryos are incubated until day 10 when they are sacrificed and their limbs, stained to visualize the aberrant pattern (d). The normal digit pattern *(234)* is duplicated to a *432234* mirror-image pattern with three additional digits (marked by asterisks).

solution of 2 *N* sodium hydroxide. Apply about 6 V to the circuit, then rapidly and repeatedly dip the needle into the hydroxide solution until the tungsten wire forms a pointed, fine needle. All surgical manipulations can be carried out on a clean bench; a sterile hood is not necessary. Surgical tools used should be rinsed in 70% ethanol.

Step 1: Preparation of Embryos

Fertile eggs from White Leghorn chickens are purchased from a breeder (in the United States, Spafas, Norwich, CT, is a reliable supplier). The eggs can be stored at 10–14° for a few days. Prior to incubation, the eggs are equilibrated to room temperature and subsequently incubated in a horizontal position. To position eggs horizontally, cut triangular holes into a styrofoam plate so that eggs do not tumble. It is important that the incubator maintains a constant temperature of 37.5–38° and a humidity of about 60%.

After about 65 hr of incubation the eggs are taken out of the incubator and flipped twice by 180° (to detach the embryo from the inner shell membrane). Position the egg horizontally on a ring made of modeling clay. Using a dissecting needle, make a small hole at the blunt end of the egg. Use the same tool to create a 3- to 4-mm-wide opening in the shell above the embryo, which floats on top. Care must be taken to ensure that only the calcified part of the shell is removed, otherwise the embryo may be damaged. This is achieved by chiseling and chipping away small shell fragments with the dissecting needle. After the hole is generated, the underlying opalescent membrane is carefully punctured. The embryo will immediately detach from the shell membrane and sink, and an airspace will form. The opening is covered with Scotch tape and, with the help of a pair of small scissors, extended to a hole about 12 mm in diameter. The embryo should now be in the center of the hole. If not, the egg can be slightly rocked until the embryo is centered.

Using the scheme of Hamburger and Hamilton,[3] the embryo is staged. The embryos should be around Hamburger–Hamilton stage 17. Embryos can be synchronized by leaving them at room temperature for up to 5 hr. After staging, the opening is covered with piece of tape, and the embryo is returned to the incubator. Further incubation for about 15–18 hr will bring the embryos to stage 20, when they appear as schematically illustrated in Fig. 1a.

Step 2: Impregnation of Beads with all-trans-Retinoic Acid

All retinoid work should be carried out in yellow light (e.g., General Electric F40 GO). AG1-X2 ion-exchange beads of approximately 200–250 μm diameter are selected under a dissecting microscope. Approximately 15 beads are placed by forceps into a 1.5-ml microcentrifuge polypropylene tube and vigorously shaken in a microtube shaker for 20 min at room temperature in 200 μl of solution of all-*trans*-retinoic acid in dimethyl sulfoxide. The concentration of retinoic acid typically used ranges from 1 to 100 μg/ml, depending on the purpose of the experiment. After removing the retinoid solution, the beads are rinsed briefly in 200 μl PBSP. The rinsing solution is replaced with 200 μl of fresh PBSP, and the beads are washed for 10 min, with shaking. This wash is repeated, after which the beads are placed into a 35-mm petri dish, from which they can be picked out individually using fine forceps and implanted into the embryo.

Local Release of Other Retinoids. AG1-X2 resin beads have successfully been used by our laboratory for the release of a series of natural or synthetic carboxylated retinoids.[4,5,6] AG1-X2 beads are also appropriate for releasing uncharged retinoids such as retinol, retinal, or retinyl esters.[2] However, the capacity of the beads for uncharged retinoids is only about 1/50 of that for retinoic acid, requiring a corresponding increase in the soaking concentration. In addition, uncharged retinoids are released within 4 hr, making the beads less suitable for experiments requiring a prolonged treatment. This problem can be alleviated by using beads made of XAD-7 resin (Amberlite, Rohm and Haas Company, Philadelphia, PA), which provide a longer-lasting release.[2]

Step 3: Implantation of the Bead at Anterior Wing Bud Margin

An egg containing a stage 20 embryo is taken from the incubator, and the tape seal is removed. The egg is examined under the dissecting microscope, which is well illuminated with a low-voltage lamp that has a heat filter and a green filter (to increase the contrast). In the region of the right wing bud, serosa and amnion membranes are torn away using two watchmaker's forceps. With the help of an electrolytically sharpened tungsten needle, a slit is cut between the apical ectodermal ridge (AER) and the underlying mesenchyme (Fig. 1b). This microsurgery requires a good quality dissection microscope with a magnification range from ×10 to ×50. The ridge is then carefully lifted up from the mesenchyme to create a space for the bead. Using watchmaker's forceps, a bead is gently maneuvered into the space, where it is held in place by the AER. This is quite a delicate procedure, but with some practice it is possible to implant a bead in less than 30 sec. The egg is then sealed with tape and returned to the incubator. The duration of retinoic acid treatment can readily be controlled by simply removing the bead any time after implantation. This is done by making a slit in the wing bud tissue that has healed over the bead and hooking the implant out with a tungsten needle.

[4] G. Eichele and C. Thaller, *J. Cell Biol.* **105,** 1917 (1987).
[5] G. Eichele, C. Tickle, and B. M. Alberts, *J. Cell Biol.* **101,** 1913 (1985).
[6] M. B. Sporn, A. B. Roberts, N. S. Roche, H. Kagechika, and K. Shudo, *J. Am. Acad. Dermatol.* **15,** 756 (1986).

Other Sites of Implantation

Implantation to Other Sites in the Wing Bud. Beads can also be implanted to the dorsal surface of the wing bud.[7] A hole of the appropriate size (200 μm) is made (at Hamburger–Hamilton stages 20–24) by removing a cube of tissue using a tungsten needle. A retinoid-soaked bead is then pushed into the hole, and the egg is sealed and returned to the incubator. It goes without saying that beads can also be implanted into the leg bud, either underneath the AER or into the dorsal surface, although these operations are usually a bit more difficult.

Implantation of Beads into Somites or Presomitic Mesoderm of Early Embryos.[8] Although originally developed to investigate the effects of retinoids on limb development, targeted delivery can also be used to examine the effects of retinoids on other developing organ systems in the chick embryo. When retinoid-soaked beads are placed in unsegmented mesoderm, there are a variety of effects on the developing vertebral column, including abnormal segmentation and the formation of hemi-vertebra, extra ribs, and reduced tails. In addition, beads that deliver retinoids to early embryos (Hamburger–Hamilton stages 8–16) will also alter the pattern of limb development before the limb has grown out. In these experiments, retinoids are applied to presumptive leg bud, rather than wing bud, regions, as placing the implant too close to the developing heart increases the mortality rate of embryos. For implantation at stages 8 to 13, place an AG1-X2 bead of 100 μm diameter directly on the vitelline membrane over the lateral plate about 1 mm away from the tail. This way, the bead will end up in the presumptive leg bud region. For embryos at stages 14 to 16, remove the vitelline membrane and, with the help of a tungsten needle, make a small hole into the flank either in the unsegmented lateral plate mesoderm or directly into one of the somites. To enhance the usually weak contrast between somites, a drop of a diluted India ink solution or sterile Nile blue sulfate solution (0.01% in chick Ringer's solution) is added to the embryo.

Step 4: Staining of Developed Embryos to Reveal Pattern of Altered Cartilage Pattern

After 10 days of incubation, embryos are removed with blunt forceps from the egg and sacrificed by decapitation. The specimens are fixed overnight in 5% trichloroacetic acid and stained for about 8 hr in Alcian Green. Excess dye is removed by washing the embryo for 6 hr in acid

[7] C. Tickle and A. Crawley, *Roux's Arch. Dev. Biol.* **197,** 27 (1988).
[8] S. E. Wedden, J. R. Ralphs, and C. Tickle, *Development* **103,** 31 (1988).

ethanol. After dehydration in absolute ethanol for 6 hr, the specimen is dropped into methyl salicylate, which clears the embryo and allows its skeletal pattern to be analyzed under a dissecting microscope. Typically, retinoic acid-soaked beads placed at the anterior wing bud margin will result in the formation of additional digits (Fig. 1d). High concentrations of retinoic acid give rise to a defective beak. In order to quantify the effects of retinoids on the digit pattern, duplicated wings are scored by assigning a value to the most anterior digit. If this digit is an additional digit *2,* the score is 1. If the most anterior digit is a *3* or *4,* then the score is 2 and 4, respectively. The scores of n specimens treated with a particular dose are added, multiplied by 100, and divided by $4n$. The resulting figure is the "percentage respecification value." These values, which range from 0 to 100%, are plotted against the concentration of retinoid in the soaking solution to yield a dose–response curve. Alternatively, the mean number of additional digits can be plotted against the soaking concentration.

Facial defects can be quantified in a similar way using a "severity of beak defect index."[9] The beaks are assigned a score (0–5) depending on the amount of shortening of the upper beak and the degree of clefting of the primary palate. For example, normal beaks score 0, and a score of 3 is given when the upper beak is drastically reduced and the right nostril (nearest the retinoid source in the wing bud) is missing. When the upper beak is completely absent, the beak scores 5. For a group of experimental embryos the mean severity of beak defect index can be calculated. These values are then plotted against the concentration of retinoid in the soaking solution to give a dose–response curve.

Analysis of Locally Applied Retinoids by Chromatography

Sustained release of retinoids is valuable for investigating the metabolism of retinoids in embryos or in parts of embryos.[10] It is important, though, to use radioactive retinoids for such studies in order to be able to distinguish the metabolites that are formed from the pool of unlabeled endogenous retinoids.[11]

Step 1: Local Application of Radiolabeled Retinoids

Five to ten AG1-X2 or XAD-7 beads are soaked in 5–10 μl of a DMSO solution of all-*trans*-[^{3}H]retinol (or all-*trans*-[^{3}H]retinal and all-*trans*-[^{3}H]retinoic acid) in a 500-μl microcentrifuge tube.[10] DMSO and excess

[9] S. E. Wedden and C. Tickle, *J. Craniofacial Genet. Dev. Biol. Suppl.* **2,** 169 (1986).
[10] C. Thaller and G. Eichele, *Development* **103,** 473 (1988).
[11] C. Thaller and G. Eichele, *Nature (London)* **327,** 625 (1987).

label are removed by 3 brief PBSP washes, and the bead is implanted as described above. Retinoic acid and retinol are commercially available from New England Nuclear (Boston, MA) and Amersham (Arlington Heights, IL) at a specific activity of 50 to 60 Ci/mmol, whereas retinal must be synthesized from retinoic acid as described.[12] At a specific activity of 50 to 60 Ci/mmol, approximately 20 to 40 implants are necessary to generate the amounts of metabolites that are required for subsequent chemical identification.

Step 2: Extraction of Tissue

After incubating the embryos for the desired time span (usually a few hours), the beads are removed, and the embryos are dissected out of the egg and rinsed in ice-cold stabilizing buffer (350 mg EDTA, 500 mg sodium ascorbate, 300 mg Na_2SO_4, 10 ml of 10× PBS, water to 100 ml, and the pH adjusted to 7.4; prepare this buffer immediately before use or store frozen at $-20°$). The tissue of interest [e.g., limb buds (Fig. 1c), facial primordia] is dissected out of the embryo in ice-cold stabilizing buffer under a dissecting microscope and collected into microcentrifuge vials kept on dry ice. To the frozen tissue add 200 μl of stabilizing buffer, 100 μl of a saturated Na_2SO_4 solution, 40 μl ethanol, 40 μl of 2 mg/ml butylated hydroxytoluene (antioxidant) in ethanol, and 20 μl of a nonradioactive carrier retinoid cocktail (e.g., 100 ng each of all-*trans*-retinol, all-*trans*-retinal, all-*trans*-retinoic acid, 13-*cis*-retinoic acid, and 4-oxo-all-*trans*-retinoic acid) that also serves as an internal standard. The sample is sonicated and extracted with 800 μl of ethyl acetate/methyl acetate (8 : 1, v/v) mixture by shaking vigorously on a microcentrifuge tube shaker for 30 min. The extraction is repeated twice more. The combined organic phases are evaporated to dryness at room temperature with nitrogen gas, and the residue is dissolved in 20 μl of the eluents used for subsequent HPLC analysis. This extraction procedure routinely results in retinoid recoveries exceeding 90%.

Step 3: Chromatographic Identification and Quantification of Retinoid Metabolites

Chemical identification of the metabolites generated in the embryo requires a combination of sequential HPLC fractionations in which the unknown metabolites must cochromatograph with authentic internal standards. Since amounts of metabolites generated in the embryo are insufficient for spectroscopic analyses, it is advisable to back up the HPLC studies

[12] C. D. B. Bridges and R. A. Alvarez, this series, Vol. 81, p. 463.

with one or more chemical derivatizations, followed by HPLC analysis of the derivatives. We have used the following isocratic systems for the analysis of metabolites generated in the chick embryo (to optimize separation of retinoid metabolites the solvent composition may have to be somewhat adjusted to the particular column used).[10,12]

System 1. Column: C_{18} reversed-phase column (e.g., Microsorb C_{18}, 5-μm particle size, Rainin Instruments, Woburn, MA). Solvent A: Acetonitrile/methanol/2% aqueous acetic acid (6:2:2, v/v/v), flow rate 1.3 ml/min. Solvent A is suitable to separate a broad range of naturally occurring retinoids. Solvent B: Methanol/acetonitrile/water/acetic acid (80:10:9:1, v/v/v/v), flow rate 1.3 ml/min. Solvent B separates retinoic acid, retinoic acid methyl ester, and retinal.

System 2. Column: Straight-phase column (e.g., Microsorb Silica 5 μm particle size, Rainin Instruments). Solvent C: *n*-Hexane/dioxane (95:5, v/v), flow rate 2 ml/min. Solvent C separates retinol and retinal. Solvent D: 1:1 mixture of eluents C and *n*-hexane/acetonitrile/acetic acid (99.5:0.4:0.1, v/v/v), flow rate 2 ml/min. Solvent D separates retinoic acid from retinal.

The tissue extract is first fractionated on a reversed-phase column using solvent A. This results in base-line separation of the internal standard retinoid cocktail described above. Fractions of 1.3-ml volume are collected, and an aliquot is counted to determine the elution profile of the radioactive metabolites. Prior to further analysis, the fractions of interest are pooled and transferred from solvent A to the appropriate solvent in the following way: Add 2 ml HPLC-grade water to the sample fractions and extract the aqueous phase 3 times with 2 ml *n*-hexane each. The combined organic phases are evaporated to dryness, and the residue is immediately dissolved in eluents (for further chromatography), methanol (for methylation), or in acetonitrile (for acetylation).

Methylation of all-trans-Retinoic Acid. Convert retinoic acid to its methyl ester by dissolving the sample in 50 μl methanol and treating it for 30 min on ice with 200 μl of an ether solution of diazomethane. The diazomethane solution is best generated using the *N*-methyl-*N'*-nitronitrosoguanidine (MNNG)–diazomethane kit from Aldrich (Milwaukee, WI). The methyl ester is subsequently chromatographed in solvent system B.

Acetylation of all-trans-Retinol. Acetylate retinol in 100 μl acetonitrile by adding 20 μl dimethylaminopyridine (0.1 mg/ml in acetonitrile) and 50 μl acetic anhydride. The reaction is carried out on ice for 1 hr.[13] Retinyl acetate can be chromatographed using either solvent B or C.

[13] M. L. Gubler and M. I. Sherman, *J. Biol. Chem.* **260**, 9552 (1985).

[22] Biosynthesis of 3,4-Didehydroretinol and Fatty Acyl Esters of Retinol and 3,4-Didehydroretinol by Organ-Cultured Human Skin

By HANS TÖRMÄ and ANDERS VAHLQUIST

Introduction

Vitamin A plays an important role in the maintenance of the normal structure and function of the epidermis, the outermost part of the skin. Without vitamin A, hyperkeratinization will occur. Conversely, hypervitaminosis A prevents terminal differentiation of the epidermal keratinocytes. Unlike most other human tissues, epidermis contains significant amounts of both retinol (vitamin A_1) and 3,4-didehydroretinol (vitamin A_2), and it can produce fatty acyl esters of each of these compounds.[1,2] The structures of the compounds are shown in Fig. 1. The biological function of 3,4-didehydroretinol is obscure. The esters may serve a dual function, providing local storage of vitamin A and also protection against retinol toxicity. The metabolism of vitamin A in epidermis has recently been studied in organ-cultured skin,[1] cell cultures of epidermal keratinocytes,[3] and subcellular fractions of epidermis.[2] This chapter presents an organ culture technique specially adapted for studies on the formation of 3,4-didehydroretinol and retinyl esters from all-*trans*-retinol.

Materials and Methods

Reagents and Equipment.

Potassium hydroxide 14 *M*
Potassium phosphate buffer, 0.2 *M*, pH 7.4
Phosphate-buffered saline (PBS)
Bovine serum albumin (BSA), essentially fatty acid-free
Retinol-binding protein (RBP) prepared from human plasma[4]
Butylated hydroxytoluene (BHT)
Dulbecco's modified Eagle's medium (Flow Labs, Ayershire, UK)
Chloroform, ethanol, methanol, and hexane (HPLC-grade) (Rathburn Chemicals, Peeblesshire, Scotland)

[1] H. Törmä and A. Vahlquist, *J. Invest. Dermatol.* **85,** 498 (1985).
[2] H. Törmä and A. Vahlquist, *J. Invest. Dermatol.* **88,** 398 (1987).
[3] K. E. Creek, C. S. Silverman-Jones, and L. M. De Luca, *J. Invest. Dermatol.* **92,** 283 (1989).
[4] H. Törmä and A. Vahlquist, *Arch. Dermatol. Res.* **276,** 390 (1984).

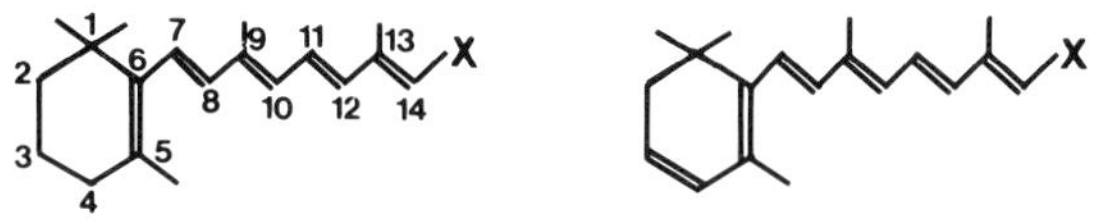

All-trans-retinol $X = CH_2OH$ 3,4-Didehydroretinol
Retinyl esters $X = CH_2$-O-acyl 3,4-Didehydroretinyl esters

FIG. 1. Structures of retinoids discussed in the text.

Potter–Elvehjem homogenizer
Castroviejo keratotome or razor blade
Tissue culture plate (3.5 ml/well)
Glass tubes with Teflon-lined screw caps
Scintillation vials
Scintillation cocktail
CO_2 incubator

Retinoids. all-*trans*-Retinol was obtained from Sigma (St. Louis, MO). 3,4-Didehydroretinol and Ro 12-0586 [all-*trans*-9-(4-methoxy-2,3,6-trimethylphenyl)-3,7-dimethyl-2,4,6,8-nonatetraen-1-ol, i.e., the TMMP analog of retinol] are available from Hoffmann-La Roche (Basel, Switzerland). Retinyl and 3,4-didehydroretinyl esters are prepared by reacting 1 mg of retinol or 3,4-didehydroretinol in pyridine (0.2 ml) with the appropriate acyl chloride (10 mg) for 2 hr.[5] all-*trans*-[11,12(*n*)-^{3}H]Retinol (specific activity 40–60 Ci/mmol) was obtained from Radiochemical Centre, Amersham.

Preparation of Skin Samples

Excised skin (e.g., from mammary reduction surgery) is cooled and prepared for organ culture within 60 min. Sheets of skin mainly composed of epidermis and a thin layer of papillary dermis are prepared by the use of a Castroviejo keratotome set at 0.3 mm, or by cutting with a razor blade parallel to the skin surface (shave biopsy). The skin specimens are cut into smaller pieces (disks) measuring approximately 0.5 cm^2.

Organ Culture

The skin samples are organ cultured on tissue culture plates (1 ml medium/well) using the free floating model.[4] L-Glutamine (0.58 mg/ml), penicillin (60 μg/ml), gentamycin (50 μg/ml), and amphotericin B (50

[5] H. S. Huang and D. S. Goodman, *J. Biol. Chem.* **240,** 2839 (1965).

units/ml) are included in the medium. An incubation temperature of 32° is preferred, because this is the physiological temperature of the outer part of the skin. The humid atmosphere consists of 5% CO_2 in air. Using this method, the skin remains viable for more than 50 hr.[4] Normally, we culture the skin for 24 hr, but in the case of prolonged periods the medium is exchanged every 24 hr. Before addition of the radiotracer, the skin samples are allowed to equilibrize in the tissue culture medium for 1 hr.

all-*trans*-[^{3}H]Retinol diluted with unlabeled retinol is added to the medium at a final concentration of 1 μM. The tracer is either supplied with BSA (1 μCi/0.1 mg of protein/well) or coupled to retinol-binding protein (RBP).[4] The rate of 3,4-didehydro[^{3}H]retinol and [^{3}H]retinyl ester formation is higher when free tracer retinol is added than with the more physiological retinol–RBP complex. The culture is terminated by washing the skin sample twice in 5 ml of cold BSA-containing PBS (1%, w/v) and then twice in PBS. The incubated skin is separated into dermis and epidermis using fine curved forceps after heating the sample to 54° for 45 sec. The samples can be stored at $-70°$ until required.

Analysis

Detection of Newly Formed 3,4-Didehydro[^{3}H]retinol in Epidermis

The incubated epidermis is subjected to alkaline hydrolysis (20 min at 80°) in a mixture of 0.5 ml ethanol, 0.1 ml potassium hydroxide, 20 μg BHT, and 7.5 ng of Ro 12-0586 (internal standard). Alkaline hydrolysis improves the recovery of vitamin A compounds from the tissue and saponifies the vitamin A esters, thereby optimizing chromatographic detection of small amounts of radioactive products. Hydrolysis is stopped by cooling on ice for 5 min and adding 0.5 ml of water. The neutral retinoids are extracted twice with 4 ml hexane, and the hexane layers are transferred to a conical glass tube and evaporated under nitrogen. The retinoids are redissolved in methanol and subjected to reversed-phase high-performance liquid chromatography (HPLC). At our laboratory the HPLC system consists of a Beckman 110 pump in combination with an LDC 1203 UV monitor with fixed wavelength (360 nm). The monitor outlet is connected to a fraction collector (FRAC-100, Pharmacia, Uppsala, Sweden). A Nucleosil 5-μm ODS column (20 cm × 4.6 mm i.d.) is isocratically eluted with acetonitrile/water (80:20) at a flow rate of 1.0 ml/min (system A). The eluate is collected in scintillation vials (1-min fractions) and subjected to liquid scintillation counting after addition of 6 ml scintillation cocktail Emulsifier scintillator 299 (Packard Instruments, Downers Grove, IL).

As an example, after organ culture for 22 hr with [^{3}H]retinol, 3,4-didehydro[^{3}H]retinol is the main neutral metabolite produced (Fig. 2). Boiling

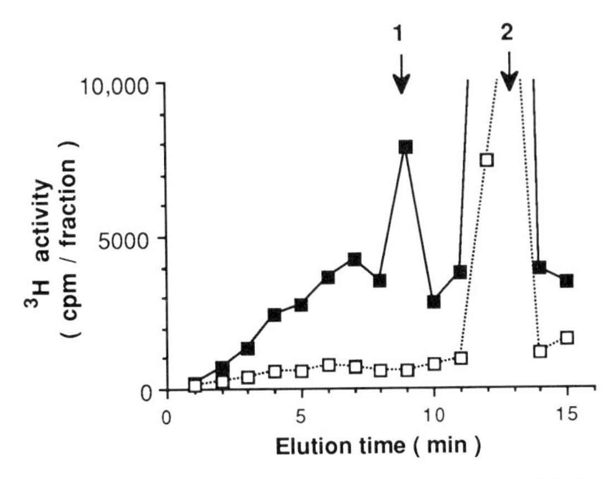

FIG. 2. HPLC of radioactive vitamin A compounds in human skin incubated for 22 hr with [^{3}H]retinol. Extraction with hexane after alkaline hydrolysis. Normal skin (■) and heat-treated skin (□). Fractions (1 ml) were collected and the ^{3}H activity monitored by liquid scintillation counting. Peak 1, 3,4-Didehydroretinol; peak 2, all-*trans*-retinol.

the skin sample for 10 min abolishes the formation of metabolites, thus disclosing enzyme-mediated reactions (Fig. 2). The product can be identified by its coelution in different HPLC systems with authentic 3,4-didehydroretinol either as such or after anhydro derivatization with ethanolic HCl (Fig. 3).[1] The formation of 3,4-didehydro[^{3}H]retinol is related to the protein content by a biuret technique using bovine serum albumin as internal standard.[6]

Analysis of Radioactive Retinyl and 3,4-Didehydroretinyl Esters

The incubated epidermis is lyophilized overnight and mechanically homogenized in phosphate buffer (50–75 mg tissue/ml) using a Teflon–glass Potter–Elvehjem homogenizer. Although extraction is not complete by this procedure, alkaline hydrolysis must be avoided in order to preserve the vitamin A esters. The sample is extracted twice with 4 ml of hexane after addition of ethanol to the buffer (45%, v/v). The combined extracts are evaporated to dryness, redissolved in 50 μl of methanol, and subjected to HPLC using the same C_{18} column as described above. Methanol at a flow rate of 1.8 ml/min is used as the mobile phase (system B). The eluate is collected in 1-min fractions in glass tubes.

Because no HPLC system currently available adequately resolves all the various esters of retinol and 3,4-didehydroretinol, fractions containing vitamin A esters must be saponified and rechromatographed in system A

[6] A. Vahlquist, H. Törmä, O. Rollman, and E. Anderson, this volume [18].

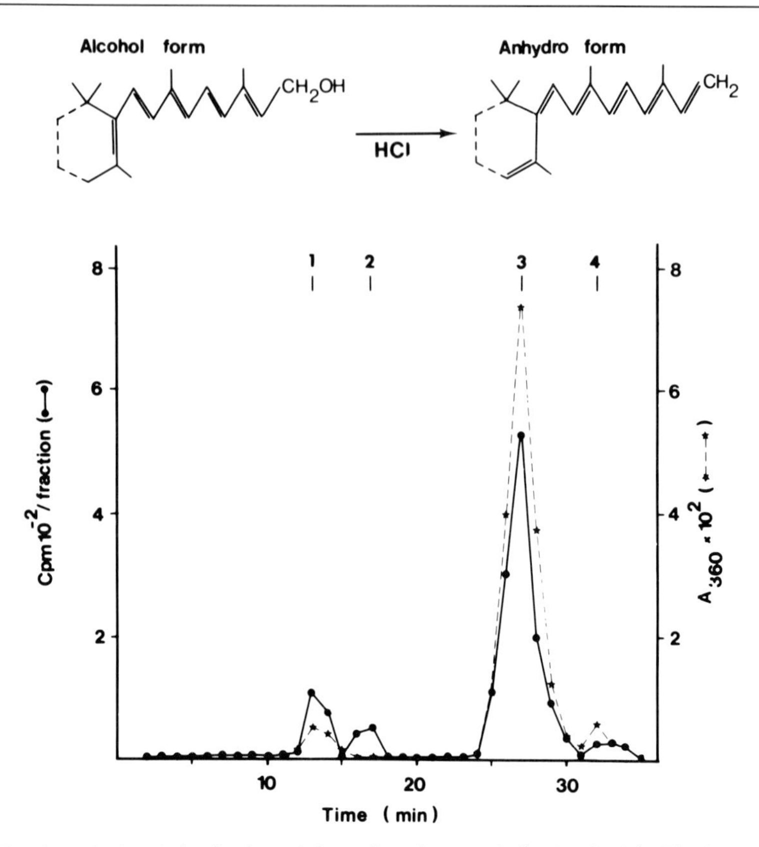

FIG. 3. Anhydro derivatization of the radioactive metabolite (peak 1) in Fig. 2. Peak 1, 3,4-Didehydroretinol; 2, retinol; 3, 3,4-didehydroanhydroretinol; 4, anhydroretinol. [Reprinted from H. Törmä and A. Vahlquist, Biosynthesis of 3-dehydroretinol (vitamin A_2) from all-*trans*-retinol (vitamin A_1) in human epidermis. *J. Invest. Dermatol.* **85** (No. 6), 498, © by Williams & Wilkins, 1985.]

(see above) for quantitation of the radioactivity associated with retinol and 3,4-didehydroretinol, respectively. Each fraction is evaporated under nitrogen and redissolved in a mixture of 0.5 ml ethanol, 0.1 ml potassium hydroxide, and 1 mg BSA (10 min, 80°). The retinoids are extracted twice with hexane after addition of 0.5 ml water and then subjected to HPLC (system A) as described above. Combined data from the two steps of analysis are used to construct the chromatogram (Fig. 4); the positions of several of the radioactivity peaks coincide with those of authentic retinyl and 3,4-didehydroretinyl esters. As a rule, the linoleate esters of retinol and 3,4-didehydroretinol predominate over the palmitate/oleate esters, similar to the situation *in vivo* (H. Törmä, 1989, unpublished observation). In order to evaluate esterification quantitatively, the amounts of retinyl esters

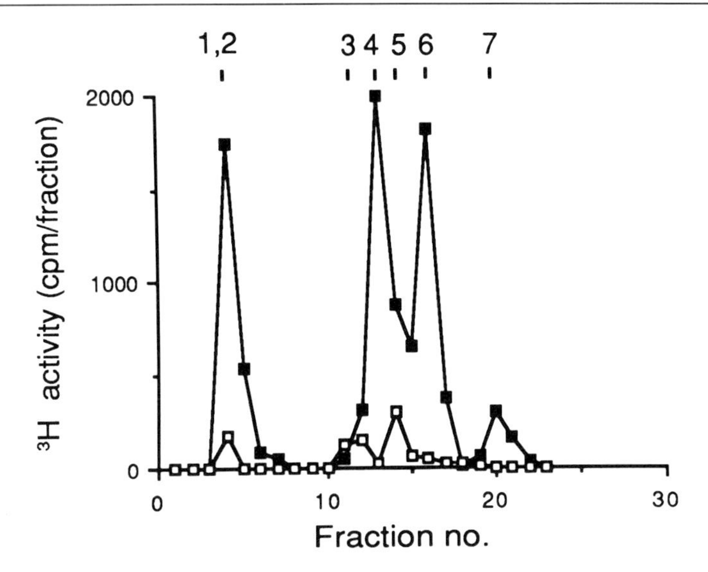

FIG. 4. HPLC of ^{3}H-labeled retinoids extracted from epidermis 48 hr after starting culture. The skin sample was first incubated for 24 hr with [^{3}H]retinol–RBP (1 μM) in the medium, then without tracer for a further 24 hr. The column was developed with tetrahydrofuran/methanol (8:92) at a flow rate of 1.5 ml/min. The eluate was collected in 1-min fractions and then analyzed for [^{3}H]retinol (■) and 3,4-didehydro[^{3}H]retinol (□) as described in the text. Numbers denote positions for authentic standards: 1, 3,4-Didehydroretinol; 2, retinol; 3, 3,4-didehydroretinyl myristate/linoleate; 4, retinyl myristate/linoleate; 5, 3,4-didehydroretinyl oleate/palmitate; 6, retinyl oleate/palmitate; 7, retinyl stearate. [Reprinted by permission of Elsevier Science Publishing Co., Inc., from H. Törmä and A. Vahlquist, Vitamin A esterification in human epidermis: A relation to keratinocyte differentiation. *J. Invest. Dermatol.* **94**, 132–138 (1990).]

are related to the protein content of the homogenized epidermis determined by an ordinary biuret after alkaline hydrolysis of the sample.[6]

Discussion

Short-term organ culture provides a physiological model for *in vitro* studies of vitamin A metabolism in human skin. An even closer resemblance to the *in vivo* situation can be obtained if the skin sample is mounted, dermal side down, on a Millipore filter placed on top of the well,[4] which allows for entry of nutrients and tracer to the epidermis solely through the basal membrane zone. A drawback is the much slower production of metabolites compared to the immersed system. The retinol metabolites formed in cultured epidermis are identical to those endogenously present in the skin, and under steady-state conditions the relative amounts

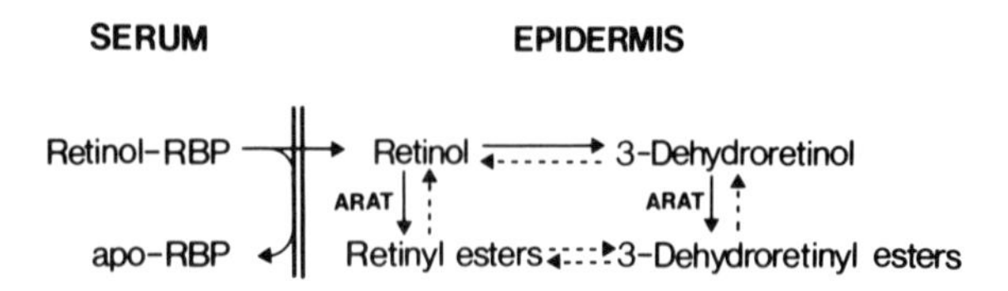

FIG. 5. Major metabolic pathways of retinol in human epidermis. ARAT, Acyl-CoA:retinol acyltransferase. Dashed lines indicate possible but not established pathways. 3-De-hydroretinol = 3,4-Didehydroretinol.

of radioactivity in the various retinoid fractions are proportional to those in epidermal retinoids *in vivo*. Another way of improving the yield of radioactive metabolites instead of [^{3}H]retinol is to remove the tracer from the medium during a 12–24 hr postincubation period.

Figure 5 illustrates some of the metabolic pathways of vitamin A in human epidermis. Some of the presumed metabolites, for example, retinal and retinoic acid, are difficult to detect in organ-cultured skin, probably because they are rapidly further metabolized and ultimately released to the medium as water-soluble conjugates or oxidized products (H. Törmä and A. Vahlquist, 1988, unpublished observation). The formation of retinyl and 3,4-didehydroretinyl esters has recently been reproduced in a cell-free system using human epidermal microsomes supplied with [^{14}C]palmitoyl-coenzyme A(CoA). Evidence suggests that the reaction is catalyzed by an acyl-CoA:retinol acyltransferase with an unusually low pH optimum.[7] Because aberrations in the epidermal formation of 3,4-didehydroretinol and various vitamin A esters probably occur in connection with certain disease states, further studies of vitamin A metabolism in organ-cultured human skin are needed.

Acknowledgements

We thank Marcia Skogh for kindly revising the English text. Supported by the Swedish Medical Research Council (Project No. 03x-07133), the Edvard Welander Foundation, and the Finsen Foundation.

[7] H. Törmä and A. Vahlquist, *J. Invest. Dermatol.* **94,** 132 (1990).

[23] Retinoid-Sensitive Cells and Cell Lines

By BRAD AMOS and REUBEN LOTAN

In this chapter we present a tabulated listing of retinoid-sensitive cells and cell lines. The listing is based on a literature search of a database containing citations from 1983 until May, 1989; additional references are derived from previous reviews. We have made every attempt to cite the primary reference. In all cases only effects that are due to a retinoid alone are included; synergistic or antagonistic effects with other agents are excluded from the list owing to space limitations. Only sources that demonstrated an effect of the retinoid on growth or differentiation are compiled.

RETINOID-SENSTIVE CELLS AND CELL LINES

Cell type and origin	Growth		Differentiation	Ref.[e]
	AD[a]	AID		
Neural crest- and brain-derived cells				
Neuroblastoma cells				
Mouse				
C1300	↓	ND[b]	↑[c]	*1*
S-20	↓	↓	↑	*2*
NB2a	↓	ND	↑	*3*
N18TG-2	↓	ND	ND	*4*
Human				
LA-N-1	↓	↓	↑	*5*
LA-N-2	↓	↓	↑	*6*
LA-N-5	↓	↓	↑	*6*
CHP 100	↓	↓	↑	*6*
CHP 126	ND	ND	↑	*7*
CHP 134	↓	↓	↑	*6*
SK-N-SH	ND	ND	↑	*6*
SK-N-DZ	↓	ND	ND	*8*
SK-N-BE	ND	ND	↑	*8*
SK-N-LE	ND	ND	↑	*8*
KA	↓	↓	↑	*6*
SH-SY5Y	ND	ND	↑	*9*
SMS-SAN	ND	ND	↑	*7*
IMR-32	ND	ND	↑	*10*
SMS-KCNR	↓	ND	↑	*10*

(continued)

RETINOID-SENSTIVE CELLS AND CELL LINES *(continued)*

Cell type and origin	Growth		Differentiation	Ref.[e]
	AD[a]	AID		
Glioma cells				
Human				
LG	↓	↓	ND	*11*
D54-MG	↓	↓	ND	*11*
KE	↓	ND	ND	*11*
EFC-2	↓	ND	ND	*11*
MM normal brain	↓	ND	ND	*11*
Rat				
C6	↓	ND	↑	*12, 13*
C6BU-1	↓	ND	ND	*4*
T-MG1	ND	↓	ND	*14*
Astrocytoma cells				
Human				
U 343 MG-A	↓	ND	↑	*15*
Retinoblastoma cells				
Human				
Y-79	↓	ND	ND	*16*
RBLA3	ND	ND	↑	*17*
RBLA14	ND	ND	↑	*17*
Melanoma cells				
Mouse				
S91-clone 2	↓	↓	↑	*18*
S91-clone M3	↓	↓	↑	*18*
MEL 11-A	↓	ND	ND	*19*
S91 CCL 53.1	ND	↓	ND	*20*
B16 parental	↓	ND	ND	*21*
B16 clone 9	↓	ND	ND	*21*
B16 clone 12	↓	ND	ND	*21*
B16 clone 14	↓	ND	ND	*21*
B16 clone 15	↓	ND	ND	*21*
B16-F1	↓	↓	ND	*18, 22*
B16-F10	↓	↓	ND	*21, 22*
B16 F10 Lr-6	↓	↓	ND	*21, 22*
B16O10	↓	ND	ND	*21*
B16B10n	↓	ND	ND	*21*
B16C3	ND	↓	ND	*23*
K1735-P	↓	↓	↑	*24*
Human				
A375	↓	↓	ND	*25*
Hs939	↓	↓	ND	*25*
Hs294	↑	=	ND	*25*
Hs 852	=	↓	ND	*25*
SH4	=	↓	ND	*25*

(continued)

RETINOID-SENSTIVE CELLS AND CELL LINES *(continued)*

Cell type and origin	Growth AD[a]	Growth AID	Differentiation	Ref.[e]
LiBr	ND	ND	↑	*26*
HXG-2	↓	ND	↑	*27*
C8146c	ND	↓	ND	*28*
JCT-Mel 2	↓	ND	↑	*29*
RPMI 7931	ND	ND	↑	*8*
MIRW P	↓	ND	↑	*30*
MIRW clone A6	↓	ND	↑	*30*
MIRW clone A9	↓	ND	↑	*30*
MIRW clone A15	↓	ND	↑	*30*
Mesenchymal cells				
Hematopoietic cells				
Mouse				
BW5147 lymphoma	↓	ND	ND	*31*
S49 lymphoma	↓	ND	ND	*31*
EL4 lymphoma	↓	ND	ND	*31*
L1210 leukemia	↓	ND	ND	*31*
S194 myeloma	↓	ND	ND	*31*
P3 myeloma	↓	ND	ND	*31*
RAW 117 lymphoma	↓	ND	ND	*31*
M1 myeloid	ND	ND	↓	*32*
Friend erythroleukemia	ND	ND	↑	*33*
P388D1 macrophage-like	↓	ND	↑	*34*
J774.2 macrophage-like	↓	ND	↑	*34*
WEHI-265 macrophage-like	↓	ND	↑	*34*
WEHI-3 macrophage-like	↓	ND	=	*34*
PV-5 macrophage-like	↓	ND	=	*34*
MHY206 hybridoma	↓	ND	ND	*35*
Human				
RPMI 1788 B cells	↓	ND	ND	*31*
Daudi Burkitt's lymphoma	↓	ND	ND	*31*
KG-1 myeloblast	↓	ND	ND	*36*
ML-1 myeloblast	↓	ND	↑	*37*
RFD-2 myeloblast	ND	ND	↑	*38*
HL-60 promyelocyte	ND	ND	↑	*39*
HL-60/MRI promyelocyte	ND	ND	↑	*40*
U-937 monoblast	ND	ND	↑	*41*
THP-1 promonocytic	ND	ND	↑	*42*
Namalva lymphoblast	↓	ND	ND	*43*
Normal bone marrow	↓	ND	↓	*44*
Muscle cells				
Human				
RMS rhabdomyosarcoma	ND	ND	↑	*45*
Rat				
BA-HAN-1C rhabdomyosarcoma	↓	ND	↑	*46*

(continued)

RETINOID-SENSTIVE CELLS AND CELL LINES *(continued)*

Cell type and origin	Growth		Differentiation	Ref.[e]
	AD[a]	AID		
Bone- and cartilage-derived cells				
Rat				
ROS 17/2.8 osteosarcoma	ND	ND	↑	*47*
ROS 17/2 osteosarcoma	ND	ND	↑	*47*
RCJ 3.2T.1 osteosarcoma	ND	ND	↑	*47*
RCJ 3.2.4.1 CAM osteosarcoma	ND	ND	↑	*47*
RCJ 3.2CE2.1 osteosarcoma	ND	ND	↑	*47*
UMR-106-01 osteoblast-like	ND	ND	↑	*48*
Chicken				
Chondrocytes	↓	ND	↓	*49*
Human				
Te85 osteosarcoma	↓	↓	ND	*22*
Hs781 osteosarcoma	↓	↓	ND	*22*
Hs791 osteosarcoma	↓	↓	ND	*50*
Hs705 chondrosarcoma	↓	↓	ND	*50*
Fibroblastic cells				
Mouse				
3T3 fibroblasts	↓	ND	ND	*51*
3T6 fibroblasts	↓	ND	ND	*51*
AKR-2B mouse embryo	↓	ND	ND	*52*
AKR-MCA chemically transformed	↓	ND	ND	*52*
L929 chemically transformed	ND	↓	ND	*23*
UV-2337P fibrosarcoma	ND	↓	ND	*22*
Rat				
NRK-SA6				
Clone 536-3-1 untransformed	↑	ND	ND	*53*
Clone 536-3-9 transformed	ND	↑	ND	*53*
Rabbit				
Dermal fibroblasts	↓	ND	ND	*54*
Hamster				
BHK fibroblasts	↓	ND	ND	*31*
DES-4 v-*src* transformed	ND	↓	ND	*55*
DES-4 H-*ras* transformed	ND	↑	ND	*55*
Chinese hamster ovary (CHO)	↓	ND	ND	*56*
Human				
WI-38 adult lung	↓	ND	ND	*57*
VA13A SV40 transformed	↓	ND	ND	*57*
IMR-90 adult lung	↓	ND	ND	*57*
WISH transformed placental	↓	ND	ND	*43*
F1 foreskin	↓	ND	ND	*58*
HT 1080 fibrosarcoma	↓	ND	ND	*31*
Normal adult skin	↑	ND	ND	*59*

(continued)

RETINOID-SENSTIVE CELLS AND CELL LINES *(continued)*

Cell type and origin	Growth		Differentiation	Ref.[e]
	AD[a]	AID		
Epithelial cells				
Mammary cells				
Mouse				
Mm5mT carcinoma	↓	ND	ND	*31*
M12 carcinoma	↓	ND	ND	*31*
DD3 carcinoma	↓	ND	ND	*31*
Rat				
13762 NF carcinoma	↓	ND	ND	*31*
DMBA No. 8 carcinoma	↓	ND	ND	*31*
R3230AC carcinoma	↓	ND	ND	*31*
Human				
734 B carcinoma	↓	ND	ND	*25*
Hs578T carcinoma	↓	ND	ND	*25*
MDA-MB-157 carcinoma	↓	ND	ND	*25*
SK-BR-3 carcinoma	↓	ND	ND	*25*
MCF-7 carcinoma	↓	ND	ND	*60*
T47D carcinoma	↓	ND	ND	*61*
ZR-75-B carcinoma	↓	ND	ND	*62*
MDA-MB-231 carcinoma	↓	ND	ND	*63*
Rat				
Rama 25 carcinoma	ND	ND	↑	*64*
Upper respiratory tract				
Rat				
RE-149 normal esophageal	ND	ND	↓	*65*
Rabbit				
RbTE normal tracheal	ND	ND	↓	*66*
Human				
NHBE normal bronchial	ND	ND	↓	*67*
1483 SC[d]	↓	↓	ND	*68*
183A SC	=	↑	ND	*68*
UMSCC-10A SC	↓	ND	ND	*69*
UMSCC-19 SC	↓	ND	ND	*69*
UMSCC-22B SC	↓	ND	ND	*69*
UMSCC-30 SC	↓	ND	ND	*69*
UMSCC-35 SC	↓	ND	ND	*69*
Lung cells				
Human				
RH-SCL-L11 SC	↓	↓	ND	*70*
RH-SCL-L10 SC	↓	↓	ND	*70*
LP-9 mesothelial	ND	ND	=	*71*
N417d small cell carcinoma	ND	↓	ND	*72*
SK-MES SC	ND	↓	ND	*72*
P3 SC	ND	↓	ND	*72*

(continued)

RETINOID-SENSTIVE CELLS AND CELL LINES *(continued)*

Cell type and origin	Growth		Differentiation	Ref.[e]
	AD[a]	AID		
P6 adenocarcinoma	ND	↓	ND	*72*
P7 adenocarcinoma	ND	↓	ND	*72*
P8 adenocarcinoma	ND	↓	ND	*72*
H23 adenocarcinoma	ND	↓	ND	*72*
SUT adenocarcinoma	ND	↓	ND	*72*
Lu-134-B-S small cell carcinoma	ND	ND	↑	*73*
Mouse				
Lewis lung carcinoma	↓	ND	ND	*74*
Colorectal cells				
Human				
DLD-2 carcinoma	=	↓	↑	*75*
COLO 205 carcinoma	↓	ND	ND	*76*
SW 620 carcinoma	↓	ND	ND	*76*
HT 29 carcinoma	↓	ND	ND	*76*
MOSER carcinoma	↓	ND	ND	*77*
Rat				
IEC-6 small intestinal	↓	ND	ND	*78*
Skin cells				
Human				
Conjunctival keratinocytes	ND	ND	↓	*79*
Epidermal cells	ND	ND	↓	*79*
SCC-13 SC	ND	ND	↓	*80*
SCC-12 SC	ND	ND	↓	*81*
SCC-15 SC	ND	ND	↓	*81*
SqCC/Y1 SC	↓	ND	↓	*82*
SCC-4 SC	ND	ND	↓	*83*
SCC-9 SC	ND	ND	↓	*83*
SCC-12F2 SC	ND	ND	↓	*83*
12B2 SC	ND	ND	↓	*83*
13B2 SC	ND	ND	↓	*83*
Mouse				
Epidermal cells				
JB6 clone 41 untransformed	↓	ND	ND	*84*
JB6 clone aT7 untransformed	↓	ND	ND	*84*
Guinea pig				
Epidermal cells	↑	ND	ND	*85*
Other cells				
Embryonal carcinoma cells				
Mouse				
F-9	↓	↓	↑	*86*
PC-13	ND	ND	↑	*87*
PCC4-aza1	ND	ND	↑	*88*
PCC4-aza 1R	ND	ND	↑	*88*

(continued)

RETINOID-SENSTIVE CELLS AND CELL LINES *(continued)*

Cell type and origin	Growth AD[a]	Growth AID	Differentiation	Ref.[e]
Nulli-SCCI	ND	ND	↑	*88*
6050 AJ	ND	ND	↑	*88*
P19	ND	ND	↑	*89*
311	ND	ND	↑	*90*
IH5 teratocarcinoma	ND	ND	↑	*91*
Human				
Tera-2	↓	ND	↑	*92*
YK teratoma	↓	ND	ND	*93*
Other cell types				
Mouse				
L cells	↓	ND	ND	*94*
LPA transformed L cells	↓	ND	ND	*95*
TM3 Leydig cells	↓	ND	ND	*96*
TM4 Sertoli cells	↑	ND	ND	*97*
ST123 adipocytes	ND	ND	↓	*98*
Rat 3T3-F442A adipocytes	ND	ND	↓	*99*
Bovine aortic endothelial cells	↑	ND	ND	*100*
NHIK 3025 hyperplastic cervix	↓	ND	ND	*101*
HeLa carcinoma	↓	↓	ND	*22, 23*
Rat NB II bladder tumor	↑	ND	ND	*102*

[a] AD, Anchorage-dependent growth; AID, anchorage-independent growth.
[b] ND, Not determined.
[c] Arrows pointing down indicate suppression; arrows pointing up indicate stimulation.
[d] SC, Squamous carcinoma.
[e] *Key to References: (1)* R. Lotan and G. L. Nicolson, *J. Natl. Cancer Inst.* **59,** 1717 (1977); *(2)* N. Prashad, D. Lotan, and R. Lotan, *Cancer Res.* **47,** 2417 (1987); *(3)* T. B. Shea, I. Fischer, and V. Sapirstein, *Dev. Brain Res.* **21,** 307 (1985); *(4)* H. Higashida, N. Miki, M. Ito, T. Iwata, and K. Tsukida, *Int. J. Cancer* **33,** 677 (1984); *(5)* N. Sidell, *J. Natl. Cancer Inst.* **68,** 589 (1982); *(6)* N. Sidell, A. Altman, M. R. Haussler, and R. C. Seeger, *Exp. Cell Res.* **148,** 21 (1983); *(7)* M. Tsokos, S. Scarpa, R. A. Ross, and T. J. Triche, *Am. J. Pathol.* **128,** 484 (1987); *(8)* L. Helson and C. Helson, *J. Neuro-Oncol.* **3,** 39 (1985); *(9)* S. Palman, A. Ruusala, L. Abrahmsson, M. E. K. Mattson, and T. Esscher, *Cell Differ.* **14,** 135 (1984); *(10)* C. J. Thiele, C. P. Reynolds, and A. M. Isreal, *Nature (London)* **313,** 405 (1985); *(11)* W. K. A. Yung, R. Lotan, P. L. Lee, D. Lotan, and P. A. Steck, *Cancer Res.* **49,** 1014 (1989); *(12)* S. K. Chapman, *Life Sci.* **26,** 1359 (1980); *(13)* I. Fischer, C. E. Nolan, and T. B. Shea, *Life Sci.* **41,** 463 (1987); *(14)* E. Helseth, G. Unsgaard, A. Dalen, and R. Vik, *Cancer Immunol. Immunother.* **26,** 273 (1988); *(15)* J. T. Rutka, S. J. DeArmond, J. Giblin, J. R. McCulloch, C. B. Wilson, and M. L. Rosenblum, *Int. J. Cancer* **42,** 419 (1988); *(16)* A. Kyritsis, G. Joseph, and G. J. Chader, *J. Natl. Cancer Inst.* **73,** 649 (1984); *(17)* E. Bogenmann, *Int. J. Cancer* **38,** 883 (1986); *(18)* R. Lotan, G. Giotta, E. Nork, and G. L. Nicolson, *J. Natl. Cancer Inst.* **60,** 1035 (1978); *(19)* M. R. Haussler, C. A. Donaldson, M. A. Kelly, D. J. Mangelsdorf,

G. T. Bowden, W. J. Meinke, F. L. Meyskens, and N. Sidell, *Biochim. Biophys. Acta* **803,** 54 (1984); *(20)* M. D. Bregman, E. Peters, D. Sander, and F. L. Meyskens, Jr., *J. Natl. Cancer Inst.* **71,** 927 (1983); *(21)* R. Lotan and G. L. Nicolson, *Cancer Res.* **39,** 4767 (1979); *(22)* R. Lotan, D. Lotan, and A. Kadouri, *Exp. Cell Res.* **141,** 79 (1982); *(23)* L. D. Dion, J. E. Blalock, and G. E. Gifford, *Exp. Cell Res.* **117,** 15 (1978); *(24)* R. Lotan, D. Lotan, and B. Amos, *Exp. Cell Res.* **177,** 284 (1988); *(25)* R. Lotan, *Cancer Res.* **39,** 1014 (1979); *(26)* J. A. Werkmeister, T. Triglia, I. R. Mackay, J. P. Dowling, G. A. Varigos, G. Morstyn, and G. F. Burns, *Cancer Res.* **47,** 225 (1987); *(27)* Y. Yongshan and W. Stanley, *Cancer Genet. Cytogenet.* **31,** 253 (1988); *(28)* M. D. Bregman, E. Peters, D. Sanders, and F. L. Meyskens, Jr., *J. Natl. Cancer Inst.* **71,** 927 (1983); *(29)* E. Hoal, E. L. Wilson, and E. B. Dowdle, *Cancer Res.* **42,** 5191 (1982); *(30)* F. L. Meyskens, Jr., and B. B. Fuller, *Cancer Res.* **40,** 2194 (1980); *(31)* R. Lotan and G. L. Nicolson, *J. Natl. Cancer Inst.* **59,** 1717 (1977); *(32)* K. Takenaga, M. Hozumi, and Y. Sakagami, *Cancer Res.* **40,** 914 (1980); *(33)* L. C. Garg and J. C. Brown, *Differentiation* **25,** 79 (1983); *(34)* R. Goldman, *J. Cell. Physiol.* **120,** 91 (1984); *(35)* M. Bosma and N. Sidell, *J. Cell. Physiol.* **135,** 317 (1988); *(36)* D. Douer and H. P. Koeffler, *J. Clin. Invest.* **69,** 277 (1982); *(37)* R. W. Craig, O. S. Frankfurt, H. Sakagami, K. Takeda, and A. Blach, *Cancer Res.* **44,** 2421 (1984); *(38)* J. Fontana, M. Munoz, and J. Durham, *Leuk. Res.* **9,** 1127 (1985); *(39)* T. R. Breitman, S. E. Selonick, and S. J. Collins, *Proc. Natl. Acad. Sci. U.S.A.* **77,** 2936 (1980); *(40)* M. Imaizumi, J. Uozumi, and T. Breitman, *Cancer Res.* **47,** 1434 (1987); *(41)* I. L. Olsson and T. R. Breitman, *Cancer Res.* **42,** 3924 (1982); *(42)* K. Mehta and G. Lopez-Berenstein, *Cancer Res.* **46,** 1388 (1986); *(43)* M. F. Bourgeade and F. Besacon, *Cancer Res.* **44,** 5355 (1984); *(44)* E. C. Bradley, F. W. Ruscetti, H. Steinberg, C. Paradise, and C. Blaine, *J. Natl. Cancer Inst.* **71,** 1189 (1983); *(45)* A. J. Garvin, W. S. Stanley, D. D. Bennet, J. S. Sullivan, and D. A. Sens, *Am. J. Pathol.* **125,** 208 (1986); *(46)* H. E. Gabbert, C. D. Gerharz, H. K. Biesalski, R. Engers, and C. Luley, *Cancer Res.* **48,** 5264 (1988); *(47)* P. M. Petkovich, J. N. M. Heersche, J. E. Aubin, A. E. Grigoriades, and G. Jones, *J. Natl. Cancer Inst.* **78,** 265 (1987); *(48)* N. C. Parteridge, J. J. Jeffrey, L. S. Ehlich, S. L. Teitelbaum, C. Fliszar, H. G. Welgus, and A. J. Kahn, *Endocrinology* **120,** 1956 (1987); *(49)* R. Hein, T. Krieg, P. K. Mueller, O. Braun-Falco, *Biochem. Pharmacol.* **33,** 3263 (1984); *(50)* R. Thein and R. Lotan, *Cancer Res.* **42,** 4711 (1982); *(51)* A. M. Jetten, M. E. R. Jetten, S. S. Shapiro, and J. J. Poon, *Exp. Cell Res.* **119,** 289 (1979); *(52)* N. M. Hoosein, D. E. Brattain, M. K. McKnight, and M. G. Brattain, *Exp. Cell Res.* **175,** 125 (1988); *(53)* A. M. Jetten and R. H. Goldfarb, *Cancer Res.* **43,** 2094 (1983); *(54)* R. A. Harper and T. Burgoon, *Cell Biol. Int. Rep.* **6,** 163 (1982); *(55)* A. M. Jetten, J. C. Barrett, and T. M. Gilmer, *Mol. Cell. Biol.* **6,** 3341 (1986); *(56)* M. K. Haddox and D. H. Russell, *Cancer Res.* **39,** 2476 (1979); *(57)* B. M. Stanulis-Praeger, C. H. Jacobus, and A. E. Nutall, *Nutr. Cancer* **8,** 171 (1986); *(58)* A. Lacroix, G. D. L. Anderson, and M. E. Lippman, *Exp. Cell Res.* **130,** 339 (1980); *(59)* R. A. Harper and C. R. Savage, *Endocrinology* **107,** 2113 (1980); *(60)* H. Ueda, T. Takenawa, J. C. Millan, M. S. Gesell, and D. Brandes, *Cancer* **46,** 2203 (1980); *(61)* N. T. Wetherall and C. M. Taylor, *Eur. J. Cancer Clin. Oncol.* **22,** 53 (1986); *(62)* A. Lacroix and M. E. Lippman, *J. Clin. Invest.* **65,** 586 (1980); *(63)* L. D. Fraker, S. A. Halter, and J. T. Forbes, *Cancer Res.* **44,** 5757 (1984); *(64)* F. C. Paterson and P. S. Rudland, *J. Cell. Physiol.* **124,** 525 (1985); *(65)* J. I. Rearick, G. D. Stoner, M. A. George, and A. M. Jetten, *Cancer Res.* **78,** 5289 (1988); *(66)* J. I. Rearick and A. M. Jetten, *J. Biol. Chem.* **261,** 13898 (1986); *(67)* J. I. Rearick, T. W. Hesterberg, and A. M. Jetten, *J. Cell. Physiol.* **133,** 573 (1987); *(68)* R. Lotan, P. G. Sacks, D. Lotan, and W. K. Hong, *Int. J. Cancer* **40,** 224 (1987); *(69)* R. Lotan, J. S. Kim, M. Maarouri, S. P. Schantz, and W. K. Hong, *Cancer Bull.* **39,** 93 (1987); *(70)* L. Olsson, O. Behnke, and

H. R. Soresen, *Int. J. Cancer* **35,** 189 (1985); *(71)* K. H. Kim, V. Stellmach, J. Javors, and E. Fuchs, *J. Cell Biol.* **105,** 3039 (1987); *(72)* M. Munker, R. Munker, R. E. Saxton, and H. P. Koeffler, *Cancer Res.* **47,** 4081 (1987); *(73)* T. Terasaki, Y. Shimosato, T. Nakajima, M. Tsumuraya, H. Ichinose, T. Nagatsu, and K. Kato, *Cancer Res.* **47,** 3533 (1987); *(74)* L. J. Wilfoff, E. A. Dulmadge, and D. H. Chopra, *Proc. Soc. Exp. Biol. Med.* **163,** 233 (1980); *(75)* R. M. Niles, S. A. Wilhelm, P. Thomas, and N. Zamcheck, *Cancer Invest.* **6,** 39 (1988); *(76)* G. Keri, A. Balogh, I. Telpan, and O. Csuka, *Tumor Biol.* **9,** 315 (1988); *(77)* K. M. Mulder and M. G. Brattain, *Int. J. Cancer* **42,** 64 (1988); *(78)* W. Sasak, A. Herscovics, and A. Quaroni, *Biochem. J.* **201,** 359 (1982); *(79)* E. Fuchs and H. Green, *Cell* (Cambridge, Mass.) **25,** 617 (1981); *(80)* S. M. Thacher, E. L. Coe, and R. Rice, *Differentiation* **29,** 82 (1985); *(81)* K. H. Kim, F. Schwartz, and E. Fuchs, *Proc. Natl. Acad. Sci. U.S.A.* **81,** 4280 (1984); *(82)* M. Reis, S. W. Pitman, and A. C. Sartorelli, *J. Natl. Cancer Inst.* **74,** 1015 (1985); *(83)* A. Rubin and R. M. Rice, *Cancer Res.* **46,** 2356 (1986); *(84)* H. L. Gensler, L. M. Matrisian, and G. T. Bowden, *Cancer Res.* **45,** 1922 (1985); *(85)* E. Christophers, *J. Clin. Invest. Dermatol.* **63,** 450 (1974); *(86)* S. Strickland and V. Mahdavi, *Cell (Cambridge, Mass.)* **15,** 393 (1978); *(87)* E. D. Adamson, S. J. Grunt, and C. F. Graham, *Cell (Cambridge, Mass.)* **17,** 469 (1979); *(88)* A. M. Jetten, M. E. R. Jetten, and M. I. Sherman, *Exp. Cell Res.* **124,** 381 (1979); *(89)* M. W. McBurney, E. M. V. Jones-Villeneve, M. K. S. Edwards, and P. J. Anderson, *Nature (London)* **299,** 165 (1982); *(90)* Y. Ogiso, A. Kume, Y. Nishimune, and A. Matsushiro, *Exp. Cell Res.* **137,** 365 (1982); *(91)* E. D. Adamson, S. Strickland, M. Tu, and B. Kahan, *Differentiation* **29,** 68 (1985); *(92)* P. W. Andrews, *Dev. Biol.* **103,** 285 (1984); *(93)* Y. Kikuchi, E. Momose, I. Kizawa, M. Ishida, H. Sunaga, K. Mukai, K. Seki, and K. Kato, *Cancer Res.* **44,** 2952 (1984); *(94)* L. D. Dion, J. E. Blalock, and G. E. Gifford, *J. Natl. Cancer Inst.* **58,** 795 (1977); *(95)* J. Koziorowska, K. Paczek, and J. Tautt, *Exp. Cell Res.* **150,** 97 (1984); *(96)* J. Mather, *Biol. Reprod.* **23,** 243 (1980); *(97)* J. Mather and D. M. Phillips, *in* "Retinoids and Cell Differentiation" (M. Sherman, ed.), p 17. CRC Press, Boca Raton, Florida, 1986; *(98)* M. Sato, K. Shudo, and A. Hiragun, *J. Cell. Physiol.* **135,** 179 (1988); *(99)* J. Pairault and F. Lasnier, *J. Cell. Physiol.* **132,** 279 (1987); *(100)* G. Melnykovych and K. K. Clowes, *J. Cell. Physiol.* **109,** 265 (1981); *(101)* O. Bakke and T. Espevik, *Exp. Cell Res.* **180,** 20 (1989); *(102)* R. Tchao and J. Leighton, *Invest. Urol.* **16,** 476 (1979).

Section II

Nutrition, Tissue and Immune Status, and Antioxidant Action

A. Nutrition

Articles 24 and 25

B. Tissue and Immune Status

Articles 26 through 28

C. Antioxidant Action

Articles 29 and 30

[24] Preparation of Vitamin A-Deficient Rats and Mice

By JOHN EDGAR SMITH

Introduction

The production of vitamin A-deficient animals is a relatively simple process. However, four very important things must be remembered: (1) start with animals with low vitamin A reserves, (2) use a diet nutritionally adequate in all nutrients other than vitamin A, (3) be certain that the diet is free of vitamin A, and (4) maintain strict sanitation.

Diet

Although it is still sold, the vitamin A-deficient diet formulated by the U.S.P. Convention in 1936[1] is not suitable for use. This diet is deficient in choline, vitamin B_6, calcium, manganese, zinc, copper, and iodine in addition to vitamin A.

The American Institute of Nutrition is currently considering a reformulation of its AIN-76 rodent diet.[2] In the meantime, minor modifications of the AIN-76A[3,4] diet will provide a diet suitable for studying vitamin A deficiency in rats or mice. These modifications include (1) removing all sources of vitamin A from the diet; (2) adding 500 *u*g of vitamin K per kg of diet;[4] (3) using a 50/50 mixture of glucose and starch rather than sucrose as the source of carbohydrate; (4) adding 1 g of ascorbic acid per kg of diet;[5,6] and (5) use peanut oil or soybean oil as the source of lipids. The suggested formulation of the diet is shown in Table I.

A solid diet can be prepared by including agar as 2.4% of the diet.[7] The agar will replace an equal amount of the glucose. After all of the other ingredients have been well mixed, dissolve the agar (24 g) in 1000 ml of water at 80° and then thoroughly mix the agar solution with 976 g of the other ingredients. Spread the slurry in layers about 2.5 cm thick in shallow

[1] U.S.P. Convention, "The Pharmacopoeia of the United States of America," 11th Ed., Mack Printing, Easton, Pennsylvania, 1936.

[2] P. G. Reeves, *Nutr. Notes* **25**(2), 6 (1989).

[3] J. G. Bieri, G. S. Stoewsand, G. M. Briggs, R. W. Phillips, J. C. Woodard, and J. J. Knapka, *J. Nutr.* **107,** 1340 (1977).

[4] J. G. Bieri, *J. Nutr.* **110,** 1726 (1980).

[5] J. Mayer and W. A. Krehl, *J. Nutr.* **35,** 523 (1948).

[6] W. E. Rogers, Jr., *Am. J. Clin. Nutr.* **22,** 1003 (1969).

[7] M. H. Green, J. B. Green, and K. C. Lewis, *J. Nutr.* **117,** 694 (1987).

TABLE I
MODIFICATION OF AIN-76A DIET FOR PRODUCTION OF VITAMIN A DEFICIENCY[a,b]

Ingredient	% (w/w)
Vitamin-free casein[c]	20.0
DL-Methionine	0.3
Cellulose-type fiber	5.0
Peanut oil or soybean oil[d]	5.0
Cornstarch	32.5
Glucose	32.4
AIN-76 mineral mix[a]	3.5
Vitamin A-free AIN-76A vitamin mix[a,b,e]	1.0
Choline bitartrate	0.2
Ascorbic acid	0.1
	100.0

[a] J. G. Bieri, G. S. Stoewsand, G. M. Briggs, R. W. Phillips, J. C. Woodard, and J. J. Knapka, *J. Nutr.* **107,** 1340 (1977).

[b] J. G. Bieri, *J. Nutr.* **110,** 1726 (1980).

[c] Before use, commercial vitamin-free casein is refluxed with 2.5 volumes (w/v) of 95% ethanol and 0.0025 volumes of concentrated HCl for 8 hr [P. R. Sundaresan, V. G. Winters, and D. G. Therriault, *J. Nutr.* **92,** 474 (1967)]. The casein is then washed with ethanol and air-dried on a Büchner funnel.

[d] Pure peanut oil and pure soybean oil have negligible vitamin A activity.

[e] The modified vitamin mix contains the following per kg of vitamin mix: thiamin-HCl, 600 mg; riboflavin, 600 mg; pyridoxine-HCl, 700 mg; nicotinic acid, 3 g; *D*-calcium pantothenate, 1.6 g; folic acid, 200 mg; D-biotin, 20 mg; cyanocobalamin, 5 mg; all-*rac*-α-tocopheryl acetate, 5000 IU as stabilized powder; vitamin D_3, 2.5 mg; menadione sodium bisulfite, 50 mg; and powdered sucrose to make 1000.0 g.

pans. Cut the diet into 2.5-cm squares, and then freeze and lyophilize the diet. The solid cubes substantially reduce the wastage of the diet.

A control diet can be prepared by adding 1200 to 3600 retinol equivalents (RE) of retinyl esters per kg of diet. A gelatin-stabilized form of the vitamin is preferred.

Animals

Ideally the animals to be depleted will have very modest liver vitamin A stores. The typical accumulation of vitamin A in the liver of young rats is

TABLE II
ACCUMULATION OF VITAMIN A IN YOUNG RATS[a]

Age (days)	Liver Vitamin A (μg)
0	3 ± 1
20	44 ± 3
21	51 ± 3
25	79 ± 6
29	150 ± 19

[a] Data derived from three separate experiments.

shown in Table II. The transfer of vitamin A to the young rat through the milk is quite modest compared to the later accumulation.[8] When the young rats begin to eat the diet of the dam the liver vitamin A stores increase quite rapidly. Rats are typically weaned on Day 21, but weaning on Day 19 or 20 may shorten the depletion period by 2 weeks or more. The liver stores deplete at a rate of about 2 *u*g/day.

Commercial specific-pathogen-free rat colonies are frequently fed diets that contain 12,300 RE (41,000 IU) of vitamin A/kg. As a result weanling rats may have more than 200 μg of vitamin A in their livers by Day 20. A few suppliers have responded to the complaints of vitamin A researchers and have lowered the vitamin A content, but many have not. Asking specific questions about the vitamin A content of the diet fed to the colony avoids erroneous results. Most major suppliers will provide the names of investigators who are doing vitamin A research with rats.

Mice also are difficult to deplete of their vitamin A stores.[9] Pups weaned from dams consuming commercial chow diets may require 1 year to become fully depleted.

If the animals to be depleted are mice or specific-pathogen-free rats with high vitamin A stores, the studies should be done with the second generation. Pregnant females fed the vitamin A-free diet (Table I) from the 10th day of gestation through the lactation period will produce pups with low vitamin A stores. Either rat or mouse pups produced in this manner should reach the growth plateau by approximately 7 weeks of age.[9] Some commercial suppliers will provide pregnant female rats with timed pregnancies based on vaginal smears or sperm plugs. Timed pregnancies based only on exposure to the male are not suitable.

[8] A. J. Lamb, P. Apiwatanaporn, and J. A. Olson, *J. Nutr.* **104,** 1140 (1974).
[9] S. M. Smith, N. S. Levy, and C. E. Hayes, *J. Nutr.* **117,** 857 (1987).

Production of Vitamin A Deficiency

Rats should be housed in hanging cages with one animal per cage. Tray or box type cages with wood chip bedding are also satisfactory. To reduce infection, the cages should be positioned at least 50 cm above the floor. The air close to the floor has a much higher bacterial population than the air higher in the room. If possible, no other animals should be housed in the same room. Definitely, no new animals should be placed in the room after the depletion has been started. The introduction of new pathogens will cause a severe weight loss and possibly death in the vitamin A-deficient animals. This will occur even if all animals are specific-pathogen-free. The animals should be provided with a continuous supply of fresh water and the vitamin A-free diet (Table I).

In the early phase of the depletion the young rats should be petted, so that they enjoy being handled. If severely deficient rats are afraid or dislike being handled, they are very apt to bite. A few extra minutes playing with the young rats can make them much easier to handle later in the study. Mice are more timid and do not respond to handling.

As the animals become depleted, they usually become dissatisfied with their diet. This dissatisfaction may be expressed in several ways including urination on the food, excessive wasting of the diet, or refusal to eat the diet. Many of these problems can be avoided by using the solid diet described above. A few chunks of diet can be offered each day in a feeder suspended from the top of the cage. Wastage of the diet will be substantially reduced, and clean food will be available if the animal should decide to eat the diet later in the day.

The weight of the rats should be followed on a regular basis. In our laboratory we typically weigh rats on Monday, Wednesday, and Friday. After each weighing, the mean daily weight gain of the rats fed the vitamin A-free diet is compared with the gain of control animals. The daily weight gain is a more sensitive indicator of vitamin A status than total body weight. Individual animal weight gains are also examined. Any animal that fails to have an adequate weight gain is closely examined. If the animal shows any sign of disease other than vitamin A deficiency, it is removed from the colony and sacrificed. Any animal that appears healthy but consistently has a very low rate of gain is also sacrificed. Dwarf rats have very unusual vitamin A metabolism, so they should not be used in experiments with normal animals. After 3–4 weeks on the vitamin A-free diet the rats may start to show a slight reduction in weight gain compared to control rats.

In rats, the serum vitamin A levels can also be monitored as an indication of vitamin A depletion. Blood samples may be obtained from a prick of a tail vein or from the subclavian venous plexus of anesthetized rats.[10] Tail vein pricks bleed better if they are not too large or deep. In mice, blood can be obtained from the retroocular venous sinus with a Pasteur pipette.[11] However, this procedure is not suitable for routine sampling, and excessive bleeding can occur. The small volumes of serum or plasma obtained by these methods can be analyzed by high-performance liquid chromatographic (HPLC) methods.[12] Alternately, the correction-formula fluorometric method[13] is suitable to analyze the small volumes of serum for total vitamin A content. For many studies, serum vitamin A levels below 2 μg/dl signify an adequate stage of deficiency to conduct the experiment.

In the interval between 4 and 6 weeks on the diet, the animals should show a marked reduction in the growth rate, known as the growth plateau. For growth bioassays vitamin A deficiency has been classically defined as the fifth consecutive day when the weight gain is 1 g/day or less. Around this time the animals will extend their claws each time they are touched, and a few animals may show paralysis of the front legs. These are relatively consistent symptoms of the deficiency. If the sanitation for the colony is adequate the classic eye symptoms of xerophthalmia will not be observed. Animals at this stage of deficiency experience considerable discomfort. Therefore, these extreme stages of deficiency should not be allowed to develop unless it is absolutely necessary. If the animals are the second generation on the vitamin A-free diet as described above the growth plateau may occur much earlier.

The time required for individual animals to reach the growth plateau may differ by 2 to 3 weeks. Retinoic acid can be used to bring the animals to deficiency at a uniform time. As the rats individually reach the growth plateau, they are given a single daily dose of 50 μg of retinoic acid per day to maintain good health and to enable them to resume a normal growth rate. When all animals have reached the growth plateau, they are maintained on retinoic acid for 1 additional week. After the retinoic acid is discontinued the rats will stop growing within 3–4 days. Although other investigators have used much lower levels of retinoic acid, we have found that either 50 μg/day as a single dose or 12 mg/kg of diet provides more uniform animals, which are closer in weight to the control animals.

[10] W. A. Phillips, W. W. Stafford, and J. Stuut, Jr., *Proc. Soc. Exp. Biol. Med.* **143,** 733 (1973).

[11] W. J. Herbert *in* "Handbook of Experimental Immunology" (D. M. Weir, ed.), 3rd Ed., p. A4.1. Blackwell, London, 1978.

[12] J. G. Bieri, T. J. Toliver, and G. L. Catignani, *Am. J. Clin. Nutr.* **32,** 2143 (1979).

[13] J. N. Thompson, P. Erdody, and W. B. Maxwell, *Biochem. Med.* **8,** 403 (1973).

Suspension of Retinoic Acid in Vegetable Oil

To prepare a solution of retinoic acid in soybean oil, place exactly 1 g of retinoic acid in a 1-liter round-bottomed flask along with 0.5 g of 2,6-di-*tert*-butyl-*p*-cresol (BHT) and 100 ml of peroxide free diethyl ether. Test the ether for peroxides by placing a drop on commercial potassium iodide–starch test paper. If the test paper develops a brown spot that turns blue after adding a drop of water the ether contains peroxides.[14] The peroxides can be removed by passing the ether over a column of activated alumina.[15] After the retinoic acid and BHT are dissolved in the ether, add 498.5 g of soybean oil to the flask. Bubble either nitrogen or argon through the solution for about 20 min. After the oil is saturated with the inert gas, remove the rest of the ether under reduced pressure in a rotary evaporator at 40°. This process usually takes 5–6 hr. All of these procedures should be done under subdued gold lighting and whenever possible in the dark. One drop of this solution should contain roughly 50 *u*g of retinoic acid. To prepare 1 kg of diet containing 12 mg of retinoic acid substitute 6 g of this solution for an equal amount of the vegetable oil (see Table I).

Administration of Retinoids

The simplest way to give an exact amount of retinoic acid is by dropping the solution directly onto the tongue. Hold the rat in your left hand with your palm over its back and your ring finger and little finger positioned below its left front leg. Place your left index finger on the top of its head and gently pull back the skin along the jaws with the thumb and middle finger. The rat should open its mouth and extend its tongue as the skin tightens. If the rat does not extend its tongue, gently touch the tongue and it will almost always be extended. Drop the solution on the back of the tongue, and the rat should swallow immediately. Use a calibrated dropping pipette (medicine dropper) to administer the oil. The weight of the drop will be very uniform if the pipette is held in a vertical position. The weight of a drop of oil will be approximately 25 mg, but the exact weight will be a function of the size of the orifice of the pipette and the surface tension of the oil. If you must use a more complicated device, it must be a positive displacement system. Most pipetting devices are very inaccurate when pipetting oils. Also remember that lingual lipase is elaborated by the Ebner's glands at the base of the tongue in response to contact with lipids. The efficiency of absorption of a retinoid in vegetable oil will be 10 to 20%

[14] F. Feigl and V. Anger, "Spot Tests in Organic Analysis," 7th Ed., Elsevier, Amsterdam, 1966.

[15] W. Dasler and C. D. Bauer, *Ind. Eng. Chem. Anal. Ed.* **18,** 52 (1946).

greater when the oil is placed on the tongue than when it is given by stomach tube.[16,17]

When the quantity of retinoic acid to be given is less critical or when it is to be given for a long time, it should be mixed in the diet.

Other Depletion Schemes

Rapid, Synchronous Vitamin A Deficiency

Although rats depleted by the above procedure above show severe signs of vitamin A deficiency, the tissues of the body still contain significant amounts of retinol or retinyl esters. Continuous long-term supplementation with retinoic acid actually seems to spare these modest reserves of retinol and retinyl esters. A procedure to deplete these reserves was developed by Lamb, Olson, and co-workers.[8,18] The rats are depleted as described above until they reach the growth plateau. The rats are then fed a diet supplemented with 2 ug of retinoic acid per g for 18 days. The retinoic acid supplementation is then discontinued for 10 days. The process is then repeated with 18 days of supplementation followed by 10 days of depletion for at least 3 cycles. After 3 cycles the withdrawal of retinoic acid should produce a cessation of growth within 24 hr and weight loss by 48 hr. Although this procedure results in a very rapid onset of deficiency symptoms after the withdrawal of retinoic acid, the animals are subjected to considerable stress during the periods without retinoic acid supplementation.

Vitamin A Deficiency Prior to Weaning

Rogers and Bieri[19] developed a procedure to produce vitamin A-deficient rat pups in the first or second week of life. Weanling female rats were depleted of their vitamin A stores essentially as described above and were maintained on 20 ug of retinoic acid/day. When they became sexually mature the females were mated with normal males. During pregnancy and lactation the dose of retinoic acid was increased to 50 μg/day. An additional supplement of 1 ug of retinyl acetate/day was given orally through the critical stages of fetal differentiation (Days 9 through 21). This level of retinyl acetate was successful during the first and second pregnancies, but 2 ug of retinyl acetate/day was required during the third pregnancy. The

[16] S. R. Ames, personal communication (Nov. 26, 1968).

[17] J. E. Smith and R. Borchers, *J. Nutr.* **102,** 1017 (1972).

[18] M. A. Anzano, A. J. Lamb, and J. A. Olson, *J. Nutr.* **111,** 496 (1981).

[19] W. E. Rogers and J. G. Bieri, *Proc. Soc. Exp. Biol. Med.* **132,** 622 (1969).

pups were delivered on Day 22 or 23 of gestation. In some litters the deficiency symptoms appeared in the first week of life, whereas in other litters the deficiency did not appear until the second week. The deficient pups typically were uncoordinated and exhibited trembling. In the next few days the front legs became paralyzed. By giving 5 μg of retinoic acid per day the visible symptoms of vitamin A deficiency were reversed.

Lamb *et al.*[8] have suggested that at least 5 ug of retinyl acetate is required during the critical period between Days 9 and 21. However, we have observed normal birth with female rats given 8 RE/day/kg body weight and no retinoic acid.[20] This is substantially less than 5 μg retinyl acetate/day.

Chronic, Low-Level Vitamin A Deficiency

In humans vitamin A deficiency symptoms usually develop in response to stress after a chronic low vitamin A intake. Green *et al.*[7] have developed a model in which a rat with a very small vitamin A reserve slowly depletes these reserves. This animal model should be satisfactory to study the effects of stress on vitamin A depletion. The rats were depleted of vitamin A as described above until their serum vitamin A levels indicated that their liver vitamin A reserves were nearly depleted. The rats were then maintained of a diet that contained 0.1 RE of vitamin A activity per gram. This provided about 2 ug retinol/rat/day. These rats had disposal rates between 2 and 3 μg of retinol/day. The plasma retinol levels were between 5 and 10 μg/dl. Although the vitamin A reserves were slowly being depleted the animals remained in good health and exhibited a reasonable growth rate. Subjecting these animals to stress should greatly increase the utilization of vitamin A.

Sanitation

The most important thing to remember in the production of vitamin A-deficient rats is sanitation. Vitamin A-deficient rats maintained in a germ-free isolator have survived up to 5 times longer than vitamin A-deficient rats in an open environment.[21] Adhering to the following guidelines for sanitation will greatly improve the success of the experiment: (1) Provide fresh food and water daily. (2) Remove urine and fecal material daily. (3) Clean the room daily. (4) Sterilize the cages and cage racks at least weekly. (5) Keep the animals separated from other animals.

[20] Y. I. Takahashi, J. E. Smith, M. Winick, and D. S. Goodman, *J. Nutr.* **105,** 1299 (1975).
[21] J. G. Bieri, E. G. McDaniel, and W. E. Rogers, Jr., *Science* **163,** 574 (1969).

[25] Use of Food Composition Tables for Retinol and Provitamin A Carotenoid Content

By KENNETH L. SIMPSON

Introduction

A number of countries and organizations such as the Food and Agricultural Organization (FAO) of the United Nations have issued food composition tables. The tables are meant to cover the foods consumed by a particular population. Although the older tables mainly report raw foods, more recently attention has been given to food as eaten. The tables generally give a single value rather than a range of values. This has led to problems because of the large variation in carotenoid content owing to different varieties of fruits and vegetables, as well as factors such as sunlight, rainfall, soils, and harvest maturity. It has also been shown that even the methods of analysis may contribute to uncertainty in the contribution of the carotenoids to the vitamin A values. Most values in the tables represent the analysis of a number of samples taken from various regions and genetic lines. The single value reported may not be a mean but a weighted mean. This is illustrated by the reported fat content of the avocado in the United States: the values reported[1] give figures from two states but also a third, weighted value representing the fact that 90% of avocados come from California and that these fruits are higher in fat content.

Food composition tables are used by dietitians, nutritionists, food scientists, and anthropologists to establish the level of nutrition for a given population. The tables and more recent data have been used to create a computer program for nutrient analysis of an individual. The point of this chapter is to stress the limitations in the nutritional tables and make recommendations for improvements.

Agricultural Practices

Variation in nutrient content of fruits and vegetables can be caused by a number of agricultural practices or conditions.[2] This list includes genetic manipulation, maturity at harvest time, and fertilization, as well as other

[1] S. E. Gebhardt, R. Cutrufell, and R. H. Matthews, "Composition of Foods—Fruits and Fruit Juices—Raw Processed, Prepared." Agriculture Handbook 8-9, USDA Human Nutrition Information Service, Washington, D.C., 1982.

[2] R. S. Harris, *in* "Nutritional Evaluation of Food Processing" (R. S. Harris and E. Karmas, eds.), p. 33. AVI, Westport, Connecticut, 1975.

factors that are less controllable such as sunlight, reliable rainfall, topography, soils, and location. Of these factors, the major contribution to carotene content has come through genetic manipulation largely because more carotene has made the product more marketable. Carrots have been developed with 5 times the β-carotene content, and the high Beta-tomato has 12 times the β-carotene content. Sweet potatoes can have varying amounts of carotene.

What can be accomplished technologically may, however, run counter to consumer acceptance. Yellow tomatoes or yellow tomato products are not readily accepted. In some parts of Latin America white sweet potatoes having small amounts of carotenes are more acceptable than colored varieties. More light generally leads to greater carotene production in tomatoes. Tomatoes ripened at temperatures above 30° produce less lycopene and are thus yellow. Vegetables and fruits are often harvested prior to maturity because of distribution problems or, as in the case of the bitter mellon, because of cultural practices. Generally the carotene content is highest at maturity.

The nutritional tables may not reflect recent improvements in breeding or the other variations caused by various agricultural practices.

Analysis of Carotenoids

The preferred methods for retinol analysis according to Arroyave *et al.*[3] are high-performance liquid chromotography (HPLC), the Bessey–Lowry method, and colorimetry using trifluoroacetic acid. Many of the data in the nutritional tables were determined using the Carr Price $SbCl_3$ method. The analysis is simpler for retinol because it is a single compound. Where esters of retinol are found the sample can be saponified.

The analysis of the carotenoids is complex because over 500 have been isolated. Certainly, only a few are encountered in any one system. Likewise, a few dozen of these are potential precursors of vitamin A, but again, less than about five are found in any one tissue. The main method of analysis for carotenoids has been that recommended by the Association of Official Analytical Chemists (AOAC). This method is simple and fairly rapid. The less polar carotenes are separated on a column, from the more polar xanthophylls. The assumption is made that the nonpolar fraction is mainly composed of β-carotene, and quantification is based on the light absorption at a given wavelength. The basic problem is that there are xanthophylls which have vitamin A activity and carotenes which have

[3] G. Arroyave, C. O. Chichester, H. Flores, J. Glover, L. A. Mejia, J. A. Olson, K. L. Simpson, and B. A. Underwood, "Biochemical Methodology for the Assessment of Vitamin A Status." IVACG-Nutrition Foundation, Washington, D.C., 1982.

some activity or none at all. Also, xanthophyll esters are not polar and are not separated from β-carotene. We would thus recognize three groups: the first, characterized by spinach, in which β-carotene is the major carotene and the xanthophylls are easily removed on the column; the second group in which the carotene profile is complex; and a third group, found mainly in fruits, in which esters are present. Table I shows the effect that α-carotene has on the analysis of carrots and the effect of β-cryptoxanthin esters on the analysis of fruits. HPLC has been shown by several authors to separate α- from β-carotene (see Zakaria *et al.*[4]). As can be seen from Table I, the AOAC method often overestimates the vitamin A values, particularly when the profile of the carotenes is complex.

Conversion Factors

The traditional way of expressing vitamin A potency has been the use of the international unit (IU). More recently, the retinol equivalent (RE) has been used. The basic calculations are as follows (Food and Nutrition Board[5]):

$$\begin{aligned} 1\ \text{RE} &= 1\ \mu\text{g retinol} \\ &= 6\ \mu\text{g } \beta\text{-carotene} \\ &= 12\ \mu\text{g other provitamin A carotenoids} \end{aligned}$$

$$\begin{aligned} 1\ \text{IU} &= 0.3\ \mu\text{g retinol} \\ &= 0.344\ \mu\text{g retinol acetate} \\ &= 0.6\ \mu\text{g } \beta\text{-carotene} \\ &= 1.2\ \mu\text{g other provitamin A sources} \end{aligned}$$

Example

$$\text{RE} = \frac{\text{No. of IU of retinol}}{3.33} = \frac{\text{No. of IU of carotene}}{10}$$

$$\text{RE} = \frac{\mu\text{g } \beta\text{-carotene}}{6} = \frac{\mu\text{g other provitamin A compounds}}{12}$$

The actual calculations depend on the sources and are expressed in terms of carotene, whereas foods such as egg yolks and milk, which have both retinol and carotene, would include both sources. The term β-carotene equivalent includes the amount of β-carotene and the equivalent amount of vitamin A activity of other carotenoids.

These calculations assume that retinol is completely absorbed and that

[4] M. Zakaria, K. Simpson, P. R. Brown, and A. Krstulovic, *J. Chromatogr.* **176,** 109 (1979).

[5] Food and Nutrition Board, "Recommended Dietary Allowances." National Research Council, National Academy of Science, Washington, D.C., 1980.

TABLE I
PROVITAMIN A CONTENT[a] OF SELECTED VEGETABLES AND FRUITS[b]

Method of analysis	Group 1		Group 2		Group 3	
	Kang-Kong	Spinach	Carrot	Pumpkin	Loquat	Papaya
AOAC	3.27	2.06	13.39	10.03	2.10	1.73
Stepwise solvent gradient elution	—	—	10.58	4.84	—	—
Saponification and stepwise solvent gradient elution	—	—	—	—	1.20	0.59
HPLC	3.32	1.99	10.83	5.09	1.04	0.52
Major carotenoids	α-Carotene	α-Carotene	α-Carotene, β-carotene	α-Carotene	α-Carotene, α-cryptoxanthin	

[a] From K. L. Simpson and S. C. S. Tsou, *in* "Vitamin A Deficiency and Its Control" (J. C. Bauernfeind, ed.), p. 461. Academic Press, Orlando, Florida, 1986.
[b] Milligrams of α-carotene equivalent/100 g.

β-carotene is partly converted to retinol and partly absorbed. This is quite complex and is covered elsewhere in this volume. What is germain to this chapter is that the Recommended Dietary Allowance (RDA) is based on either IU or RE. In the United States, the assumption has been made that retinol and β-carotene each supply 50% of the 5000 IU RDA. More recently the 1000 RE RDA was calculated as follows: 750 μg coming from retinol (2500 ÷ 3.33) and 250 RE coming from β-carotene (2500 ÷ 10).[5] This would mean that an all-vegetarian diet of 5000 IU (1.0 RDA) would be 0.5 RDA 5000 ÷ 10 = 500 RE.

Food Composition Tables

Tsou *et al.*[6] lists some vegetable and fruit values from various national and regional tables: 11 fruits and vegetables from Taiwan, Japan, the Philippines, and the United States, as well as a regional table from the

[6] S. C. S. Tsou, J. Gershon, K. L. Simpson, and C. O. Chichester, *in* "Human Nutrition: Better Nutrition, Better Life" (V. Tanphaithetri, W. Dahlan, U. Suphakarn, and A. Valyasevi, eds.), p. 179. Proceedings of the 4th Asian Congress on Nutrition, Aksornsmai Press, Bangkok, 1984.

FAO. A large variation is seen in the papaya: 47 IU, Japan (425 IU Philippines) and 1750 IU, United States. This large variation may be due to the difference in β-carotene content of the yellow versus the red papaya, which contains mainly vitamin A-inactive lycopene. The tables do not indicate the type of fruit analyzed. Green and red peppers also show a great variation: green peppers, Taiwan 4000 IU, FAO 2717, Philippines 260; red peppers, Japan 1100, United States 21,600; persimmon, Japan 65, United States 2710. In some cases the difference is due to the method of calculation. In the United States the International Unit (IU) is arrived at by dividing micrograms by 0.6. In Japan, the factor is 1.8; thus, the IU in Japan is one-third the value given in tables from the United States. However, the Japanese RDA is 2000, rather than 5000 IU given for men in the United States.[7]

In a large country such as Brazil or the United States, a single value for the vitamin A content of a fruit or vegetable may not be representative of samples from all sections of the country. This is especially true in Brazil, where very large variations in climate, rainfall, soil, etc., exist. Closer inspection of the Brazilian data shows that only four of the values in the current table are determined on domestic fruits and vegetables; the rest are from foreign fruits and vegetables.[8] The current tables could be very much improved if the provitamin A values were listed and if individual values were given where large variations exist.

Conclusion

Food composition tables are used extensively to assess the nutritional status of a population. Most of the tables report a single value which seems to be open to question when we look at the variation among the national and regional values for the same fruit or vegetable. In some cases the difference may be in the analysis or the calculation of the IU. Another major difference seems to be due to the variety and conditions native to that country. Clearly there is uncertainty regarding these values which is not resolved by the information given in the tables. There is a need for accurate food composition tables, especially in areas where there is a vitamin A deficiency, in order to assess nutritional status and to select crops.

Acknowledgments

Rhode Island Agriculture Experiment Station contribution No. 2503.

[7] Resource Council Science and Technology Agency, "Standard Tables of Food Composition in Japan," 4th Ed., Rep. 7. Resource Council, Tokyo, 1980.

[8] G. Franco, "Nutricao—Text Basico e Tabela de Composicao Quimica dos Alimentos," 6th Ed. Livraria Athencv, Rio de Janeiro, 1982.

[26] Biochemical and Histological Methodologies for Assessing Vitamin A Status in Human Populations

By BARBARA A. UNDERWOOD

Introduction

Developing quantitative methodologies for determining the relative vitamin A nutritional status of humans by reasonably noninvasive biochemical or histochemical techniques has challenged the biomedical community. Vitamin A nutriture exists as a continuum between clinically evident deficiency and toxicity. The signs and symptoms, and the corresponding biochemical alterations in blood, have been well described at the extremes of the continuum.[1,2] However, the intermediate states along the continuum that could identify the vitamin A status of an individual or population as marginal, adequate, or excessive are not overt and have been difficult to quantitatively measure because vitamin A can exist in animal tissues at concentrations disproportionate to levels in blood or urine. The level found in blood at intermediate levels of nutriture is regulated and relatively insensitive. The vitamin normally is not excreted in urine, except as a variety of metabolites whose concentrations are not parallel to body stores, and people do not savor providing tissue biopsy specimens for non-disease-related assessment purposes.

Marginal vitamin A status has been arbitrarily defined by assigning distinguishing cutoff concentrations to levels in blood, breast milk, or tissues.[3] Usually the cutoff value is chosen to reflect the lower range of the usual distribution curve of populations where clinical malnutrition and vitamin A deficiency are rare. Or, as in the case for liver, the cutoff concentrations are selected to provide a calculated protective period before deficiency symptoms would be expected to appear.[4] Static measures of vitamin A concentrations at the selected cutoff values, however, can only suggest the risk that physiological functions might be compromised in populations. However, they fail to indicate whether physiological functions are, in fact, compromised in individuals.

Functional measures are an alternative and more dynamic means of

[1] B. A. Underwood, *in* "The Retinoids" (M. B. Sporn, A. B. Roberts, and D. S. Goodman, eds.), Vol. 1, p. 381. Academic Press, New York, 1984.

[2] J. C. Bauernfeind, "The Safe Use of Vitamin A." IVACG, Washington, D.C., 1980.

[3] G. Arroyave, C. O. Chichester, H. Flores, J. Glover, L. A. Mejia, J. A. Olson, K. L. Simpson, and B. A. Underwood, "Biochemical Methodology for the Assessment of Vitamin A Status." IVACG, Washington, D.C., 1982.

[4] J. A. Olson, *Am. J. Clin. Nutr.* **45,** 704 (1987).

describing marginal vitamin A status applicable to individuals as well as populations. Of those currently available for use in humans only one measures vitamin A directly, the relative dose response test (RDR). This test, performed on blood, is an indication of marginal liver reserves. The other functional measures, dark adaptation or a history of night blindness, and a reduction in goblet cells in ocular or other tissues, are indirect measures that are subject to other influences. For example, systemic and localized disease states can confound interpretation of quantitatively derived, indirect data.

Static Measures of Vitamin A Status

Isotope Dilution to Estimate Total Body Reserves

Principle. Over time, newly administered, labeled vitamin A uniformly mixes with vitamin A stores in the body. The dilution of the label in blood, relative to the amount administered, can be used to calculate the total body pool size, assuming that a constant proportion is stored in the liver. This methodology is the only one that directly and quantitatively measures relative vitamin A status. It has not been extensively used in humans because it requires the use of stable isotopes and sophisticated instrumentation, namely, HPLC and gas chromatography–mass spectrometry,[5] which are not widely available for population studies.

Method. A noted amount of labeled vitamin A tracer, such as tetradeuterated retinol, is administered, and one or two blood samples are drawn over a period of about 10–50 days. The ratio of unlabeled to labeled retinol is calculated [(H/D) − 1], and a formula is applied to calculate pretreatment total liver reserve of vitamin A (TLR):

$$\mathrm{TLR} = \mathrm{dose} \times F(Sa[(\mathrm{H/D}) - 1]$$

The calculation is based on the dose given and a series of assumptions about the efficiency of storage of the orally administered dose ($F = 0.5$), the ratio of specific activities of retinol in serum to liver ($S = 0.65$), and the fraction of the labeled dose remaining in the liver at the time of blood sampling (a).

Comments. Although this procedure, after additional refinements and validation, may be the most accurate one for measuring from the marginal through most of the adequate range of liver reserves,[5] the technical requirements for conducting it are likely to limit its use to individual assessment

[5] H. C. Furr, O. Amédée-Manesme, A. J. Clifford, H. R. Bergen III, A. D. Jones, D. P. Anderson, and J. A. Olson, *Am. J. Clin. Nutr.* **49**, 713 (1989).

and treatment purposes, excluding it as a practical method for population-based surveys.

Liver Concentrations

Principle. The concentration of vitamin A in the liver estimates vitamin A status, assuming that the liver retains a constant percentage (about 90%) of the total body reserve. However, liver biopsy performed solely for assessment of vitamin A content, in the absence of suspected pathology, is not justified ethically. This restricts the use of this methodology in humans to situations where biopsies are indicated for diagnostic purposes or on specimens obtained postmortem.

Method. A liver sample of approximately 1–5 g is cut from the central portion of the right lobe (or a few milligrams obtained by needle biopsy).[6] The sample is accurately weighed, ground with anhydrous sodium sulfate and a suitable solvent such as 50% (v/v) glycerol in water,[7] chloroform,[8] or dichloromethane[6] in a ratio of about 1:10:15. The sample can safely be left at room temperature overnight in the solvent. This extracts lipids including vitamin A. After decanting the extract, the residue is reextracted twice, the extracts are brought to volume, and an aliquot is analyzed for vitamin A content by a suitable analytical technique.

Comments. Vitamin A is not uniformly distributed throughout the liver. To minimize within-individual fluctuations, it is preferable to standardize the site for sampling. However, some claim that this variation is insufficient to warrant the extra precautions.[6] Vitamin A in liver is quite stable in the solvent extractant at room temperature if protected from light. It is also stable for extended periods in frozen tissue when protected from dehydration and light.

Blood Levels

Principle. Serum levels of vitamin A are the most commonly used biochemical measure of vitamin A status. This is true even though it is well known that serum levels of individuals are regulated by physiological controls which are only partially understood and highly individual. These levels become predictive of the status of an individual only when body reserves have been critically depleted or overfilled. On the other hand, serum distribution curves based on random sampling can be useful in comparing the status of populations relative to the occurrence of clinical

[6] O. Amédée-Manesme, H. C. Furr, and J. A. Olson, *Am. J. Clin. Nutr.* **39,** 315 (1984).
[7] H. Flores and C. R. C. de Araújo, *Am. J. Clin. Nutr.* **40,** 146 (1984).
[8] J. A. Olson, D. B. Gunning, and R. A. Tilton, *Am. J. Clin. Nutr.* **39,** 903 (1984).

vitamin A deficiency and malnutrition, dietary intake patterns, and/or other socioeconomic and ecological characteristics.

Method. A blood sample is obtained and analyzed by one of several analytical techniques outlined elsewhere in this volume. Preferably, the sample is obtained from individuals who have fasted overnight or for at least 4–5 hr. There is disagreement about the magnitude of the influence of the nonfasting state on circulating levels.[9] Blood samples should be protected from light, heat, and air exposure; blood-filled vacutainers retained in an ice-cooled, opaque container until centrifugation are best under survey conditions. If not immediately analyzed, the serum or plasma should be stored at -20° or lower, with minimal exposure to air. This can be assured by flushing the head space in the tube with an inert gas, such as nitrogen or helium. Under these conditions of storage, little loss of the vitamin occurs over more than 2 years.[10]

Comments. Serum values of populations are best expressed as distribution curves that are specific for age (<6 years, 6–11 years) and sex groups (12–17 years, 18–44 years, >45 years).[11] Serum levels are age-related before puberty and sex-related thereafter. Values are lower for females than males during the fertile years, then catch up postmenopause. A shift to the right in the lower end of the age/sex-specific distribution curves following an intervention is evidence of an improved vitamin A status in the population.

Functional Measures of Vitamin A Status

Functional tests reflect whether the vitamin is sufficiently available for normal participation in its known metabolic activities. The recognized physiological functions of vitamin A that are useful for assessment purposes are generation of the visual pigment rhodopsin and maintenance of the differentiation of cells originating from the epithelium, including the production of goblet cells. The response in circulating holo-RBP [retinol bound to its specific binding protein (RBP)], after providing an additional exogenous source of vitamin A, is a measure of whether the vitamin is available in the circulation from liver reserves at its appropriate homeostatic level. These known functions of, or responses to, vitamin A have

[9] M. H. C. Barreto-Lins, F. A. C. S. Campos, M. C. N. A. Azevedo, and H. Flores, *Clin. Chem.* **34,** 2308 (1988).

[10] N. E. Craft, E. D. Brown, and J. C. Smith, Jr., *Clin. Chem.* **34,** 44 (1988).

[11] S. M. Pilch (ed.), "Assessment of the Vitamin A Nutritional Status of the U.S. Population Based on Data Collected in the Health and Nutrition Examination Surveys." Life Science Research Office, Federation of American Societies for Experimental Biology, Bethesda, Maryland, 1985.

been the basis for development of specific and sensitive field-applicable methods to measure subclinical vitamin A status.

Night Blindness

Principle. Rhodopsin is generated when the protein opsin in the rods of the retina combines with a cis isomer of retinaldehyde. The complex is split in response to light, yielding opsin and a trans isomer. After isomerization of the trans to the cis isomer, recycling of the process occurs at a rate that normally maintains the level of rhodopsin so that visual accommodation is not prolonged when going from brightly to dimly lighted conditions. When the retinol supply is limited, the rate of regeneration of rhodopsin after bleaching by exposure to bright light is impaired. The final dark-adaptation threshold can be quantitatively measured. In practical terms, individuals who are deficient in vitamin A have difficulty seeing when going from bright to dimly lighted situations, that is, they are "night blind."

Methods. In clinical settings psychophysical measurements (dark-adaptation threshold) and electrophysiological measurements (electroretinograms) are available to measure directly the level of rhodopsin and its rate of regeneration.[12] Modified instruments and procedures have been used in field situations among older children and adults who are able to respond actively to visual discrimination tests subsequent to bright light (bleaching) exposure. These tests, however, have been insufficiently reliable when applied among preschool-aged and younger children from deprived environments, where vitamin A deficiency is most likely to occur.

In the latter situation, a history of night blindness elicited from the mother or responsible adult has been the most successful approach.[13] This is true, however, only in cultures where a specific word or words exist that characterize the condition. Such a word usually, but not always, exists in areas with endemic vitamin A inadequacy. For example, in Bangladesh the Bengali term is *rat kana,* the Indonesian and Sundanese terms are *buta ayam* or *kotokeun* (chicken blindness) respectively, in Tamil Nadu, South India, the term is *malai ken* (evening eyes), and in Zambia it is *kafifi.* Investigators must carefully determine under each local situation whether an appropriate term exists and decide if it can be reliably used in an historical interview.

Where local terms are not used, alternate strategies to indirectly test dark adaptation may be useful. For example, a child taken from the bright

[12] H. Ripps, *Invest. Ophthalmol. Visual Sci.* **23,** 588 (1982).

[13] A. Sommer, G. Hussaini, Muhilal, I. Tarwotjo, D. Susanto, and J. S. Saroso, *Am. J. Clin. Nutr.* **33,** 887 (1980).

outdoor sun into a darkened room may be asked to find the waiting mother or sibling, or to locate a favorite toy. The time required to achieve this task is compared to that needed by a similarly aged child who is not deficient.

Differentiation and Morphology of Epithelial Tissues

Principle. When mucus-secreting epithelial tissues receive too little vitamin A, they develop a metaplasia which in some cases is characterized by keratinization (e.g., skin). In other epithelia, keratinization does not occur or is minimal (e.g., intestinal mucosa, lungs, and conjunctiva), but the number of goblet cells declines and in severe deficiency may be absent. In addition, the morphology of the epithelial cells may be altered in deficiency. In the conjunctiva they take on a squamous or flattened appearance with small nuclei and an expanded proportion of cytoplasm.[14] Various modifications of a test in humans to evaluate the presence or absence of conjunctival goblet cells and the morphology of the epithelial cells are referred to in the literature as conjunctival impression cytology (CIC), ocular impression cytology (OIC), and impression cytology with transfer (ICT).

Conjuctival or Ocular Impression Cytology with Staining of Cells on Filter Paper. A training manual is available that describes the CIC technique in detail.[15] Briefly, the lower lid of the eye is retracted, and a strip of cellulose ester filter paper of 0.45-μm pore size is touched to the lower, outer (temporal) portion of the conjunctiva for 3–5 sec and then gently peeled away. The paper must flatten down on the epithelium to obtain sufficient contact for cells to adhere. This can be accomplished by careful touch application of the filter paper[16] or by lightly touching the paper with a blunt-ended glass rod, or a soft tubing extending from the outer arm of tweezers used to hold the filter paper strip. (Recently, a disk applicator with a suction apparatus has been advocated.) The filter paper strip, with adhering cells, is placed in a fixative solution (75 ml 95% ethanol, 25 ml distilled water, 5 ml glacial acetic acid, 5 ml 37% formaldehyde) in a tightly closed vial. The specimen is stable in the solution and can be stained at any time after 20 min of exposure.

Staining is based on the principle that periodic acid–Schiff (PAS)

[14] D. L. Hatchell and A. Sommer, *Arch. Ophthalmol.* **102,** 1389 (1984).

[15] ICEPO, "Training Manual: Assessment of Vitamin A Status by Impression Cytology." Data Center for Preventive Ophthalmology, The Johns Hopkins University, Baltimore, Maryland, 1988.

[16] C. L. Kjolhede, A. M. Gadomski, J. Wittpenn, J. Bulux, A. R. Rosas, N. W. Solomons, K. H. Brown, and M. R. Forman, *Am. J. Clin. Nutr.* **49,** 490 (1989).

reagent causes the acid mucopolysaccharide, which is a principal component of goblet cells and their mucoid secretions, to stain a bright pink. Sodium metabisulfite fixes the stain, and Harris hematoxylin counterstains the epithelial nuclei.[17] After washing away the fixative, specimens are sequentially subjected to a series of baths containing 0.5% periodic acid for 2 min, Schiff reagent diluted 1:1 with distilled water for 8 min, 0.5% sodium metabisulfite for 2 min, and Harris hematoxylin for 15–20 sec. (Care must be taken to determine the precise timing to avoid under- or overstaining and to assure that the water washes between each bath are thorough.) The stained strips are then rigorously dehydrated by subjecting them to a 95% ethanol wash, a 100% ethanol wash (reusable up to 10 times), and a final wash in fresh 100% ethanol. (Methanol can substitute for ethanol.) The dehydrated strips are placed in xylene until transparent, then mounted on a glass slide using Permount or a comparable mounting media that is devoid of moisture. The strips must not be allowed to dry after removal from the xylene and before mounting, and the mounting medium must be water-free. If these precautions are not observed, the transparent filter paper strips become opaque and unreadable by microscopy. Properly mounted transparent strips are evaluated using a simple microscope. Mounted specimens are stable on storage.

Impression Cytology with Transfer, or Staining of Cells Transferred from Filter Paper to Glass Slides. ICT is a simplified modification of the above technique.[18] After the cells are secured on the filter paper strip as described above, they are transferred using light, finger pressure directly to a cleaned glass slide, placed in 95% alcohol as a fixative for 15 min, and stained in a single step for 20 min using a mixture of carbon–fuchsin (Ziehl–Nielsen solution) and 0.2% Alcian blue in 5% acetic acid. After water washing, the nuclei and cytoplasm of epithelial cells and the nuclei of goblet cells stain pink with carbon–fuchsin. The acid mucous substance stains blue with Alcian blue. Provided care is exercised to assure that sufficient cells are transferred to the slide, the procedure is quick; moreover, the cells without the filter paper are easier to view microscopically, and the method is less expensive in terms of both reagents needed and personnel time required for processing.[19] Slides can be stored protected in a box without mounting for many months. In fact, application of a mounting medium may leach the stain from the cells during storage.

Certain precautions are necessary to assure adequate transfer of cells

[17] G. Natadisastra, J. R. Wittpenn, K. P. West, Jr., Muhilal, and A. Sommer, *Arch. Ophthalmol.* **105,** 1224 (1987).

[18] O. Amédée-Manesme, R. Luzeau, C. Carlier, and A. Ellrodt, *Lancet* **1,** 1263 (1987).

[19] R. Luzeau, C. Carlier, A. Ellrodt, and O. Amédée-Manesme, *Int. J. Vitam. Nutr. Res.* **58,** 166 (1987).

from the filter paper to the slide. Slides should be cleaned with alcohol and dry. Ambient temperatures should be below 35°. After obtaining the filter paper impression and just before transfer, the slide should be slightly moistened by an exhaled breath, then immediately the strip should be applied and lightly pressed with the thumb, without allowing the paper to move, and the filter paper carefully peeled away. After transfer the paper strip can be discarded or put into fixative as a backup in the event that the transfer was insufficient. In the latter case, the filter paper can be stained by the longer procedure described above. The transferred cells on the dry slide are put into the 95% alcohol fixative.

Comments. Irrespective of the processing procedure used for staining, the histological assessment is only as good as the initial impression obtained from the conjunctiva; in other words, an adequate number of cells must adhere to the filter paper for direct reading or transfer to the slide. Experience has shown unreliable results from children under 3 years of age, from those who are crying profusely so that the paper does not make close contact with the conjunctiva, and when ocular infections coexist.[20,21] Interpretation of the slides also requires clearly defined criteria of normal and abnormal cytology as well as careful standardization of readers.[22] The specificity and sensitivity of the test for assessing marginal vitamin A status requires confirmation in the setting where it will be used. Current studies indicate that conjunctival infections, including trachoma, keratoconjunctivitis of a variety of etiologies, and perhaps severe malnutrition and systemic infections may limit sensitivity and even specificity.

Relative Dose Response Test for Insufficient Hepatic Stores

Principle. Owing to a chronically inadequate dietary supply, liver reserves of vitamin A become progressively depleted. At a point along the downward continuum, when the liver reserve is depleted below a critical threshold, the rate of release of the remaining reserve is diminished. Concurrently, conservation mechanisms are invoked to increase the efficiency of utilization among tissues and to maintain the level in blood circulating to the target tissues. Synthesis of the carrier protein RBP continues, resulting in accumulation of an hepatic pool of preformed RBP. Providing an exogenous source of vitamin A causes the release of holo-RBP at a level

[20] C. L. Kjolhede, A. M. Gadomski, J. Wittpenn, J. Bulux, A. R. Rosas, N. W. Solomons, K. H. Brown, and M. R. Forman, *Am. J. Clin. Nutr.* **49,** 490 (1989).

[21] G. Natadisastra, J. R. Wittpenn, Muhilal, K. P. West, Jr., L. Mele, and A. Sommer, *Am. J. Clin. Nutr.* **48,** 695 (1988).

[22] A. M. Gadomski, C. L. Kjolhede, J. Wittpenn, J. Bulux, A. R. Rosas, and M. R. Forman, *Am. J. Clin. Nutr.* **49,** 495 (1989).

and in a characteristic time course relative to the amount of accumulated preformed carrier protein.[23,24]

Method. Individuals suspected to be marginally deficient in vitamin A based on moderately low serum levels (15–30 μg/dl) and a dietary or disease history that suggests a prolonged inadequate intake, but who may not have clinical signs or symptoms, are given a small dose of vitamin A after at least a 4- to 5-hr fast. A fasting blood sample of 2–3 ml is taken by venipuncture. (Finger or heel prick bloods can be used if the analyses are adapted to microtechniques, such as HPLC). A liquid solution containing about 450–1000 μg retinyl palmitate in an oily solution or aqueous dispersion is given orally directly onto the tongue. The exact amount given should be such as to not overload the system with retinyl esters but sufficient to assure complete mobilization of preformed accumulated hepatic RBP. If high-performance liquid chromatography (HPLC) is the analytical technique, retinol is distinguished from esters, and, therefore, this is not a problem. Care must be taken to assure that the child swallows the dose. Children are then given a small meal that consists of foods with some fat and minimal vitamin A sources, such as a slice of bread with about 25 g cheese and a small glass of fruit juice. Nothing but water or non-vitamin A-containing juices and snacks are allowed for the next 5 hr. After 5 hr, a second blood sample of about 2–3 ml (or a capillary tube from a finger/heel stick) is obtained.

The blood is analyzed for vitamin A by a suitable analytical technique. The RDR is calculated as follows:

$$\mathrm{RDR} = \frac{\text{vitamin A}_5 - \text{vitamin A}_0}{\text{vitamin A}_5} \times 100$$

where vitamin A_5 is the serum vitamin A level 5 hr after dosing and vitamin A_0, the fasting serum vitamin A level. An RDR exceeding 20% is considered to be positive and indicative of inadequate hepatic stores of vitamin A, that is, marginal vitamin A status.

Recently, a modification of the procedure (MRDR) that uses dehydroretinol (vitamin A_2) and requires a single blood sample at 5 hr postdosing has been tested in rats,[25] in healthy children,[26] and in those suspected of

[23] J. D. Loerch, B. A. Underwood, and K. C. Lewis, *J. Nutr.* **109,** 778 (1979).

[24] L. M. De Luca, J. Glover, J. Heller, H. A. Olson, and B. A. Underwood, *in* "Guidelines for the Eradication of Vitamin A Deficiency and Xerophthalmia: VI. Recent Advances in the Metabolism and Function of Vitamin A and Their Relationship to Applied Nutrition." IVACG, The Nutrition Foundation, Washington, D.C., 1979.

[25] S. A. Tamumihardjo and J. A. Olson, *J. Nutr.* **118,** 598 (1988).

[26] S. A. Tamumihardjo, P. G. Koellner, and J. A. Olson, *J. Nutr.* in press (1990).

being vitamin A-deficient.[27] Dehydroretinol is a biologically active naturally occurring form of vitamin A but is uncommon in most food supplies. A single dose (0.35 μg/kg body weight) of dehydroretinyl acetate dissolved in corn oil is fed under conditions as described for the RDR. The calculated dehydroretinol to retinol (DR/R) ratio is determined on a single blood sample obtained 5 hr postdosing. A tentative DR/R ratio for a satisfactory vitamin A status of 0.03 has been proposed.[26,27]

The approach is limited at present because dehydroretinol is not readily available from commercial sources and must be chemically synthesized. In addition, an HPLC is required for the analysis.

Comments. Currently, the validation data that are available indicate that a positive RDR occurs when hepatic stores are below 20 μg/g. Mild to moderate protein–energy malnutrition (PEM) does not interfere with the test, and even in severe PEM a single test is valid. In severe PEM repeat testing should not be attempted for several days to allow sufficient time to elapse for resynthesis of RBP. A positive RDR will revert to normal after sufficient vitamin A has been given to restore hepatic stores above the critical level of depletion (about 20 μg/g).

Summary

In recent years, new biochemical and histological methodologies have been developed for assessing vitamin A nutritional status in humans at subclinical levels of nutriture. Insensitive static blood levels no longer are the only practical assessment parameter. Some of the newer functional methodologies require additional testing of their sensitivity and specificity under a variety of conditions existing in human populations and that frequently are associated with an inadequate vitamin A status. Some of these conditions could confound the interpretation when only a single assessment method is applied.

[27] S. A. Tamumihardjo, Muhilal, Y. Yuniar, D. Permaesih, Z. Sulaiman, D. Karyadi, and J. A. Olson, *J. Nutr.* in press (1990).

[27] Characterization of Immunomodulatory Activity of Retinoids

By DENISE A. FAHERTY and ADRIANNE BENDICH

Introduction

Historically, specific immunity in mammals is regarded as comprising two distinct responses to antigenic challenge: humoral and cell-mediated immune responses. Humoral immunity is B-lymphocyte-mediated, although it is now well recognized that the production of antibodies in response to many antigens is T-lymphocyte-dependent. Cellular immunity is T-lymphocyte-mediated and can be divided into two main effector mechanisms: cytolytic T lymphocyte (CTL) responses and delayed-type hypersensitivity (DTH) responses. Retinoids have been reported to affect both humoral and cellular immune responses.[1-3]

CTL and DTH responses are considered to be important in the rejection of transplanted organs.[4,5] The role of humoral immune responses in transplant rejection is less clear.[6] As part of an ongoing program to assess the potential utility of retinoids in models of transplantation, we have tested retinoids for their ability to affect defined cellular and humoral responses in mice. The methodologies for the following models are included: the cytolytic response to an allogeneic tumor cell line,[7] the humoral response to sheep erythrocytes,[8] and the DTH response to a chemical sensitizing agent, oxazolone.[9]

General Considerations

The initial screening of compounds to detect immunomodulatory activity usually involves *in vitro* assay systems. However, there are certain classes of compounds, such as the retinoids, which cannot be readily

[1] J. B. Barnett, *Int. Arch. Allergy Appl. Immunol.* **72,** 227 (1983).
[2] R. Lotan and G. Dennert, *Cancer Res.* **39,** 55 (1979).
[3] D. Katz, S. Mukherjee, J. Maisey, and K. Miller, *Int. Arch. Allergy Appl. Immunol.* **82,** 53 (1987).
[4] D. W. Mason and P. J. Morris, *Annu. Rev. Immunol.* **4,** 119 (1986).
[5] D. Steinmuller, *Transplantation* **40,** 229 (1985).
[6] M. W. Baldwin, L. C. Paul, F. H. J. Claas, and M. R. Daha, *in* "Transplantation: Approaches to Graft Rejection" (H. T. Meryman, ed.), p. 41. Alan R. Liss, New York, 1986.
[7] K. T. Brunner, J. Mauel, J.-C. Cerottini, and B. Chapuis, *Immunology* **14,** 181 (1968).
[8] N. K. Jerne and A. A. Nordin, *Science* **140,** 405 (1963).
[9] G. L. Asherson and W. Ptak, *Immunology* **15,** 405 (1968).

evaluated *in vitro* owing to certain physical properties. For example, retinoids are relatively lipophilic and thus poorly soluble in water. Consequently, these compounds are insoluble in most vehicles used in *in vitro* test systems. In many instances, retinoids precipitate out of solution when added to culture media. The solubility characteristics can result in equivocal data from laboratory to laboratory and, indeed, from assay to assay. Thus, optimal characterization of the immunomodulatory activity of retinoids requires *in vivo* assays. However, the bioavailability and absorption characteristics of individual retinoids can differ considerably. In addition, the route of administration can also influence the effects seen.[10] Thus, monitoring of retinoid blood levels is recommended as an index of comparisons between experiments.[11] Reproducibility of results is enhanced by the consistent use of the same vehicle and route of administration for all studies. In this chapter, emphasis is placed on the immunomodulatory effects of retinoids suspended in a corn oil vehicle and administered intraperitoneally.

Materials and Methods

Animals

Six- to eight-week-old, female, C57Bl/6 mice can be obtained from Jackson Laboratories (Bar Harbor, ME). Male CD-1 mice, at 6 weeks of age, can be obtained from Charles River (Wilmington, MA). C57Bl/6 mice are housed four or five to a cage, and CD-1 mice are housed eight to a cage in a barrier facility maintained at a temperature of 21–22° on a 12-hr light/dark schedule. Mice are provided with autoclavable rodent laboratory chow (Purina Mills, St. Louis, MO) and acidified water *ad libitum*. Mice are acclimated for at least 1 week after receipt prior to commencement of experiments.

Administration of Retinoids

Retinoids are administered intraperitoneally on a daily basis to the mice, beginning on the day of antigen administration and continuing to the day prior to the assay. A suspension of the retinoid is prepared by sonication of the compound mixed with corn oil. Concentrations are adjusted to administer the desired amount of retinoid in a 0.1-ml volume. Fresh stock solutions are prepared weekly and refrigerated in amber vials

[10] J. G. Allen and D. P. Bloxham, *Pharmacol. Ther.* **40,** 1 (1989).
[11] C. J. L. Bugge, L. C. Rodriguez, and F. M. Vane, *J. Pharm. Biomed. Anal.* **3,** 269 (1985).

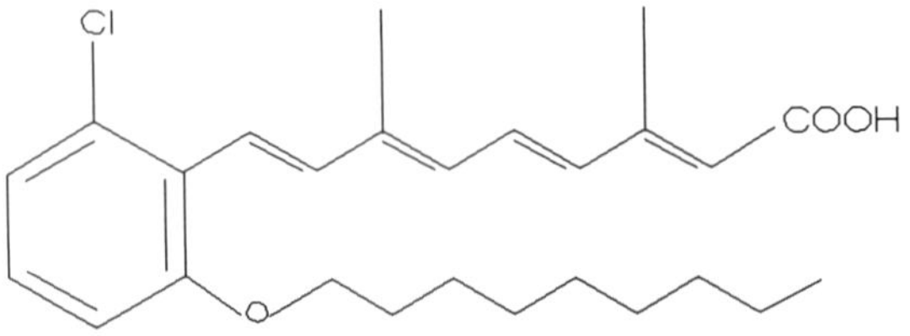

FIG. 1. Structure of Ro 23-5023.

during storage. Care should be taken to avoid exposure of the stock solutions to light, as most retinoids are light-sensitive.

The retinoid Ro 23-5023 (Fig. 1) was used in the experiments reported here. The results provide an example of retinoid-induced inhibition in each model.

In Vivo Cytolytic T Lymphocyte Response

The *in vivo* CTL response assay measures the ability of the mice to mount a cell-mediated immune response to foreign MHC antigens. Inject each C57Bl/6 (H-2^b) female mouse with 3×10^6 viable P815 mastocytoma cells (H-2^d, see following section on the maintenance of this cell line) intraperitoneally (i.p.) in 0.1 ml of phosphate-buffered saline (PBS; Gibco, Grand Island, NY). Approximately 10 days later, sacrifice the mice and assay the splenocytes for cytolytic T-cell activity in a 4-hr chromium release assay.

In order to prepare suspensions of splenocytes, passage each spleen individually through a steel mesh to prepare a single cell suspension in PBS. Allow the tissue debris to settle and remove the supernatant containing the splenocytes. Centrifuge the cell suspension at 300 *g* for 10 min at 5°, discard the supernatant, and resuspend the cell pellet in PBS. After an additional wash, resuspend the cells in 2 ml of assay medium consisting of minimal essential medium (MEM) without L-glutamine or $NaHCO_3$ (Gibco), 10% (v/v) heat-inactivated fetal bovine serum (FBS; Sterile Systems, Logan, UT), 2 m*M* L-glutamine (Gibco), 1 m*M* sodium pyruvate (Gibco), 0.1 m*M* MEM nonessential amino acids (Gibco), 1 ng/ml sodium bicarbonate solution (Gibco), 50 μg/ml gentamicin solution (Gibco), and 5×10^{-5} *M* 2-mercaptoethanol (Sigma, St. Louis, MO). Count the viable splenocytes using a hemacytometer or a Coulter counter (Hialeah, FL) and dilute in assay medium to 16×10^6 cells/ml.

Prepare a series of four 2-fold dilutions of the splenocytes. Plate 100 μl of each dilution, in triplicate, into 96-well U-bottom plates (Costar #3799, Cambridge, MA). Add 100 μl of the ^{51}Cr-labeled P815 mastocytoma target cell suspension (see below) to each well of the 96-well plates containing splenocytes to yield effector to target cell ratios of 160:1, 80:1, 40:1,

TABLE I
EFFECT OF DAILY ADMINISTRATION OF Ro 23-5023 ON CYTOTOXICITY RESPONSES

Treatment[a] (mg/kg)	Percent specific lysis, effector to target ratio[b]		
	160:1	80:1	40:1
Vehicle	43 ± 10	46 ± 10	37 ± 7
3	3 ± 2	1 ± 1	1 ± 1
1	2 ± 1	4 ± 1	3 ± 1
0.3	8 ± 8	5 ± 4	4 ± 3
0.1	29 ± 5	26 ± 4	20 ± 1

[a] Mice were administered the indicated doses of Ro 23-5023 intraperitoneally on a daily basis in 0.1 ml of corn oil.

[b] Effector cells were tested against P815 target cells in a standard 4-hr chromium-51 release assay at the indicated effector to target ratios. Data represent means ± S.D. of four mice per group.

20:1, and 10:1. Establish an additional 6 wells on each plate with 0.1 ml of assay medium plus labeled target cells to determine spontaneous release of chromium. The incorporation of label in 0.1 ml of target cells is used as the maximum counts per minute (cpm). Incubate the plates at 37° in a 5% CO_2 in air incubator for 4 hr. Harvest the supernatants with a Skatron (Sterling, VA) harvesting system and count in a γ counter. Alternatively, the supernatants can be collected manually with a multichannel pipettor and counted in a γ counter.

Determine the percent specific cytotoxicity by the following formula:

$$\% \text{ cytotoxicity} = 100 \left(\frac{\text{cpm/sample} - \text{spontaneous release (cpm)}}{\text{maximum cpm} - \text{spontaneous release (cpm)}} \right)$$

The effect of retinoid treatment on the generation of the CTL response can be evaluated by comparing the percent specific cytotoxicity by effector cells from retinoid-treated mice to the cytotoxicity by effector cells from vehicle-treated controls. Data from a representative experiment are shown in Table I. In this experiment, only three effector to target ratios were assayed.

Alternatively, the data can be converted to lytic units.[12] The specific release values are plotted versus the log of the effector to target ratios. The number of splenocytes necessary to obtain 30% specific lysis is defined as

[12] J.-C. Cerottini and K. T. Brunner, *in* "*In Vitro* Methods in Cell-Mediated Immunity" (B. Bloom and P. Glade, eds.), p. 369. Academic Press, New York, 1971.

one lytic unit. Comparison of the number of lytic units per million splenocytes from control and retinoid-treated mice can then be made.

Maintenance and Labeling of P815 Mastocytoma Target Cell Line

The P815 mastocytoma cell line (American Type Culture Collection, Rockville, MD), which is used as the alloantigen in the CTL assay, can be propagated *in vitro* or *in vivo*. For *in vitro* maintenance, the cell line is passaged twice weekly at cell densities of 1 to 5×10^3 cells/ml in RPMI 1640 medium (Gibco) supplemented with 10% FBS (Sterile Systems) and 50 μg/ml gentamicin solution (Gibco). For optimal results, the cells are harvested from tissue culture at a maximum cell density of 1×10^6 cells/ml. Cells grown to a higher density are less immunogenic and have high spontaneous release of chromium. For *in vivo* maintenance, the cells are passaged 3 times a week in the peritoneal cavities of DBA/2 mice. Approximately 2 to 8×10^6 cells are injected per mouse. The cells are harvested by peritoneal lavage with PBS (Gibco) and should not exceed 1.5×10^8 cells per animal for optimal results.

The P815 cells are labeled with chromium-51 as follows. Collect the target cells by centrifugation at 300 *g* for 10 min, at 5°. Incubate 7.5×10^6 cells for 2 hr at 37° in 1 ml of assay medium containing 100 μCi sodium [Cr^{51}]chromate (specific activity 250–500 mCi/mg; Amersham, Arlington Heights, IL). Collect the labeled target cells by centrifugation at 300 *g* and wash twice by centrifugation through 1 ml of undiluted FBS. Resuspend the cells in 2 ml of assay medium and count using a hemacytometer. Adjust the cells to a concentration of 1×10^5 cells/ml and add to the plates containing effector cells.

In Vivo Antibody Response

The *in vivo* antibody response assay measures the ability of mice to mount a humoral immune response to a particulate antigen, sheep erythrocytes (SRBC; Colorado Serum Company, Denver, CO). Immunize C57Bl/6 mice with SRBC by injecting each animal i.p. with 2×10^8 SRBC in 0.2 ml PBS. Five days after SRBC administration, prepare single cell suspensions from individual spleens by placing the spleen in a 10×35 mm tissue culture dish containing 5 ml of RPMI 1640 medium (Gibco) and disrupting the tissue with the plunger of a 3-cm^3 syringe. Draw the splenocytes through a 26-gauge needle into a 5-cm^3 syringe, transfer the cells to a polypropylene centrifuge tube, and wash them once by centrifugation at 300 *g* for 10 min, at 5°. Count the number of viable cells with either a hemacytometer or a Coulter counter. Suspend the cells at 2×10^6/ml in RPMI medium and assay for the number of hemolytic plaque-forming cells (pfc) using a modification of the Jerne hemolytic pfc assay.[8] Briefly, mix 50 μl of the splenocyte suspension with 50 μl of a 30% (v/v) suspension of SRBC in RPMI 1640 and 0.6 ml of a solution of 0.7% (w/v) SeaPlaque agarose (FMC Bioproducts, Rockland, ME) in RPMI 1640. Pour this mixture into a 10 × 35 mm tissue culture dish containing a 2 ml base layer of solidified 0.7% (w/v) agarose in RPMI 1640. After overnight incubation at 37° in a 5% CO_2 (v/v) in air incubator, the hemolytic plaques are developed by adding 0.5 ml of 5% (v/v) guinea pig serum (Pel-Freeze, Rogers, AR) in PBS. The number of plaque-forming cells are enumerated visually on duplicate plates prepared from each splenocyte suspension. A comparison of the responses from the retinoid-treated mice versus the vehicle-treated mice can then be made. Data from a representative experiment are shown in Fig. 2.

In Vivo Delayed-Type Hypersensitivity Response

The *in vivo* DTH response assay measures the ability of mice to mount an antigen-specific inflammatory response to a chemical sensitizing agent. The sensitizing agent, 4-ethoxymethylene-2-phenyloxazol-5-one (oxazo-

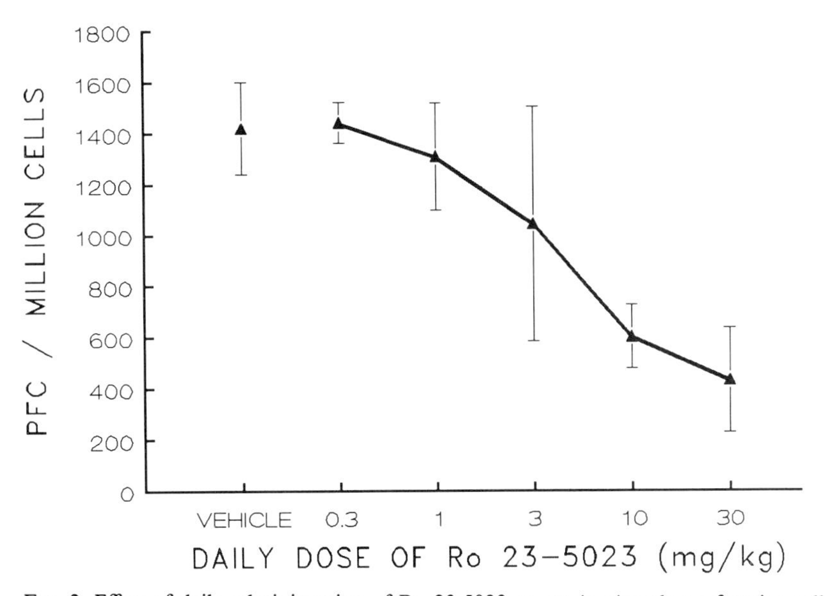

FIG. 2. Effect of daily administration of Ro 23-5023 on an *in vivo* plaque-forming cell response. Mice were administered the indicated doses of the retinoid intraperitoneally in 0.1 ml of corn oil. The response of each animal was determined in duplicate in a Jerne plaque assay. Data are means ± S.D. of five mice per group.

lone), was purchased from Sigma. Male CD-1 mice, in groups of eight, are sensitized by a single topical application of 25 μl of a 16 mg/ml solution of oxazolone in acetone applied to the scrotum. One week later, mice are challenged by application of 25 μl of a 8 mg/ml solution of oxazolone in acetone to the outer surface of the ear. Control animals are challenged in the same manner with acetone (basal ear weights). Twenty-four hours later, the mice are sacrificed and a punch biopsy of the challenged ear is taken, using a number six Keyes dermal punch (Arista Surgical Supply, New York, NY). The ear punches are weighed to the nearest milligram. It is important to weigh each group of ear punches immediately because fluid is rapidly lost, which can result in lower weights. A comparison of the responses from the retinoid-treated mice versus the vehicle-treated mice can then be made. Data from a representative experiment are shown in Fig. 3.

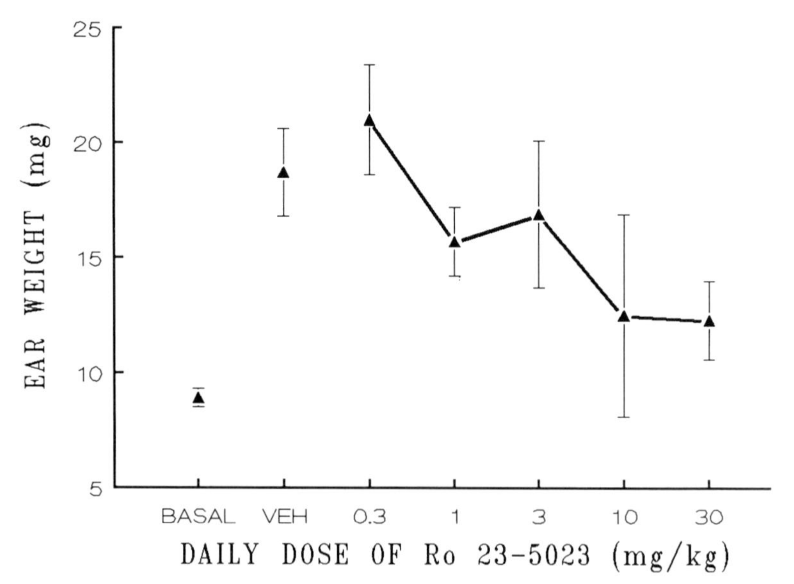

FIG. 3. Effect of daily administration of Ro 23-5023 on the *in vivo* delayed-type hypersensitivity response to oxazolone. Mice were administered the indicated doses of the retinoid intraperitoneally in 0.1 ml of corn oil. Twenty-four hours after challenge, ear punch biopsies were taken and weighed to the nearest milligram. Data are means ± S.D. of eight mice per group.

Concluding Remarks

Characterizing the activity of a retinoid in these three model systems will permit an interpretation of the relative capacity of the compound of interest to affect both cellular and humoral immune responses. The compounds with activity can then be analyzed in animal models of immune disease states. The discovery of compounds which selectively inhibit either cellular or humoral immune responses will also aid in the determination of which effector functions contribute to the various diseases of immune dysfunction.

[28] Immunotrophic Methodology

By KATHLEEN M. NAUSS, A. CATHARINE ROSS, and SALLY S. TWINING

Introduction

Interest in the immunoregulatory properties of vitamin A stems from three types of observations: the susceptibility of vitamin A-deficient humans and animals to infection, the association of vitamin A deficiency with depressed immune responses, and the stimulating effects of excess vitamin A, retinoic acid, and synthetic retinoids on various immune parameters.[1-3] A large number of *in vitro* and *in vivo* immunologic assays have been used to assess immunocompetence in these studies. In this chapter, the tests used most frequently to evaluate immunologic status in vitamin A-deficient or retinoid-supplemented human and animal subjects are discussed. These assays have been described in detail in earlier volumes of this series[4] and in specialized handbooks.[5-7] The reader is referred to

[1] M. E. Gershwin, R. S. Beach, and L. S. Hurley, "Nutrition and Immunity." Academic Press, Orlando, Florida, 1985.

[2] K. M. Nauss, *in* "Vitamin A Deficiency and Its Control" (J. C. Bauerenfiend, ed.), p. 207. Academic Press, Orlando, Florida, 1986.

[3] R. Chandra, "Nutrition and Immunology" Alan R. Liss, New York, 1988.

[4] Immunochemical Techniques, this series, Vol. 70 (1980); Vol. 73 (1981); Vol. 74 (1981); Vol. 84 (1982); Vol. 92 (1983); Vol. 93 (1983); Vol. 108 (1984); Vol. 116 (1985); Vol. 121 (1986); Vol. 132 (1986); Vol. 150 (1987).

[5] D. M. Weir (ed.), "Handbook of Experimental Immunology," Vols. 1–3. Blackwell, Oxford, 1989.

[6] N. R. Rose, H. Friedman, and J. L. Fahey, "Manual of Clinical Laboratory Immunology," 3rd Ed. American Society of Microbiology, Washington, D.C., 1986.

[7] L. M. Bradley, *in* "Selected Methods in Cellular Immunology" (B. B. Mishell and S. M. Shigii, eds.), p. 153. W. H. Freeman and Co., New York, 1980.

these sources and to those indicated in the text for details regarding specific procedures. Our purpose is to (1) provide an overview of the methodologies available to identify retinoid-associated immunosuppression or immune enhancement, (2) discuss interpretation of the data, and (3) indicate the strengths and limitations of the different assays when applied to nutritional or pharmacological studies.

Models

The effects of vitamin A deficiency on immunocompetence have been examined in human subjects in epidemiological surveys, as well as in clinical studies and in experimental animals. It is difficult to interpret many field studies because of the confounding effects of concurrent protein–energy malnutrition (PEM), infection, or the failure to control for such variables as sanitation and socioeconomic factors. In a limited number of clinical trials, an immunopotentiating effect of retinoids or high doses of vitamin A has been reported; in these studies, however, collection of immunology data was often secondary to determination of the therapeutic effect of the test agent.

Animal models bypass some of these difficulties, because they permit regulation of nutrient status and utilization of proper controls. Well-defined experimental animal models allow the investigator not only to identify whether cell-mediated, humoral, phagocytic, or nonspecific responses have been affected but also to develop hypotheses for the role of vitamin A in the immune response. Immune function studies have been conducted in vitamin A-deficient rats, chickens, rabbits, and swine. The mouse model, although attractive for immunology studies, has not been used extensively because induction of vitamin A deficiency in this species requires initiation of a vitamin A-deficient diet during fetal development. The immune status of most animal species is very sensitive to vitamin A deficiency. Immune response impairment often precedes other early indicators of vitamin A deficiency, such as growth retardation or the development of ocular lesions.[2]

Vitamin A deficiency has been induced in experimental rats by two methods: by elimination of all forms of the vitamin from the diet of rapidly growing animals[8] or by elimination of retinol while providing cyclic retinoic acid supplementation.[9] The deprivation method, although mimicking the development of human vitamin A deficiency, requires the use of

[8] G. Wolf, *in* "Nutrition and the Adult: Micronutrients" (R. B. Alfin-Slater and D. Kritchevsky, eds.), p. 97. Plenum, New York, 1980.

[9] A. J. Lamb, P. Apiwatanaporn, and J. A. Olson, *J. Nutr.* **104,** 1140 (1974).

pair-fed controls because of the anorexia and decreased weight gain associated with consumption of a vitamin A-deficient diet. Such controls are often omitted, but are necessary because PEM and other single-nutrient deficiencies also lead to impaired immunocompetence. It has not yet been established whether the two methods of inducing vitamin A deficiency have similar effects on the immune system.

Investigators need to be sensitive to the fact that the impact of vitamin A deprivation on the immune system is dependent on the stage of deficiency and the response being examined. Even before rats reach the weight plateau, which is generally regarded as the first clinical sign of vitamin A deficiency, serum retinol levels are low, and the splenic lymphocyte transformation response to T-cell mitogens is decreased.[10] As the weight plateau progresses, spleen cellularity decreases and the splenic lymphocyte response to B-cell mitogens diminishes, but the mitogen-induced responses of the peripheral lymphoid organs actually increase.[11]

The confounding effect of opportunistic infections is a serious problem in animals that have moderate to severe vitamin A deficiency. Special attention needs to be given to animal housing and the microbial environment. A precipitous weight loss in a vitamin A-deficient animal is usually a sign of infection. The clinical status of the animals, especially any indications of ocular or other types of infectious disease, should be carefully noted.

There is an extensive literature describing immunostimulation associated with treatment of animals with pharmacological doses of vitamin A or retinoids. These effects include enhanced antibody responses, increased lymphoid tissue reactivity, accelerated graft-versus-host reactions, as well as stimulated cell-mediated and cytotoxic responses.[2,12] The biological relevance of some of these reports has been called into question because of the acute toxicity associated with high doses of systemically administered vitamin A. Both increases and decreases of immune responsiveness have been reported following supplementation of animal diets with excess, but nontoxic, levels of vitamin A.[2,13] The magnitude and direction of the immune response in vitamin A-supplemented animals are dependent, in part, on the dose of the retinoid and the treatment protocol (continuous feeding versus bolus administration).[13] Investigators studying the effects of vitamin A or retinoid supplementation also need to consider the immunomolula-

[10] K. M. Nauss, D. A. Mark, and R. M. Suskind, *J. Nutr.* **109,** 1815 (1979).

[11] K. M. Nauss, C.-C. Phua, L. Amborgi, and P. M. Newberne, *J. Nutr.* **115,** 909 (1985).

[12] G. Dennert, *in* "The Retinoids" (M. B. Sporn, A. B. Roberts, and D. S. Goodman, eds.), Vol. 2, p. 373. Academic Press, Orlando, Florida, 1984.

[13] A. Friedman and D. Sklan, *J. Nutr.* **119,** 790 (1989).

tory properties of the vehicle and the timing of the immune function tests relative to the supplementation.

Tests of Immune Function

Multiple tests have been developed to evaluate the complex cellular reactions involved in the expression of immunity. These assays commonly include immunopathology assessment, tests of cellular and humoral immune function, tests of nonspecific immunity, and tests of resistance to tumor cells, microorganisms, or other pathogens.

Immune dysfunction can be evaluated *in vivo* by testing for alterations in the ability of an animal to resist infection or tumor cell challenge. Susceptibility to experimental infection with bacteria, parasites, and viruses has been reported in vitamin A-deficient animals, but interpretation of many of the earlier studies is compromised because comparisons were made to *ad libitum* rather than to pair-fed controls.[2,14]

Investigators interested in the immunosuppressive or immunopotentiating effects of nutrients such as vitamin A have often focused their efforts on individual tests of *in vivo* or *in vitro* immune function, without considering the relevance of the assay, or the results, to overall immune competence. It is instructive in this regard to consider the efforts of toxicologists, who have made impressive advances in developing valid and meaningful strategies for assessing chemical-induced immunotoxicity.[15,16] The National Toxicology Program (NTP) has evaluated a number of different murine host-resistance assays and has chosen four models for its validation assays: challenge with syngeneic tumor cells (PYB6 and B16F10), bacterial challenge (*Listeria monocytogenes* and *Streptococcus* species), viral challenge (influenza virus), and parasite challenge *(Plasmodium yoelli)*.[15] These models were selected for technical considerations and because current information on the mechanisms involved in the host response indicates that different components of the immune system mediate the responses to these agents. The NTP has developed a large data base on these four host-resistance challenge assays as well as *in vitro* assays which may be useful to investigators interested in retinoid-induced modulation of immune defenses.

[14] N. S. Scrimshaw, C. E. Taylor, and J. E. Gordon, *W.H.O. Monogr. Ser.* **57** (1968).

[15] M. I. Luster, A. E. Monson, P. T. Thomas, M. P. Holsapple, J. D. Fenters, K. L. White, Jr., L. D. Lauer, D. R. Germolec, G. J. Rosenthal, and J. H. Dean, *Fundam. Appl. Toxicol.* **10,** 2 (1988).

[16] D. Trizio, D. A. Basketter, P. A. Botham, P. H. Graepel, and C. Lambré, S. J. Magda, T. M. Pal, A. J. Riley, H. Ronneberger, N. J. Van Sittert, and W. J. Bontinck, *Food Chem. Toxicol.* **6,** 527 (1988).

Immunopathology

Routine immunopathology procedures may be very useful in assessing the immunomodulating effect of specific nutrients. Such procedures include hematology (complete blood count and differential) and assessment of the weight, cellularity, and histology of lymphoid organs. Hematological parameters are very sensitive to vitamin A status. A significant drop in the number of circulating lymphocytes is associated with mild vitamin A deficiency, a condition that is rapidly reversed with vitamin A supplementation.[11]

Peripheral blood and bone marrow, spleen, lymph node, and thymus cell suspensions can be typed and quantified using fluorescently labeled (fluorescein isothiocyanate, phycoerythrin, or tetramethylrhodamine isothiocyanate) monoclonal antibodies directed toward cell-specific antigen clusters (CD). Single and/or double labeling experiments with human cells can distinguish between T (CD3) and B (CD20) cells, natural killer cells (CD56), monocytes ($CD14^+$, $Leu8^+$), macrophages (CD14, $Leu8^-$, or $CD11b^-$), and polymorphonuclear leukocytes ($Leu8^+$, $CD11b^+$). In addition, activation state (CD25), B-cell subsets,[17] and T-cell subsets (mitogenic, $CD3^+$; cytotoxic, $CD8^+$ and $CD11b^-$; supressor, $CD8^+$ and $CD11b^+$; helper, $CD4^+$ and $CD45R^-$; and inducer, $CD4^+$ and $CD45R^+$) can be identified.[18-20] (Mouse monoclonal antibodies to the above human antigens are available from various suppliers including Becton Dickinson Immunocytometry Systems, Sunnyvale, CA; Ortho Diagnostic Systems, Raritan, NJ; and Coulter Immunology, Hialeah, FL. Monoclonal antibodies to mouse and rat monocyte, B-, and T-cell antigens are available from Becton Dickinson and from Bioproducts for Science, Indianapolis, IN, respectively.) Cell populations can be counted manually using fluorescence microscopy; however, this is a time-consuming process. In contrast, 5,000 to 10,000 cells can be quantified rapidly on the basis of fluorescence intensity, size, and granularity using a flow cytometer equipped with fluorescence detectors, in addition to detectors for forward-angle and right-angle light scatter.[17-23] These fluorescence techniques can be combined

[17] E. Sherr, D. C. Adelman, A. Saxon, M. Gilly, R. Wall, and N. Sidell, *J. Exp. Med.* **168,** 55 (1988).

[18] R. B. Lal, L. J. Edison, and T. M. Chused, *Cytometry* **9,** 213 (1988).

[19] T. A. Fleisher, G. E. Marti, and C. Hagengruber, *Cytometry* **9,** 309 (1988).

[20] R. Festin, B. Björklund, and T. H. Tötterman, *J. Immunol. Methods* **101,** 23 (1987).

[21] R. R. Watson, M. D. Yahya, H. R. Darban, and R. H. Prabhala, *Life Sci.* **43,** xiii (1988).

[22] B. S. Edwards and G. M. Shopp, *Cytometry* **10,** 94 (1989).

[23] S. M. Smith, N. S. Levy, and C. E. Hayes, *J. Nutr.* **117,** 857 (1987).

with rosetting[24] or "panning,"[25] using specific antibodies to identify or separate cell populations.

Microscopic examination of tissue sections by a pathologist may yield information about nutrient effects on thymus-dependent and thymus-independent areas of the spleen, thymus, and lymph nodes. Because there is a high degree of variability among subjects, and even among different sections from the same tissue, these procedures require a large sample size to acquire valid data.

Cell-Mediated Immunity

Both *in vivo* and *in vitro* assays have been used to assess retinoid-mediated cell-mediated immunity (CMI) in human subjects and animal models.

Delayed Hypersensitivity Response. The delayed hypersensitivity response (DHR), the classic clinical test for assessing the *in vivo* CMI status in humans, measures the ability of a subject to respond to new antigens (primary response) or to antigens to which the subject has been previously exposed (secondary response). Intradermal injection of antigen initiates a complex series of reactions including stimulation of sensitive T cells to produce lymphokines and inflammatory mediators, which, in turn, attract phagocytic cells to the injection site, resulting in an inflammatory response. Inconsistent reports on the effect of vitamin A deficiency on the DHR in human subjects may be due, in part, to the complications of PEM.[26-28] Oral retinoid therapy has been reported to increase *in vivo* responses to primary and recall antigens.[29]

The methods developed to quantify DHR in animal models have generally not proved to be as sensitive or reproducible as those used in humans. Radioisotopic procedures that compare monocyte influx at a challenge site (e.g., rodent ear) to the influx at a control site (the opposite ear) are generally superior to methods that depend on assessment of the size of an inflammatory wheal.

Three different *in vitro* tests are commonly used to assess different aspects of CMI function in retinoid-deficient or retinoid-supplemented subjects: the lymphocyte proliferation assay, the mixed leukocyte response, and the cytotoxic T-lymphocyte response. Retinol, retinoic acid, and other retinoids can also be added directly to *in vitro* cultures. Their effects,

[24] B. C. Elliot and H. F. Pross, this series, Vol. 108, p. 49.

[25] L. J. Wysocki and V. L. Sato, *Proc. Natl. Acad. Sci. U.S.A.* **75,** 2844, 1978.

[26] V. T. Jayalakshmi and C. Gopalan, *Indian J. Med. Res.* **46,** 87 (1958).

[27] C. Bhaskaram and V. Reddy, *Br. Med. J.* **3,** 522 (1975).

[28] K. H. Brown, M. M. Rajan, J. Charkraborty, and K. M. A. Aziz, *J. Clin. Nutr.* **33,** 212 (1980).

[29] R. A. Fulton, P. Souteyrand, and J. Thivolet, *Dermatologica* **165,** 568 (1982).

however, are influenced by a number of experimental variables, particularly the cell type, retinoid concentration, and toxicity, and whether the culture is being maintained at optimal conditions. In the hands of careful investigators, *in vitro* cultures are powerful tools for unraveling the mechanisms of retinoid interactions with different components of the immune system.

Lymphocyte Proliferation Assay. A number of investigators have demonstrated the sensitivity of the mitogen-induced lymphocyte proliferation assay to vitamin A status or retinoid supplementation.[2] Lymphocytes isolated from peripheral blood or lymphoid organs are stimulated by polyclonal activators (mitogens) to undergo blastogenesis and proliferation. For technical reasons, investigators conducting studies in human subjects, or in large animals, generally use lymphocytes prepared from blood samples by density centrifugation, whereas those studying rodents rely on either crude cell suspensions from the spleen, thymus, or regional lymph nodes or suspensions from which red blood cells have been removed by brief hypotonic lysis with buffered ammonium chloride. Specific methods for separating and purifying populations of lymphoid cells, as well as details regarding the mitogen-induced lymphocyte proliferation assay, have been described in earlier volumes in this series[4] as well as in other basic manuals of immunologic methods.[5-7] The most commonly used mitogens in the lymphocyte proliferation assay are the plant lectins concanavalin A (Con A) and phytohemagglutinin (PHA), which stimulate different populations of T cells,[30] and the bacterial product lipopolysaccharide (LPS), which is mitogenic for B lymphocytes. There are, however, marked species differences in the sensitivity of lymphocytes to mitogenic stimulation.

The magnitude of the mitogen-induced lymphocyte proliferation response is generally measured by the cellular uptake of [*methyl*-^{3}H]thymidine and its incorporation into cellular DNA. Given the assumption that [*methyl*-^{3}H]thymidine incorporation reflects the rate of cellular proliferation, radiolabeled thymidine should be added as a pulse and its concentration should not be rate-limiting.[31] The latter consideration is often overlooked and becomes significant if retinoid treatment alters cellular thymidine synthesis and, hence, thymidine pool size and specific radioactivity of [*methyl*-^{3}H]thymidine. Although there is a typical dose and time response for each mitogen, considerable variation exists among laboratories depending on the cell type, cell concentration, mitogen concentration, serum source, and other culture variables. When comparing the lymphocyte proliferation response of cells from control and retinoid-treated animals, investigators should examine the response at mitogen concentrations

[30] J. D. Stobo and W. E. Paul, *J. Immunol.* **110,** 362 (1973).
[31] C. F. Beyer and W. E. Bowers, *J. Immunol.* **119,** 2120 (1977).

above and below the optimal dose level for control cultures, because nutrient deprivation or excess can shift the dose–response curve. The mitogenic response can be expressed two ways, either as the counts per minute (cpm) of the stimulated cells minus the cpm of the unstimulated cells, or as the stimulation index (SI), which is equal to (cpm stimulated − cpm unstimulated)/cpm unstimulated. The SI, although frequently used, is very sensitive to small changes in unstimulated cpm. It is recommended that a full data set be presented.

Mixed Lymphocyte Response. Mixtures of lymphocytes from two genetically dissimilar individuals or animals are capable of stimulating each other to proliferate. Usually one-way stimulation is obtained by inactivating the stimulator cells with mitomycin C or irradiation. Equal numbers of stimulator and responder cells are cultured for 3 to 5 days, and stimulation is quantified by measuring incorporation of [*methyl*-^{3}H]thymidine into DNA. Appropriate controls to measure nonspecific activity and the efficacy of the inhibition of the stimulator cells should be included. Because the allogeneic mixed lymphocyte response (MLR) is a measure of histoincompatibility, it is used in clinical settings to predict host responses to transplantation and as a general indicator of immunocompetence. Luster *et al.*[15] found that the MLR was generally a sensitive indicator of chemical-induced CMI immunosuppression in mice; however, the assay has not been used to any great extent in studies of retinoid-induced immunomodulation.

Cytotoxic T-Lymphocyte (CTL) *Response.* Retinoid treatment has been shown to stimulate the CTL response in animal models.[12,32,33] CTL can be elicited *in vivo* by stimulator cells that carry new surface antigens or *in vitro* by culturing spleen cells with appropriate stimulator cells that have been irradiated to prevent proliferation. In animal models CTL are generally found in the spleen and in the peritoneal fluid. Cytotoxicity is evaluated in *in vitro* assays by measuring the release of radioisotope label such as chromium-51 from the target cells.

Humoral Immunity

A number of investigators have reported effects of either poor retinoid status or retinoid supplementation on antibody production.[2,12] We discuss here the most common methods to assess the humoral immune response, which include measuring serum antibody concentrations (titer), determining the number of antibody-producing lymphocytes, and assessing the proliferative potential of B lymphocytes *in vitro.*

Serum Antibody Response. Serum antibody titer provides a convenient

[32] R. Lotan and G. Dennert, *Cancer Res.* **39,** 55 (1979).

[33] G. Dennert, C. Crowley, J. Kouba, and R. Lotan, *JNCI, J. Natl. Cancer Inst.* **62,** 89 (1979).

measure of a response to immunization or of the level of "natural," preexisting antibodies of a particular class. Often, an implicit assumption is that titer reflects antibody production; more accurately, it represents a balance between antibody secretion and clearance, which could be affected differently with changes in retinoid status. A major advantage of measuring serum antibody titer is the possibility of making serial measurements so as to conduct kinetic or response studies in individuals or populations.

Excellent references such as the *Manual of Clinical Laboratory Immunology*[6] provide numerous specific methods to assess antibody concentrations, including those based on complement fixation, cytolytic reactions, neutralization of toxins by antibody, nephelometry of the antigen–antibody complex, radioimmunoassay (RIA), and enzyme-linked immunosorbent assays (ELISA). For any of these methods, preliminary experiments are required to validate the assay with new antigens. These include tests of sensitivity to determine the working range of the assay and tests of specificity for the antigen of interest. Each methodologic approach has its strengths or limitations: for instance, methods based on complement fixation require careful standardization of complement; nephelometric assays can be sensitive to serum turbidity such as could be present after retinoid treatment; the results of competitive binding assays such as RIA and ELISA methods reflect antibody–antigen affinity as well as antibody amount, each of which may change following immunization; RIA and ELISA methods offer the possibility to investigate isotype specificity by utilizing a second antibody specific for an isotype class or subclass.

For any method, it is advisable to prepare one or more pooled serum "standards" for inclusion in each analytical run to ensure comparability over time, and this is critical for kinetic measurements. The investigator should conduct competition experiments to determine specificity and to evaluate the extent of antibody cross-reactivity. It is desirable to obtain a preimmunization sample from each subject to test for preexisting antibodies. In the case of RIA or ELISA procedures, the investigator should demonstrate that the addition of soluble immunogen competes with labeled or coating antigen for binding to antibodies. Studies with irrelevant antigens which should not compete, or related but nonidentical antigens which may cross-react, are useful in defining the specificity of the humoral response under investigation.

Hemolytic Plaque-Forming Assay. Determining the number of lymphocytes secreting antibody to a particular antigen remains a mainstay of experimental immunologic testing. The hemolysis-in-gel method of Jerne and associates[34,35] is based on the principle that antibody secreted by

[34] H. K. Jerne and A. A. Nordin, *Science* **26,** 504 (1963).
[35] H. K. Jerne, C. Henry, A. A. Nordin, H. Fugi, A. M. C. Kores, and I. Lefkovits, *Transplant. Rev.* **18,** 130 (1974).

plasma cells will combine with antigen-coated target cells (red blood cells) in an agar or agarose matrix, and after addition of complement, target cells will lyse to form visible plaques against a background of unlysed red cells. Although the method has most often been applied to antibody production by splenic lymphocytes, other tissue sources from which a whole cell suspension containing B cells can be prepared are also amenable to investigation. Generally, the number of plaque-forming cells (pfc) correlates well with serum antibody titer. Determination of pfc is versatile, with application to (1) assessing the response of intact animals to immunization, (2) determining the response to antigen restimulation *in vitro* of lymphocytes from antigen-primed animals, and (3) conducting studies in which unprimed lymphocytes are immunized *in vitro*. The latter two forms of study are most useful for determining the requirement or effects of additional cellular or soluble factors on antibody production. Data may be referenced to the number of cells per organ or to a reference population (e.g., per 10^8 spleen cells).

An advantage of the pfc assay is its sensitivity—individual immunocompetent plasma cells can be counted. When the assay is used to assess the immune response to red blood cells as antigen, no derivatization of the target cells is needed. Determination of pfc is, however, versatile for use with other antigens, provided these can be stably coated to, or coupled with, the target cell. For instance, we have used chromium chloride as an agent to couple bacterial antigens to sheep red blood cells.[36] The investigator can assess the primary (IgM) immune response using the "direct" version of the pfc assay in which pentavalent IgM directly binds complement.

Among the disadvantages of pfc determination is that kinetic studies in the same animal are usually precluded. Determination of pfc is essentially a bioassay and therefore is subject to day-to-day variations, often arising from variations in target cells, efficiency of antigen–red cell coupling, or efficacy of complement (generally, guinea pig serum or another serum source of complement). A small preliminary trial a few days before a major experiment is most helpful. For the same reasons, it is important to design retinoid studies such that all treatment groups are assessed simultaneously to ensure that pfc counts are truly comparable. In addition, as with any test that is read by observation, the pfc assay has an inherent element of subjectivity which is likely to become important when using antigens that produce small plaques. To minimize observer bias, we prepare coded slides and apportion replicate slides for counting by independent observers.

[36] A. M. G. Pasatiempo, M. Kinoshita, C. E. Taylor, and A. C. Ross, *FASEB J.* **4**, 2518 (1990).

Appropriate controls are required to show specificity: each experiment should include nonimmunized control animals to assess nonspecific antibody production; this is critical when red cells are used as immunogen. Testing against heterologous red cells is also useful in this situation. When antigen-coupled target cells are employed, the lymphocytes of each animal should also be tested against the same target cells that remain underivatized to assess the nonspecific background response to the cells themselves. That plaque-forming cells produce antibody specific for the coating antigen can be demonstrated by showing that addition of small amounts of soluble antigen inhibit lysis of antigen-coated red cells.

Consistent with a complex biological response and the monoclonal theory of antibody production, the quantitative pfc response can be quite variable, even with good design and within an inbred rodent strain. It is worthwhile to determine whether the response of the population is normally distributed to determine the appropriate statistical test. In the case of some pfc responses, the logarithm of pfc number is normally distributed and can be used for statistical testing.

Lipopolysaccharide-Stimulated Lymphocyte Transformation Response. The proliferative response to LPS, a mitogen capable of stimulating B lymphocytes directly, can be useful to assess retinoid-dependent maturation and proliferation of these cells.

Phagocyte Response

Investigators have studied the effect of retinoids on phagocytic function by adding these molecules *in vitro* to phagocyte cultures[37–42] or by manipulating the dietary level of the retinoid.[43–46] Retinoids added *in vitro* can activate phagocytes or modify the response to other stimulators. *In vivo* studies of dietary retinoid excess or deficiency permit examination of retinoid effects on the synthesis and maturation of phagocytic cells.

[37] M. Wolfson, E. S. Shinwell, M. Zvillich, and B. Rager-Zisman, *Clin. Exp. Immunol.* **72,** 505 (1988).

[38] B.-I. Coble, C. Dahlgren, L. Molin, and O. Stendahl, *Acta Derm. Venereol.* **67,** 481 (1987).

[39] H. Hemilä and M. Widstörm, *Scand. J. Immunol.* **21,** 227 (1985).

[40] S. Pontremoli, E. Melloni, M. Michetti, O. Sacco, F. Salamino, B. Sparatore, and B. L. Horecker, *J. Biol. Chem.* **261,** 8309 (1986).

[41] J. E. Lochner, J. A. Badwey, W. Horn, and M. L. Karnovsky, *Proc. Natl. Acad. Sci. U.S.A.* **83,** 7673 (1986).

[42] R. W. Randall, J. E. Tateson, J. Dawson, and L. G. Garland, *FEBS Lett.* **214,** 167 (1987).

[43] S. Moriguchi, L. Werner, and R. R. Watson, *Immunology* **56,** 169 (1985).

[44] Y. Kishino, S. Takama, and S. Kitajima, *Virchows Arch. B:* **50,** 81 (1985).

[45] M. Ongsakul, S. Sirisinha, and A. J. Lamb, *Proc. Soc. Exp. Biol. Med.* **178,** 204 (1985).

[46] D. A. Norris, R. Osborn, W. Robinson, and M. G. Tonnesen, *J. Invest. Dermatol.* **89,** 38 (1987).

Sources of phagocytes include peripheral blood for nonstimulated monocytes[46] and polymorphonuclear leukocytes (PMN),[46,47] the peritoneal cavity[43,48] and alveolar spaces[44] for resting macrophages, the peritoneal cavity after injection of activators (such as thioglycolate or shellfish glycogen) for large numbers of partially stimulated PMN and macrophages,[49,50] and bone marrow for neutrophil and monocyte precursors.[51,52] For most experiments, nonstimulated cells are preferred. The appropriate cell purification procedure depends on the cell source and the experiment to be performed. Whole blood,[53] peritoneal[43] and alveolar[44] lavages, or leukocytes isolated by dextran sedimentation[39,54] can be used. Cells are identified and counted, either manually or with a flow cytometer. Relatively pure populations (96–98%) of monocytes, macrophages, or PMN can be obtained by using Ficoll–Hypaque gradients,[39,47,54] customized for the species from which the cells are obtained, or by separating monocytes and macrophages from other leukocytes on the basis of their adherence properties.[44,48]

Phagocytic cells, and other leukocytes, can be quantified using a hemocytometer with trypan blue to determine cell number and viability, followed by examination of stained cell smears for differential counts.[55] Fluorescence flow cytometry also can be used to count cells, to identify cell type based on size, shape, and binding of monoclonal antibodies, and to determine cell viability by ethidium exclusion or vital dye metabolism.[55–57]

Retinoid-induced effects can be tested by studying phagocyte maturation, morphology, and such functions such as chemotaxis, phagocytosis, killing, digestion, and secretion.[44,49,51,52,55,57] In addition, signal transduction studies address the effects of retinoids on cell receptors,[57,58] membrane

[47] J. D. Ogle, J. G. Noel, R. M. Sramkoski, C. K. Ogle, and J. W. Alexander, *J. Immunol. Methods* **115,** 17 (1988).

[48] C. Fiedler-Nagy, B. H. Wittreich, A. Georgiadis, W. C. Hope, A. F. Welton, and J. W. Coffey, *Dermatologia* **175** (Suppl.), 81 (1987).

[49] J. M. Robinson, J. A. Badwey, M. L. Karnovsky, and M. J. Karnovsky, *J. Cell Biol.* **105,** 417 (1987).

[50] S. S. Twining, K. M. Lohr, and J. E. Moulder, *Invest. Ophthalmol. Visual Sci.* **27,** 507 (1986).

[51] N. Sato, O. Takatani, H. P. Koeffler, H. Sato, S. Asano, and F. Takaku, *Exp. Hematol.* **17,** 258 (1989).

[52] E. Januszewicz and I. A. Cooper, *Br. J. Haematol.* **58,** 119 (1984).

[53] S. S. Twining, R. J. Kawczynski, and P. McFadden, *J. Cell Biol.* **108S,** 574a (1988).

[54] F. S. Southwick, G. A. Dabiri, and T. P. Stossel, *J. Clin. Invest.* **82,** 1525 (1988).

[55] D. R. Absolom, this series, Vol. 132, p. 95.

[56] R. Bjerknes, *J. Immunol. Methods* **72,** 229 (1984).

[57] R. Bjerknes, C.-F. Bassøe, H. Sjursen, O. D. Laerum, and C. O. Solberg, *Rev. Infect. Dis.* **11,** 16 (1989).

[58] T. Nakamura and H. Hemmi, *Eur. J. Haematol.* **41,** 258 (1988).

potential changes,[57] phospholipase C[41] and protein kinase C activity,[40,41,57] calcium mobilization,[40,59] arachidonic acid metabolite production,[42,48] and actin polymerization.[54,60] *In vivo*[46] and *in vitro*[46,50,61] methods for studying chemotaxis involve quantification of cell migration through a controlled pore membrane toward a chemoattractant with correction for chemokinesis.

Phagocytic function is evaluated by measuring the uptake of living or inert particles such as microorganisms,[53,56,57,62] sheep red blood cells,[43] or polystyrene microspheres.[47] These particles can be quantified by counting stained or fluorescent particles, either manually[45,55] or with a flow cytometer,[53,57,62] or by measuring radiolabeled particles with a scintillation counter.[43,55] The major problem with phagocytosis assays is distinguishing between extracellular adherent organisms and those inside the cells. Three approaches have been used to resolve this difficulty: (1) lysis of extracellular organisms with enzymes such as lysostatin for *Staphyloccus*[55] or with EDTA for *Pseudomonas aeruginosa,*[53] (2) quenching of extracellular fluorescence,[62] and (3) reacting extracellular particles with an antibody labeled with a second fluorescent molecule.[47]

Killing of phagocytosed organisms is determined by plating techniques[55] or flow cytometer methods using fluorescent compounds that bind only to killed organisms, that is, propidium iodide added to lysed cells[56] or hydroethidine added to intact cells, which is converted in viable cells to ethidium.[53] Oxygen consumption,[39] chemiluminescence,[38] and production of superoxide,[37,39] hydrogen peroxide,[37,63] and hypochlorous acid,[37,64] as well as metabolic activities associated with phagocytic killing function, can also be studied. Digestion is tested by following the loss of label from phagocytes incubated with radiolabeled[65] or fluorescently labeled[56] organisms. Secretion is studied by assaying supernatant fractions from cytochalasin B- or phorbol myristate acetate-treated cells for specific lysosomal enzymes.[40,57]

[59] M. Lopez, D. Olive, and P. Mannoni, *Cytometry* **10,** 165 (1989).

[60] K. M. K. Rao, *Cancer Lett.* **28,** 253 (1985).

[61] C. N. Ellis, S. Kang, R. C. Grekin, J. J. Voorhees, and J. Silva, Jr., *J. Am. Acad. Dermatol.* **13,** 437 (1985).

[62] J. Hed, G. Hallden, S. G. O. Johansson, and P. Larsson, *J. Immunol. Methods* **101,** 119 (1987).

[63] D. A. Bass, J. W. Parce, L. R. DeChatelet, P. Szejda, M. C. Seeds, and M. Thomas, *J. Immunol.* **130,** 1910 (1983).

[64] S. J. Weiss, R. Klein, A. Slivka, and M. Wei, *J. Clin. Invest.* **70,** 598 (1982).

[65] A. W. Segal, *Clin. Immunol. Allergy* **5,** 491 (1985).

Natural Killer Cell (NK) Assay

Natural killer cells, considered an arm of the innate immune system, were originally defined as resting lymphocytes having as their hallmark the ability to lyse certain cultured tumor cells without prior sensitization and, in contrast to cytolytic T cells, in a major histocompatibility complex-unrestricted manner. Lymphocytes with NK activity bear a number of different surface receptors and thus show phenotypic heterogeneity; they may also be of various lineage.[66,67] In addition, lymphocytes activated in mixed culture, by lymphokines or by lectins, can acquire NK-like activity.[66,68]

Methods for assessing NK activity have been described.[69,70] Generally, these are based on cytolysis of transformed cells in short-term culture. In one method,[71] target cells (murine YAC-1 lymphoma cells) are labeled with chromium-51 and incubated in microwell dishes with effector cells (rat splenocytes) at various ratios of effector to target cells. Lysis is measured as chromium-51 released into the culture supernatant after 4 hr. Various modulators may be preincubated with the effector cells prior to addition of YAC-1 cells. Cytotoxicity is expressed as a percentage: [100(experimental chromium-51 release minus spontaneous release in the absence of effector cells)/(total release determined after addition of 0.5% Triton X-100 minus spontaneous release)]. It is important to confirm that results are consistent over a range of effector to target cell ratios.

Local Immune Response

The local immune response under vitamin A deficiency has been studied using intestinal fluid[72] or bile[73] to quantify the total and specific secretory IgA and IgG present. Immunofluorescence techniques on frozen tissue sections such as small intestine and colon can also be used.[72]

[66] P. Hersey and R. Bolhuis, *Imunol. Today* **8,** 233 (1987).

[67] C. W. Reynolds and J. R. Ortaldo, *Immunol. Today* **8,** 172 (1987).

[68] R. B. Herberman, J. Hiserodt, N. Vujanovic, C. Balch, E. Lotzova, R. Bolhuis, S. Golub, L. L. Lanier, J. H. Phillips, C. Riccardi, J. Ritz, A. Santoni, R. E. Schmidt, and A. Uchida, *Immunol. Today* **8,** 178 (1987).

[69] R. B. Herberman and D. M. Callewaert, *in* "Mechanisms of Cytotoxicity by NK Cells" (R. B. Herberman and D. M. Callewaert, eds.), p. 1. Academic Press, New York, 1984.

[70] P. J. Romano, *in* "Manual of Clinical Laboratory Immunology" (N. R. Rose, H. Friedman, and J. L. Fahey, eds.), 3rd Ed., p. 72. American Society for Microbiology, Washington, D.C., 1986.

[71] P. S. Morahan, W. L. Dempsey, A. Volkman, and J. Connor, *Infect. Immun.* **51,** 87 (1986).

[72] S. Sirisinha, M. D. Darip, P. Moongkarndi, M. Ongsakul, and A. J. Lamb, *Clin. Exp. Immunol.* **40,** 127 (1980).

[73] S. Puengtomwatanakul and S. Sirisinha, *Proc. Soc. Exp. Biol. Med.* **182,** 437 (1986).

Lymphokine Production and Response

Many soluble factors are now recognized to influence immune responses either quantitatively or qualitatively. To date, relatively little is known of the effects of vitamin A on the production of, or response to, specific lymphokines. As an example, Malkovsky *et al.*[74] have shown that supplementing newborn mice with retinyl acetate results in cessation of immunologic tolerance, and interleukin-2 (IL-2) is implicated in this process. At present, most assays of lymphokine production rely on the ability of supernatants from mitogen- or antigen-stimulated lymphocytes to stimulate growth of lymphokine-dependent cell lines.[75] The specificity of such assays for a particular lymphokine is usually not absolute but may be improved by neutralizing other lymphokines with specific antibodies.[75] Purified or recombinant interleukins are becoming available, and this should result in improved standardization of bioassays; similarly, specific antibodies will allow development of immunoassays to quantify lymphokine protein. In addition, these purified or recombinant lymphokines make possible studies of the effects of exogenous soluble factors on immune responses both *in vitro* and *in vivo.*

[74] M. Malkovsky, P. B. Medawar, D. R. Thatcher, J. Toy, R. Hunt, L. S. Rayfield, and C. Dore, *Proc. Natl. Acad. Sci. U.S.A.* **82,** 536 (1985).

[75] T. R. Mosmann and R. L. Coffman, *Annu. Rev. Immunol.* **7,** 145 (1989).

[29] Antioxidant Activity of Retinoids

By MIDORI HIRAMATSU and LESTER PACKER

Introduction

Although the effects of retinoids are numerous in biological systems, among them are indications that retinoids may exert some of their actions by virtue of their acting as lipid-soluble antioxidants. Thus, epidemiological studies have suggested that, as for vitamin E, higher serum retinol concentrations are directly correlated with lower risks of cancer[1] and ischemic heart disease mortality.[2] The use of retinoic acid in many animal models of carcinogenesis has also suggested that their action may be due to antioxidant activity. Moreover, retinol has been shown to inhibit iron-de-

[1] R. Peto, R. Doll, J. D. Buckley, and M. B. Sporn, *Nature (London)* **290,** 201 (1981).

[2] K. F. Gey, G. B. Brubacher, and M. B. Stahelin, *Am. J. Clin. Nutr.* **45,** 1368 (1987).

pendent peroxidation of rat liver microsomes by Adriamycin[3] and to inhibit prostaglandin and hydroxyeicosatetraenoic acid production from arachidonic acid and bovine seminal vesicles and kidney.[4]

In evaluating a substance as an antioxidant it is important to use more than one assay, because an antioxidant may be very sensitive in certain assays but it may not react in others, owing to, for example, steric hindrance. Factors such as antioxidant solubility in water or hydrophobic environments are also important. We describe here the application of two assays.

Electron Spin Resonance

In the first assay the free radical electron spin resonance (ESR) signal of a stable free radical is quenched in the presence of an antioxidant. The concentration dependence of the 1,1-diphenyl-2-picrylhydrazyl (DPPH) free radical signal quenching is the basis for the antioxidant assay.[5] The second assay makes use of phycoerythrin fluorescence. When phycoerythrin is exposed to a continuous source of peroxy radicals the molecule forms stable products, which is accompanied by a loss of fluorescence. Inhibition of the fluorescence decay rate by a substance is used as the principle for assaying antioxidant action by this method.[6]

Experimental Methods

General Reagents and Retinoids. DPPH was purchased from Sigma Chemical Co. (St. Louis, MO). Ro 4-3870 (13-*cis*-retinoic acid), Ro 1-5488 (all-*trans*-retinoic acid), Ro 10-1670, Ro 11-1430, Ro 13-6298, Ro 13-7410, Ro 15-0778, Ro 15-1570, Ro 21-6583, Ro 22-1318, and Ro 10-9359 (Table I) were gifts from Dr. Stanley Shapiro of Roche Dermatologics Inc. (Nutley, NJ). Other chemicals and reagents were of the highest grade available from commercial suppliers. The molecular structures of the compounds tested are shown in Fig. 1.

Antioxidant Activity of Retinoids Assayed by Electron Spin Resonance Spectrometry

DPPH Radical Analysis. DPPH (200 μM) is dissolved in ethanol. Thirty microliters of this solution and 30 μl of sample dissolved in ethanol

[3] G. F. Vile and C. C. Winterbourn, *FEBS Lett.* **238,** 353 (1988).

[4] O. Halevy and D. Sklan, *Biochim. Biophys. Acta* **918,** 304 (1987).

[5] M. Hiramatsu, R. Edamatsu, M. Kohno, and A. Mori, *in* "Recent Advances in the Pharmacology of Kampo Medicines" (E. Hosoya and Y. Yamamoto, eds.), p. 120.

[6] A. N. Glazer, *FASEB J.* **2,** 2497 (1988).

TABLE I
EFFECT OF RETINOIDS ON THE ELECTRON SPIN RESONANCE SIGNAL OF 1,1-DIPHENYL-2-PICRYLHYDRAZYL (DPPH) RADICAL[a]

Common name	Ro number	Inhibition (% of control)
Furyl analog of retinoic acid	22-1318	77.3
13-*cis*-Retinoic acid	04-3870	81.7
Acitretin	10-1670	95.6
Etretin	10-9359	97.5
all-*trans*-Retinoic acid	01-5488	99.7
Arotinoid acid	13-7410	100.1
Theinyl analog of retinoic ester	21-6583	101.5
Arotinoid sulfone ester	15-1570	111.7
Motretinide	11-1430	112.3
Termarotene	15-0778	115.1
Arotinoid ester	13-6298	118.8

[a] Retinoids at 2.5 m*M* were added to a 100 μM solution of DPPH. Each value is the mean of two or three determinations.

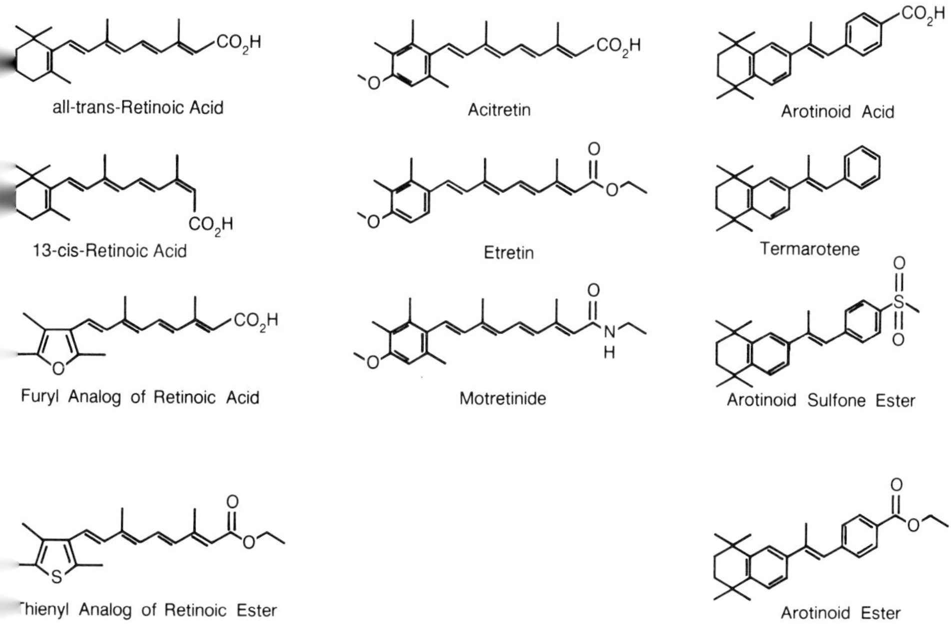

all-trans-Retinoic Acid
Acitretin
Arotinoid Acid
13-cis-Retinoic Acid
Etretin
Termarotene
Furyl Analog of Retinoic Acid
Motretinide
Arotinoid Sulfone Ester
Thienyl Analog of Retinoic Ester
Arotinoid Ester

FIG. 1. Molecular structures of retinoids.

are mixed for 10 sec, and 50 μl of the solution is transferred to a capillary tube for measurement of the DPPH concentration by ESR. For the Bruker ER200 D-SCR instrument, conditions for measurement of DPPH are as follows: 3480 ± 100 G for magnetic field, 10 mW for power, 0.5 sec for response time, 2.5 G for modulation, 2.5×10^4 for amplitude, and 0.5 sec for sweep time at ambient temperature. Manganese oxide is used as the internal standard. The internal standard is located outside of quartz ESR tube into which the capillary tube containing the DPPH solution with or without retinoids present is always inserted.

Measurement of Antioxidant Activity. The ESR spectrum of the DPPH radical shows a 5-lined spectrum (pentad signals) as indicated in Fig. 2. The control signal height intensity (in the absence of retinoids) is taken as 100%, and the percentage signal height intensity is calculated. Both Ro 22-1318 and Ro 4-3870 quench about 20% of 100 μM DPPH radical at concentration of 2.5 mM whereas Ro 10-1670, Ro 10-9359, Ro 01-5488, Ro 13-7410, Ro 21-6583, and Ro 15-1570 show no effect on DPPH radical quenching. Ro 15-0778 and Ro 13-6298 slightly increase the DPPH radical

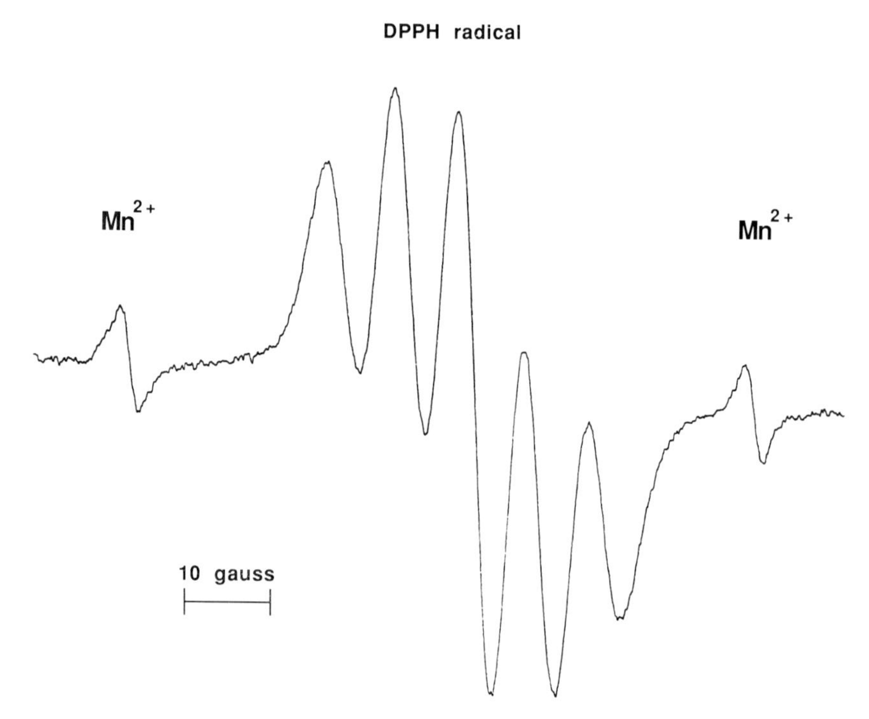

FIG. 2. ESR spectrum of the 1,1-diphenyl-2-picrylhydrazyl radical.

signal at the same concentration (Table I). However, no quenching effect on the DPPH radical is detected at concentrations of 0.0005, 0.005, 0.05, and 0.5 m*M* for the retinoids examined.

Ro 4-3870 (13-*cis*-retinoic acid) and Ro 22-1318 are the only retinoids found to slightly quench the DPPH radical at a concentration of 2.5 m*M*. Glutathione completely quenches the DPPH radical (100 μM), and the IC_{50} value for glutathione at 100 μM DPPH radical is 50 μM. The IC_{50} values of vitamins C and E for the DPPH radical (30 μM) are 5 and 70 μM, respectively.[5] Thus, the antioxidant action of Ro 4-3870 and Ro 22-1318 is much less than that of vitamins C and E, or glutathione.

Decay of Phycobiliprotein Fluorescence as Assay for Reactive Oxygen Species and Antioxidant Activity of Retinoids

This section reports on a method developed in the laboratory of A. N. Glazer (with permission). The data on antioxidant action of retinoids shown below is from unpublished results of A. Koushafar and A. N. Glazer.

Reagents and Stock Solutions

2,2′-Azobis(2-amidinopropane)-HCl (MW 267), Polysciences, Inc. (Warrington, PA, Cat. No. 8963)

Chelex 100 resin (200–400 mesh) sodium form, Bio-Rad Laboratories (Richmond, CA, Cat. No. 142-2842)

B-Phycoerythrin, from *Porphyridium cruentum,* molar extinction coefficient at 545 nm 2.41×10^6 M^{-1} cm^{-1}, Calbiochem (San Diego, CA, Cat. No. 526255)

Polysciences, Inc. (Cat. No. 18110)

R-Phycoerythrin, from *Gastroclonium coulteri,* molar extinction coefficient at 566 nm 1.96×10^6 M^{-1} cm^{-1}, Calbiochem (Cat. No. 526258) or Polysciences, Inc. (Cat. No. 18188)

75 m*M* Sodium phosphate buffer, pH 7.0, passed through Chelex 100 to remove metal ions

B-Phycoerythrin, 1.7×10^{-6} *M* in 75 m*M* sodium phosphate buffer, pH 7.0

2,2′-Azobis(2-amidinopropane)-HCl, 40 m*M* in 75 m*M* sodium phosphate buffer, pH 7.0 (freshly prepared; stored at 0°)

Plasma, preferably fresh or stored frozen at −20°

Procedure. The reaction mixture contains, in a total volume 2.0 ml, the following (listed in order of addition): 75 m*M* sodium phosphate buffer, pH 7.0, 1.772 ml; B-phycoerythrin solution, 20 μl (final concentration 1.7×10^{-8} *M*); plasma, 8 μl (1:250 dilution); and 2,2′-azobis(2-amidinopropane)-HCl, 200 μl (4 mM). The temperature should be maintained constant at 37° because constant temperatures cause thermal decay of the azo initiator and this determines the rate of peroxy radical generation. The excitation wavelength is 540 nm, and the emission wavelength is 575–580 nm (for R-phycoerythrin an excitation wavelength of 495 nm may be used if preferred).

Comments. The reactions involved in the generation of radicals and the subsequent radical reactions which result in fluorescence decay are shown below.

Generation of AAPH radicals:

$$\text{A—N=N—A} \rightarrow [\text{A}\cdot\text{N}_2\cdot\text{A}] \rightarrow 2e\text{A}\cdot + (1-e)\text{A—A} + \text{N}_2$$

where e is the efficiency of free radical generation.

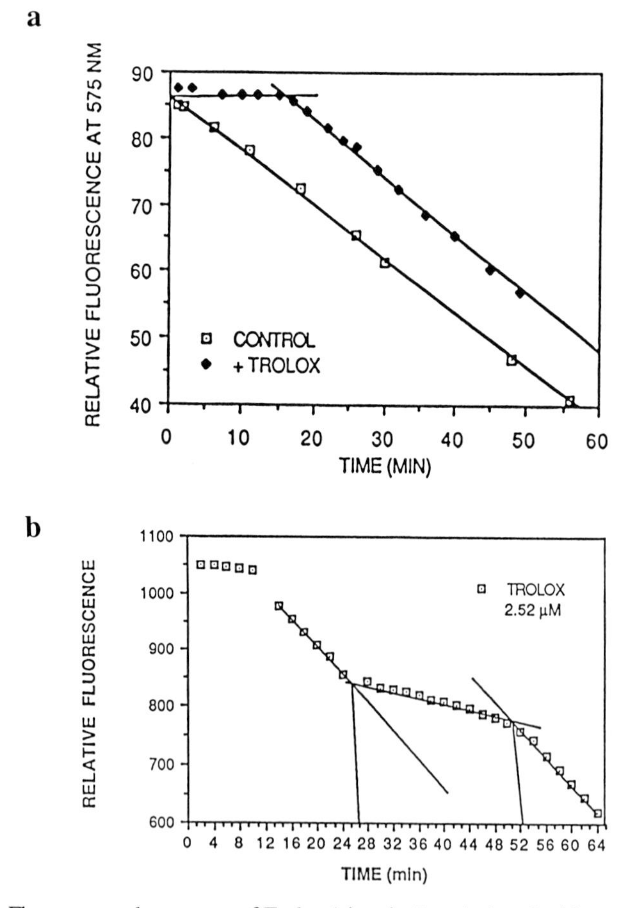

FIG. 3. Fluorescence decay assay of Trolox (vitamin E analog) antioxidant activity.

Subsequent radical reactions:

$$A\cdot + O_2 \rightarrow AO_2\cdot$$
$$AO_2\cdot + \text{R-phycoerythrin} \rightarrow \text{stable products}$$
$$AO_2\cdot + AO_2\cdot \rightarrow \text{stable products}$$

The antioxidant is added either prior to initiation of fluorescence decay or during the course of the reaction. These are both ways that can be used to assay for activity as illustrated for the very efficient antioxidant action of vitamin E using the water-soluble analog trolox (Fig. 3). Trolox consumes almost exactly 2 mol peroxy radicals per mole of antioxidant. Its high affinity and efficiency result in the slope of fluorescence decay being unchanged after the antioxidant is expended. This illustrates ideal conditions for using this assay.

With retinoids, the inhibition of the fluorescence decay is also evident. After retinoids are expended the change in slope of fluorescence decay is not reestablished to the initial value and is usually slower. This is probably

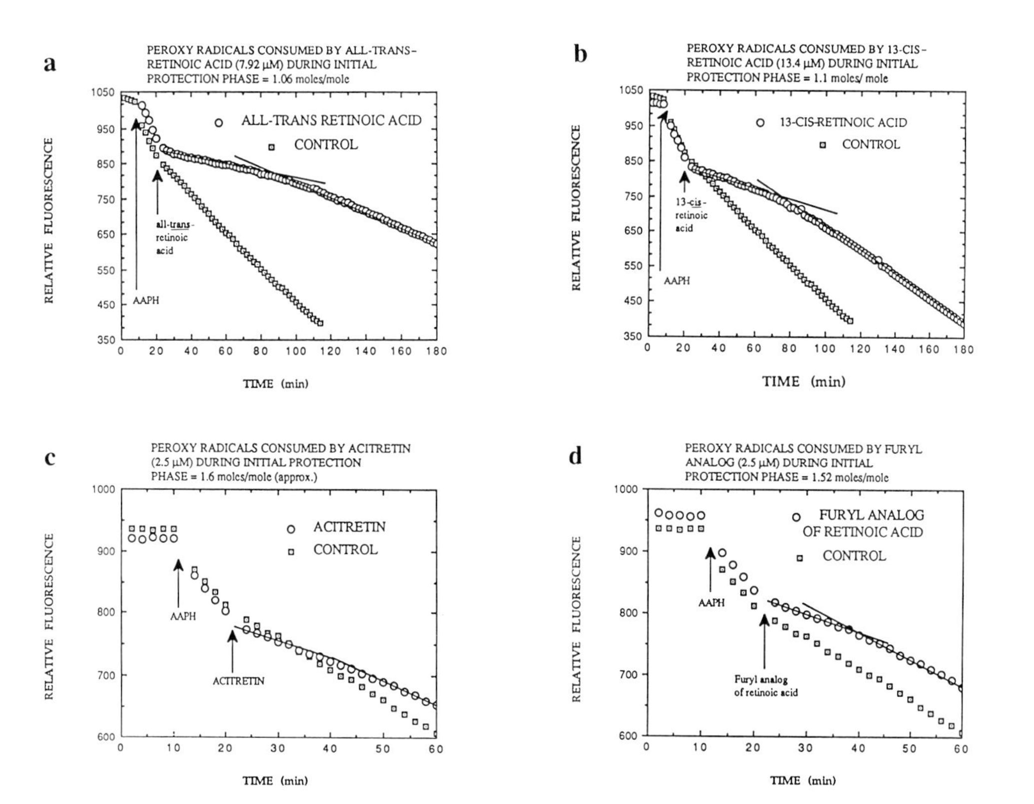

FIG. 4. Antioxidant action of retinoids measured by the fluorescence decay method.

TABLE II
REACTION STOICHIOMETRY OF ANTIOXIDANTS, USING PHYCOBILIPROTEIN FLUORESCENCE DECAY ASSAY

Substance	Amount added (μmol)	Reaction stoichiometry (Mol peroxy radicals consumed/mol)
13-*cis*-Retinoic acid	13.40	1.10
	2.52	1.04
all-*trans*-Retinoic acid	7.92	1.06
	2.52	1.05
Acitretin	2.50	1.60
Furyl analog	2.50	1.52
Trolox	2.52	2.00

because radical products (from the retinoids), which continue to be formed in the reaction mixture, have some inhibitory action on fluorescence decay. However, computer-assisted selection of the best slopes by a least-squares fit allow one to make a close approximation of the number of peroxy radicals consumed per mole of retinoid. Examples are shown in Fig. 4 for four substances. Results with 13-*cis*- and all-*trans*-retinoic acid, the furyl analog, and acitretin are summarized in Table II.

It should be noted in the assays of antioxidant activity of retinoids that it is necessary to introduce the retinoids in alcoholic solution. Thus, controls for ethanol are used. Ethanol itself has a small effect on the velocity of the fluorescence decay which must be corrected. Furthermore, it is necessary to dilute the retinoids slowly in order to ensure that they are in solution.

Among the retinoids tested, the reaction stoichiometry varies between 1 and 1.6. Hence, these compounds display quite significant antioxidant activity. All retinoids examined exhibited antioxidant activity in the same range of concentration between 2.5 and 13.4 μM in the test system described. Vitamin A (all-*trans*-retinol), but not retinoids, has also been reported to be an effective scavenger of thiyl radicals.[7]

[7] M. D'Aquino, C. Dunster, and R. L. Willson, *Biochem. Biophys. Res. Commun.* **161,** 1199 (1989).

[30] Inhibition of Microsomal Lipid Peroxidation by 13-*cis*-Retinoic Acid

By VICTOR M. SAMOKYSZYN and LAWRENCE J. MARNETT

Introduction

We have previously reported that 13-*cis*-retinoic acid is an effective peroxidase-reducing cosubstrate that undergoes hydroperoxide- or arachidonic acid-dependent oxidation by prostaglandin H synthase (EC1.14.99.1).[1,2] The mechanism of oxidation involves hydrogen atom abstraction by the peroxidase at allylic carbons, resulting in the formation of retinoid-derived carbon-centered radicals. These radicals couple with dioxygen to yield peroxyl radicals that add to other retinoid molecules, resulting in epoxidation at the 5,6-position (Fig. 1) and perhaps at other positions of the polyene chain.[2] Polyunsaturated fatty acid (PUFA)-derived peroxyl radicals, generated by reaction of PUFA hydroperoxides with hematin, also epoxidize 13-*cis*-retinoic acid at the 5,6-position.[2] However, neither retinoid- nor PUFA-derived peroxyl radicals react with retinoid molecules via H-atom abstraction reactions as evidenced by the absence of free radical chain-propagation reactions or detectable formation of 4-hydroxy-13-*cis*-retinoic acid, a metabolite generated via H-atom abstraction.[2] The ability of 13-*cis*-retinoic acid to react preferentially with peroxyl radicals by addition but not H-atom abstraction mechanisms suggested that it may be an effective antioxidant and prompted us to investigate its potential to inhibit lipid peroxidation. We have reported that 13-*cis*-retinoic acid is an effective inhibitor of ascorbate-dependent, iron-catalyzed microsomal lipid peroxidation.[3] Consistent with results obtained in systems that generate high yields of peroxyl radicals, a major microsomal oxidation product is 5,8-oxy-13-*cis*-retinoic acid,[3] which is generated via acid-catalyzed rearrangement of the 5,6-epoxide[2] (Fig. 1). The 4-hydroxy metabolite was not detected. We describe herein several methods to measure the ability of 13-*cis*-retinoic to inhibit lipid peroxidation including thiobarbituric acid and O_2 consumption assays as well as the high-performance liquid chromatographic (HPLC) method used to analyze retinoid oxidation products.

[1] V. M. Samokyszyn, B. F. Sloane, K. V. Honn, and L. J. Marnett, *Biochem. Biophys. Res. Commun.* **124**, 430 (1984).
[2] V. M. Samokyszyn and L. J. Marnett, *J. Biol. Chem.* **262**, 14119 (1987).
[3] V. M. Samokyszyn and L. J. Marnett, *Free Radical Biol. Med.* **8**, 491 (1990).

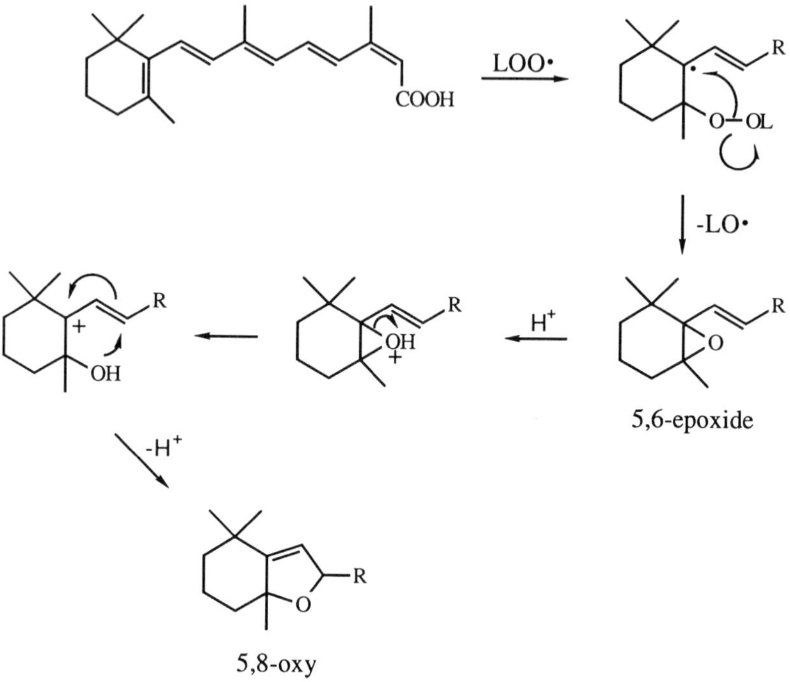

FIG. 1. Products of peroxyl radical-dependent epoxidation of 13-*cis*-retinoic acid.

Thiobarbituric Acid Assay

The thiobarbituric acid assay,[4] which measures malondialdehyde (MDA) formation, is used to quantitate lipid peroxidation. The peroxidation system used is essentially that of Orrenius *et al.*[5] consisting of rat liver microsomes (RLM, 0.5 mg microsomal protein/ml), ADP-chelated Fe(III) [4 m*M* ADP/15 μM Fe(III)], and ascorbate (1 m*M*).

Reagents

Tris-HCl buffer, 0.1 *M*, pH 7.5
Rat liver microsomes, in 0.2 *M* KCl buffer prepared from Long-Evans rats as described by Schelin *et al.*[6]
ADP (0.4 *M*)/Fe(III) (1.5 m*M*), in buffer
L-Ascorbic acid, sodium salt (40 m*M*), in buffer

[4] J. A. Buege and S. D. Aust, this series, Vol. 52, p. 302.
[5] S. Orrenius, G. Dallner, and L. Ernster, *Biochem. Biophys. Res. Commun.* **14,** 329 (1964).
[6] C. Schelin, A. Tunek, B. Jernstrom, and B. Jergil, *Mol. Pharmacol.* **18,** 529, (1980).

13-*cis*-Retinoic acid (18.75 m*M*), in dimethyl sulfoxide (DMSO)
Thiobarbituric acid solution, prepared as described by Buege and Aust[4]

Procedure

Microsomes (0.5 mg protein/ml), 10 μl of ADP/Fe(III), and 8 μl of the retinoid solution or vehicle (DMSO) are added to buffer to give a final volume of 0.975 ml. Following a 1-min preincubation, lipid peroxidation is initiated by the addition of 25 μl of ascorbate, allowed to continue for 20 min at 37°, then terminated by admixture with an equal volume of thiobarbituric acid reagent. The mixture is heated in a boiling water bath for 15 min, centrifuged at 1000 *g*, and the absorbance of the supernatant determined at 535 nm ($\epsilon = 156{,}000\ M^{-1}\ cm^{-1}$). As shown in Table I, 13-*cis*-retinoic acid, but not DMSO alone, effectively inhibits microsomal lipid peroxidation as evidenced by inhibited formation of MDA. Under these conditions, the IC_{50} is approximately 10 μM.[3]

Remarks

Because of the photoreactivity of retinoids, all procedures are carried out in the dark or under yellow light. Retinoids are also susceptible to autoxidation; thus, the retinoid should be stored under argon at −70° and stock solutions prepared fresh in argon-saturated DMSO.

Oxygen Consumption Assays

Because dioxygen is consumed during lipid peroxidation, particularly during free radical propagation reactions, uptake of O_2 is used as a complementary assay for measuring the effects of 13-*cis*-retinoic acid on ascorbate-dependent, iron-catalyzed microsomal lipid peroxidation. Oxygen uptake is measured polarographically using a Clark electrode.

TABLE I
INHIBITION OF LIPID PEROXIDATION BY 13-*cis*-RETINOIC ACID

System	Malondialdehyde formation (nmol MDA/mg RLM protein)
−DMSO	37.6 ± 0.8[a]
+DMSO (0.8%)	36.4 ± 0.9
+Retinoid (0.8% DMSO)	1.2 ± 0.0

[a] Values represent the means ± S.D. of triplicate measurements.

Procedure

Rat liver microsomes (0.5 mg protein/ml), 13 μl ADP/Fe(III), and 10 μl of retinoid solution or DMSO are added to buffer to give a final volume of 1.27 ml in a 2-ml water-jacketed cell maintained at 37° and containing a Clark electrode (Yellow Springs Instruments, Yellow Springs, OH) connected to a Gilson (Middleton, WI) 5/6 oxygraph. After a 1-min preincubation, 32 μl of the ascorbate solution is added and O_2 consumption measured. The ambient O_2 concentration in the Tris buffer at 37° is approximately 220 μ*M*. As shown in Fig. 2, 13-*cis*-retinoic acid effectively inhibits lipid peroxidation as evidenced by inhibition of O_2 consumption (Fig. 2C) compared with the complete lipid peroxidation system devoid of retinoid (Fig. 2D). The slow rate of O_2 depletion in controls containing ADP/Fe(III)/ascorbate in the absence of microsomes and absence or presence (Fig. 2A,B) of retinoid is probably due to autoxidation of ADP-chelated ferrous complexes.

Metabolite Analysis

Evidence that 13-*cis*-retinoic acid reacts with microsomal PUFA-derived peroxyl radicals, generated during lipid peroxidation, is obtained by

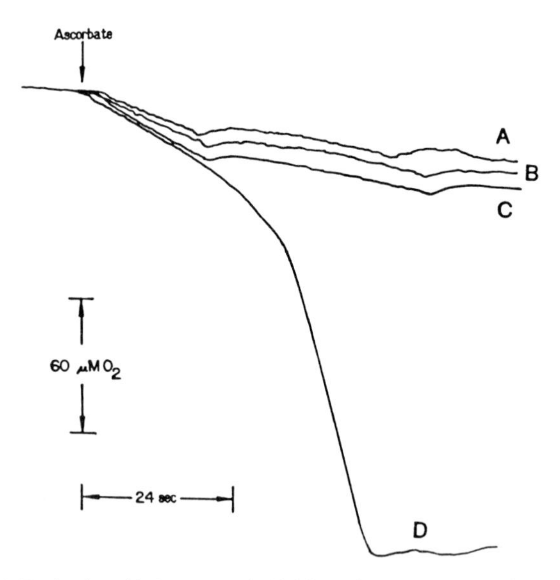

FIG. 2. 13-*cis*-Retinoic acid-dependent inhibition of O_2 uptake associated with ascorbate-dependent, iron-catalyzed microsomal lipid peroxidation. (A) ADP/Fe(III)/ascorbate; (B) ADP/Fe(III)/ascorbate plus retinoid; (C) ADP/Fe(III)/ascorbate plus microsomes and retinoid; (D) ADP/Fe(III)/ascorbate plus microsomes. (Data from Samokyszyn and Marnett.[3])

reversed-phase HPLC analysis of incubation mixtures and unambiguous identification of 5,8-oxy-13-*cis*-retinoic acid as a major oxidation product, which has previously been shown to arise via peroxyl radical addition reactions[2] (Fig. 1).

Procedure

Incubations containing microsomes and retinoid, in the presence or absence of the ADP/Fe(III)/ascorbate lipid peroxidation system, are carried out as described above for the thiobarbituric acid assay. After the 20-min incubation, reaction mixtures are acidified to pH 3 with HCl, saturated with NaCl, and metabolites extracted with HPLC-grade ethyl acetate. The solvent is removed *in vacuo,* and the residue dissolved in methanol. Chromatographic separations of metabolites are performed on a Varian Model 5060 instrument using an Altex Ultrasphere ODS (5 μm) column (25 cm $\times$ 4.6 mm i.d.) connected to a Hewlett-Packard (Avondale, PA) 1040A diode array detector interfaced with a disk-driven HP 85B computer. This detector system allows simultaneous detection of up to seven different wavelengths as well as aquisition of UV spectra.

The following solvent program is used. Reservoir A contains methanol–20 mM ammonium acetate (50:50, v/v), prepared by dissolving 1.54 g of ammonium acetate in 1000 ml of deionized water and adding 1000 ml of methanol. Add acetic acid to bring the pH to 6.65 ± 0.05. Reservoir B contains methanol–0.10 M ammonium acetate (90:10, v/v), prepared by dissolving 1.54 g of ammonium acetate in 200 ml of deionized water and adding 1800 ml of methanol. Add acetic acid to bring the pH to 6.65 ± 0.05. The flow rate is 1.2 ml/min. The gradient is programmed for 0–28 min with 20–90% B and for 28–45 min with 90% B, isocratic.

Figure 3 demonstrates the product profiles obtained when 13-*cis*-retinoic acid is incubated with microsomes in the presence (top) or absence (bottom) of the ADP/Fe(III)/ascorbate lipid peroxidation system. Several more polar metabolites as well as the all-trans isomer are generated when the retinoid is incubated in the complete lipid peroxidation system compared with microsomes alone. 5,8-Oxy-13-*cis*-retinoic acid was identified as a major metabolite on the basis of cochromatography with a chemically synthesized authentic standard, UV spectroscopy, and mass spectrometry of the methyl ester obtained by reaction of chromatographically purified metabolite with ethereal diazomethane.[3] Identification of this metabolite unambiguously demonstrates the reaction of the retinoid with PUFA-derived peroxyl radicals generated during lipid peroxidation. Consistent with results obtained in other peroxyl radical-generating systems,[2] 4-hydroxy-13-*cis*-retinoic acid ($t_r \sim 32$ min) is not detected, indicating that the

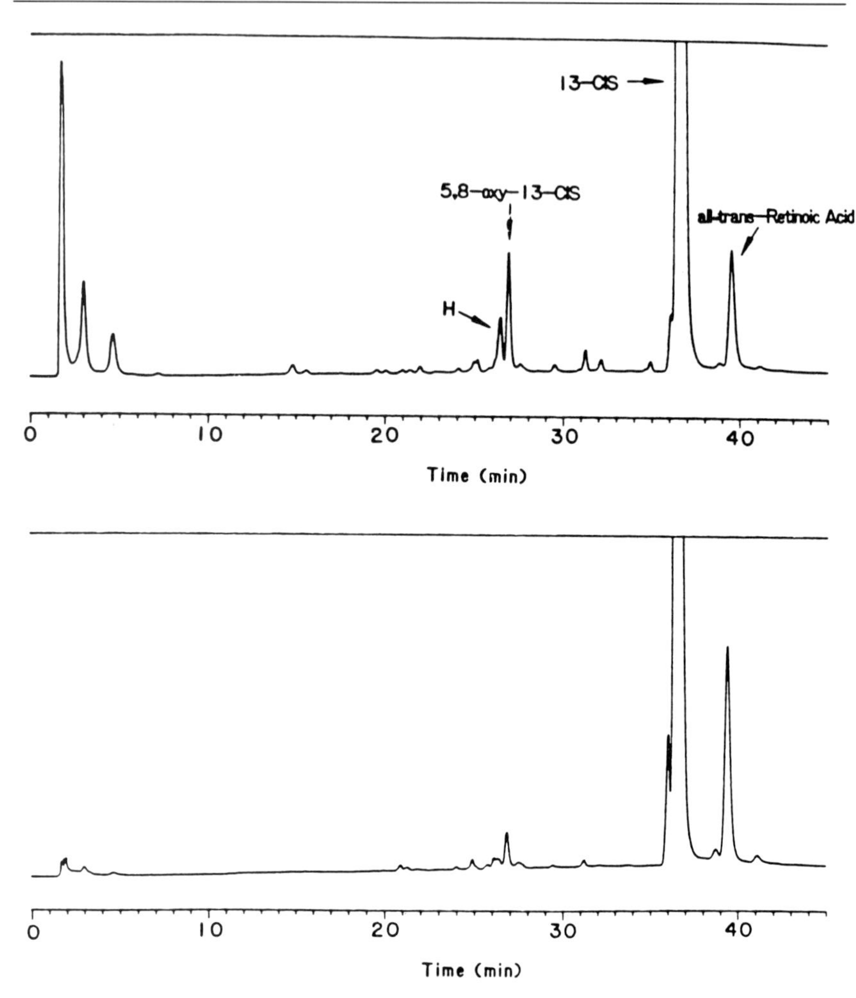

FIG. 3. Chromatographic elution profile of 13-*cis*-retinoic acid metabolites generated during ascorbate-dependent, iron-catalyzed microsomal lipid peroxidation. (Top) Retinoid plus complete lipid peroxidation system; (bottom) minus ADP/Fe(III)/ascorbate. (Data from Samokyszyn and Marnett.[3])

PUFA-derived peroxyl radicals failed to react with the retinoid via an H-atom abstraction mechanism.

Concluding Remarks

13-*cis*-Retinoic acid is an effective antioxidant as evidenced by its ability to inhibit lipid peroxidation (Table I, Fig. 2). In addition, the retinoid reacts preferentially with peroxyl radicals by addition but not H-atom abstraction reactions because 5,8-oxy-13-*cis*-retinoic acid (an epoxidation product[2]), but not the 4-hydroxy metabolite (an H-atom abstraction product[2]), is detected as a major oxidation product of lipid peroxidation. In general, peroxyl radical addition to double bonds has been shown to be favored over abstraction of allylic hydrogens when the double bond is conjugated with another vinyl group.[7,8] However, preferential reactivity by addition reactions cannot fully account for the antioxidant behavior of 13-*cis*-retinoic acid because peroxyl radical-dependent epoxidation must necessarily result in the formation of reactive alkoxyl radicals (Fig. 1). Furthermore, besides epoxidation, addition of peroxyl radicals to double bonds under aerobic conditions may also result in further generation of peroxyl radicals by reaction of dioxygen with the α-carbon-centered radical intermediates.[8] The latter mechanism may explain the dichotomous pro- and antioxidant behavior of β-carotene, an analogous conjugated polyene.

Burton[9] and Ingold have shown that β-carotene functions as an effective inhibitor of azobisisobutyronitrile (AIBN)-initiated hydrocarbon autoxidation in chlorobenzene at less than ambient O_2 tension. However, the carotenoid functions as a prooxidant at ambient or elevated O_2 concentrations. Presumably, the antioxidant behavior arises as a consequence of cross-termination reactions involving the coupling of polyene carbon-centered radicals, generated via addition of peroxyl radicals to β-carotene, with chain-propagating peroxyl radicals (and/or their carbon radical precursors). In contrast, the prooxidant effect is probably a consequence of the coupling of polyene carbon-centered radicals with dioxygen, resulting in new chains of propagation. Similarly, the AIBN-initiated autoxidation of all-*trans*-retinoic acid or its methyl ester in chlorobenzene and elevated O_2 tension is characterized by free radical propagation reactions.[10] In contrast, our results indicate that 13-*cis*-retinoic acid behaves as an antioxidant in microsomal lipid bilayers at ambient O_2 tension where the O_2 concentration is expected to be significantly higher in the lipid phase compared with the bulk aqueous phase. This is somewhat unexpected based on the behavior of 13-*cis*-retinoic acid in chlorobenzene at ambient O_2 tensions. Perhaps the microsomal phospholipid imposes solvent cage effects that favor cross-termination (and epoxidation) reactions rather than coupling of polyene-derived carbon radicals with dioxygen. Such putative solvent cage effects are consistent with the demonstrated strong interaction of retinoic

[7] F. R. Mayo, A. A. Miller, and G. A. Russell, *J. Am. Chem. Soc.* **80,** 2500 (1958).
[8] K. U. Ingold, *Acc. Chem. Res.* **2,** 1 (1969).
[9] B. W. Burton and K. J. Ingold, *Science* **224,** 569 (1984).
[10] E. I. Finkelshtein, Y. F. Rubachinskaya, and Y. I. Kozlov, *Zh. Org. Khim.* **17,** 936 (1981).

acid with phospholipid bilayers.[11,12] The various unidentified metabolites in Fig. 3 (top) may represent products arising from cross-termination reactions including radical coupling with PUFA-derived alkoxyl radicals generated during retinoid epoxidation (Fig. 1).

[11] R. G. Meeks, D. Zaharevitz, and R. F. Chen, *Arch. Biochem. Biophys.* **207,** 141 (1981).

[12] S. R. Wassall, T. M. Phelps, M. R. Albrecht, C. A. Langsford, and W. Stillwell, *Biochim. Biophys. Acta* **939,** 393 (1988).

Section III

Pharmacokinetics, Pharmacology, and Toxicology

A. Pharmacokinetics
Articles 31 through 33

B. Pharmacology
Articles 34 through 42

C. Toxicology
Articles 43 through 47

[31] Retinoids: An Overview of Pharmacokinetics and Therapeutic Value

By H. GOLLNICK, R. EHLERT, G. RINCK, and C. E. ORFANOS

Introduction

Retinoids are naturally existing substances and synthetic derivatives of retinol exhibiting vitamin A activity.[1] The first compound of this class used in clinical trials in the treatment of disorders of keratinization was vitamin A itself, but an unfavorable ratio between its rather moderate therapeutic value and frequent side effects resembling hypervitaminosis A syndrome along with the development of less toxic retinoid derivatives banned its further systemic application. The first synthetic retinoid was tretinoin (all-*trans*-retinoic acid), which soon became a valuable substance for topical treatment, but its systemic use did not reveal superiority over vitamin A. During the early 1980s, a number of retinoids have been synthesized, but only three therapeutically useful retinoids have become commercially available for oral medication: isotretinoin, etretinate, and acitretin. Some of the so-called third generation retinoids (e.g., arotinoid, arotinoid ethyl ester, arotinoid ethyl sulfone; Fig. 1) may be of future importance, but these are still in the preliminary clinical testing phase and are teratogenic, as are the others mentioned above.[2]

Bioavailability, plasma transport, and tissue distribution of retinoids vary widely and are determined predominantly by their physicochemical properties. Retinoids bind strongly to plasma proteins. In contrast to retinol, which binds to a specific serum retinol-binding protein, the main route of plasma transport for isotretinoin and acitretin is by binding to albumin, and for etretinate to lipoprotein.[3]

Isotretinoin

Following oral administration the bioavailability of isotretinoin is approximately 25%. Although interindividual variations are large, all retinoids show a marked increase of their bioavailability when taken together with food or milk.[4,5] Following a time lag of 0.5–2 hr before the drug

[1] D. S. Goodman, *N. Engl. J. Med.* **310,** 1023 (1984).
[2] W. Bollag, *Fortschr. Med.* **103,** 691 (1985).
[3] A. Vahlquist and O. Rollman, *Dermatologica* **175** (Suppl. 1), 20 (1987).
[4] W. A. Colburn, D. M. Gibson, R. E. Wiens, and J. J. Hanigen, *J. Clin. Pharmacol.* **23,** 534 (1983).
[5] J. J. DiGiovanna, E. G. Gross, S. W. McClean, M. E. Ruddel, G. Ganti, and G. L. Peck, *J. Invest. Dermatol.* **82,** 636 (1984).

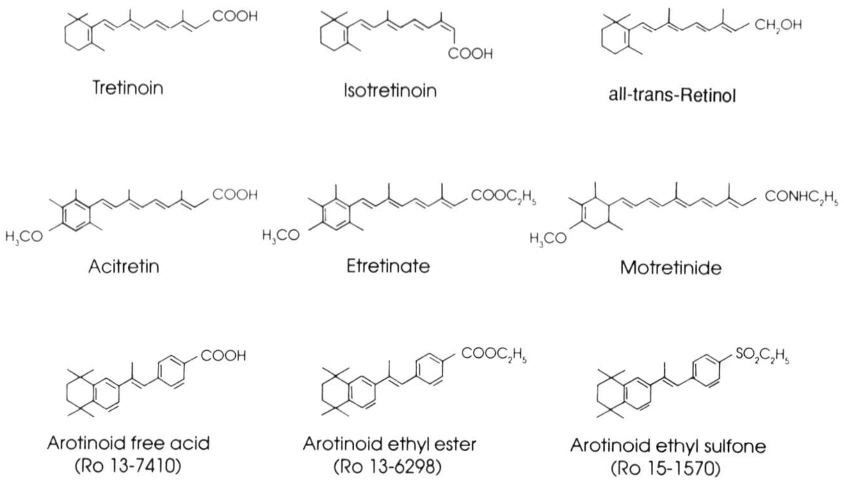

FIG. 1. Structures of retinoids.

reaches systemic circulation, peak blood concentrations occur 2–4 hr after single and multiple doses of approximately 300 ng/milliliter.[6]

The major metabolite, 4-oxo-isotretinoin, predominates over isotretinoin after 6 hr in serum and in tissues.[6,7] Owing to enterohepatic circulation, there is a biphasic course of blood concentrations. Isotretinoin is excreted in the bile in the form of conjugated metabolites, whereas 4-oxo-isotretinoin is excreted in glucoronized form also in the bile. Neither isotretinoin nor any of its possible metabolites were detected in the urine.[8] The mean elimination half-life of isotretinoin ranged from 10 to 30 hr in single and multiple dose studies, and tissue accumulation has not been a problem so far.[9] Experimental evidence suggests that isotretinoin is transferred across the placenta and is secreted into breast milk, though no such clinical studies are available.[10,11]

Etretinate

The bioavailability of etretinate is approximately 40%. Peak concentrations occur 2.5–6 hr after intake, and it takes about 30 min after oral

[6] R. K. Brazzell, F. M. Vane, C. W. Ehmann, and W. A. Colburn, *Eur. J. Clin. Pharmacol.* **24,** 695 (1983).

[7] O. Rollman and A. Vahlquist, *J. Invest. Dermatol.* **86,** 384 (1986).

[8] K. C. Khoo, D. Reik, and W. A. Colburn, *J. Clin. Pharmacol.* **22,** 395 (1982).

[9] A. Ward, R. N. Brogden, R. C. Heel, T. M. Speight, and G. S. Avery, *Drugs* **28,** 6 (1984).

[10] F. M. Vane and W. L. Bugge, *Ann. N.Y. Acad. Sci.* **359,** 424 (1981).

[11] C. L. Wang, *Proc. Am. Assoc. Cancer Res.* **18,** 45 (1977).

intake until the drug can be detected in plasma.[12] Because of the high lipophilicity of etretinate, subcutaneous fat serves as a storage tissue from which the drug is only slowly released, resulting in an elimination half-life of 80–120 days[13] after long-term administration. Etretinate has been detected 150 weeks after discontinuation of therapy.[14] In our laboratory (Table I) we were able to detect low levels of the 13-*cis*-acitretin metabolite in some of our patients, whereas etretinate and acitretin as the first and main metabolite were below the sensitivity level of the high-performance liquid chromatographic (HPLC) assay of 2 ng/ml. The residual levels of 13-*cis*-acitretin indicate the presence of etretinate in the body.[15] This is in contrast to the short elimination half-life of etretinate after a single dose (6–13 hr).[16] In animal studies etretinate has been shown to be transferred across the placenta and to be secreted into breast milk.[17] Excretion occurs via the bile and urine.[18]

Acitretin

Acitretin, the free aromatic acid, is a metabolite of etretinate. The bioavailability is approximately 60%, with a wide range of 36–95%. Peak plasma concentrations are reached approximately 4 hr after oral administration.[19] The profound advantage of the free acid metabolite acitretin as compared with its so-called parent drug, etretinate, is that it is eliminated at a much faster rate. After multiple doses its elimination half-life is about 50 hr, and after single dosing, only 2 hr. The plasma concentration of acitretin during continuous therapy with 50 mg/day is about 20 ng/ml in the pseudo-steady-state after 3–12 weeks as compared to approximately 50 ng/ml after etretinate administration.[20,77,81] It is suggested that the lower blood concentration of acitretin after oral administration is due to its rapid

[12] U. Paravicini, *in* "Retinoids: Advances in Basic Research and Therapy" (C. E. Orfanos, O. Braun-Falco, E. M. Farber, C. Grupper, M. K. Polano, and R. Schuppli, eds.), p. 13. Springer-Verlag, Berlin, Heidelberg, and New York, 1981.

[13] R. W. Lucek and W. A. Colburn, *Clin. Pharmacokinet.* **10,** 30 (1985).

[14] J. J. DiGiovanna, L. A. Zech, M. E. Ruddel, G. Gantt, G. L. Peck, *Arch. Dermatol.* **125**(2), 246 (1989).

[15] G. Rinck, H. Gollnick, C. E. Orfanos, *Lancet* (April 15), p. 845. (1989).

[16] A. Ward, R. N. Brogden, R. C. Heal, T. M. Speight, and G. S. Avery, *Drugs* **26,** 9 (1983, updated in 1986).

[17] U. Paravicini and A. Busslinger, *in* "Retinoid Therapy: A Review of Clinical and Laboratory Research" (W. J. Cunliffe and A. J. Miller, eds.), p. 11. MTP Press, Lancaster, 1984.

[18] M. Tateishi, H. Tomisawa, S. Nakamura, S. Ichihara, and H. Fukazawa, *Pharmacometrics* **24,** 339 (1982).

[19] C. J. Brindley, *Dermatologica* **178,** 79 (1989).

[20] H. Gollnick, *Dermatologica* **175** (Suppl. 1), 182 (1987).

TABLE I
LEVELS OF ETRETINATE AND ACITRETIN[a]

Patient No.	Diagnosis	Age (years)	Sex	Weight (kg)	Intake (months)	Total (mg)	Pause (months)	Etretinate (ng/ml)	Acitretin (ng/ml)	13-*cis*-Acitretin (ng/ml)
1	Psor. arth	81	F	71	6	6150	18	10	3	13
2	Psor. vulg	72	F	50	20	14,100	12	8	—	16
3	Psor. vulg	72	F	70	24	16,240	11	6	3	10
4	SCLE	47	F	47	4	4200	12	4	—	7
5	Psor. vulg	17	F	60	8	7000	12	—	—	7
6	Psor. vulg	24	M	78	4	6200	8	8	5	21
7	Icht. cong	25	M	60	78	99,450	14	—	—	—
8	Psor. vulg	63	M	75	4	6100	18	—	—	—
9	Psor. palm	27	F	54	6	4650	22	—	—	—
10	Psor. vulg	43	M	75	3	2350	8	2	—	10
11	CDLE	28	F	61	6	4200	10	5	3	13
12	Psor. vulg	66	M	75	2	3200	8	9	9	31

[a] Psor. arth, Psoriasis vulgaris with arthropathy; Psor. vulg, psoriasis vulgaris; SCLE, subacute cutaneous lupus erythematosus; Icht. cong, congenital ichthyosis; Psor. palm, psoriasis palmaris; CDLE, chronic discoid lupus erythematosus; F, female; M, male.

isomerization to 13-*cis*-acitretin, which may be less active in psoriasis than the all-trans isomer. Thirty days after cessation of acitretin therapy, acitretin plasma concentrations were below the quantification limit; however, residual concentrations of 3–4 ng/ml of its metabolite 13-*cis*-acitretin could be detected. Six weeks after therapy with 50 mg acitretin, drug concentrations of both substances were not detectable in 4 of our patients. At present, we do not know whether the 13-cis metabolite is as teratogenic as its prodrugs.

Isotretinoin

Isotretinoin and etretinate are the two retinoids currently commercially available for systemic oral therapy. Acitretin is available only in a few countries.

In acne, isotretinoin affects[21] (1) the epithelium by inhibiting excessive proliferation of keratinocytes and influencing infundibular differentiation of keratinocytes; and (2) the sebaceous glands by decreasing the size of the glands, reducing the activity of the glands, and influencing the lipid composition of sebum and comedos. Severe conglobate and nodulocystic acne including acne tetrade and acne fulminans are the main indications for oral isotretinoin therapy.[22,23] Isotretinoin is the most potent drug currently know to reduce effectively sebum secretion rates (SER),[24–26] when reliably measured with the direct gravimetric technique introduced by Lookingbill and Cunliffe.[27] Along with the reduction of the SER, a change in sebum composition can be observed, including an increase in squalene and a decrease in free fatty acids, waxes, triglycerides, and cholesterol ester.[28] Usually a 50% regression of the inflammatory acne lesions can be expected after 2–4 weeks of therapy. However, a flare-up reaction with the appearance of papulopustules derived from the expulsion of comedos is not uncommon around this time and is often misinterpreted by the patient as a worsening of the disease; but it rapidly clears when medication is continued.[29] In view of the relationship between dosage and long-term remis-

[21] C. E. Orfanos, *Hautarzt* **40,** 123 (1989).

[22] G. L. Peck, T. G. Olsen, F. W. Yoder, J. S. Strauss, D. T. Downing, and M. Pandya, *N. Engl. J. Med.* **300,** 329 (1979).

[23] G. Plewig, J. Nikolowski, and H. H. Wolff, *J. Am. Acad. Dermatol.* **6,** 766 (1982).

[24] K. King, D. H. Jones, D. C. Daltrey, and W. J. Cunliffe, *Br. J. Dermatol.* **107,** 583 (1982).

[25] D. H. Jones, A. J. Miller, and W. J. Cunliffe, *Br. J. Dermatol.* **108,** 333 (1983).

[26] M. E. Stewart, B. S. Benoit, B. S. Stranieri, M. Anna, R. P. Rapini, J. S. Strauss, and D. T. Downing, *J. Am. Acad. Dermatol.* **8,** 532 (1983).

[27] D. P. Lookingbill and W. J. Cunliffe, *Br. J. Dermatol.* **114,** 75 (1985).

[28] J. B. Schmidt, J. Spona, and G. Niebauer, *Z. Hautkrankh.* **58,** 1743 (1983).

[29] H. Gollnick, *in* "Retinoide in der Praxis" (R. Bauer and H. Gollnick, eds.), p. 79. Grosse Verlag, Berlin, 1984.

sions, we recommend in general a dose of 1.0 mg/kg/day of isotretinoin for 3 months followed by a dose of 0.2–0.5 mg/kg/day for another 3 months, but the severity of the disease the the tolerance of the patient must be considered individually.[30] Even after discontinuation of the drug further improvement of acne lesions will be observed, and if the aforementioned dose regimen is completed, no more than 20% of the patients will show a relapse in their condition. Long-term remission is not correlated with sebosuppression after treatment but is correlated with normalization of infundibular keratinization.[31]

Other acne-related dermatoses also appear to respond well to isotretinoin therapy. In gram-negative folliculitis, immunomodulatory properties of the drug may play an additional role in the mode of action of isotretinoin. Significant improvement concomitant with a reduction in pathogenic microorganisms, such as *Proteus mirabilis* or *Klebsiella* species, has been observed at dosages between 0.5 and 1.0 mg/kg/day.[24,32]

In severe recalcitrant rosacea papulopustulosa and conglobata and even in early stages of rhinophyma, as long as no fibrous connective tissue transformation has occurred, a response to isotretinoin therapy may be expected at dose levels of 0.5–1.0 mg/kg/day after 3–6 months of therapy.[23] In individual cases, relapse-free intervals of up to 15 months have been reported after isotretinoin therapy, but relapses are usually more frequent and occur earlier than in acne.[33,34] In various types of ichthyoses clinical trials with isotretinoin revealed contradictory results.[30] Also, if an improvement could be observed, after discontinuation of the drug a deterioration of the skin condition was constantly observed. Considering the toxicity of isotretinoin, continuous long-term therapy for more than 1 year cannot be recommended.

Except for antiandrogenic contraceptive preparations in female patients, no combination therapy is recommended under oral isotretinoin medication. The simultaneous administration of tetracycline[35] and vitamin A-containing preparations[36] must be avoided. Likewise, additional topical therapy with, for example, tretinoin or benzoyl peroxide as well as

[30] C. E. Orfanos, R. Ehlert, and H. Gollnick, *Drugs* **34,** 459 (1987).

[31] W. J. Cunliffe, D. H. Jones, K. T. Holland, S. Millard, and H. Albaghdadi, *in* "Retinoid Therapy" (W. J. Cunliffe and A. J. Miller, eds.), p. 255. MTP Press, Lancaster, 1984.

[32] W. D. James and J. J. Leyden, *J. Am. Acad. Dermatol.* **12,** 319 (1985).

[33] J. Nikolowski and G. Plewig, Hautarzt **31,** 660 (1981).

[34] R. A. Fulton, *In* "Retinoid Therapy" (W. J. Cunliffe and A. J. Miller, eds.), p. 315. MTP Press, Lancaster, 1984.

[35] A. R. Shalita, W. J. Cunningham, J. J. Leyden, P. E. Pochi, and J. S. Strauss, *J. Am. Acad. Dermatol.* **9,** 629 (1983).

[36] L. M. Milstone, J. McGuire, and R. C. Ablow, *J. Am. Acad. Dermatol.* **7,** 663, (1982).

overexposure to natural or artificial UV irradiation should be avoided because of increased cutaneous side effects.

Etretinate

The development of the aromatic retinoid etretinate was a real breakthrough in the therapy of psoriasis. A large number of reports have proved the successful use of etretinate in patients with psoriasis either as monotherapy, particularly in patients with pustular or erythrodermic variants of psoriasis, or as a combination therapy, preferred in the chronic plaquelike form.[16,37,38]

Even though the exact molecular mechanisms responsible for the mode of action of retinoids remain to be elucidated, ultrastructural and biochemical investigations have shown etretinate to produce the following main effects on psoriatic skin: etretinate has strong keratinolytic activity, leading to restoration of normal epidermis by causing reappearance of the granular layer, disappearance of parakeratosis, and reduction of acanthosis. Autoradiographic studies have demonstrated a decrease of the labeling index for [^{3}H]thymidine incorporation and a prolonged time of DNA synthesis in the affected and uninvolved epidermis of psoriatic patients.[39] Concerning the effects of retinoids on changes in gene expression, two possibilities should be considered: (1) retinoids may enter the cell and interact directly with chromosomal material in the nucleus; (2) the first primary action with the cell may take place on the plasma membrane, forming a second messenger which enters the nucleus and influences gene expression.[40]

Other hypotheses of retinoid action include the following: (1) on the nuclear level, steroid hormone-like action concerning transcription and protein synthesis; (2) as a coenzyme, involving retinoyl-CoA and retinoyl phosphate (glycosylation); (3) cyclic AMP-dependent protein kinase; (4) posttranslational effects on glycosylation.[41] Retinoids also appear to be dual inhibitors of both cyclooxygenase and lipoxygenase, thus explaining their antiinflammatory effects.[42] It has also been shown that some retinoids have effects on the release and metabolism of arachidonic acid by inhibiting the production of leukotriene B_4 (LTB_4) and prostaglandin E_2

[37] H. Gollnick and C. E. Orfanos, *Hautarzt* **36,** 2 (1985).

[38] H. Gollnick and C. E. Orfanos, *in* "Psoriasis" (H. H. Roenigk and H. I. Maibach, eds.), p. 597. Dekker, New York and Basel, 1985.

[39] E. Dierlich, C. E. Orfanos, H. Pullmann, and G. K. Steigleder, *Arch. Dermatol. Res.* **264,** 169 (1979).

[40] F. Chytil and D. R. Sherman, *Dermatologica* **175** (Suppl. 1), 8 (1987).

[41] P. M. Elias, *Dermatologica* **175** (Suppl. 1), 28 (1987).

[42] S. Nigam, *Dermatologica* **15** (Suppl. 1), 73 (1987).

(PGE_2).[43] *In vitro* experiments have also shown an inhibition of interleukin 1 (IL-1), which is an important mediator in mounting an immune response to antigens and in eliciting inflammation and tissue remodeling, possibly linking the antiinflammatory effects of retinoids with their immunomodulatory action.[44] Primary effects and secondarily conveyed influences on growth regulation and differentiation of epidermal keratinocytes are decisive aspects of the biological action of retinoids in general and, thus, of etretinate.[45-48]

The first clinical effects following initiation of etretinate therapy occur not earlier than 1-2 weeks after desquamation begins. Psoriatic lesions become flatter and their color changes from red to pink. After 3 weeks of therapy the centers of the lesions slowly clear, changing the psoriatic plaques to more gyrated lesions. Healing proceeds in the craniocaudal direction, and intertriginous areas and the lower extremities usually show delayed clearing. On the palms and soles therapeutic effects of etretinate therapy can be observed after 2-4 weeks after onset of therapy.[37,49] In order to enhance clinical improvement of the monotherapy, which is capable of clearing psoriasis in about 60% of patients, combined regimens, for example, etretinate and UV irradiation and suitable topical therapy such as dithranol, tar plus UV light, are recommended in chronic plaquelike psoriasis. Thus, in order to maintain a favorable risk/benefit ratio moderate doses of etretinate can be used.[50]

In contrast, for pustular types of psoriasis such as pustulosis palmaris et plantaris (Barber-Königsbeck type), generalized pustular psoriasis (Zumbusch type), and plaquelike psoriasis with pustular eruptions, etretinate monotherapy is considered the treatment of first choice, but in general at least high initial doses of 1.0 mg/kg/day are required. After cessation of the pustular eruptions, the dose can be reduced rapidly to a maintenance dose of 0.3-0.5 mg/kg/day. On the other hand, low initial doses of approximately 0.3 mg/kg/day in order to avoid additional retinoid-induced irritation of the skin should be given in psoriatic erythroderma. According to the clinical appearance of the patient and considerations of individual

[43] C. Fiedler-Nagy, B. H. Wittreich, A. Georgiadis, W. C. Hope, A. F. Welton, and J. W. Coffey, *Dermatologica* **175** (Suppl. 1), 81 (1987).

[44] U. M. Ney, I. J. Gall, R. P. Hill, D. Westmacott, D. P. Bloxham, *Dermatologica* **175** (Suppl. 1), 93 (1987).

[45] J. Lauharanta, M. Kousa, K. Kapyaho, K. Linnamaa, and K. Mustakallio, *Br. J. Dermatol.* **105,** 267 (1981).

[46] F. L. Meyskens, *Life Sci.* **28,** 2323 (1981).

[47] R. Lotan, *Biochim. Biophys. Acta* **605,** 33 (1980).

[48] R. Stadler and C. E. Orfanos, *Hautarzt* **32,** 564 (1981).

[49] R. P. Kaplan, D. H. Russell, and N. J. Lowe, *J. Am. Acad. Dermatol.* **8,** 95 (1983).

[50] N. Väätäinen, A. Hollmen, and J. E. Fräki, *J. Am. Acad. Dermatol.* **12,** 52 (1985).

TABLE II
DOSAGE REGIMEN OF ETRETINATE IN PSORIASIS

Type of psoriasis	Dosage recommendation
Chronic plaquelike psoriasis	Mean daily dose approximately 0.5 mg/kg/day over several weeks to months; use combined therapy if necessary
Psoriatic erythroderma	Low doses (~0.25 to 0.35 mg/kg/day) over several weeks; increase dosage according to clinical effectiveness and individual tolerance
Localized or generalized pustular psoriasis	High initial doses (~1.0 mg/kg/day); reduce dosage to maintenance levels after 2 to 3 weeks according to clinical effectiveness

tolerance, the dosage should be carefully increased to up to 0.7 mg/kg/day in erythrodermic patients.[37,51]

The recommended dosage regimens of etretinate in various types of psoriasis are listed in Table II.[30] The therapeutic value of etretinate in arthropathic psoriasis is still controversial. Some authors do not recommend etretinate therapy,[52] whereas others have demonstrated a clear advantage of etretinate as compared to common therapy with nonsteroidal antiphlogistic medication.[53–55]

Nonpsoriatic sterile pustulosis such as pustulosis palmaris et plantaris and the so-called Andrew's bacterid are further indications for etretinate monotherapy.[56] However, some authors believe that these pustular disorders of the palms and soles are only special types of psoriasis. Others reported on the use of etretinate in subcorneal pustulosis Sneddon–Wilkinson;[57] acrodermatitis continua Hallopeau;[58] pustular type of ery-

[51] G. Goerz and C. E. Orfanos, *Dermatologica* **157** (Suppl. 1), 38 (1978).

[52] G. Mahrle, S. Meyer-Hamme, and H. Ippen, *Arch. Dermatol.* **118,** 97 (1982).

[53] R. Hopkins, H. A. Bird, H. Jones, J. Hill, K. E. Surrall, L. Astbury, A. Miller, and V. A. Wright, *Ann. Rheum. Dis.* **44,** 189 (1985).

[54] R. Stollenwerk, H. Fischer-Hoinhes, K. Komenda, and F. Schilling, *in* "Retinoids: Advances in Basic Research and Therapy" (C. E. Orfanos, O. Braun-Falco, E. M. Farber, C. Grupper, M. K. Polano, and R. Schuppli, eds.), p. 205. Springer-Verlag, Berlin, Heidelberg, and New York, 1981.

[55] G. C. Chieregato and A. Leoni, *Acta Derm. Venerol.* **66,** 321 (1986).

[56] A. Lassus, J. Lauharanta, T. Juvakovski, and L. Karneva, *Dermatologica* **166,** 215 (1983).

[57] D. Lubach, M. Edmüller, and A. L. Rahm-Hoffmann, *Hautarzt* **31,** 545 (1980).

[58] L. H. Pearson, B. S. Allen, and J. G. Smith, *J. Am. Acad. Dermatol.* **11,** 755 (1984).

thema anulare centrifugum;[48] and in Reiter's disease, the latter especially if skin lesions on the palms and feet are present.[59]

Generally, palmoplantar keratoses of congenital and acquired origin show a good response to etretinate.[60] Successful treatment has been reported in keratosis palmoplantaris of the Unna–Thost type,[61] in keratosis palmoplantaris areata Siemens,[62] in the Papillon–Lefevre syndrome,[63] and in keratosis palmaris et plantaris mutilans Vohwinkel.[64] In keratosis palmaris et plantaris of the Voerner type, etretinate therapy has remained unsatisfactory.[65]

In hyperkeratotic rhagadiform eczema of the hands (acquired keratoderma) etretinate has been proved to be a valuable therapeutic measure. Treatment should be started at an initial dose of 0.8 mg/kg body weight/day up to a maximum of 75 mg/day. As clinical improvement occurs, the dosage should be reduced gradually until a long-term maintenance dosage of 10–20 mg/day is achieved. Usually, additional topical therapy with urea and lactic acid-containing preparations is recommended.[66]

In severe ichthyoses of various types, satisfactory results have been achieved with moderate initial doses of approximately 0.5 mg/kg/day, then continued with the lowest acceptable maintenance dosage. Considering the risk/benefit ratio of long-term etretinate therapy, even on a low-dose level, this form of systemic treatment should be limited only to severely affected patients.[67–69]

Even in cases resistant to other conventional therapy, etretinate was able to clear keratosis follicularis (Darier). Low initial doses (0.1–0.2 mg/kg/day gradually increasing up to 0.6 mg/kg body weight/day are recom-

[59] S. Marghescu, B.-W. Kock, and D. Lubach, *Hautarzt* **36,** 291 (1985).

[60] Happle, *Dermatologica* **175** (Suppl. 1), 107 (1987).

[61] G. L. Peck, E. G. Gross, and D. Butkus, *in* "Retinoids: Advances in Basic Research and Therapy" (C. E. Orfanos, O. Braun-Falco, E. M. Farber, C. Grupper, M. K. Polano, and R. Schuppli, eds.), p. 279. Springer-Verlag, Berlin, Heidelberg, and New York, 1981.

[62] C. Luderschmidt and C. Plewig, *Hautarzt* **31,** 96 (1980).

[63] W. Wehrmann, H. Traupe, and R. Happle, *Hautarzt* **36,** 173 (1985).

[64] N. Hundziker, R. Brun, and J.-P. Jeanneret, *in* "Retinoids: Advances in Basic Research and Therapy" (C. E. Orfanos, O. Braun-Falco, E. M. Farber, C. Grupper, M. K. Polano, and R. Schuppli, eds.), p. 453. Springer-Verlag, Berlin, Heidelberg, and New York, 1981.

[65] E. Haneke, *Hautarzt* **33,** 654 (1982).

[66] G. Bäurle and E. Haneke, *in* "Retinoide in der Praxis" (R. Bauer and H. Gollnick, eds.), p. 55. Grosse Verlag, Berlin, 1984.

[67] C. Blanchet-Bardon and A. Puissant, *in* "Retinoids: Advances in Basic Research and Therapy" (C. E. Orfanos, O. Braun-Falco, E. M. Farber, C. Grupper, M. K. Polano, and R. Schuppli, eds.), p. 303. Springer-Verlag, Berlin, Heidelberg, and New York, 1981.

[68] J. J. DiGiovanna and G. L. Peck, *Dermatol.* **1,** 77 (1983).

[69] R. Ruiz-Maldonado and L. Tamayo, *Dermatologica* **175** (Suppl. 1), 125 (1987).

mended; after a clinically satisfactory response, maintaining the patient on 0.25-0.5 mg/kg/day is generally sufficient.[60,70,71]

Usually, in the treatment of lichen planus of verrucous or disseminated forms including mucous involvement, marked improvement, but rarely complete clearing, has been observed with etretinate monotherapy. Therefore, we usually combine treatment with topical or systemic corticosteroids with a good to excellent response by lowering the duration of treatment and thus side effects. Especially in erosive oral lichen planus, including the bullous type, resistant to other therapies, etretinate should be considered at high initial doses of 1.0 mg/kg/day (maximum 75 mg/day) followed by lower doses (0.5 mg/kg/day) after 2-3 weeks of treatment.[72,73] Finally, etretinate therapy has been tried in a variety of skin disorders, including bullous dermatoses, lichen sclerosus et atrophicans, lichen amyloidosus, lupus erythematosis, as well as benign and even malignant neoplasias (reviewed in Orfanos *et al.* [30] and Gollnick[20]).

Indications and Precautions

When considering systemic retinoid therapy, a clear indication following a clear diagnosis and, as a rule, a preceding failure using conventional measures are required before therapy is started. Pregnancy, lactation, and severe hepatic and renal dysfunction, as well as diabetes mellitus, are absolute contraindications for oral retinoid treatment. Patients with increased serum lipid levels, neurological symptoms, and the presence of bone pain must be monitored very closely. Concomitant therapy with vitamin A, aspirin (in high doses), and tetracyclines (including minocycline) must be avoided.[74]

Contraindications and Risk Factors

Prior to retinoid therapy, risk factors such as obesity, alcohol abuse, diabetes mellitus, hyperuricemia or gout, familial lipid abnormalities, and concomitant therapies must be considered. Obligatory laboratory examinations before and regularly (approximately once a month) during therapy include measurements of muscle and liver enzymes, creatinin, uric acid, cholesterol, and triglycerides. Owing to dysmorphogenicity and teratogenicity, retinoids should not be administered to women of childbearing

[70] C. E. Orfanos, M. Kurka, and V. Strunk, *Arch. Dermatol.* **114,** 1211 (1978).

[71] M. Binazzi and E. G. Cicilioni, *Arch. Dermatol. Res.* **264,** 365 (1979).

[72] R. Schuppli, *Dermatologica* **157** (Suppl. 1), 60 (1978).

[73] R. Happle and H. Traupe, *in* "Retinoide in der Praxis" (R. Bauer and H. Gollnick, eds.), p. 35. Grosse Verlag, Berlin, 1984.

[74] R. Viraben, C. Matthew, and B. Fonton, *J. Am. Acad. Dermatol.* **13,** 515 (1985).

TABLE III
SIDE EFFECTS OF RETINOID THERAPY

Side effect	Etretinate (0.25–1.0 mg/kg) (%)	Isotretinoin 0.5 mg/kg (%)	Isotretinoin 1.0 mg/kg (%)
Cheilitis/dry lips	75–100	75	95
Dry mouth	25	20	30
Dry nose	25	35	50
Epistaxis	5	25	25
Facial dermatitis	5	30	50
Palmoplantar desquamation	40	10	20
Desquamation of skin	30	10	20
Thinning of skin	50	15	25
Skin fragility	25	15	20
Xerosis	20	30	50
Retinoid dermatitis	5	5	5
Hair loss (varying degrees)	50	10	20
Conjunctivitis	5	30	30
Itching/pruritus	15	25	25

age.[30,37] If retinoid therapy is unavoidable, strict and reliable contraception measures must be applied 4 weeks prior to retinoid therapy and 8 weeks after treatment with isotretinoin and at least 2 years after treatment with etretinate.[15,75,76]

Side Effects

If long-term retinoid therapy, especially with isotretinoin, has been introduced, radiological examinations of the skeletal system and the epiphyseal cartilage of tubular bones must be performed before and regularly (every 6–12 months) during treatment, particularly if high doses are applied. Side effects of retinoids include liver function abnormalities, serum lipid alterations, myalgias and arthralgias, bone changes, and other systemic side effects, as well as a variety of mucocutaneous side effects[30] (Table III). Usually, the side effects do not justify discontinuation of treatment. A lower dosage of the retinoid or appropriate topical treatment will, in general, lead to successful continuation of retinoid therapy.

A few patients appear to be "nonresponders" to retinoid therapy.

[75] C. E. Orfanos, *Hautarzt* **35,** 503 (1984).

[76] R. Happle, H. Traupe, Y. Bounameaux, and T. Fish, *Dtsch. Med. Wschr.* **109,** 1476 (1984).

TABLE IV
THERAPEUTIC VALUE OF RETINOIDS[a]

Retinoid	Effectiveness[b]
First generation	
Vitamin A	s
Tretinoin	s, t
Isotretinoin	s, t
Second generation	
Etretinate	s
Motretinide	t
Etretin	s
Ro 12-7554 (chlorinated)	s
Polyprenoic acid (E 5166)	s
Third Generation	
Arotinoid (Ro 15-0778)	s?, t
Arotinoid ethyl ester (Ro 13-6298)	s, t?
Arotinoid carboxylic acid (Ro 13-7410)	s
Arotinoid ethyl sulfone (Ro 15-1570)	t
Arotinoid benzofuran analog (TTNF)	—
Arotinoid indole analog (TTNI)	—
Arotinoid naphthalene anlog (TTNN)	t
Arotinoid benzothiophene analog (TTNT)	—

[a] From Ref. 20
[b] s, Systemically; t, topically effective in humans.

Whether this is due to malresorption of the drug or a different pathway of retinoid metabolism in these patients remains to be elucidated.[81]

Future Aspects of Retinoids

A derivative of etretinate, acitretin, is the free aromatic trimethylmethoxyphenyl acid (TMMP, Ro 10-1670). *In vitro* experiments have suggested similar clinical properties as compared with its parent compound, etretinate. Clinical studies do not reveal any significant differences in either the therapeutic value or the side-effect profile when acitretin is compared with etretinate in a double-blind fashion. However, there is an overall tendency in favor of etretinate because the side effects occur in higher incidence with acitretin.[77-79] The major advantage of acitretin is its short elimination

[77] H. Gollnick, R. Bauer, C. Brindley, C. E. Orfanos, G. Plewig, *et al., J. Am. Acad. Dermatol.* **19,** 458 (1988).
[78] J. M. Geiger and B. Czarnetzki, *Dermatologica* **176,** 182 (1988).
[79] Kragballe *et al., Acta. Derm. Venereol.* **69**, 35 (1989).

half-life of 2–3 days compared to etretinate (up to 120 days). Although acitretin and etretinate show the same degree of dysmorphogenicity and teratogenicity, the advantage would be that female patients of childbearing age could interrupt contraception 2 months following discontinuation of therapy as compared to the 2-year contraception period required after etretinate therapy.

Several other retinoids have been synthesized recently (Table IV). Some show interesting and promising results in experimental as well as in clinical testing.[20,80,82] Future efforts in the development and testing of new retinoids will focus on attempts to find new compounds which will be free of undesirable side effects, such as dysmorphogenicity and teratogenicity, while having more selective therapeutic effects compared to the retinoids presently available.

[80] C. E. Orfanos, R. Stadler, H. Gollnick, and D. Tsambaos, *Curr. Probl. Dermatol.* **13,** 33 (1985).
[81] H. Gollnick, G. Rinck, Th. Bitterling, and C. E. Orfanos, Z. *Hautkr.* **65** (1) 40 (1990).
[82] Shroot B., *J. Am. Acad. Dermatol.* **15** (4) 748 (1986).

[32] Experimental and Kinetic Methods for Studying Vitamin A Dynamics *in Vivo*

By MICHAEL H. GREEN and JOANNE BALMER GREEN

Introduction

Many types of kinetic studies have been done to investigate the metabolism of retinoids. We review here methods we have used to study vitamin A dynamics in the rat. Our approach has been to administer radioactive vitamin A in a physiological form, so that tracer data can be extrapolated to the tracees of interest (retinol and retinyl esters). We have applied several kinetic methods to analyze the *in vivo* data, with the goal of developing a quantitative description of whole-body vitamin A dynamics at different levels of vitamin A status. Such techniques are applicable to studies of vitamin A metabolism in different physiological or nutritional states using rats or other species, as well as to experiments on the effects of alcohol and xenobiotics on vitamin A metabolism and to those on disposition of pharmacologically useful retinoids.

Preparation of Physiological Doses of Labeled Vitamin A

As a rule, the isotopically labeled compound should be administered in a physiological form to recipient animals if one wishes to assume that subsequent kinetic data reflect the dynamic behavior of the tracee *in vivo.* In the case of vitamin A, the two primary physiological plasma transport forms, retinol carried by retinol-binding protein (RBP)/transthyretin (TTR) and retinyl esters in chylomicrons, can be conveniently labeled *in vivo* using donor animals.

[^{3}H]Retinol-Labeled Plasma

[^{3}H]Retinyl Acetate/Tween Emulsion. All procedures are done under shaded natural light or yellow lights. A disperson of [^{3}H]retinyl acetate in Tween 40 (Sigma Chemical Co., St. Louis, MO) is prepared as follows using a modification[1,2] of the method of Smith *et al.*[3] Evaporate the solvent from an appropriate amount (0.5–1 mCi) of high specific activity [^{3}H] retinyl acetate {e.g., Amersham Corp., Arlington Heights, IL; [11,12(*n*)^{3}H]vitamin A acetate; specific activity, 180 μCi/μg} or [^{3}H] retinol at 40° using a gentle stream of nitrogen. Add absolute ethanol (5 μl), cap the vial, and vortex briefly. Immediately add Tween 40 (0.2 ml; warmed to 40°) and begin to stir the mixture mechanically at 40°. Add 0.4 ml of saline (9 g NaCl/liter of water; warmed to 40°) and then 0.4 ml of distilled water (40°) by slowly injecting the liquids under the surface of the detergent via a 27-gauge needle. Purge the vial with nitrogen, cap, and stir at 40° for 30 min. We generally hold the emulsion at room temperature overnight before injection into donor rats.

Labeled Plasma. Since vitamin A deficiency is associated with increased liver levels of apo-RBP,[4] we use vitamin A-depleted rats as donors. Thus, labeled vitamin A cleared by the liver will be rapidly and maximally released as holo-RBP of high specific activity. Dietary protocols for inducing vitamin A depletion are reviewed by Smith.[5] A vitamin A-deficient rat[2] (plasma retinol <5 μg/dl) is anesthetized with diethyl ether. To ensure rapid and maximal hepatic clearance of Tween micelles containing [^{3}H] retinyl acetate, the emulsion ($\sim$1 ml) is injected into the exposed portal vein over 2–2.5 min using a 27-gauge needle; the rat is sutured and allowed to recover under a heat lamp. We have also administered the

[1] M. H. Green, L. Uhl, and J. B. Green, *J. Lipid Res.* **26,** 806 (1985).
[2] M. H. Green, J. B. Green, and K. C. Lewis, *J. Nutr.* **117,** 694 (1987).
[3] J. E. Smith, D. D. Deen, Jr., D. Sklan, and D. S. Goodman, *J. Lipid Res.* **21,** 229 (1980).
[4] Y. Muto, J. E. Smith, P. O. Milch, and D. S. Goodman, *J. Biol. Chem.* **247,** 2542 (1972).
[5] J. E. Smith, this volume [24].

emulsion into a jugular vein; however, this may result in a somewhat lower recovery of label, as will administering the dose by gastric intubation.[6]

We have found that isotope levels in plasma peak about 100 min after intravenous (i.v.) administration of the emulsion. At that time, the rat is reanesthetized, and blood is collected from the abdominal aorta or heart into a heparinized syringe; we use diethyl ether because, in our hands, this results in a maximal harvest of blood. Plasma is isolated at 4°, stored at 4° in a sterile vial under nitrogen, and injected into recipients within 1–3 days. In past experiments,[1,2,7] donor plasma was first ultracentrifuged at 1.5×10^6 *g* per min to remove any unmetabolized Tween micelles that might remain as a nonphysiological contaminant. In current experiments, less than 4% of the label in plasma floats under these conditions, so plasma is now used as obtained for *in vivo* turnover studies. Care should be taken during all procedures to avoid denaturing proteins, as labeled retinol associated with such proteins is apparently cleared rapidly (i.e., it does not show the normal kinetics of plasma retinol).

In our earlier experiments,[1,2] we showed that, using this procedure, essentially all of the plasma ^3H activity is retinol associated with the RBP/TTR complex. Recovery of injected isotope in the total plasma volume averages 17–32%, and the activity ranges from 3 to 20 μCi/ml; recovery of plasma typically averages 40–60% of total plasma volume.

Vitamin A-Labeled Chylomicrons

Chylomicrons labeled with retinyl esters can be obtained by administering labeled vitamin A to a lymph duct-cannulated donor rat (e.g., see Refs. 8 and 9). In our laboratory, a rat is anesthetized, and catheters are placed in the thoracic duct, caudal to the diaphragm (for lymph collection), and in the duodenum proximal to the entrance of the bile duct (for continuous infusion of saline and oil, and administration of the label); see Green and Green[10] for experimental details. When the rat has recovered from surgery (~2 days), [^{3}H]retinyl acetate can be administered intraduodenally in a small amount of oil (e.g., 0.2 ml of corn oil) or as a Tween emulsion (see above). Appearance of radioactivity in lymph is monitored

[6] P. A. Bank, M. E. Cullum, R. K. Jensen, and M. H. Zile, *Biochim. Biophys. Acta* **990,** 306 (1989).

[7] K. C. Lewis, Ph.D. thesis, The Pennsylvania State University, University Park (1987).

[8] R. Blomhoff, P. Helgerud, M. Rasmussen, T. Berg, and K. R. Norum, *Proc. Natl. Acad. Sci. U.S.A.* **79,** 7326 (1982).

[9] H. F. J. Hendriks, E. Elhanany, A. Brouwer, A. M. De Leeuw, and D. L. Knook, *Hepatology* **8,** 276 (1988).

[10] M. H. Green and J. B. Green, *Nutr. Res.* **8,** 1265 (1988).

at 15-min intervals for 4–5 hr. Labeled chylomicrons are isolated from selected lymph samples by ultracentrifugation[10] and recovered by tube slicing or aspiration.

We store the labeled chylomicrons at 4° under nitrogen and use the preparation for *in vivo* studies within 1–3 days. Gently disperse chylomicrons through a 22-gauge needle before administration. Based on limited experience in our laboratory and on published results,[8] nearly all (90–99%) of the labeled vitamin A in chylomicrons so produced will be in retinyl esters; recovery of isotope in lymph may range from 10 to 50%. If a specific size range of chylomicrons is not required, we recommend using whole lymph rather than isolated chylomicrons, as ultracentrifugation seems to alter a fraction of the dose so that it acts nonphysiologically. Although nonphysiological components can be handled during modeling, it is obviously better to avoid this situation if possible.

Other Methods of Preparation of Physiological Doses of Vitamin A

Although it is theoretically possible to label RBP with retinol[11] and chylomicrons with retinyl acetate or palmitate[12] *in vitro,* we are not aware that such entities have been prepared and shown to have kinetic behavior identical in recipient animals to that of *in vivo* doses. Such information would be valuable, as the use of donor animals is not feasible for all species and recoveries of tracer are relatively low for the *in vivo* methods.

For studies in humans and possibly nonhuman primates, plasma retinol could be labeled by orally administering a (stable) isotopic form of retinol, as recently done by Furr *et al.*[13] Unfortunately, the processes of concurrent appearance/disappearance of labeled chylomicrons and appearance/disappearance of labeled retinol/RBP/TTR generate data that are difficult to interpret. However, using kinetic methods currently available, data from such studies may be meaningfully interpreted if (1) the amount of dose absorbed is known, (2) tracer in early plasma samples is separated into retinol and retinyl esters, (3) the experiment is carried out long enough that tracer equilibrates with endogenous pools, and (4) the dose contains sufficient tracer for accurate measurement but a small enough mass to not excessively perturb the system.

[11] M. A. Gawinowicz and D. S. Goodman, *J. Protein Chem.* **4,** 199 (1985).

[12] D. S. Goodman, O. Stein, G. Halperin, and Y. Stein, *Biochim. Biophys. Acta* **750,** 223 (1983).

[13] H. C. Furr, A. J. Clifford, H. R. Bergen, A. D. Jones, and J. A. Olson, *Fed. Proc., Fed. Am. Soc. Exp. Biol.* **46,** 1335 (Abstr. 5946) (1987).

In Vivo Turnover Studies

Preliminaries

Fortunately, a few guidelines can be given for choosing the most ideal subjects for vitamin A kinetic studies, as well as for determining the optimal length and the optimal times for sampling. Although the kinetics of many biological systems are nonlinear, a system is effectively linearized if the entity of interest is in a steady state during the kinetic study. Then the underlying kinetic parameters (i.e., fractional transfer coefficients) are time-invariant (i.e., constant), and mathematical interpretation of system dynamics is greatly simplified. For the case of vitamin A, this means that one should use animals that are near a growth plateau and whose body stores of the vitamin (reflected primarily by liver total retinol levels) are relatively constant. To detect underlying nonlinearities in the system, the system can be studied in different steady states.

To choose the duration of the study, we conduct long-term pilot experiments to determine the approximate length of time required for the plasma tracer concentration versus time curve to reach a final slope when plotted semilogarithmically. This "terminal slope" indicates that the tracer has "equilibrated" with endogenous pools, and it is required for meaningful computation of area under the curve (see below). To fulfill this requirement, we studied vitamin A-depleted rats for 35 days and vitamin A-sufficient ones (liver vitamin A levels > 1000 μg of retinol equivalents) for 115 days;[2] the latter was apparently longer than necessary.

For a first study, times for sampling blood can be calculated [Eq. (1)] using a geometric progression

$$T_{i+1} = T_i\,([T_N/T_1]^{1/(N-1)}) \tag{1}$$

where T_1 is the time of the first sample, T_N the time of the terminal sample, T_i the time of the ith sample, and N the number of samples. For subsequent experiments, sampling times and number can be determined using an optimization procedure.[14-16]

Experimental Details

For our experiments, recipient rats are anesthetized with methoxyflurane (Pitman-Moore, Inc., Washington Crossing, NJ), and the right exter-

[14] J. J. DiStefano III, *Am. J. Physiol.* **240,** R259 (1981).

[15] J. J. DiStefano III, *Fed. Proc., Fed. Am. Soc. Exp. Biol.* **39,** 84 (1980).

[16] M. Berman, W. F. Beltz, P. C. Greif, R. Chabay, and R. C. Boston, "CONSAM User's Guide." PHS Publication number 1983-421-132:3279, U.S. Government Printing Office, Washington, D.C., 1983.

nal jugular vein is exposed.[17] An accurately known amount of labeled dose (0.3–1 g) is injected via a 27-gauge needle; the wound is closed, and anesthesia is removed. For the experiment cited above,[2] doses containing approximately 13 and 22 μCi provided adequate levels of radioactivity in terminal plasma samples (0.1 ml) of vitamin A-depleted and -sufficient rats, respectively; these doses contained 100 and 160 ng of retinol and perturbed the plasma retinol mass only minimally (<7 and 2%, respectively).

After dose administration, serial blood samples (0.1–0.25 ml) are collected at predetermined times until the rats are sacrificed. The first sample should be taken as soon as the animal is awake and moving around (~10 min after injection). To minimize blood loss, small samples are collected at early times when plasma radioactivity is high. Plasma samples are obtained from a caudal blood vessel after nicking the tail with a #10 scalpel blade; unanesthetized rats are free-moving on the investigator's lap during blood sampling. With practice, blood can be so obtained with minimal distress to the rat. Blood is collected into 250-μl microcentrifuge tubes containing Na_2EDTA (final concentration ~4 μM), and plasma samples are aliquoted as soon as practical. Samples are purged with nitrogen and stored at −16°. Some samples (including one collected before dose administration) are analyzed for retinol mass as well as radioactivity. Accurately weighed aliquots of the injected dose are also taken for analysis.

After collection of the final blood sample, rats are anesthetized with methoxyflurane. It is important to collect the last blood sample before anesthetizing the animal because anesthesia apparently causes a change in plasma volume that results in a change in plasma retinol concentration. If organs are to be collected, the whole body is perfused by circulating 0.9% saline containing 0.2% of the antioxidant D-isoascorbic acid (Sigma) or Hanks' balanced salt solution buffered to pH 7.4 with HEPES (~150 ml at 40 ml/min) through a butterfly needle inserted in the left cardiac ventricle to an exit incision in the right auricle.[2] Organs are excised, weighed, frozen, freeze-dried, reweighed, and refrozen for analysis. Frequently, the remaining carcass (minus contents of GI tract and urinary bladder) is also retained.

Analytical Methods

Methods for extracting and analyzing retinoids are described in detail elsewhere.[18] For our studies, plasma samples and aliquots of the injected

[17] S. Renaud, *Lab. Anim. Care* **19**, 664 (1969).

[18] A. C. Ross, this series, Vol. 189, p. 81, and H. C. Furr, this series, Vol. 189, p. 85.

doses are extracted using a modification[2] of the procedure of Thompson *et al.*[19] If retinol mass is to be determined, samples are spiked with an internal standard. We use 20 μl of a solution of retinyl acetate (Sigma) in absolute ethanol (3–4 ng/μl). Add 1 ml of absolute ethanol, 5 ml of HPLC-grade hexane that contains 5 μg/ml of the antioxidant butylated hydroxytoluene (BHT; Sigma), and 1 ml of distilled water. Vortex and centrifuge at low speed. Aliquot the upper lipid-containing phase for high-performance liquid chromatography (HPLC) (see Ref. 2 for conditions used in our laboratory) and/or for liquid scintillation spectrometry. To determine radioactivity, solvent is evaporated and samples are solubilized in scintillation solution. We count all samples twice to an accuracy of 1.5% (i.e., a total of 40,000 accumulated counts). Sample counts per minute (cpm) are converted to disintegrations per minute (dpm).

We have used two methods for extraction of retinoids from tissues. To determine mass or radioactivity in total retinol (i.e., unesterified plus esterified retinol), a modification[1] of the saponification procedure described by Thompson *et al.*[19] is used. A lingering problem with this method is finding an appropriate internal standard for quantifying masses. One promising candidate brought to our attention by Dr. H. Furr (Iowa State University, Ames) is 15,15-dimethyl retinol.[20] When both unesterified and esterified retinol are of interest, we use the hexane/2-propanol procedure of Hara and Radin[21,22] developed specifically by K. C. Lewis as part of his work in our laboratory at Penn State and described by Green *et al.*[2] Compared to the commonly used tissue lipid extraction method of Folch *et al.*,[23] this procedure uses solvents that are minimally toxic, and it is technically easier and quicker. In brief, spike aliquots (0.15–3 g) of freeze-dried or fresh tissue with retinyl acetate or another appropriate internal standard. Add 18 ml of hexane/2-propanol (3 : 2, v/v) that contains 5 μg/ml of BHT; vortex. Add 9 ml of sodium sulfate wash (66.7 g/liter of water); vortex and centrifuge. Aspirate the upper lipid-containing phase into a volumetric flask. Wash the aqueous phase 2 times with 4 ml of hexane/2-propanol (7 : 2, v/v) containing BHT, and pool the upper phases with the first extract. Aliquot for radioactivity and/or for HPLC. Using this procedure, recovery of retinoids added to liver samples ($n = 9$) averaged $98 \pm 0.8\%$ (mean $\pm$ S.D.) for retinyl acetate, $95 \pm 2.4\%$ for retinol, and $101 \pm 2.9\%$ for retinyl palmitate.[2] We have also used this extraction method for eyes and adrenals, for freeze-dried kidneys, small intestine, testes, and lungs, and for samples of ground or homogenized carcass.

[19] J. N. Thompson, P. Erdody, R. Brien, and T. K. Murray, *Biochem. Med.* **5,** 67 (1971).
[20] P. Tosukhowong and J. A. Olson, *Biochim. Biophys. Acta* **529,** 438 (1978).
[21] A. Hara and N. S. Radin, *Anal. Biochem.* **90,** 420 (1978).
[22] N. S. Radin, this series, Vol. 72, p. 5.
[23] J. Folch, M. Lees, and G. H. Sloane Stanley, *J. Biol. Chem.* **226,** 497 (1957).

Kinetic Analyses

The effort required to understand and correctly apply mathematical modeling to tracer kinetic data is well justified by the information and insights that are obtained. We recommend several excellent sources as an introduction to tracer kinetic theory and methods.[24-31] Newcomers to this arena may wish to ponder some wise words of Albert Einstein: "Things should be made as simple as possible, but not simpler."

In order to choose the appropriate kinetic approach for studying the dynamic behavior of vitamin A or other retinoids, one must first formulate the experimental goals. To determine specific kinetic parameters, one can apply the relatively straightforward techniques of empirical modeling, in which plasma tracer data are fit to a multiexponential equation and parameters are calculated based on this equation and other information about the system. Alternatively, if the aim is to develop a whole-body compartmental model which includes individual organs, model-based compartmental analysis is appropriate. We have applied both methods to data on vitamin A dynamics *in vivo* and briefly review the approaches here.

Preliminary Calculations

To calculate the fraction of the dose in plasma at each time, individual animal data (dpm/ml) at each time are normalized to the estimated plasma tracer concentration at the time of injection: tracer concentration (dpm/ml) at time 0 = dpm injected ÷ estimated plasma volume (ml), where plasma volume is approximated as body weight (g) times 0.038 ml of plasma/g of body weight.[32] The fraction of the dose in organs is calculated as dpm in total organ divided by dpm injected. If an appreciable number of blood samples is removed early in the study, then the denominator should be corrected for dpm removed in previous blood samples.

Ideally, kinetic analysis is done on the data for individual animals. This

[24] J. J. DiStefano III and E. M. Landaw, *Am. J. Physiol.* **246,** R651 (1984).
[25] E. M. Landaw and J. J. DiStefano III, *Am. J. Physiol.* **246,** R665 (1984).
[26] M. Berman, *Prog. Biochem. Pharmacol.* **15,** 67 (1979).
[27] R. A. Shipley and R. E. Clark, "Tracer Methods for *In Vivo* Kinetics." Academic Press, New York, 1972.
[28] J. A. Jacquez, "Compartmental Analysis in Biology and Medicine," 2nd Ed. The University of Michigan Press, Ann Arbor, Michigan, 1985.
[29] E. R. Carson, C. Cobelli, and L. Finkelstein, "The Mathematical Modeling of Metabolic and Endocrine Systems: Model Formulation, Identification, and Validation." Wiley, New York, 1983.
[30] M. Berman, *in* "Lipoprotein Kinetics and Modeling" (M. Berman, S. M. Grundy, and B. V. Howard, eds.), Chap. 1. Academic Press, New York, 1982.
[31] E. Gurpide, "Tracer Methods in Hormone Research." Springer-Verlag, New York, 1975.
[32] L. Wang, *Am. J. Physiol.* **196,** 188 (1959).

is easy in the case of empirical modeling. For model-based compartmental analysis, the usual approach is to sacrifice groups of rats at predetermined, optimized times, average the data at each time, then combine all data into a data set for modeling. Unfortunately, there are theoretical problems associated with interpretation of the models developed using this approach which should be recognized.[28]

Empirical Modeling

Also called compartmental analysis, multiexponential modeling, input/output analysis, models of data, or the back door approach to compartmental analysis, empirical modeling is well described in the literature;[24,25,27,31,33] it has been applied to many entities, including cholesterol,[34–36] thyroid hormone[37] and retinol.[2,38]

For retinol, data are collected on the kinetics of [^{3}H]retinol in plasma after injection of [^{3}H]retinol-labeled plasma. To obtain accurate estimates of kinetic parameters, experiments must be carried out long enough that the tracer equilibrates with endogenous pools of the tracee, reflected graphically for an interchanging system in a steady state by the attainment of a "terminal slope" for the semilogarithmic plot of the plasma tracer response curve. Data on fraction of dose in plasma versus time for each animal are fit to a multiexponential equation [Eq. (2)] of the form

$$y_t = \sum_{i=1}^{n} I_i e^{-g_i t} \tag{2}$$

When we fit data on plasma retinol dynamics in rats with low, marginal, or adequate vitamin A status using this method, the data were described by three- or four-component exponential equations. This indicates that plasma acts as one kinetic compartment which exchanges with two or three kinetically distinct extravascular compartments. In fitting the data to an exponential equation with a maximum of four components, eight parameters (four exponential coefficients and four constants) have to be identified. We thus collect approximately three samples per parameter at times that the data are most sensitive to these parameters. More sophisticated optimization methods are available.[14,15]

[33] E. Gurpide, J. Mann, and E. Sandberg, *Biochemistry* **3,** 1250 (1964).

[34] F. R. Smith, R. B. Dell, R. P. Noble, and D. S. Goodman, *J. Clin. Invest.* **57,** 137 (1976).

[35] D. S. Goodman, F. R. Smith, A. H. Seplowitz, R. Ramakrishnan, and R. B. Dell, *J. Lipid Res.* **21,** 699 (1980).

[36] W. Perl and P. Samuel, *Circulation Res.* **25,** 191 (1969).

[37] J. J. DiStefano III, M. Jang, T. K. Malone, and M. Broutman, *Endocrinology* **110,** 198 (1982).

[38] K. C. Lewis, M. H. Green, and B. A. Underwood, *J. Nutr.* **111,** 1135 (1981).

Fitting data to Eq. (2) requires nonlinear regression methods[25,39] and an appropriate weighted, nonlinear least-squares regression computer program. We use the Simulation, Analysis, and Modeling (SAAM) computer program[40] and its interactive version, CONSAM,[16] which were developed by the late Mones Berman and colleagues at the National Institutes of Health. Fitting data to exponential equations using SAAM is well described in the SAAM tutorials[41] developed by D. M. Foster, R. C. Boston, J. A. Jacquez, and L. A. Zech.

Using the exponential constants (I_i) and coefficients (g_i) from Eq. (2), many interesting kinetic parameters describing vitamin A dynamics can be calculated. To do so, first set y_t at time 0 [Eq. (2)] equal to 1 (i.e., the fraction of dose in plasma at time 0 equals 1). This is done [Eq. (3)] by calculating

$$H_i = I_i / \sum_{i=1}^{n} I_i \tag{3}$$

Next calculate [Eq. (4)] the area under the plasma tracer response curve (AUC_p) as

$$AUC_p = \int_0^\infty y_t / y_{t_0} = \int_0^\infty \sum_{i=1}^{n} H_i e^{-g_i t} = \sum_{i=1}^{n} H_i / g_i \tag{4}$$

The plasma-derived fractional catabolic rate (FCR_p), or the rate of vitamin A utilization as a fraction of the plasma retinol pool, is calculated [Eq. (5)] as

$$FCR_p = 1/AUC_p \tag{5}$$

The minimal rate of irreversible utilization of vitamin A, or the disposal rate (DR_p), is calculated [Eq. (6)] as

$$DR_p = FCR_p M_p \tag{6}$$

where M_p is the plasma retinol pool calculated from the mean plasma retinol concentration and an estimate of plasma volume. Plasma volume (V_p) can be calculated [Eq. (7)] as

$$V_p = (\text{Mean body weight} \times 0.038) \div \sum_{i=1}^{n} I_i \tag{7}$$

[39] H. J. Motulsky and L. A. Ransnas, *FASEB J.* **1,** 365 (1987).

[40] M. Berman and M. F. Weiss, "SAAM Manual." DHEW Publication No. (NIH) 78-180, U.S. Government Printing Office, Washington, D.C., 1978.

[41] Available from D. M. Foster, Resource Facility for Kinetic Analysis (RFKA). University of Washington, Seattle, Washington.

The denominator in Eq. (7) corrects the initial estimate of plasma volume and is based on the isotope dilution principle (i.e., $y_{t_0} = 1.0$; if the extrapolated value for $y_{t_0} = \Sigma I_i$ is >1.0, plasma volume was initially overestimated and if <1.0, it was underestimated). The disposal rate calculated by Eq. (6) will not include any newly absorbed vitamin A which is irreversibly utilized before it enters the plasma as retinol; this would presumably be a small fraction of total utilization. In a steady state, the disposal rate will equal the production rate; for vitamin A, the latter is the rate of absorption of dietary vitamin A. The fractional transfer coefficient for plasma retinol (L_p; the fraction of plasma retinol leaving the plasma reversibly or irreversibly per unit time) is computed [Eq. (8)] as

$$L_p = \sum_{i=1}^{n} H_i g_i \tag{8}$$

and the plasma retinol transfer (turnover) rate (R_p) is calculated [Eq. (9)] as

$$R_p = M_p L_p \tag{9}$$

Other kinetic parameters that can be computed include the plasma retinol mean transit time or turnover time [$\bar{t}_p$; the mean of the distribution of time a retinol molecule spends in the plasma during a single transit; Eq. (10)], plasma mean residence time [$\bar{T}_p$; the mean exposure time of the plasma to retinol, or the mean of the distribution of time retinol spends in the plasma before irreversibly leaving the plasma; Eq. (11)], the plasma retinol recycling number [ν_p; the number of times on average that a retinol molecule cycles back to the plasma before irreversibly leaving the plasma; Eq. (12)], and the plasma-derived mean sojourn time [MST_p; the average time a vitamin A molecule spends in the body from the time it first enters the plasma until it irreversibly leaves the plasma; Eq. (13)]. MST_p is thus less than the system mean sojourn time, or the total time a vitamin A molecule is in the body before irreversible utilization. One can also calculate the mean recycle time [$\bar{t}_{pp}$; the mean of the distribution of time a traceable vitamin A molecule spends outside the plasma before cycling back to the plasma; Eq. (14)]; the total traced mass of vitamin A [M_T; Eq. (15)]; and the traced extravascular mass of vitamin A [M_{EV}; Eq. (16)]. By irreversible loss, we mean conversion of retinol to retinoic acid or other polar metabolites that cannot chemically recycle to retinol plus loss of label owing to chemical isomerization plus any loss of retinol in urine, feces, or desquamated cells. These kinetic parameters can be computed by standard methods as

$$\bar{t}_p = 1/\sum_{i=1}^{N} H_i g_i = 1/L_p \tag{10}$$

$$\bar{T}_p = 1/FCR_p = AUC_p \qquad (11)$$

$$v_p = (\bar{T}_p/\bar{t}_p) - 1 \qquad (12)$$

$$MST_p = \sum_{i=1}^{n} (H_i/g_i^2)/\sum_{i=1}^{n} H_i/g_i \qquad (13)$$

$$\bar{t}_{pp} = (MST_p - \bar{T}_p)/v_p \qquad (14)$$

$$M_T = MST_pDR \qquad (15)$$

$$M_{EV} = M_T - M_p \qquad (16)$$

Finally, if certain assumptions are made, still other useful kinetic parameters can be calculated. For a discussion of this approach, referred to as interval analysis, see DiStefano and Landaw.[24]

Model-Based Compartmental Analysis

Perhaps the most challenging type of kinetic analysis is model-based compartmental analysis.[26,30] In contrast to empirical modeling, we call this the front door approach to compartmental analysis, since the investigator formalizes a model at the outset, then adjusts the structure and associated kinetic parameters (i.e., fractional transfer coefficients) to arrive at a description which is compatible both with the known physiology and biochemistry of the system and with current data. Development of such a model requires much patience and many, many hours, as well as a thorough understanding of the system and tracer kinetic theory.

The SAAM and CONSAM programs mentioned above[16,40] were specifically developed, and are continually being refined, for compartmental analysis of biologically interesting systems; they are available, along with supporting documentation, from the RFKA.[41] Current versions can handle up to 75 compartments and 75 adjustable parameters. The programs have been used to generate models describing the metabolism of lipoproteins, minerals, blood-borne fuels, and retinol.[1,7,42,43]

Model-based compartmental analysis using SAAM/CONSAM is distinguished by several important features: information from the literature and from previous work can be incorporated during the modeling process to describe and constrain the system; the modeling process is iterative in that each experiment is designed based on what was learned or hypothesized in previous work; the programs can accommodate data (tracer and tracee) on any tissue subsystem as well as plasma, urine, and feces; the user

[42] M. H. Green and J. B. Green, *Fed. Proc., Fed. Am. Soc. Exp. Biol.* **46,** 1011 (Abstr. 4047) (1987).

[43] M. H. Green, J. B. Green, T. Berg, K. R. Norum, and R. Blomhoff, *FASEB J.* **2,** A1094 (Abstr. 4636) (1988).

can simulate aspects of the system which cannot be easily measured; and the programs do not require that experiments be conducted in a steady state.

It is not so easy to systematically describe the sequence of events in model-based compartmental analysis as to enumerate the equations used in empirical modeling. To get started, one may want to enlist the help of an experienced biological modeler who has used this approach. If no previous compartmental models are available in the literature, the first step is to conceptualize one, assign quantitative estimates to the processes described, and then design and carry out an appropriate kinetic study.[14] It is useful to develop a mechanistic model including algebraic descriptions of underlying biological processes. This will help prevent misinterpretation of kinetic information when model predictions are mapped back to the real system. The structure and parameter values of the starting model [the fractional transfer coefficients, $L(I,J)$ or the fraction of the mass of compartment J transferred to compartment I per unit time] are adjusted in light of the data until a good fit is obtained between the observed values and the model predictions calculated by SAAM. Technical aspects of this process are well described.[41,44] Final parameter values describing the model, and estimates of their statistical uncertainties, are calculated by weighted nonlinear least-squares regression. Kinetic parameters and state variables (e.g., rates of transfer between compartments, compartment masses, time and recycling characteristics) can then be estimated.

As noted previously, it is usual to develop a whole-body model using data from a large number of animals sacrificed at different times (the superrat approach[25]). In fact such a model does not provide the actual arithmetic mean of kinetic parameters for individual animals, but rather gives central tendencies and thus estimates for the dynamics expected in a theoretical average rat.

Several features of the SAAM/CONSAM programs are particularly useful for modeling complex biological systems. One is the forcing function[44] which uncouples individual organs from the rest of the system. SAAM and CONSAM can also be used to develop parallel models when before-treatment and after-treatment data are collected, if there are data from different experimental groups, or if multiple tracers are administered. Such examples are described in the SAAM tutorials.[41]

[44] D. M. Foster and R. C. Boston, *in* "Compartmental Distribution of Radiotracers" (J. S. Robertson, ed.), Chap. 5. CRC Press, Boca Raton, Florida, 1983.

Concluding Remarks

We began to apply model-based compartmental analysis to whole-body retinol dynamics in 1981. Based on conventional wisdom at that time,[45] we postulated a five-compartment starting model. We then conducted an *in vivo* turnover study in rats with marginal vitamin A status; short- and long-term tracer and tracee data were collected for plasma, liver, kidneys, and the rest of the carcass. Data were not compatible, however, with the initial model; we used SAAM/CONSAM to arrive at a "simplest" model with 11 physiological compartments.[1] Some parameters were well defined (i.e., statistical uncertainties were low); others, for example, those describing liver vitamin A dynamics, were not. This study not only generated some new and thought-provoking concepts about retinol metabolism, but it suggested improvements in experimental design and hypotheses to be tested in subsequent and ongoing experiments. It is clear that the application of model-based compartmental analysis will continue to add to our understanding of the complex dynamics of whole-body retinol metabolism.

Acknowledgments

Work in the laboratory of M. H. Green was supported by U.S. Department of Agriculture Competitive Research Grants 81-CRCR-1-0702 and 88-37200-3537. We acknowledge Barbara A. Underwood who introduced us to the exciting field of vitamin A dynamics through a collaborative research project.

[45] B. A. Underwood, J. D. Loerch, and K. C. Lewis, *J. Nutr.* **109,** 796 (1979).

[33] Quantification of Embryonic Retinoic Acid Derived from Maternally Administered Retinol

By DEVENDRA M. KOCHHAR

Introduction

Hypervitaminosis A during pregnancy is probably teratogenic in humans as it certainly is in animals.[1-3] It is generally assumed that the human embryo is safe as long as the mother does not overdose herself, but

[1] S. Q. Cohlan, *Science* **117,** 535 (1953).
[2] F. W. Rosa, A. L. Wilk, and F. O. Kelsey, *Teratology* **33,** 355 (1986).
[3] Anonymous, *Teratology* **35,** 269 (1987).

the question is, What constitutes a vitamin A "overdose"? From a survey of malformed human babies whose mothers ingested isotretinoin [Accutane (Hoffmann-La Roche), 13-*cis*-retinoic acid] a derivative of retinol, it turned out that even small quantities of the drug resulted in teratogenesis.[2,4,5]

Several questions can be addressed in experimental animals which may yield not only initial estimates of the risk associated with vitamin A-induced teratogenesis but also some information on mechanisms. These include the following: What is the identity of the chemical form(s) of vitamin A [retinol, retinal, retinyl esters, retinoic acid (RA), or metabolites] that poses the greatest teratogenic threat to the embryo? Is there a threshold dose of retinol beyond which teratogenic outcome becomes certain? At the threshold dose, is there a relationship of the levels of retinol (or of its teratogenic metabolites) in maternal circulation and the embryonic target tissues? Are there species differences in transplacental pharmacokinetics of retinol and other retinoids? Using retinol as an example, the following describes an experimental approach and methodologies suitable to address these questions.

The aim is to quantitate teratogenicity in mice after exposure to a stepwise increase in an oral dose of retinol in order to correlate this response with the patterns of formation and distribution of the parent compound and its individual metabolites in the maternal and embryonic compartments. Some requirements for such an experiment should be noted. The degree and the type of teratogenic effects vary with the stage of embryonic development; this is so even if the dose of the test chemical is held constant. Hence, the onset of pregnancy should be timed with some precision so as to obtain consistency in the teratogenic response. Another requirement is the selection of one or more distinctive end point(s) from among a variety of embryonic defects produced by highly embryotoxic agents such as retinoids. It is preferable that the end point selected for measuring the teratogenic response should be a specific and a direct effect of the chemical rather than a nonspecific pharmacologic change in the dam. Therefore, care should be taken to ensure that the treatment does not result in overt maternal toxicity. Although indirect effects are of considerable importance in safety evaluations, they may not be so useful as the direct effects in understanding the mechanisms of retinoid action. For the same reason the selected defect should be such that it can be unambiguously and rapidly ascertained in a large number of exposed fetuses without

[4] E. J. Lammer, D. T. Chen, R. M. Hoar, N. D. Agnish, P. J. Benke, J. T. Braun, C. J. Curry, P. M. Fernohoff, A. W. Girixte, L. T. Lott, J. M. Richard, and S. C. Sun, *N. Engl. J. Med.* **313,** 837 (1985).

[5] J. C. Kraft, H. Nau, E. Lammer, and A. Olney, *N. Engl. J. Med.* **321,** 262 (1989).

the need for elaborate sample processing or extensive anatomic expertise. These are some of the criteria we have used in planning the following experimental protocols.

Dosing of Animals

Four doses of retinol are chosen for examination: 0, 10, 100, and 200 mg/kg maternal body weight. Retinol (all-trans-retinol, Sigma Chemical Co., St. Louis, MO) is dissolved in absolute ethanol (10 mg/ml) from which dilutions are made fresh in soybean oil on the day of use. The volume of vehicle is held constant at 5 ml/kg body weight.[6] Thus, a mouse weighing 40 g receives 0.2 ml of ethanol/oil vehicle containing either 0, 0.4, 4, or 8 mg retinol. Precautions against exposure of retinoid preparations to light, air, and heat are observed.

Teratology

Five pregnant females are dosed orally for each time point, using an animal feeding needle (Perfektum, 18 gauge × 2 inches, from Popper & Sons, Inc., New Hyde Park, NY) attached to a 1-ml syringe. For the teratology study, dams are intubated once on the morning (10 am) or evening (10 pm) of Day 10 of gestation (designated as Days 10 and 10.5, respectively; day of vaginal plug is designated Day 0 of gestation), or on the morning of Day 11.

The dose and treatment regimens are based on previous studies where the teratologic outcome is known for a number of diverse retinoids.[7-10] A single 100 mg/kg dose of an active retinoid affects virtually every mouse embryo exposed on Days 10 or 11 of gestation, and these embryos exhibit pronounced reduction of limb bones and are born with a cleft palate. If exposed earlier in gestation, on Days 8 or 9, the embryos frequently die and are resorbed or survive with a pattern of malformations quite different and anatomically more complex than that obtained with the chosen regimen. Embryos at more advanced developmental stages (Days 13 and 14) become increasingly resistant to the teratological effects of retinoids.

The dams are sacrificed by cervical dislocation 1 day before delivery, on Day 18, to prevent loss of litters through cannibalism. The fetuses are

[6] A. Kistler and W. B. Howard, this volume [44]–[46].

[7] D. M. Kochhar and J. D. Penner, *Teratology* **36,** 67 (1987).

[8] B. Zimmerman, D. Tsambaos, and H. Sturje, *Teratog. Carcinog. Mutagen.* **5,** 415 (1985).

[9] D. M. Kochhar, J. C. Kraft, and H. Nau, *in* "Pharmacokinetics in Teratogenesis" (H. Nau and W. J. Scott, eds.), p. 173. CRC Press, Boca Raton, Florida, 1987.

[10] W. S. Webster, M. C. Johnston, E. J. Lammer, and K. K. Sulik, *J. Craniofacial Genet. Dev. Biol.* **6,** 211 (1986).

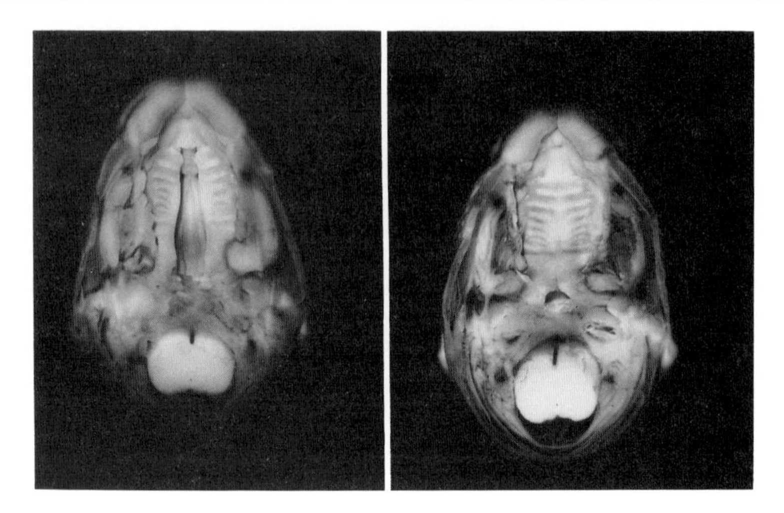

FIG. 1. Cleft palate (left) in a Day 18 mouse fetus is compared with the complete secondary palate of a normal fetus. The specimens were fixed in Bouin's fixative and the mandible removed to confirm the palatal abnormality.

weighed and examined fresh under a dissection microscope or fixed and prepared for subsequent examination.[7,11] A typical example of cleft palate in a Day 18 mouse fetus after fixation in Bouin's solution is shown (Fig. 1). For examples of limb reduction defects, see Kistler and Howard.[11]

Dose-dependent frequencies of malformed fetuses are expressed as the percentage of total number of live fetuses obtained after exposure on either Day 10, 10.5, or 11 (Fig. 2). All fetuses exposed to the 10 mg/kg dose of retinol survive without any malformations. About one-quarter of fetuses are affected at the 100 mg/kg dose; the frequencies of cleft palate among the survivors varies from 7% after exposure on Day 10 to 28% after Day 11; some of these affected fetuses also show a mild degree of limb bone reductions. The highest dose of retinol (200 mg/kg) results in virtually 100% incidence of not only cleft palate but also severe shortening of limb bones (phocomelia). Of importance is that even at the highest dose the embryonic mortality (resorbed or dead fetuses) is only 8%, which is only slightly greater than that observed in control litters.[12]

Pharmacodynamics

Our aim is to correlate maternal/fetal levels of retinol derivatives with teratogenesis. Each pregnant female yields only a single blood sample at

[11] A. Kistler and W. B. Howard, this volume [44]–[46].

[12] D. M. Kochhar, J. D. Penner, and M. A. Satre, *Toxicol. Appl. Pharmacol.* **96,** 429 (1988).

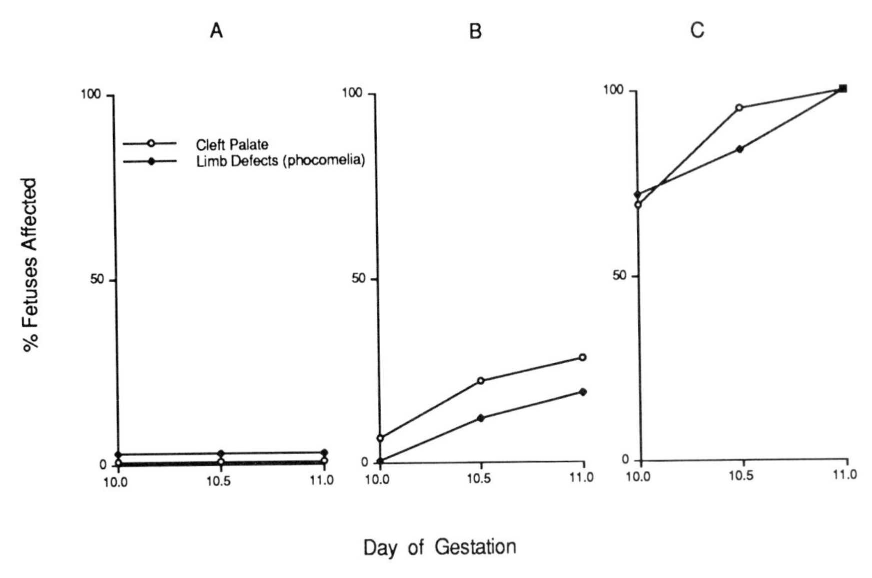

FIG. 2. Teratologic effects of an oral dose of retinol (A, 10 mg/kg; B, 100 mg/kg; C, 200 mg/kg) given to pregnant ICR mice on either Day 10, 10.5, or 11 of gestation. The fetuses were examined for anomalies on Day 18 of gestation. The frequency of occurrence of cleft palate and limb reduction defects is given as the percentage of the total surviving fetal populations. No teratologic effects are associated with the 10 mg/kg dose, whereas the 100 mg/kg dose is mildly teratogenic when given on Days 10.5 and 11. The 200 mg/kg dose is strongly teratogenic, resulting in virtually 100% frequencies of cleft palate and limb defects.

the time of removal of her litter. Although serial samples can be taken from an individual animal, controls must be included to assess the resultant effects of maternal stress or hemodynamic change, either of which may influence not only teratogenesis but also the pharmacokinetic parameters themselves.

Maternal blood, embryo, and placental samples are obtained at designated time intervals. The blood is drawn from the inferior vena cava of an ether-anesthetized animal into a heparinized syringe, after which the animal is sacrificed by cervical dislocation and its implantation sites removed.[13] Each embryo is removed from its placenta, divested of embryonic membranes, and examined for its proper developmental stage designation before processing for the extraction of retinoids. Each mouse litter consists of an average of 12 embryos, and a Day 11 embryo weighs on an average about 30 mg (wet weight) and contains about 40 μg protein. A sample size of about 200 mg wet weight of embryos is generally adequate

[13] M. A. Satre and D. M. Kochhar, *Dev. Biol.* **133,** 529 (1989).

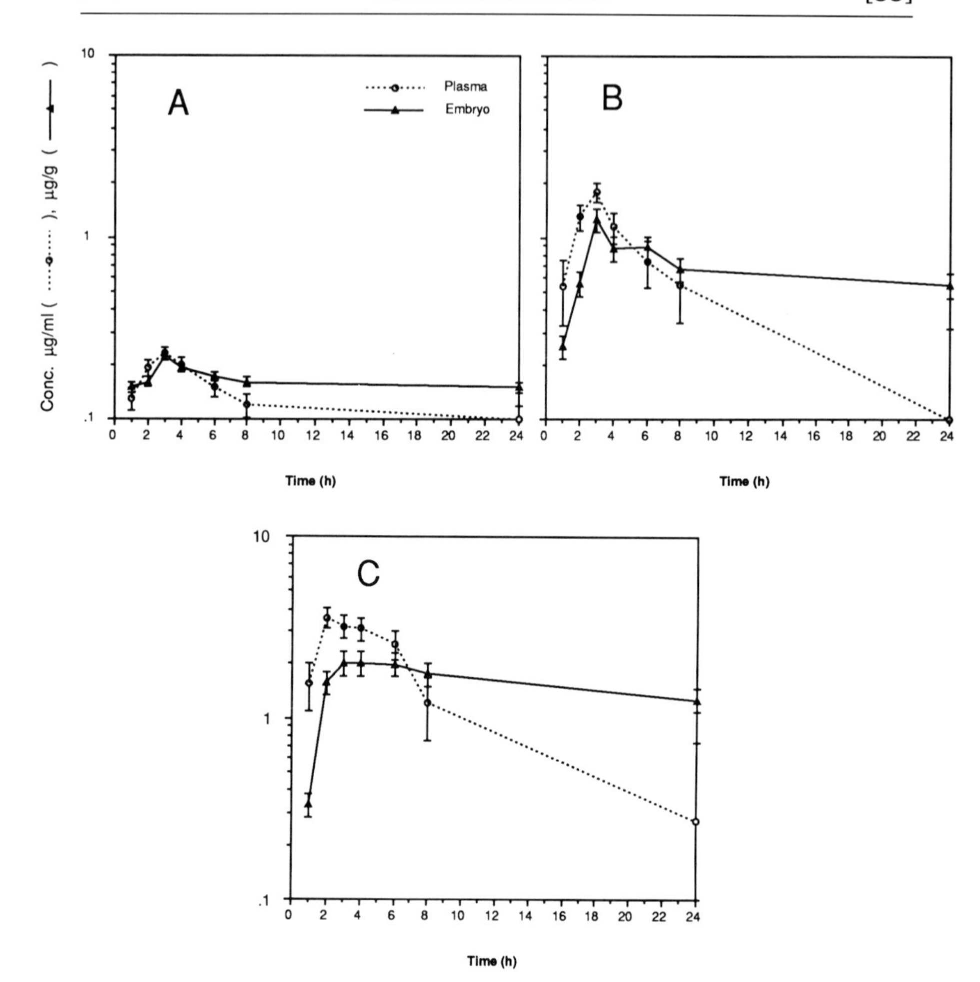

FIG. 3. Retinol concentrations in the embryo and in maternal circulation during the first 24 hr after intubation of Day 11 pregnant mice with nonteratogenic (10 mg/kg, A) or teratogenic (100 mg/kg, B; 200 mg/kg, C) doses of retinol. Although the magnitude of increase in retinol concentration is roughly proportional to the dose in either compartment, the higher levels persist longer in the embryo than in the maternal circulation.

for measurement; therefore, each litter provides an average of two replicate samples.

The sample collection, processing, extraction, and quantification of retinoids are essentially similar to protocols reported elsewhere.[12,13] An internal standard, Ro 13-7410 (Hoffmann-La Roche, Inc., Nutley, NJ), is added to the ethanolic extracts of the sample prior to analysis by high-performance liquid chromatography (HPLC). Retinoid concentrations are calculated from the integrated peak areas with reference to standard curves obtained by analyzing known amounts of authentic retinoids.

We have found that in pregnant, untreated mice the circulating retinol levels normally decline by midgestation to about one-fourth the level of nonpregnant animals in our ICR colony.[12,13] Intubation of the animals on Day 11 of gestation with the subteratogenic 10 mg/kg dose raises slightly the retinol levels in circulation without unduly influencing the embryonic level (Fig. 3A). The presence of RA isomers is detectable after this dose, but their amounts are similar to the levels normally encountered in untreated control animals.

Intubation of the dams with the higher doses results in prompt and proportional increase in plasma levels. The peak is reached within 2–3 hr with either the 100 mg/kg or the 200 mg/kg dose followed by a gradual decline to near normal levels by 24 hr (Fig. 3B,C). Transfer to the embryo occurs fairly rapidly, and the peak is attained simultaneously with the plasma. The peak concentrations in embryo after 100 and 200 mg/kg doses are, respectively, 1 and 2 μg/g wet weight (Table I). The notable feature is that these high retinol levels in the embryo persist for a prolonged period of

TABLE I
PEAK CONCENTRATIONS (C_{max}) OF RETINOL AND DERIVATIVES FOLLOWING A SINGLE, ORAL DOSE OF RETINOL TO DAY 11 PREGNANT MICE

Retinoid	Retinol dose (mg/kg)	Maternal plasma (ng/ml ± S.E.)	Embryo (ng/g ± S.E.)	Placenta (ng/g ± S.E.)
Retinol	0	53 ± 15	148 ± 25	140 ± 32
	10	235 ± 12	222 ± 19	698 ± 72
	100	1790 ± 410	1250 ± 490	4850 ± 1510
	200	3560 ± 610	1980 ± 490	7100 ± 1200
all-*trans*-Retinoic acid	0	<10	<10	<10
	10	<10	17 ± 5	28 ± 12
	100	412 ± 56	245 ± 41	820 ± 45
	200	1200 ± 220	630 ± 70	1270 ± 60
4-Oxoretinoic acid	0	0	0	0
	10	0	0	0
	100	148 ± 20	118 ± 22	106 ± 30
	200	990 ± 60	690 ± 210	570 ± 160
13-*cis*-Retinoic acid	0	<10	0	0
	10	<10	0	0
	100	72 ± 16	83 ± 21	41 ± 5
	200	1090 ± 150	440 ± 40	1080 ± 80

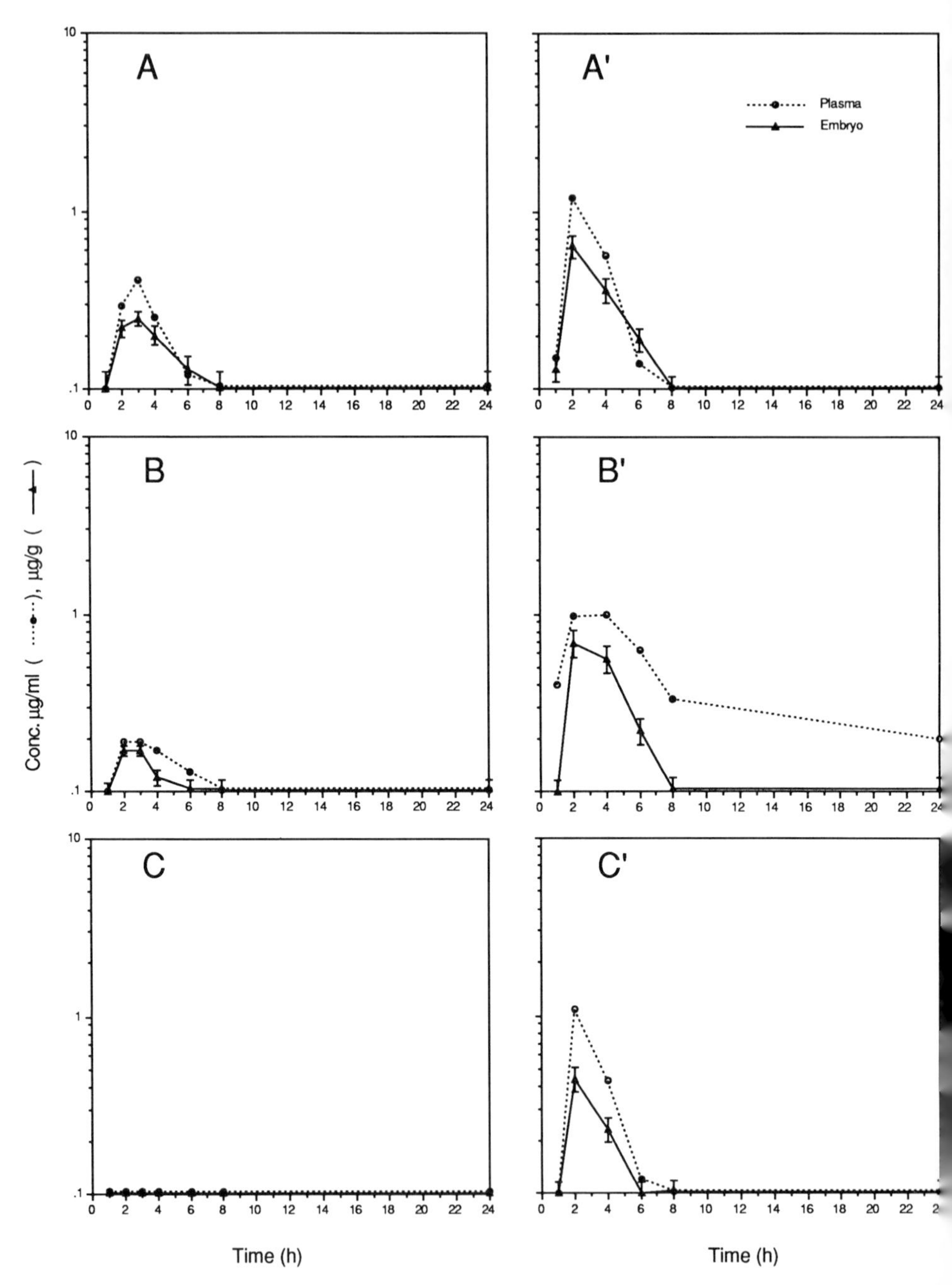

FIG. 4. Derivatives of retinol in the embryo and in maternal circulation during the first 24 hr after intubation of Day 11 pregnant mice with the low (100 mg/kg, A, B, C) or the high (200 mg/kg, A′, B′, C′) teratogenic dose of retinol. The derivatives are all-*trans*-RA (A, A′), 4-oxo-RA (B, B′), and 13-*cis*-RA (C, C′). The derivatives make a rapid and virtually simulta-

time, at least for the first 24 hr after dosing, and constitute a 5- to 10-fold increase in the normal control levels.

The formation and distribution of retinol derivatives in the maternal plasma and the embryo following the two teratogenic dose levels are shown (Fig. 4). The data are from the same samples as processed for retinol determination. The major metabolites in the embryo after the 100 mg/kg dose are all-*trans*-RA and 4-oxo-RA, whereas 13-*cis*-RA is a minor metabolite (Fig. 4A,B,C). The quantitative pattern of these metabolites changes after the higher 200 mg/kg dose (Fig. 4A′B′C′). Here, 4-oxo-RA and 13-*cis*-RA show a disproportionate increase in amount so that peak concentrations of the three metabolites in the embryo are 0.63, 0.69, and 0.44 μg/g for, respectively, all-*trans*-RA, 4-oxo-RA, and 13-*cis*-RA (Table I). Irrespective of the dose of retinol, all derivatives in the embryo decline in concentration at a rate which is much faster than that of retinol itself, a fact suggesting that the metabolic capacity of the embryo to generate the derivatives may be minimal. Because retinol is present at fairly high concentrations in the embryo after either dose, the larger difference in their teratogenic potency may be due to embryonic exposure to an excess of active metabolites formed after the 200 mg/kg dose.

Acknowledgments

The research reported here is supported by National Institutes of Health Grant HD-20925.

neous appearance in the maternal circulation and embryo, reaching near peak concentrations by 2 hr. Subsequent decline in levels is also fairly rapid except for 4-oxo-RA (B, B′). Although all-*trans*-RA is quantitatively the major metabolite after the low teratogenic dose, the high teratogenic dose (200 mg/kg) yields fairly similar levels of both all-*trans*-RA and its 4-oxo derivatives.

[34] Effect of Retinoids on Sebaceous Glands

By STANLEY S. SHAPIRO and JAMES HURLEY

Introduction

Retinoids are modulators of epithelial differentiation in many cell types[1] including sebaceous glands. Retinoids have a profound effect on sebaceous gland activity, reducing sebaceous gland size and sebum secretion in humans and animals.[2-5]

The wide distribution of sebaceous glands in mammalian skin has resulted in the development of numerous animal models of sebaceous gland activity. Because the sebaceous gland is androgen-sensitive, these models were developed and often used to study endocrine responses. In the recent past, several of these models have been utilized by investigators to assess the effects of retinoids on sebaceous glands. Representative models include the nonspecialized sebaceous glands of rat skin[6] and hamster ear,[7] as well as the easily accessible and enlarged specialized sebaceous glands, such as the rat preputial gland,[8] hamster costovertebrae (flank) organ,[9] and the gerbil ventral gland.[10]

The mechanism(s) involved in retinoid inhibition of sebaceous glands remains unknown. In studying the effect of retinoids in animal models, one cannot automatically extrapolate the results to humans. Different drug effects are commonly seen in different models. In addition to interspecies differences relating to absorption, distribution, and metabolism, there may be more fundamental differences involving unknown factors in sebaceous gland regulation. This chapter reviews the rat model, as well as the hamster ear and flank organ models, as useful systems to assess retinoid activity.

[1] S. S. Shapiro, *in* "Retinoids and Cell Differentiation" (M. I. Sherman, ed.), p. 29. CRC Press, Boca Raton, Florida, 1986.

[2] J. A. Goldstein, A. Socha-Szott, R. J. Thomsen, P. E. Pochi, A. R. Shalita, and J. S. Strauss, *J. Am. Acad. Dematol.* **6,** 760 (1982).

[3] E. C. Gomez, *J. Am. Acad. Dermatol.* **6,** 746 (1982).

[4] G. L. Peck, T. G. Olsen, D. Butkus, M. Pandya, J. Arnaud-Battandier, E. G. Gross, D. B. Windhorst, and J. Cheripko, *J. Am. Acad. Dermatol.* **6,** 735 (1982).

[5] J. S. Strauss and A. M. Stranieri, *J. Am. Acad. Dermatol.* **6,** 751 (1982).

[6] A. Archibald and S. Shuster, *Br. J. Dermatol.* **82,** 146 (1970).

[7] G. Plewig and C. Luderschmidt, *J. Invest. Dermatol.* **68,** 171 (1977).

[8] J. Mesquita-Guimaraes and A. Coimbra, *Arch. Dermatol. Res.* **270,** 325 (1981).

[9] E. C. Gomez, *J. Invest. Dermatol.* **75,** 68 (1981).

[10] D. D. Thiessen, F. E. Regnier, M. Rice, M. Goodwin, N. Isaacks, and N. Lawson, *Science* **184,** 83 (1974).

Similarities and differences with selected retinoids across models are emphasized.

Materials and Methods

Sebum Inhibition in Rats

Retinoid inhibition of sebum secretion in rats may be carried out using mature males or females, or, alternatively, castrated immature males treated with testosterone.[11,12] The use of immature, castrated, testosterone-stimulated male rats as opposed to mature intact rats eliminates some variation and raises the baseline level of sebum secretion. If castrated animals are not available, mature males (at least 3 months old) may be used.

Sebum Inhibition in Testosterone-Stimulated Castrated Rats. Sprague-Dawley male rats are obtained from the Charles River Breeding Laboratory (Wilmington, MA) at an age of 21 days. Castration is performed at 22 days of age. One week after surgery, the rats are lightly anesthetized with Metofane inhalation anesthetic (Pitman-Moore, Inc., Washington Crossing, NJ) and then scrubbed with a bristle brush in a lukewarm 0.2% lauryl sulfate detergent solution to remove any dirt and surface "sebum." The detergent solution is rinsed off with running tap water, and the rats are hand-dried with paper towels. One day after the washing, the drug-treated groups of rats are given testosterone propionate (TP) in sesame oil (100 μg/0.2 ml/day, subcutaneously), and various doses of the test compounds are administered orally by intubation in propylene glycol (0.2 ml/day). Rats in a second group receive only TP in sesame oil (100 μg/0.2 ml/day, subcutaneously) and propylene glycol (0.2 ml/day, orally). A third group of rats is treated with the carrier vehicles and serves as the control group. The rats are treated for 14 days and then sacrificed on the Day 15 by CO_2 inhalation.

Sebum Inhibition in Mature Rats. Sprague-Dawley male rats are obtained from the Charles River Breeding Laboratory at an age of 3 months. The rats are lightly anesthetized with Metofane and washed as described above. Test compounds are dissolved in propylene glycol and administered orally by intubation in a fixed volume of 0.2 ml for 28 consecutive days. The rats are washed two additional times, as described above, on Days 14 and 21. On Day 29, the rats are sacrificed by CO_2 inhalation.

[11] S. Shuster and A. J. Thody, *J. Invest. Dermatol.* **62,** 172 (1974).
[12] F. J. Ebling and J. Skinner, *Br. J. Dermatol.* **92,** 321 (1975).

Quantitation of Sebum Secretion. The extent of retinoid-induced inhibition of sebum secretion can be calculated by two different procedures.

Surface sebum from skin and fur. Surface sebum from the skin and fur is removed following sacrifice by total body immersion of the rats in 300 ml of acetone with mixing on a magnetic stirrer for 2 min. An aliquot of the solvent bath (0.1 ml) is removed, placed on a tared aluminum weighing pan, and dried at room temperature for 18 hr. The solid residue is weighed on an electronic microbalance. Inhibition of the testosterone-stimulated increase in sebum secretion compared to the vehicle-treated controls is used as the parameter of response. Each group contains eight to ten rats. From the mean sebum weights, the percent inhibition of sebum secretion is calculated by the following formula:

$$\% \text{ Inhibition} = \frac{\text{sebum wt (TP alone)} - \text{sebum wt (TP with test compound)}}{\text{sebum wt (TP alone)} - \text{sebum wt (vehicle control)}} \times 100$$

Sebum from fur. The secretion of sebum can also be monitored by measuring extractable lipid from the fur coat of the rats, thereby avoiding sacrifice and allowing the animals to be cycled into other experiments if needed. The amount of sebum on the fur is quantitated by lipid extraction from a fixed amount of hair.[12] The mid-dorsal region is shaved with an electric hair clipper. One gram of the shaved fur is placed in a glass flask with 50 ml of acetone and stirred with a magnetic stirrer for 3 min. The acetone is removed and saved, and a second 50-ml volume of acetone is added to the flask and stirred for an additional 3 min. Both acetone solutions are combined, and a 0.1-ml aliquot is removed, placed on a tared aluminum weighing pan, and dried at room temperature for 18 hr. The solid residue is weighed on an electronic microbalance. From the mean sebum weights, the percent reduction of sebum secretion in mature (non-testosterone-treated) rats is calculated by the following formula:

$$\% \text{ Reduction} = \frac{\text{sebum wt of controls} - \text{sebum wt of drug treated}}{\text{sebum wt of controls}} \times 100$$

Hamster Flank Organ Weight

The hamster flank organ is commonly used to assess sebaceous gland function and has been fully described by Gomez.[13] Retinoid inhibition of

[13] E. C. Gomez, *in* "Models in Dermatology" (H. I. Maibach and N. J. Lowe, eds.), Vol. 2, p. 100. Karger, Basel, 1985.

flank organ weight may be carried out using mature males or castrated immature males stimulated with testosterone. The use of immature castrated hamsters stimulated with testosterone eliminates some variation and increases the baseline size of the flank organs. It also eliminates the need for the 14-hr light/10-hr dark photoperiod required when using mature hamsters, in which androgen levels are a function of the light cycle.

Flank Organ Response in Testosterone-Stimulated Castrated Hamsters. Male, Golden Syrian hamsters are obtained from the Charles River Breeding Laboratory at 6 weeks of age and castrated. The animals are used 1 week after surgery. Hamsters are administered TP in sesame oil, 100 μg/0.2 ml/day subcutaneously. Concurrently, test compounds are administered orally by intubation in propylene glycol in a fixed volume of 0.2 ml/day for 14 consecutive days. Alternatively, test compounds may be evaluated for topical activity by applying the compounds directly to the flank organs of shaved hamsters. Testosterone propionate dissolved in chloroform (5 μg/0.05 ml) is applied topically to each flank organ. Test compounds are also dissolved in chloroform and administered topically to each flank organ in a volume of 50 μl daily concurrently with TP. Hamsters are treated for 14 consecutive days. On Day 15, the hamsters are sacrificed by CO_2 inhalation.

Flank Organ Response in Mature Male Hamsters. Male, Golden Syrian hamsters are obtained from the Charles River Breeding Laboratory at an age of 3 months and maintained in stainless steel cages with a photoperiod of 14 hr of light and 10 hr of darkness. Hamsters are treated orally by intubation or topically for 4 weeks. Drugs are administered orally in propylene glycol at a fixed volume of 0.2 ml/hamster/day. Drugs are administered topically in acetone at a fixed volume of 50 μl/flank organ/day. On the day after the last drug administration, the hamsters are sacrificed by CO_2 inhalation, and the dorsal skin is shaved with an electric hair clipper, exposing the pigmented paired flank organs. A flap of skin, containing the paired flank organs, is removed from the back by cutting around the perimeter of the flank organs with a pair of surgical scissors. The flank organs are removed and weighed.

Quantitation of Flank Organ Size. The inhibition of flank organ size by retinoids may be quantitated by simply weighing the organ or by planimetry of histological sections of the sebaceous glands.

Flank organ weight. A flap of skin, containing the paired flank organs, is removed following sacrifice from the back by cutting around the perimeter of the flank organs with a pair of surgical scissors. The flank organs are removed from the skin flap by holding the flap with a pair of mouse tooth forceps while the glands are gently dissected out with a scalpel. Flank organs are weighed on an electronic balance. The percent inhibition is calculated by the following formula:

$$\% \text{ Inhibition} = \frac{\text{flank organ wt (TP alone)} - \text{flank organ wt (TP with test compound)}}{\text{flank organ wt (TP alone)} - \text{flank organ wt (vehicle control)}} \times 100$$

Sebaceous gland area. At the conclusion of the experiment, the flank organs are removed from the skin by cutting around the perimeter of each gland, placed in glass scintillation vials, and fixed with 10% buffered formalin. Twelve cross sections are made through each flank organ and stained with a standard hematoxylin and eosin stain. Sebaceous gland area is determined by planimetry and expressed in square microns.

$$\% \text{ Reduction} = \frac{\text{gland area of controls} - \text{gland area of drug treated}}{\text{gland area of controls}} \times 100$$

Hamster Ear Sebaceous Gland

The ventral sebaceous glands of Syrian hamster ears have been thoroughly described by Plewig and Luderschmidt.[7] The glands are large and resemble human sebaceous glands. Retinoid inhibition of sebaceous gland size in hamster ears may be evaluated in either of two ways: cross-section analysis[7] or the whole-mount analysis method described by Matias and Orentreich.[14]

Hamster Sebaceous Gland Response. Golden Syrian male hamsters are obtained from the Charles River Breeding Laboratory at 3 months of age and maintained in stainless steel cages with a photoperiod of 14 hr of light and 10 hr of darkness. Test compounds are administered orally by intubation or topically to the ears for 4 weeks. Drugs are administered orally in propylene glycol at a fixed volume of 0.2 ml/hamster/day. Drugs are administered topically in acetone at a fixed volume of 100 μl/ear/day. At the conclusion of the study, the hamsters are sacrificed by CO_2 inhalation.

Quantitation of Sebaceous Gland Size

Planimetry of cross sections. After the hamsters have been sacrificed, both ears are removed by cutting across the base of the ear with surgical scissors. The ears are placed into glass scintillation vials and fixed with 10%

[14] J. R. Matias and N. Orentreich, *J. Invest. Dermatol.* **81**, 43 (1983).

buffered formalin v/v. Twelve sagittal sections are made and stained with standard hematoxylin and eosin processing. At least 6 sections per ear are examined, and only ventral glands are measured. Planimetry is used to determine the sebaceous gland size, which is expressed in square microns. To avoid reading tangential cuts through the sebaceous gland, only those sections that have an intact infundibular canal are measured. From the mean sebaceous gland area, the percent reduction is calculated as follows:

$$\% \text{ Reduction} = \frac{\text{gland area of controls} - \text{gland area of treated}}{\text{gland area of controls}} \times 100$$

Whole-mount analysis. A rapid, alternative method for evaluating the size of the sebaceous glands has recently been described by Matias and Orentreich.[14] After the hamsters have been sacrificed, both ears are removed by cutting along the base of the ear with surgical scissors. The ventral side of the ear is separated from the dorsal side by gently pulling the dorsal side away from the supporting cartilage. One pair of small forceps is used to hold the ventral side stationary, while, at the same time, the dorsal side is gently pulled away with a second pair of forceps, starting at the base of the ear and working toward the apical end. The dorsal side of the ear is discarded, and the ventral side is placed into a petri dish and covered with 0.9% saline (w/v) to prevent the tissue from drying out. Any remaining cartilage is then gently scraped away from the inner surface of the ear with the blunt end of a small scissors. The scraped ears are rinsed with saline and placed in scintillation vials with 10 ml of 0.1% Sudan Black B in propylene glycol (w/v). The ears are stained for 18 hr at room temperature, rinsed with 85% propylene glycol/15% distilled water (v/v) and stored overnight in 85% propylene glycol/15% formalin (v/v) at 5°. The next day, a 4-mm biopsy punch is taken through the ears along the midline, 6 mm from the apical end. The ear punch is placed, inner side facing up, on a glass microscope slide. At least six sebaceous glands per ear punch are measured using a digital planimetry system. From the mean sebaceous gland area, the percent reduction is calculated as follows:

$$\% \text{ Reduction} = \frac{\text{gland area of controls} - \text{gland area of treated}}{\text{gland area of controls}} \times 100$$

Discussion

The results of selected retinoids in three *in vivo* models are shown in Figs. 1–3. In the testosterone-treated, immature rat model (Fig. 1), temarotene is about 50 times more active than isotretinoin (13-*cis*-retinoic acid) and etretinate and 10 times more active than acitretin. Figure 2 demon-

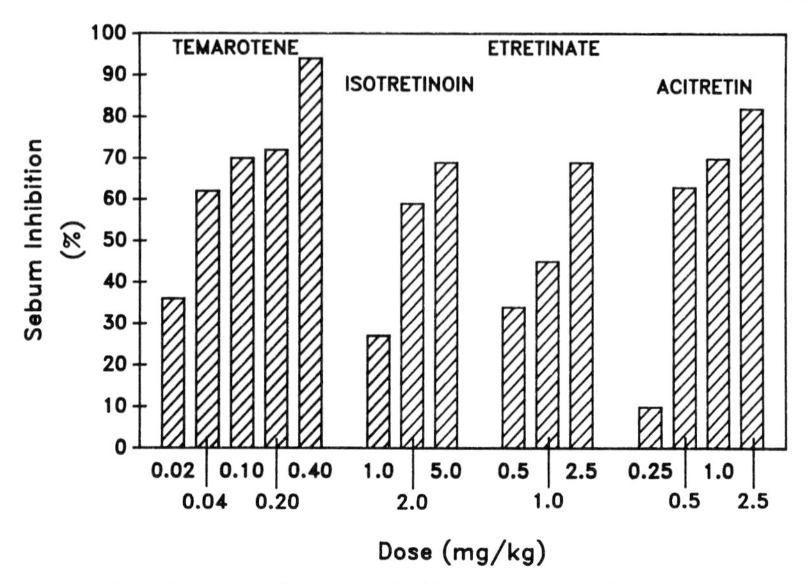

FIG. 1. Effect of orally administered retinoids on sebum secretion in rats. Drugs were administered to testosterone-stimulated, castrated rats as described in the text.

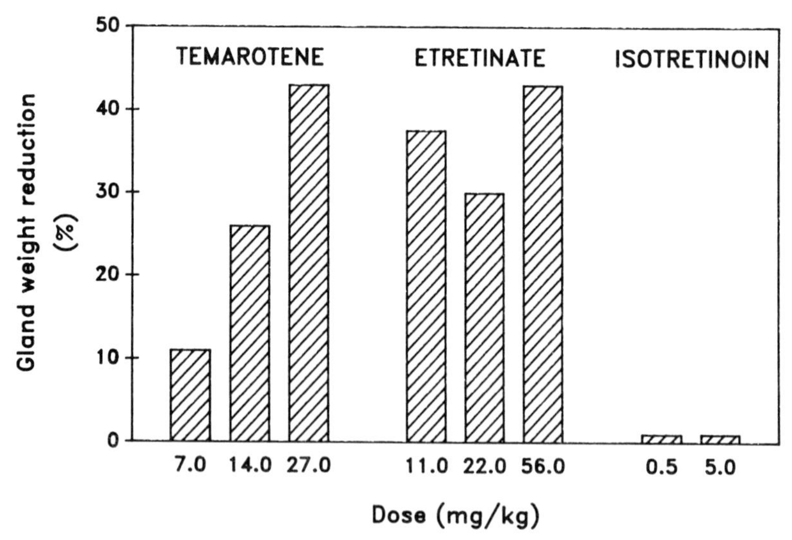

FIG. 2. Effect of orally administered retinoids on hamster flank organ weight. Drugs were administered to testosterone-stimulated, castrated hamsters as described in the text.

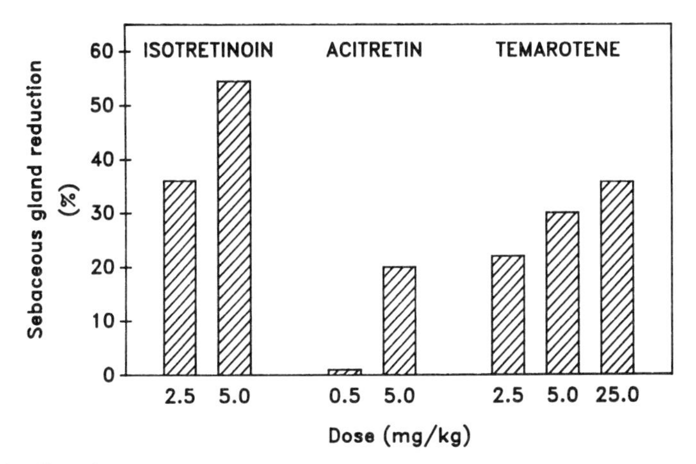

FIG. 3. Effect of retinoids on hamster ear sebaceous gland size. Drugs were administered to mature male animals as described in the text.

strates that temarotene and etretinate are orally active in the testosterone-treated immature hamster flank organ model, whereas isotretinoin is not active orally in this model. It must be emphasized that in the immature models the inhibition of testosterone-induced sebaceous gland hyperplasia is measured, not the decrease in size of a fully developed gland. In the hamster ear model (Fig. 3), isotretinoin was more active than temarotene and acitretin.

In the clinical situation, isotretinoin is orally active in reducing sebaceous gland size and sebum production.[2,4,5] No clinical studies of sebum suppression have been carried out with acitretin; however, etretinate, which is converted to acitretin *in vivo,* has been evaluated and was found to have minimal effects on sebum secretion.[2] Temarotene was found to have little or no effect on sebum secretion in human volunteers.[15,16]

When the results of the three *in vivo* models are compared to the clinical results, the hamster ear model appears to be the most predictive model for retinoid activity in acne. This evaluation is limited by the small number of retinoids that have been tested orally in the clinic for the treatment of acne.

[15] J. S. Strauss, W. P. Davey, S. J. Denton, and A. M. Stranieri, *Arch. Dermatol. Res.* **280,** 152 (1988).

[16] F. M. Vane, S. S. Chari, S. S. Shapiro, K. M. Nordstrom, and J. J. Leyden, *in* "Acne and Related Disorders" (R. Marks and G. Plewig, eds.), p. 183. Martin Dunitz, London, 1989.

[35] Retinoid Effects on Sebocyte Proliferation

By THOMAS I. DORAN and STANLEY S. SHAPIRO

Introduction

Owing to the unavailability of a true acne animal model, a number of assays based on sebum suppression in animals have been developed for antiacne drug discovery.[1-3] These models were predictive for isotretinoin [13-*cis*-retinoic acid, Accutane (Hoffman-La Roche)], which is currently used in acne therapy. However, it now appears that for at least one class of retinoids, namely, arotinoids, such models may not be predictive clinically.[4-6]

There has been considerable activity directed toward the maintenance of sebaceous glands *in vitro* as well as the growth of isolated sebaceous cells. A number of reports have appeared on the metabolic and hormonal activities of human sebaceous glands *in vitro*[7-10] and on the growth of isolated human sebaceous cells *in vitro*.[11-13] Additionally, rodent sebaceous cells, derived from the preputial gland, have been used for *in vitro* metabolic and hormone studies.[14-18] Recently, we reported on the growth of human

[1] E. C. Gomez, *J. Am. Acad. Dermatol.* **6,** 746 (1982).
[2] E. C. Gomez, *J. Invest. Dermatol.* **76,** 68 (1981).
[3] F. J. Ebling and J. Skinner, *Br. J. Dermatol.* **92,** 321 (1975).
[4] A. Boris, J. Hurley, C. Q. Wong, K. Comai, and S. Shapiro, *Arch. Dermatol. Res.* **280,** 246 (1988).
[5] J. S. Strauss, W. P. Davey, S. J. Denton, and A. M. Stranieri, *Arch. Dermatol. Res.* **280,** 152 (1988).
[6] F. M. Vane, S. S. Chari, S. S. Shapiro, K. M. Nordstrom, and J. J. Leyden, *in* "Acne and Related Disorders: An International Symposium" (R. Marks and G. Plewig, eds.), p. 183. London, Martin Dunitz, 1989.
[7] T. Kealey, C. M. Lee, A. J. Thody, and T. Coaker, *Br. J. Dermatol.* **114,** 181 (1985).
[8] D. M. Cassidy, C. M. Lee, M. F. Laker, and T. Kealey, *FEBS Lett.* **200,** 173 (1986).
[9] B. Middleton, I. Birdi, M. Heffrom, and J. R. Marsden, *FEBS Lett.* **231,** 59 (1988).
[10] M. E. Sawaya, L. S. Honig, L. D. Garland, and S. L. Hsia, *J. Invest. Dermatol.* **91,** 101 (1988).
[11] M. Karasek, *In Vitro* **22,** 22A (Abstr. 46) (1986).
[12] M. Johnson and M. Karasek, *J. Invest. Dermatol.* **88,** 496 (Abstr.) (1987).
[13] L. Xia, C. Zouboulis, and C. E. Orfanos, *J. Invest. Dermatol.* **93,** 315 (1989).
[14] J. E. R. Potter, L. Prutkin, and V. R. Wheatley, *J. Invest. Dermatol.* **72,** 120 (1979).
[15] V. R. Wheatley, J. E. R. Potter, and G. Lew, *J. Invest. Dermatol.* **73,** 291 (1979).
[16] V. R. Wheatley and J. L. Brind, *J. Invest. Dermatol.* **76,** 293 (1981).
[17] J. L. Brind, D. Marinescu, E. C. Gomez, V. R. Wheatley, and N. Orentreich, *J. Endocrinol.* **100,** 377 (1984).
[18] R. L. Rosenfield, *J. Invest. Dermatol.* **92,** 751 (1989).

sebaceous cells *in vitro* and their responsiveness to a number of retinoids.[19] This chapter describes a functional human sebocyte assay for evaluating the sebosuppressive activity of retinoids. Based on the results with a limited number of retinoids, this assay appears to correlate with clinical results in acne.

Isolation of Cells and Culture Conditions

Human sebaceous cells are isolated by the method of Karasek.[11] Briefly, tissue removed during cosmetic facelift surgery is washed in several changes of phosphate-buffered saline (PBS), and the top 0.4-mm section containing the epidermis and some of the dermis is removed by a Castroviejo keratotome. The second 0.4-mm section is used as the source of human sebaceous cells. This 0.4-mm section is placed in a solution of 10 mg/ml dispase in Dulbecco's modified Eagle's medium (DMEM) containing penicillin/streptomycin and 10% calf serum (v/v) for 30 min at 37°. The tissue is then placed in 0.3% trypsin/1% EDTA (w/v) in PBS for 15 min at 37° followed by 3 washes in PBS. The tissue is placed in medium containing serum (see below) and scraped vigorously with a scalpel blade. The released cells are counted in a hemacytometer and plated down onto a bed of mitomycin C-treated 3T3 fibroblasts.[20] Sebocytes are plated at a density of $10^4/cm^2$ onto 6-well plates for retinoid experiments. 3T3 fibroblasts are plated at a density of 2×10^4 cells/cm^2.

Cells are cultured in Iscove's medium containing 2% human serum (defibrinated human plasma) (v/v), 8% fetal calf serum (Hyclone Labs, Logan, UT) (v/v), penicillin/streptomycin, glutamine, and dexamethasone (4 μg/ml). Cultures are given fresh medium every 2 days.

Retinoid Preparation

Retinoids are stored as solids at $-20°$. Stock solutions are prepared in dimethyl sulfoxide (DMSO) at 10^{-2} *M* under reduced light and stored under nitrogen at $-20°$ for the duration of the experiment. Stock solutions are checked by high-performance liquid chromatography (HPLC) for stability. Retinoid solutions in medium are prepared each time the cells are given fresh medium.

Determination of Sebocyte Growth Suppression

Growth suppression experiments are terminated before the control cultures reach confluency. Cultures are washed with 0.03% EDTA (w/v) in

[19] T. I. Doran and R. Baff, *J. Invest. Dermatol.* **90,** 554 (Abstr.) (1988).

[20] J. G. Rheinwald and H. Green, *Cell (Cambridge, Mass.)* **6,** 331 (1975).

PBS to remove 3T3 cells. The remaining colonies are trypsinized for 15–20 min to generate a single cell suspension and counted.

Growth of Sebocytes *in Vitro*

The isolation of dermal facial tissue using a keratotome yields a sebaceous gland-enriched section. The sebaceous cells isolated from this tissue grow slowly at first but reach confluency in 10–14 days after seeding at densities of 2×10^4 cells/cm^2 onto a previously established 3T3 fibroblast monolayer. Sebaceous cells, termed sebocytes, can also be grown in KGM medium (Clonetics Corp., San Diego, CA), but the final cell density is lower than that achieved in serum-containing medium. The inclusion of 2% human serum (v/v) or dexamethasone is not necessary for sebocyte growth, but it results in a higher cell density as compared to human serum/dexamethasone-deficient cultures. Hydrocortisone is not able to substitute for dexamethasone as a growth supplement. Additionally, 2% human serum (v/v), as well as the absence of hydrocortisone in the medium, appears to be inhibitory to human keratinocytes.

Sebocytes can be cultured on a polymerized bovine type I collagen, with or without 3T3 cells. However, without 3T3 cells present, dermal fibroblasts eventually proliferate and inhibit sebocyte colony growth. Additionally, the final density of sebocytes grown on collagen without a feeder

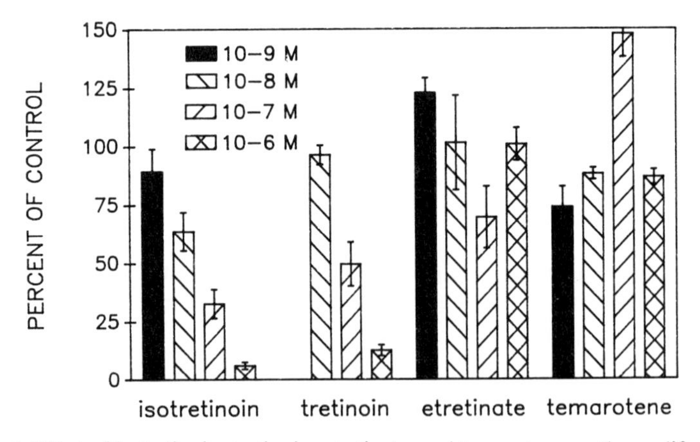

FIG. 1. Effect of isotretinoin, tretinoin, etretinate, and temarotene on the proliferation of human sebocytes *in vitro*. Sebaceous cell were cultured *in vitro* for 9 to 14 days, with retinoids being added 1–2 days after sebocytes were seeded onto the plates. Cell counts (±S.E.M.) are given as the percentage of cells in control (vehicle-treated) cultures and are from duplicate determinations.

TABLE I
RELATIVE ACTIVITIES OF RETINOIDS ON SEBOCYTE PROLIFERATION

Compound	Hypervitaminotic dose[a] (mg/kg)	ED_{50}[b] (μM)	Therapeutic index[c]
Isotretinoin	400	0.05	8000
Tretinoin	80	0.10	800
Etretinate	50	Inactive	ND[d]
Temarotene	>400	Inactive	ND

[a] From W. Bollag, *J. Am. Acad. Dermatol.* **9,** 797 (1983).
[b] ED_{50} values for sebocytes are derived from data given in Fig. 1.
[c] Sebocyte therapeutic index was calculated by dividing the hypervitaminotic dose by the sebocyte ED_{50}.
[d] Not determined owing to inactivity of compound.

layer ($5 \times 10^4/cm^2$) is considerably less than that seen with 3T3 cells present ($9 \times 10^4/cm^2$).

Effect of Retinoids on Sebocyte Proliferation

In an effort to determine the effect that compounds active in several skin disorders have on sebocyte proliferation, sebocytes were grown in the presence of several retinoids (Fig. 1). Isotretinoin was effective in suppressing sebocyte proliferation in a dose-dependent fashion with an ED_{50} of 50 n*M*. Tretinoin (all-*trans*-retinoic acid) also suppressed sebocyte proliferation but was not as effective as isotretinoin, with an ED_{50} of 0.1 μM. Etretinate and temarotene were inactive in suppressing sebocyte proliferation. The ED_{50} values and sebocyte therapeutic indices are given in Table I.

Summary

The human sebocyte model offers several advantages over the current animal models. Foremost among these is the correlation of *in vitro* activity with clinical results, which was not true for arotinoids in the animal models. It is also possible to study several parameters (total cell number, [^{3}H]thymidine uptake, protein and lipid composition/synthesis, hormone response, receptor regulation, etc.) in the same system.

The proliferation of isolated sebocytes is inhibited by retinoids, such as isotretinoin and tretinoin, which are known to be clinically active in human acne. Sebocytes are not responsive to the arotinoid temarotene, which is active in the aforementioned animal models and against dimethylbenz[a]anthracene (DMBA)-induced rat mammary carcinoma[21,22] but inactive clinically in acne.[5,6] Additionally, this model is not responsive to etretinate, a compound known to be active in psoriasis but inactive in acne.[23] The *in vitro* model is, therefore, more predicative of clinical efficacy than the animal models alone.

[21] W. Bollag and H. R. Hartmann, *Eur. J. Cancer Clin. Oncol.* **23,** 131 (1987).
[22] K. Teelmann and W. Bollag, *Eur. J. Cancer Clin. Oncol.* **24,** 1205 (1988).
[23] J. A. Goldstein, A. Socha-Szott, R. J. Thomsen, P. E. Pochi, A. R. Shalita, and J. S. Strauss, *J. Am. Acad. Dermatol.* **6,** 760 (1982).

[36] Effects of Retinoids on Human Sebaceous Glands Isolated by Shearing

By TERENCE KEALEY

Introduction

Isotretinoin, or 13-*cis*-retinoic acid, an effective treatment for acne vulgaris,[1] markedly inhibits human sebum secretion *in vivo*.[2] This is associated, histologically, with a considerable shrinkage of human sebaceous glands,[3] which is not necessarily surprising, as sebum secretion by sebaceous glands is holocrine.[4,5] However, the molecular mechanisms that underlie the effects of retinoids on human sebaceous glands remain to be fully discovered.[6]

Isolated human sebaceous glands should provide the most obvious *in vitro* model for studying the molecular mediators of retinoid action on the glands, but their isolation has not proved easy. They have been isolated by microdissection,[7] but this is a difficult technique of low yield. The glands

[1] G. L. Peck, T. Olsen, F. Yoder, J. S. Strauss, D. T. Downing, M. Pandya, D. Butkus, and J. Arnaud-Battandier, *N. Engl. J. Med.* **300,** 329 (1979).
[2] D. Jones, K. King, A. Miller, and W. Cunliffe, *Br. J. Dermatol.* **108,** 333 (1983).
[3] G. L. Peck, T. Olsen, D. Butkus, M. Pandya, J. Arnaud-Battandier, E. G. Cross, D. B. Windhorst, and J. Cheripko, *J. Am. Acad. Dermatol.* **6,** 735 (1982).
[4] E. H. Epstein and W. L. Epstein, *J. Invest. Dermatol.* **46,** 453 (1966).
[5] G. Plewig and E. Christophers, *Acta Derm. Venereol.* **54,** 177 (1974).
[6] M. Robertson, *Nature (London)* **330,** 420 (1987).
[7] J. B. Hay and M. B. Hodgins, *J. Endocrinol.* **79,** 29 (1978).

are small and white, so they are hard to see against the white dermal collagen of skin biopsies. The glands, moreover, are delicate and easily damaged. These problems have afflicted the microdissection of all the human skin glands and appendages, including human eccrine sweat glands,[8,9] apocrine sweat glands,[10] and hair follicles.[11]

In 1983, Okada *et al.*[12] and Kealey,[13] working independently of each other, showed that the collagenase digestion of human skin biopsies yields viable human eccrine sweat glands. But in 1984 we showed that shearing, which had been a preparatory step in collagenase digestion, was itself sufficient.[14] It would also isolate viable human sebaceous glands,[15] apocrine sweat glands,[16] and hair follicles.[17] The yields are high: a sliver of chest skin removed during cardiac surgery will provide 40 sebaceous glands and 100 eccrine sweat glands within 2 hr of biopsy, and an axillary sliver of skin removed during lymph node biopsy for staging carcinoma of the breast will yield 40 or more apocrine sweat glands within 10 min of biopsy. Shearing, which is no more than the repeated incision of skin biopsies with a pair of scissors, is so improbable a technique that it requires detailed description.

Shearing

Human skin is obtained at surgical operations. We find the sagittal, frontal, midline incisions that cardiothoracic surgeons employ for cardiac operations to be the best source of human sebaceous glands and eccrine sweat glands. This is partly because sebaceous glands are concentrated in the upper trunk (as well, of course, as in the face) and partly because chest surgeons tend to make longer incisions than their abdominal colleagues. Chest skin, moreover, does not age very fast.

Shearing works best on young skin. In donors up to the age of about 20, all sites yield clean glands. As skin ages, however, the glands become more firmly attached to the surrounding dermal collagen. This is presumably a

[8] P. M. Quinton, *Nature (London)* **301,** 421 (1983).
[9] K. Sato and F. Sato, *J. Clin. Invest.* **73,** 1763 (1984).
[10] K. Sato and F. Sato, *Br. J. Dermatol.* **103,** 235 (1980).
[11] R. Frater, *Aust. J. Biol. Sci.* **36,** 411 (1983).
[12] N. Okada, Y. Kitano, and T. Morimoto, *Arch. Dermatol. Res.* **275,** 130 (1983).
[13] T. Kealey, *Biochem. J.* **212,** 143 (1983).
[14] C. M. Lee, C. J. Jones, and T. Kealey, *J. Cell Sci.* **72,** 259 (1984).
[15] T. Kealey, C. M. Lee, A. J. Thody, and T. Coaker, *Br. J. Dermatol.* **114,** 181 (1986).
[16] J. H. Barth, J. Ridden, M. Philpott, M. J. Greenall, and T. Kealey, *J. Invest. Dermatol.* **92,** 333 (1989).
[17] M. R. Green, C. S. Clay, W. T. Gibson, T. C. Hughes, C. G. Smith, G. E. Westgate, M. White, and T. Kealey, *J. Invest. Dermatol.* **87,** 768 (1986).

consequence of the cross-linking that characterizes aging collagen.[18] Older glands, therefore, are not so clean on isolation, and they may need to be cleared of collagen strands by careful teasing with fine microforceps. But chest skin tends not to age so quickly as abdominal skin, perhaps because it is less stretched, and so it is a preferred tissue. Male sebaceous glands, like all the male skin glands and appendages, are bigger than those of the female, and so male patients are preferred.

Before obtaining skin, ethical committee permission must be obtained, as must the consent of the patient and, of course, the cooperation of the surgeon. These are not generally difficult to achieve, because the removal of a sliver of skin from the incision neither affects wound healing nor produces a worse scar.

The incision of skin must be made with a scalpel. Diathermy may damage the glands. The skin sliver, which may be up to $\frac{1}{8}$ inch wide, should penetrate down to the subcutaneous fat, and may extend the length of the incision. The sliver should then be placed into a suitable buffer, such as Krebs–Ringer bicarbonate-buffered medium, phosphate-buffered saline, or Earle's salts. If media are bicarbonate-buffered, the containers should be air-tight. Bicarbonate-buffered media are to be preferred, because they seem to promote better retention of glandular lipogenesis. Perhaps this is because of the bicarbonate dependence of lipogenesis.

The medium should be at room temperature. Ice-cooled medium seems to impair glandular viability, perhaps for the same reason that adipocytes are killed at 4°, namely, that the intracellular fat or sebum solidifies and so ruptures the cell. Either because of the bicarbonate dependence of lipogenesis or because of the temperature sensitivity of the glands, it has been found[19,20] that glandular lipogenesis *in vitro* is much greater after overnight maintenance at 37° in bicarbonate-buffered medium under 5% CO_2 as described below than in freshly isolated glands.

In the laboratory, the skin is trimmed of subcutaneous fat, washed several times in buffer to elute the bactericidal solution that surgeons apply to the skin, and then sheared. Shearing involves repeated cutting of the skin with scissors. The skin is first chopped into small pieces ($\frac{1}{8} \times \frac{1}{8} \times \frac{1}{8}$ inch), and the pieces are placed in a small volume of buffer (which cannot be bicarbonate buffered as it will be exposed to the air for some time). The skin/buffer suspension is then repeatedly chopped with scissors until it looks like porridge. This may take up to 5 min (unless one's hand feels tired, one has not done it enough). Aliquots of the suspension are then

[18] D. J. Prockop, K. I. Kivirikko, L. Tuderman, and N. A. Fuzman, *N. Engl. J. Med.* **301,** 13 and 77 (1979).

[19] D. M. Cassidy, C. M. Lee, M. F. Laker, and T. Kealey, *FEBS Lett.* **200,** 173 (1986).

[20] B. Middleton, I. Birdi, M. Heffson, and J. R. Marsden, *FEBS Lett.* **231,** 59 (1988).

poured into disposable plastic petri dishes, diluted with additional buffer to facilitate microscopy, and placed under a binocular microscope. At magnification ×10, with incident light, the isolated glands can be easily recognized (see Fig. 1). They are picked out gently with fine microforceps (we use Dumont No. 5) and placed in fresh buffer.

Shearing works because of the periglandular capsule of fibroblasts. As Fig. 2 illustrates, skin glands *in situ* are surrounded by very thin fibroblastic cytosolic projections, separated by collagen fibers, and these layers shear over each other under the pressure of scissors. Soft skin, such as that of the neonatal rat, requires the use of blunt scissors because the blades of sharp scissors tend to cut through the skin cleanly, rather than shear it. Human skin is sufficiently tough to necessitate the use of sharp scissors even to shear it. Because the shearing occurs through the capsule, the underlying gland pops out intact, and with no apparent electron micrographic evidence of damage.

Glands isolated by shearing appear to be viable as determined by a wide range of criteria. There include light and electron microscopy;[15] ATP, ADP, and AMP contents;[15] and the rates and patterns of lipogenesis.[19,20]

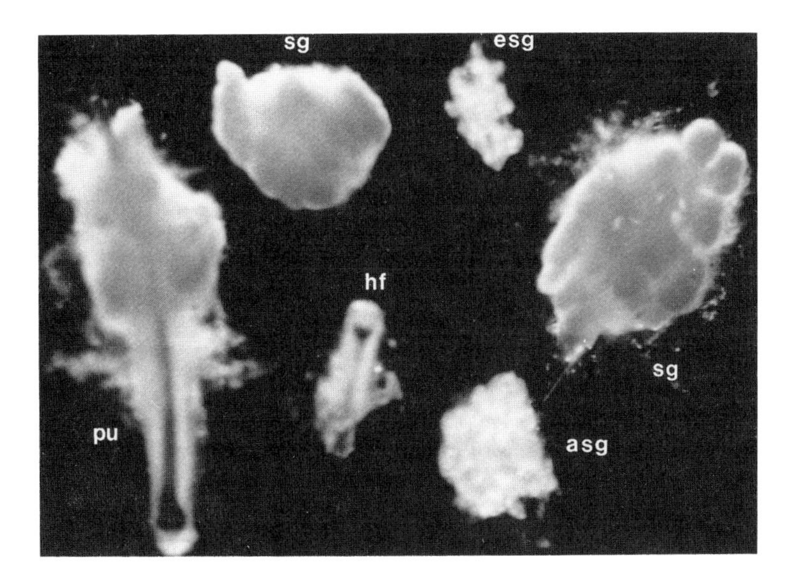

FIG. 1. Isolated human skin glands and appendages. One pilosebaceous unit (pu), two sebaceous glands (sg), one hair follicle (hf), an apocrine sweat gland (asg), and an eccrine sweat gland (esg) can be seen. Magnification: ×14. [Reproduced by permission from *Acne and Related Disorders* (R. Marks and G. Plewig, eds.), Martin Dunitz, London, 1989.]

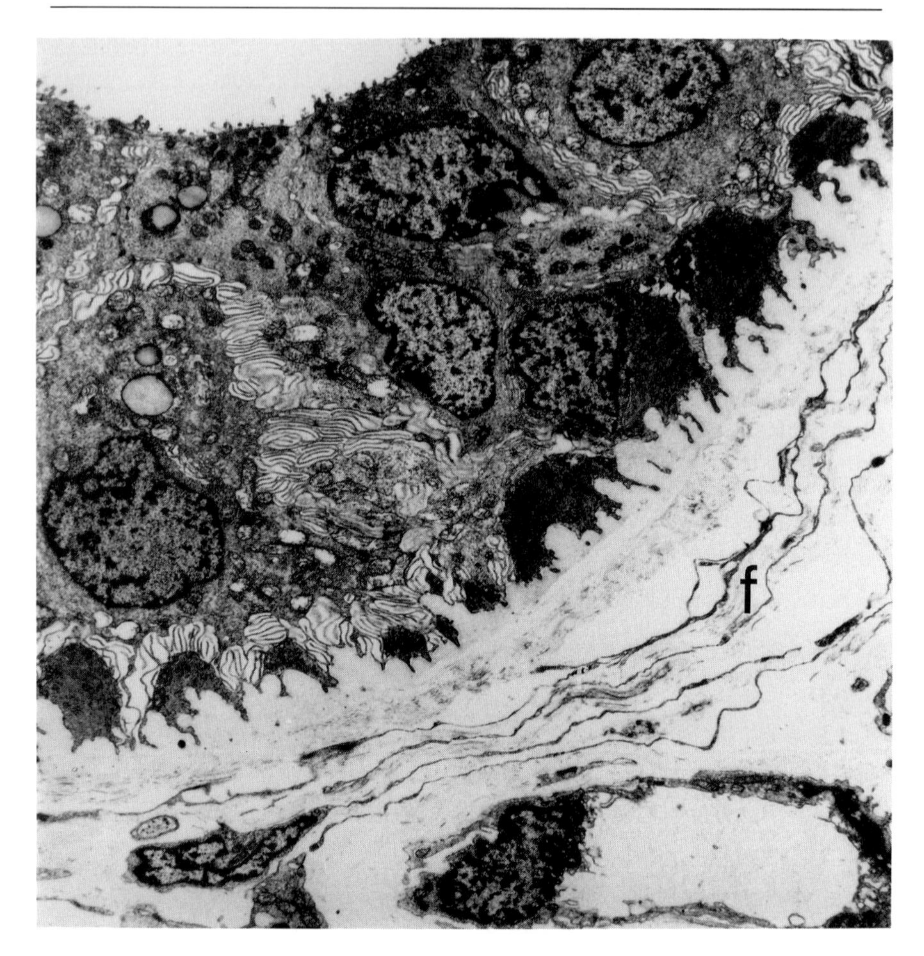

FIG. 2. Electron micrograph of a typical human skin gland or appendage (in this case an eccrine sweat gland) *in situ.* Outside the basal lamina, three or four connective rings of fibroblasts (f) can be seen. Shearing occurs between them. Magnification: ×2,800. [Electron micrograph provided by Dr. C. J. Jones. Reproduced by permission from *Epithelia: Advances in Cell Physiology and Cell Culture* (C. J. Jones, ed.), Kluwer Academic, Lancaster, 1989.]

Effects of Isotretinoin on Human Sebaceous Glands

Sebum secretion is holocrine; sebocytes originate at the periphery of the sebaceous gland lobule; and as they grow they accumulate lipid and move centrally until, after 7 to 28 days,[4,5] the cell dissolves, liberating sebum into the sebaceous duct. It is probable, therefore, that isotretinoin, which inhibits sebum secretion dramatically *in vivo,*[2] acts by inhibiting sebocyte cell turnover rather than by inhibiting lipid synthesis specifically. The reasons for this conclusion are as follows: (1) isotretinoin shrinks sebaceous glands as determined histologically,[3] and, in a holocrine gland, size and secretion rates are directly proportional;[15] (2) the isotretinoin-induced inhibition of lipogenesis *in vivo* is not immediate, as it takes up to 4 weeks for maximal effect,[2] which is compatible with an inhibition of the turnover of cells whose transit time is 7 to 28 days;[4,5] and (3) the retinoids are known to regulate the rates of division of a large number of different cell types.[21]

To study retinoid effects on sebaceous glands *in vitro,* therefore, it is necessary to culture the glands. We chose organ maintenance rather than cell culture so as to retain the native architecture of the sebocytes, cell–cell interaction, and cell–connective tissue interaction as much as possible. The retinoids can have paradoxical effects on cell turnover, such as stimulating primary cell outgrowths[22] while inhibiting established cell lines,[23] and so we hoped that organ maintenance would help conserve the physiological actions of isotretinoin.

Sebaceous glands are maintained overnight at 37°, on nitrocellulose filters, pore size 0.45 μm, floating on 5 ml Williams E medium supplemented with 2 m*M* L-glutamine, 100 units/ml penicillin and streptomycin, and 2.5 μg/ml Fungizone (Gibco, Grand Island, NY), 10 μg/ml insulin, 10 μg/ml transferrin, 10 ng/ml hydrocortisone, 10 ng/ml epidermal growth factor, 10 ng/ml sodium selenite, 3 n*M* triiodothyronine, trace elements (Gibco), 10 μg/ml prostaglandin E_1, and 10 μg/ml bovine pituitary extract and buffered in a humidified 95% air/5% (*v/v*) CO_2 atmosphere.[24] The medium is supplemented where appropriate with 1 μM isotretinoin and/or 3 μM testosterone. Isotretinoin is dissolved in dimethyl sulfoxide (DMSO), the final concentration of which does not exceed 0.02% (*v/v*). Where isotretinoin is absent from control experiments, DMSO is added to 0.02% (*v/v*), and where testosterone is absent, ethanol is added to 0.002% (*v/v*).

For assays of lipogenesis from 2 m*M* [*U*–^{14}C]acetate, five glands are incubated in 300 μl of bicarbonate-buffered saline in humidified air/CO_2 (95% : 5%, *v/v*) at 37° as described before.[15] Lipids are extracted into chloroform/water and identified by thin-layer chromatography.[25] [*U*–^{14}C]Acetate uptake into lipid is linear over 24 hr. The rates of protein and DNA synthesis from 500 μM [*U*–^{14}C]leucine and 3 μM[*methyl*-^{3}H]thymidine

[21] M. B. Sporn, A. B. Roberts, and D. S. Goodman (eds.) "The Retinoids." Vols. 1 and 2. Academic Press, Orlando, Florida, 1984.

[22] E. Christophers, *J. Invest. Dermatol.* **63,** 450 (1974).

[23] D. M. Kochhar, J. T. Dingle, and J. A. Lucy, *Exp. Cell Res.* **52,** 591 (1968).

[24] J. Ridden and T. Kealey, *J. Cell Sci.* **95**, 125 (1990).

[25] M. F. Cooper, A. J. Thody, and S. Shuster, *Biochim. Biophys. Acta* **360,** 193 (1974).

TABLE I

ORGAN MAINTENANCE OF HUMAN SEBACEOUS GLANDS[a]

Conditions of maintenance	[U-^{14}C]Acetate uptake (nmol/gland/hr)	[*methyl*-^{3}H] Thymidine uptake (fmol/mg wet weight/hr)	[U-^{14}C]Leucine incorporation (pmol/mg wet weight/hr)	DNA content (μg/gland)	Protein content (μg/gland)	Wet weight (mg/gland)
Freshly isolated	569.2 ± 60.3	173.7 ± 45.6	170.3 ± 15.2	2.19 ± 0.25	29.5 ± 3.8	3.1 ± 0.2
No additions	94.2 ± 21.2[b]	159.0 ± 49.6	175.9 ± 20.3	2.03 ± 0.22	25.9 ± 2.8	2.4 ± 0.2[e]
Testosterone (3 μM)	266.3 ± 48.2[b,c]	112.4 ± 39.8	158.1 ± 12.9	2.46 ± 0.27	23.9 ± 2.8	2.8 ± 0.3
Isotretinoin (1 μM) plus testosterone (3 μM)	49.3 ± 21.3[b,d]	28.4 ± 8.1[b]	144.6 ± 14.7	1.45 ± 0.18[e]	19.7 ± 2.6[e]	1.9 ± 0.2[b]

[a]Human sebaceous glands were maintained on Williams E medium at 37° under 5% CO_2/95% air (v/v), supplemented as described in the text. The measurements made on "freshly isolated" glands were made on glands that had been maintained overnight. All other measurements were made after a further 7 days of maintenance, either with no further additions to the medium or with 3 μM testosterone with or without 1 μM isotretinoin. All assays were performed as described in the text. Results are presented as means ± S.E.M. All the studies were made on male glands. With exception of acetate uptake ($n = 5$ subjects), all the observations were made on glands isolated from 12 subjects. Each observation per subject was made on at least 5 glands. The metabolic studies were made after incubation with the radioactive precursor for 3 hr.

[b]Significant reduction as determined by Student's t test, relative to freshly isolated glands ($p < .001$).

[d]Significant difference from testosterone treatment ($p < .01$).

[e]Significance as determined by Student's t test, relative to freshly isolated glands ($p < .05$).

are determined by the incorporation of radioactivity into trichloroacetic acid (TCA)-precipitable material as described before.[14] Uptake is linear over 18 hr.

Table I shows that over maintenance for 1 week, the rate of new sebocyte formation as determined by [*methyl*-^{3}H]thymidine uptake into TCA-precipitable material was unchanged. [*methyl*-^{3}H]Thymidine autoradiography (not illustrated) confirmed that this was restricted, on maintenance, to the basal sebocytes. Testosterone (3 μM) *in vitro* did not significantly alter this rate, but the further addition of 1 μM isotretinoin markedly inhibited new cell division. This was reflected (Table I) in the isotretinoin-induced reductions in sebaceous glandular wet weight and their contents of DNA and protein. This shows that the organ-maintained human sebaceous gland provides a good *in vitro* model for the actions of isotretinoin. Isotretinoin at 1 μM was employed because a dose–response curve (not illustrated) showed that concentration to be the minimum one to provide maximal inhibition.

The current model is not, however, perfect. Table I shows that on maintenance for 1 week, rates of lipogenesis fall to one-sixth of the control values. Control experiments show that this is a progressive fall. Histological experiments show that the new cells which arise on maintenance do not, indeed, differentiate fully, in that they fail to demonstrate an accumulation of lipid. Instead, they appear to keratinize. Nonetheless, some *in vitro* differentiation control can be seen, in that 3×10^{-6} M testosterone stimulates lipogenesis 3-fold, whereas 10^{-6} M isotretinoin markedly reduces lipogenesis by one-fifth. The effect of isotretinoin on lipogenesis is reversible, moreover, in that further maintenance for 1 week in the absence of isotretinoin, but in the presence of 3×10^{-6} M testosterone as the sole addition to the medium, causes a significant ($p < .01$) 3-fold rise in the rate of lipogenesis to 146.8 ± 31.9 pmol acetate/gland/hr (mean $\pm$ S.E.M.; $n = 4$ subjects).

We are currently trying to improve the model. In particular, we are trying different media and a lower calcium concentration, and we are testing the effects of a range of different growth factors such as the transforming growth factors (TGF). Recent testosterone dose curves show that 10^{-9} M, which is physiologically more appropriate, is also more stimulatory of lipogenesis *in vitro* than 3×10^{-6} M. Control experiments show that fetal calf serum inhibits lipogenesis in sebaceous glands *in vitro;* however, we have yet to determine if fetal calf serum inhibits sebocyte turnover or whether its associated free fatty acids inhibit acetyl-CoA carboxylase (EC 6.4.1.2).[26]

[26] A. L. Miller, M. E. Geroch, and H. R. Levy, *Biochem. J.* **118,** 645 (1970).

In conclusion, therefore, we have shown that human sebaceous glands maintained as organs in serum-free medium provide a good *in vitro* model for studying the effect of isotretinoin. The incomplete differentiation of sebocytes, however, requires that the model be further improved.

Acknowledgments

I thank my colleagues Mr. John Ridden, Mr. David Ferguson, and Dr. David Cassidy for their collaboration. I thank Dr. Stan Shapiro and Dr. R. Thieroff-Ekerdt for their invaluable advice and Ms. Rose Maxwell for the typing. This work was supported by the Medical Research Council (U.K.) and Roche. During the course of this research, I was a Wellcome Senior Clinical Research Fellow.

[37] Retinoid Modulation of Phorbol Ester Effects in Skin

By GERARD J. GENDIMENICO, ROBERT J. CAPETOLA, MARVIN E. ROSENTHALE, JOHN L. MCGUIRE, and JAMES A. MEZICK

Introduction

The phorbol ester 12-*O*-tetradecanoylphorbol 13-acetate (TPA) causes a number of biological effects when topically applied to normal skin. A prominent early response induced by TPA is inflammation, manifested as an increase in vascular permeability resulting in edema, which peaks 5 to 6 hr after TPA is applied.[1] The eicosanoids prostaglandin E_2 (PGE_2)[2] and 5-, 8-, 12-, and 15-hydroxyeicosatetraenoic acids[3,4] (HETE) are elevated in TPA-treated skin. Inhibitors of cyclooxygenase and lipoxygenase enzymes suppress edema formation.[5] Epidermal hyperplasia, owing to an increase in the number of epidermal cell layers, occurs maximally 48 to 72 hr after TPA is applied to skin.[6] At the biochemical level, the epidermal hyperplasia is preceded by an increase in ornithine decarboxylase (ODC) (EC

[1] A. Janoff, A. Klassen, and W. Troll, *Cancer Res.* **30,** 2568 (1970).

[2] C. L. Ashendel and R. K. Boutwell, *Biochem. Biophys. Res. Commun.* **90,** 623 (1979).

[3] M. Gschwendt, G. Fürstenberger, W. Kittstein, E. Besemfelder, W. E. Hull, H. Hagedorn, H. J. Opferkuch, and F. Marks, *Carcinogenesis* **7,** 449 (1986).

[4] S. M. Fischer, J. K. Baldwin, D. W. Jasheway, K. E. Patrick, and G. S. Cameron, *Cancer Res.* **48,** 658 (1988).

[5] R. P. Carlson, L. O'Neill-Davis, J. Chang, and A. J. Lewis, *Agents Actions* **17,** 197 (1985).

[6] A. N. Raick, *Cancer Res.* **33,** 1096 (1973).

4.1.1.17) activity, which peaks 4 to 5 hr after TPA application.[7] The induction of ODC leads to increased levels of the polyamines putrescine and spermidine in the epidermis.[8] Epidermal DNA synthesis also increases as early as 12 to 15 hr after TPA is applied.[8]

One of the most well-known effects of TPA is its activity as a potent tumor promoter. In the two-stage model of mouse skin carcinogenesis,[9] skin is treated once with an initiator, most typically 7,12-dimethylbenz[*a*]-anthracene, and then TPA is applied multiple times. Benign epithelial papillomas can develop within 6 weeks of TPA treatment, some of which may develop into invasive carcinomas.

Retinoids antagonize many of the effects of TPA in the skin. They inhibit the induction of ODC[10,11] and suppress the first wave of DNA synthesis.[10,12,13] Although these two parameters are inhibited by retinoids, the TPA-induced epidermal hyperplasia is not reduced by retinoid treatment.[12,13] The production of skin tumors is suppressed by retinoids when they are applied topically prior to TPA application.[14] Retinoids also cause papillomas to regress when administered to mice with papillomas 3 to 4 mm in size.[15] Retinoids also have been reported to interfere with the metabolism of arachidonic acid[16] and thus have the potential to antagonize the cutaneous inflammation induced by TPA.

The purpose of this chapter is to review convenient assay methods for retinoids against phorbol ester-induced effects in the skin. Methods will be described for measuring TPA-induced edema, ODC activity, and DNA synthesis. Readers interested in assay methods for retinoid effects on tumor promotion should consult the works of Verma *et al.*[14] or Mayer *et al.*[15]

[7] A. K. Verma and R. K. Boutwell, *Cancer Res.* **37,** 2196 (1977).

[8] E. G. Astrup and J. E. Paulsen, *Carcinogenesis* **2,** 545 (1981).

[9] A. K. Verma, *in* "Models in Dermatology, Volume 2: Dermatopharmacology and Toxicology" (H. I. Maibach and N. J. Lowe, eds.), p. 313. Karger, Basel and New York, 1985.

[10] E. G. Astrup and J. E. Paulsen, *Carcinogenesis* **3,** 313 (1982).

[11] A. K. Verma, H. M. Rice, B. G. Shapas, and R. K. Boutwell, *Cancer Res.* **38,** 793 (1978).

[12] A. K. Verma, *in* "Retinoids: Advances in Basic Research and Therapy" (C. E. Orfanos, O. Braun-Falco, E. M. Farber, C. Grupper, M. K. Polano, and R. Schuppli, eds.), p. 117. Springer-Verlag, Berlin, Heidelberg, and New York, 1981.

[13] G. J. Gendimenico, X. Nair, P. L. Bouquin, and K. M. Tramposch, *J. Invest. Dermatol.* **93,** 363 (1989).

[14] A. K. Verma, B. G. Shapas, H. M. Rice, and R. K. Boutwell, *Cancer Res.* **39,** 419 (1979).

[15] H. Mayer, W. Bollag, R. Hänni, and R. Rüegg, *Experientia* **34,** 1105 (1978).

[16] C. Fiedler-Nagy, B. H. Wittreich, A. Georgiadis, W. C. Hope, A. F. Welton, and J. W. Coffey, *Dermatologica* **175** (Suppl. 1), 81 (1987).

Methods

Phorbol Ester-Induced Ear Edema

Albino male or female CD-1 mice, 7–9 weeks of age, are used. A 0.005% (w/v) TPA solution is prepared using acetone as the vehicle. For topical studies, the retinoids are dissolved directly in the 0.005% TPA solution. For oral and systemic dosing, retinoids are usually suspended in oily vehicles, such as soybean or sesame oil. Because many retinoids are degraded by light and oxygen, it is best to prepare them fresh under yellow or dim lights immediately before use.

A 20-μl volume of the 0.005% TPA solution (vehicle control) or the 0.005% TPA/retinoid solution is applied to the dorsal left ear of the mouse. The right ear is not treated. For the positive control, a glucocorticoid such as dexamethasone can be used, which is also dissolved and applied in the 0.005% TPA solution. The mice are sacrificed by CO_2 inhalation at 5.5 hr after TPA application, and the left and right ears are removed. A 7-mm punch biopsy is removed from each ear using a dermal punch. The biopsies are weighed, and the difference in ear weight between the right ear and the left ear is calculated.

Antiinflammatory effects of compounds are evident as an inhibition of the increase in ear weight of the TPA-treated ear. The dose-related inhibition of ear edema by topically applied all-*trans*-retinoic acid and 13-*cis*-retinoic acid is shown in Fig. 1.

Ornithine Decarboxylase and DNA Synthesis Assays

Animals. For the ODC and DNA synthesis assays, hairless mice (Skh-1 or HRS/J strains) 6 to 8 weeks of age are the most convenient animals to use because the lack of hair allows for easy topical treatment of dorsal trunk skin with TPA or retinoids. The lack of hair also simplifies the subsequent preparation of skin for biochemical analyses. Female hairless mice are preferable because, in group housing, male mice fight and damage each other's dorsal trunk skin.

Treatment. The methods for treatment of animals with test materials is essentially the same for the ODC and DNA synthesis assays. The dorsal trunk skin of mice is treated topically with 0.017 μmol of TPA in a volume of 0.1 ml. Acetone or ethanol are typically used as vehicles for TPA. These vehicles can also be used for retinoids that are to be administered topically. For oral and systemic dosing, retinoids are usually suspended in oily vehicles, such as soybean or sesame oil. Because many retinoids are degraded by light and oxygen, it is best to prepare them fresh under yellow or dim lights immediately before use.

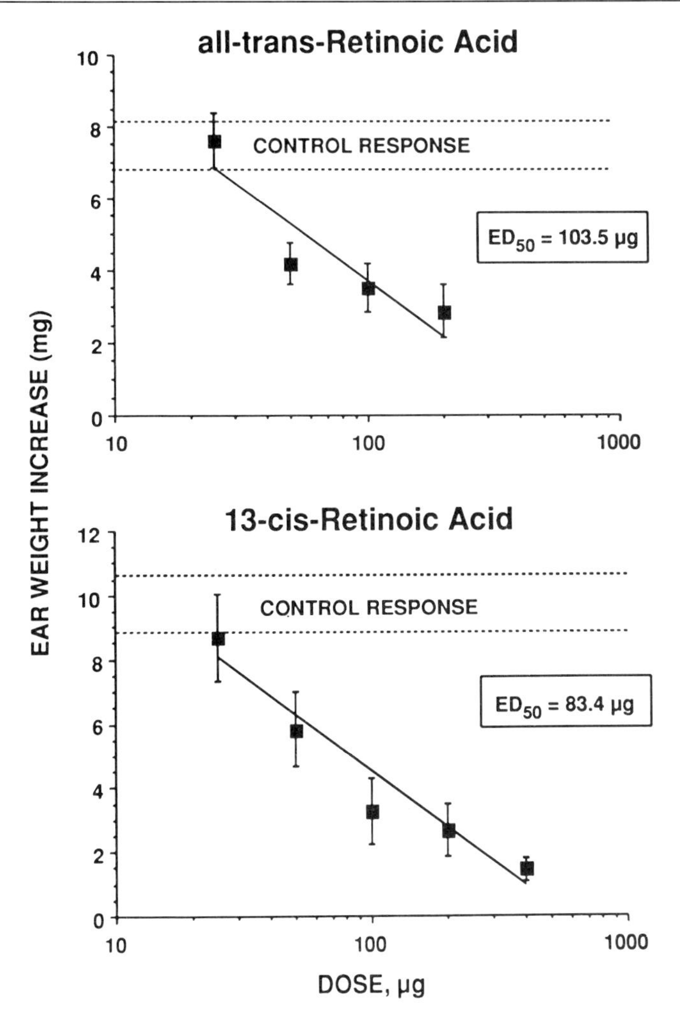

FIG. 1. Dose-related inhibition of TPA-induced ear edema by topically applied all-*trans*-retinoic acid and 13-*cis*-retinoic acid. The mean increases in ear weights (mg) ± S.E. for the vehicle controls in the all-*trans*-retinoic acid and 13-*cis*-retinoic acid groups were, respectively, 7.46 ± 0.65 and 9.67 ± 0.87. Each group contained 10 mice. The ED_{50} values were determined by least-squares linear regression analysis.

The retinoids are generally administered 1 hr before TPA is applied to assure adequate delivery of the retinoid into the epidermis. It has been reported that retinoids will also suppress ODC and DNA synthesis when they are topically applied as late as 1 hr after TPA is applied.[7,13] The time of sacrifice of mice after TPA is applied depends on the parameter being measured. ODC activity peaks at 4 to 5 hr,[7] whereas the midpoint of the first peak in DNA synthesis occurs at 15 hr.[10,12]

After mice are sacrificed, the skins are excised and placed on ice. To facilitate separation of the epidermis from the dermis, each skin is stapled to a piece of index card, and an area of skin is scored with a 1-inch-diameter arch punch. The skin, attached to the card, is immersed in 55° water for 30 sec and cooled briefly in ice water. The epidermis is scraped off within the scored area with a scalpel blade while the skin is resting on a chilled surface. Epidermis is processed for evaluation of ODC activity or DNA synthesis as described below.

Measurement of Ornithine Decarboxylase Activity. The method described by Connor and Lowe[17] is suitable for measuring ODC activity in hairless mouse skin. The epidermis is homogenized in a solution containing 50 m*M* sodium phosphate buffer, 0.4 m*M* pyridoxal 5-phosphate, and 5 m*M* dithiothreitol (DTT) with a Brinkman Polytron tissue grinder. The homogenates are centrifuged at 30,000 *g* at 4° for 30 min, and the supernatant is used for the assay. It is reported that the supernatants can be frozen at −70° for several months without loss of ODC activity.[17]

The assay of ODC is based on the release of $^{14}CO_2$ from L-[^{14}C]ornithine. The supernatant, in a volume up to 0.2 ml, is incubated at 37° for 1 hr with 50 m*M* sodium phosphate buffer, pH 7.2, 5 m*M* DTT, 0.4 m*M* pyridoxal 5-phosphate, 1 m*M* ornithine, and 0.5 μCi L-[^{14}C]ornithine in a final volume of 0.3 ml. The reaction can be carried out directly in a scintillation vial using a modification of the procedure described by Beaven *et al.*[18] The reaction mixture components are placed in a 1.5-ml plastic microcentrifuge tube, and the plastic tube is placed in a glass, screw-capped scintillation vial. Before adding the plastic tube, a piece of filter paper soaked in a CO_2 trapping reagent, such as Protosol, NCS, or Hyamine hydroxide, is placed in the scintillation vial. The vials are capped and the incubation started. To stop the reaction, the vials are opened and 0.5 ml of 2 *M* citric acid is added to the microcentrifuge tube. The scintillation vial is recapped and incubated an additional 60 min at 37°. The plastic tube is removed after the second incubation, and scintillation fluid is added to the glass vial containing the filter paper. Either the Lowry method or the Bio-Rad assay (Bio-Rad Laboratories, Richmond, CA) can be used for determining the amount of protein in the epidermal extract. The ODC activity is expressed as nanomoles $^{14}CO_2$ released per milligram protein per hour.

Measurement of DNA Synthesis. The method used for measurement of

[17] M. J. Connor and N. J. Lowe, *Cancer Res.* **43,** 5174 (1983).

[18] M. A. Beaven, G. Wilcox, and G. K. Terpstra, *Anal. Biochem.* **84,** 638 (1978).

DNA synthesis is that described by Gendimenico *et al.*[13] One hour before sacrifice, each mouse is given 25 μCi of [^{3}H]thymidine, 20 Ci/mmol by intraperitoneal injection in 0.25 ml of 0.9% (w/v) NaCl. The epidermis is homogenized with a Brinkman Polytron tissue grinder in 2.0 ml of ice-cold 0.5 *N* perchloric acid (PCA). The tubes are kept in an ice bath during homogenization. The acid-insoluble material containing DNA is isolated by centrifugation at 800 *g* at 4° and washed 2 times with 0.5 *N* PCA. The acid-soluble fractions are discarded. The DNA is hydrolyzed by resuspending the washed acid-insoluble material in 1.5 ml of 0.5 *N* PCA and heating at 90° for 30 min. The tubes are chilled in an ice bath and the precipitate sedimented by centrifugation at 800 *g* at 4° for 10 min. The supernatant, which contains hydrolyzed DNA, is removed. An aliquot of the supernatant is used to quantitate the amount of [^{3}H]thymidine incorporated into DNA by liquid scintillation spectrometry. An aliquot of the supernatant is also used to quantitate the total amount of DNA using the diphenylamine colorimetric method of Burton[19] or a shortened version of Burton's method.[20] The degree of [^{3}H]thymidine incorporation is expressed as counts per minute (cpm) per microgram DNA.

Comments

Measurements of TPA-induced ear edema, ODC activity, and DNA synthesis provide convenient methods to assess the effects of retinoids on phorbol ester-mediated events in the skin. The antiinflammatory effects of retinoids in the TPA ear edema model may be due to their action as inhibitors of arachidonic acid metabolism. all-*trans*-Retinoic acid and 13-*cis*-retinoic acid are reported to act as inhibitors of snake venom phospholipase A_2 and to inhibit the release and metabolism of arachidonic acid in calcium ionophore-stimulated rat peritoneal macrophages.[16] In TPA-treated mouse skin, phospholipase A_2 activity is elevated[21] and would provide a source of free arachidonic acid for metabolism into PGE_2 and HETE.

A number of structurally diverse retinoids are reported to be active at inhibiting epidermal ODC activity.[11,22,23] Dose-related effects of the retinoids were observed when they were applied topically 1 hr before TPA was

[19] K. Burton, this series, Vol. 12B, p. 163.

[20] G. J. Gendimenico, P. L. Bouquin, and K. M. Tramposch, *Anal. Biochem.* **173,** 45 (1988).

[21] E. Bresnick, G. Bailey, R. J. Bonney, and P. Wightman, *Carcinogenesis* **2,** 1119 (1981).

[22] M. I. Dawson, P. D. Hobbs, R. L. Chan, W.-R. Chao, and V. A. Fung, *J. Med. Chem.* **24,** 583 (1981).

[23] M. I. Dawson, R. L.-S. Chan, K. Derdzinski, P. D. Hobbs, W.-R. Chao, and L. J. Schiff, *J. Med. Chem.* **26,** 1653 (1983).

applied.[11,22] The ED_{50} values for all-*trans*-retinoic acid and 13-*cis*-retinoic acid at inhibiting ODC activity were reported to be 0.12 and 0.24 nmol, respectively.[11] Some retinoids, administered orally at single doses 1 hr before TPA was applied, also caused inhibition of ODC activity.[11] The inhibition of ODC induction by all-*trans*-retinoic acid is reported to be due to reduced transcription of two epidermal ODC mRNAs.[24]

Fewer retinoids have been examined for activity against TPA-induced DNA synthesis.[10,12,13] Representative retinoids from the alicyclic, aromatic, and arotinoid classes caused dose-related suppression of TPA-stimulated DNA synthesis.[13] The ED_{50} values reported were as follows: all-*trans*-retinoic acid, 0.0013%; 13-*cis*-retinoic acid, 0.017%; etretinate, 0.017%; and arotinoid ethyl ester, 0.0001%.[13]

[24] S. Kumar, J. deVellis, N. J. Lowe, and D. P. Weingarten, *FEBS Lett.* **208,** 151 (1986).

[38] Retinoid Effects on Photodamaged Skin*

By GRAEME F. BRYCE and STANLEY S. SHAPIRO

Introduction

Chronic exposure of the skin to sunlight or ultraviolet radiation causes severe damage to the underlying connective tissue, typically manifested in elastosis and alterations in other extracellular matrix macromolecules.[1-4] It has come to be accepted that such actinic damage is a major cause of the more rapid "aging" of exposed areas of the skin compared to sun-protected areas.[5] The UV-irradiated hairless mouse is now the species of choice in many laboratories for a variety of photobiological investigations. The action spectrum and time course for UV-induced erythema in this mouse are

*Portions reprinted by permission of Elsevier Science Publishing Co., Inc., from "Retinoic Acids Promote the Repair of the Dermal Damage and the Effacement of Wrinkles in the UVB-Irradiated Hairless Mouse," by G.F. Bryce, N.J. Bogdan, and C. C. Brown, *The Journal of Investigative Dermatology*, Vol. 91, pp. 175-180. Copyright © 1988 by The Society for Investigative Dermatology, Inc.

[1] L. H. Kligman, F. J. Akin, and A. M. Kligman, *J. Invest. Dermatol.* **78,** 181 (1982).
[2] W. M. Sams, J. G. Smith, and P. G. Burk, *J. Invest. Dermatol.* **43,** 467 (1964).
[3] J. M. Knox, E. G. Cockerrel, and R. G. Freeman, *JAMA, J. Am. Med. Assoc.* **179,** 630 (1962).
[4] A. M. Kligman, *JAMA, J. Am. Med. Assoc.* **210,** 2377 (1969).
[5] R. M. Lavker, *J. Invest. Dermatol.* **73,** 59 (1979).

similar to the sunburn response in humans;[6] UV-induced connective tissue damage is also similar to that observed in humans[1] and has been described in detail.[7] Once considered to be irreversible, this damage has been observed to undergo a substantial degree of spontaneous repair; on cessation of the UV irradiation, a band of new dermal tissue is laid down in the immediate subepidermal region[1,8] compressing the old elastotic tissue. Thus, the measure of repair is reflected in the width of this "zone of reconstruction."[9] Of particular interest was the finding that topically applied all-*trans*-retinoic acid, in a dose-dependent manner, greatly augmented the rate of the repair process.

Chronic exposure of hairless mice to UVB radiation (290–320 nm), in amounts that cause dermal elastosis and damage, leads to the appearance of wrinkles.[10] These take the form of regularly spaced furrows on the dorsal surface. Because these wrinkles do not disappear on gentle stretching, they are distinguished from the array of regularly spaced lines seen on nonirradiated animals at sites where the excess skin normally lies in folds. The term "permanent wrinkle" has been applied to such deep wrinkles in sun-exposed human skin.

Animal Treatment Schedules

Hairless mice (female, HRS/J strain, Jackson Laboratories, Bar Harbor, ME; 5–7 weeks old at the start of the experiments) are housed under yellow light and irradiated 3 times per week with a bank of eight Westinghouse Sunlamps (FS40) placed about 20 cm above the animals. The radiation dose is controlled by an International Light Model IL844A Phototherapy Exposure Control and a Model SEE240 detector. The UVB dosing schedule is such that individual doses, seldom exceeding 0.06 J/cm^2, cause distinct erythema but no burning or scarring. There is significant elastosis, detectable by histology, after a total dose of about 3.5–4.0 J/cm^2 (accumulated over a period of 5–6 months); this can be confirmed by measurements of elastin in whole skin by means of a radioimmunoassay for desmosine, an elastin-specific amino acid found in hydrolyzates of elastin and considered to be a reliable index of total elastin content.

[6] C. A. Cole, R. E. Davies, P. D. Forbes, and L. C. D'Aloisio, *Photochem. Photobiol.* **37,** 623 (1983).

[7] L. H. Kligman and A. M. Kligman, *in* "Models in Dermatology" (H. Maibach and N. J. Lowe, eds.), p. 59. Karger, Basel, 1985.

[8] L. H. Kligman, F. J. Akin, and A. M. Kligman, *J. Invest. Dermatol.* **81,** 98 (1983).

[9] L. H. Kligman, H. D. Chen, and A. M. Kligman, *Connect. Tissue Res.* **12,** 139 (1984).

[10] G. F. Bryce, N. J. Bogdan, and C. C. Brown, *J. Invest. Dermatol.* **91,** 175 (1988).

Desmosine Content

Desmosine is determined by a modification of the procedure of King *et al.*[11] using an antiserum kindly provided by Dr. Barry Starcher (University of Texas Health Center, Tyler, Texas). A 4-mm punch of whole skin, free from subcutaneous fat, is heated at 105° for 48 hr in a sealed tube containing 1 ml of constant boiling HCl (Pierce, Rockford, IL). After evaporation to dryness, the residue is taken up in 1 ml water and stored at −70° until assay. The tracer is I^{125}-labeled desmosine Bolton–Hunter reagent (Elastin Products, Inc., Pacific, MO 63069), which is prepared according the manufacturer's instructions. Standard curves are set up covering a range of 0.02 to 5.0 ng desmosine, and appropriate amounts of sample are added to a dilution of antiserum which binds about 40% of the tracer. After incubation overnight at 4°, bound desmosine is separated by precipitation with 50% ammonium sulfate and quantitated by γ counting. Calculations of desmosine content can be done by conventional methods using a program written to interpolate values from the linearized standard curve. Results are expressed as nanograms desmosine in 1 mm^2 of skin. The sensitivity of the assay run with about 3000 dpm of freshly prepared tracer is of the order of 0.02 ng.

Typically, with the described irradiation schedules, desmosine increases by about 2-fold after a dose of approximately 3.5 J/cm^2 of UVB (6.7 ± 0.4 versus 3.5 ± 0.6 ng/mm^2, $p < .01$), which is consistent with the histological appearance. When the irradiation is discontinued, no further changes in desmosine are observed either in control (7.3 ± 0.6) or treated groups (6.6 ± 0.4); this is also consistent with the histological appearance. Whole skin (dermis plus epidermis) is used as it has been shown that epidermis contains no detectable desmosine.

Retinoid Treatment

To effect repair of the dermal damage, the UVB irradiation is discontinued, and the animals are divided into groups of approximately eight and treated 3 times per week with various concentrations of the retinoids dissolved in acetone. Stock solutions are made up fresh every week in subdued light at concentrations such that the dose is delivered in a 100-μl volume and applied topically with a plastic pipette to an area of about 10 cm^2 on the back of the animal. All dosing is done under yellow light. A control group treated with acetone alone is included. After 10 weeks of treatment the animals are sacrificed and skin samples taken and processed by standard methods.

[11] G. S. King, V. S. Mohan, and B. C. Starcher, *Connect. Tissue Res.* 7, 263 (1980).

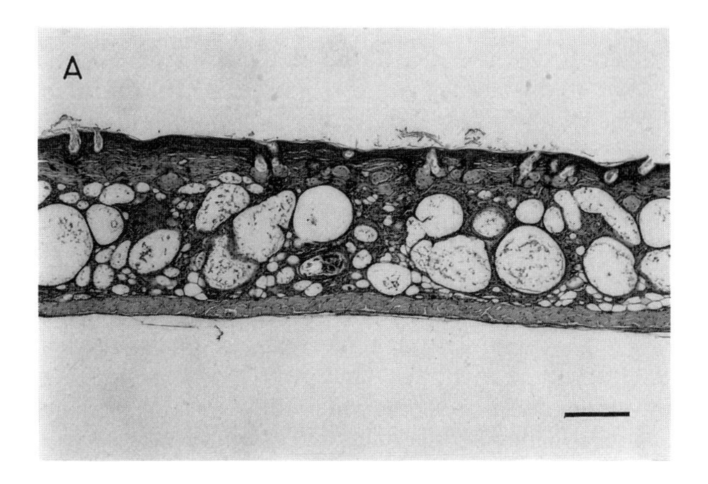

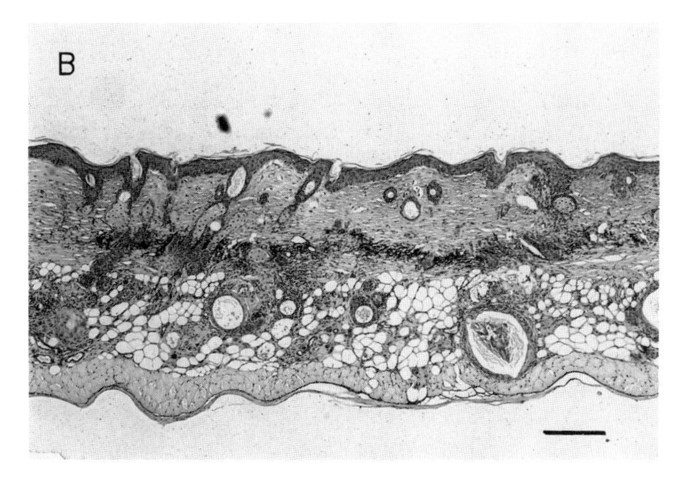

FIG. 1. Histological sections of tissue from UVB-irradiated hairless mice treated with either vehicle (A) or 100 μg 13-*cis*-retinoic acid (B). Luna stained, $\times$40. Bar: 200 μm.

Histology

Two-centimeter strips of dorsal skin, taken laterally across the center of the irradiated (and treated) area, are fixed in 10% formalin, embedded in paraffin, and sectioned at 6 μm. Hematoxylin/eosin is used for routine examination of the tissue; elastin fibers are stained with Luna's aldehyde fuchsin[12] and collagen by Van Gieson. A typical example of tissue from a

[12] L. H. Kligman, *Am. J. Dermatopathol.* **3,** 199 (1981).

treated animal is shown in Fig. 1 (Luna stained). In this model, repair is defined by the appearance of a normalized dermis extending from the epidermis down to the layer of compressed elastin. The extent of repair is reflected by the width of this zone. In these studies, because the width of the zone varies considerably, the area of the zone on a standard length of histological section is measured by an image analyzer. Twenty-four fields, each of which, at magnification ×100, represents a section of tissue 0.57 mm long, are examined and the area calculated. The results are reported as the average area per field in square millimeters. Thus, the average width of the repair zone can be obtained by dividing this average area by 0.57 mm. Areas of substantial repair are interspersed with areas with no discernible repair. This focal nature of the effect is reflected in large standard deviations and increments which sometimes do not reach statistical significance. All microscopic fields are included in the calculations of average area (or width). Data are analyzed by Student's *t*-test.

Wrinkle Measurements

Skin replicas are taken of the lower dorsal area by means of SILFLO (Flexico Developments Ltd., Potters Bar, England). The animals are anesthetized with Chloropent (Fort Dodge Laboratories, Fort Dodge, IA) (0.1 ml/animal, intraperitoneally). An adhesive electrode ring (Novametrix, Wallingford, CT) with a 1-cm-diameter hole is placed on the surface of skin that had been irradiated and/or treated, and the SILFLO is applied according to the manufacturer's instructions. It is possible to place the adhesive ring on the same part of each animal's back by positioning it relative to the base of the tail. Wrinkles appear in these impressions as ridges and cast a shadow when illuminated with low-angle light. A characteristic of the wrinkling pattern is the occurrence of furrows in a regularly spaced array about 2–3 mm apart (Fig. 2). The extent of wrinkling is assessed using this line pattern by assigning values (the wrinkle index), on a scale of 0 to 4, to the width of shadow, which is proportional to the depth of the wrinkle, with 0 representing the absence of wrinkling and 4 the greatest degree of wrinkling. Samples are read in a blind manner by two independent observers, and the results are pooled. In some instances impressions can be taken from the same animal at separate times; the period between sampling should be about 2 weeks, which allows time for the mild irritation caused by the procedure to subside.

Effect of Retinoic Acid Treatment

Figure 1 illustrates the histological appearance of skin from UVB-irradiated animals treated with either vehicle (A) or 100 μg of 13-*cis*-retinoic

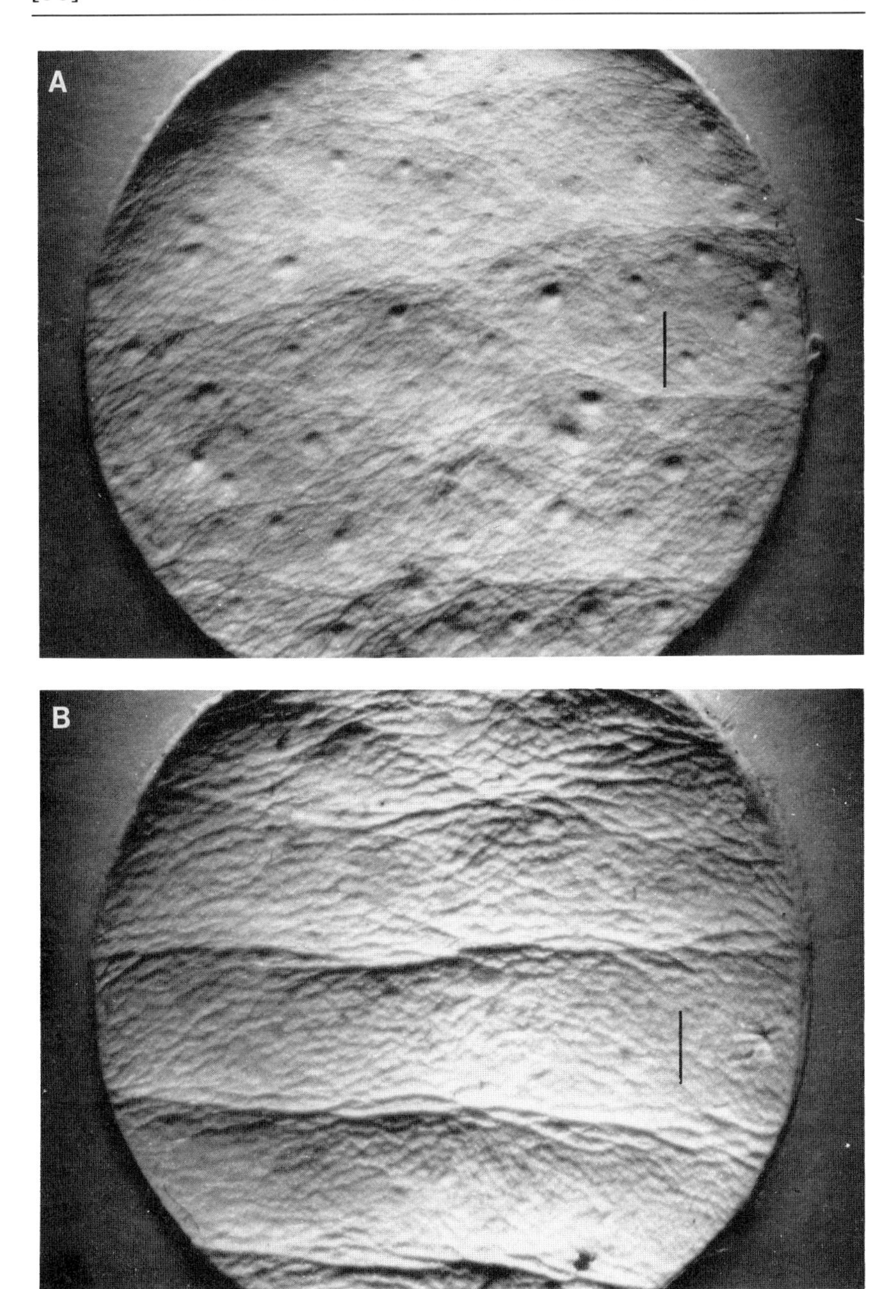

FIG. 2. Skin surface impressions of hairless mice taken from unirradiated skin (A) and from skin irradiated with 4 J/cm^2 UVB (B). Bar: 1 mm.

acid (B). There is negligible "repair" in the vehicle-treated tissue, and the elastin is found over the entire dermis, particularly at the epidermal/dermal junction. In contrast, the "repair zone" in the treated tissue varies in width from about 100 to 220 μm. In the hairless mouse, the nonirradiated epidermis is about 16 μm thick and increases to about 32 μm after UVB irradiation. Further increases are induced by the action of retinoids, with values typically in the range of 35–50 μm (Fig. 1). In some instances the epidermis has projections into the dermis, usually associated with sebaceous glands, but their low frequency does not significantly affect the overall measurements of the repair zone width. Total skin thickness (dermis plus epidermis) is about 440 μm in nonirradiated animals, increasing to about 570 μm after UVB irradiation. Retinoid treatment causes further thickening to values in the range of 800–1000 μm. Thus, despite epidermal hyperproliferation, the major cause of the skin thickening is the dermal fibroplasia.

all-*trans*-Retinoic acid and 13-*cis*-retinoic acid produce dose-related increments in the area of the dermal repair zone (Fig. 3). 13-*cis*-Retinoic acid appears less potent but, at equivalent doses, less irritating. The maximum response may be higher for the 13-cis isomer by virtue of the higher dose that could be used. The width of the repair zone increases from about 20 μm in controls to about 70 μm at the highest doses of retinoids (Fig. 3).

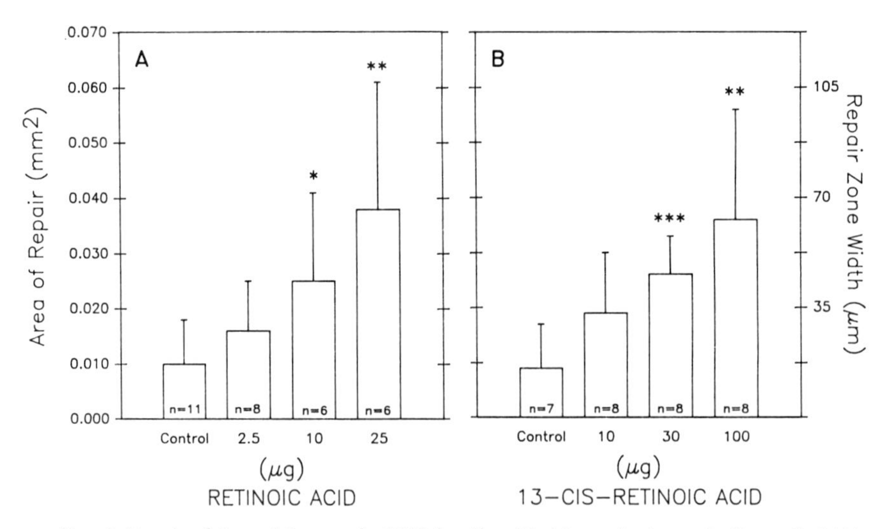

FIG. 3. Repair of dermal damage in UVB-irradiated hairless mice by topically applied (A) all-*trans*-retinoic acid and (B) 13-*cis*-retinoic acid. Animals were treated 3 times per week for 10 weeks. Numbers within the bars denote the group sizes. *$p < .05$, **$p < .01$, ***$p < .001$.

all-*trans*-Retinoic acid is very toxic to the animals at doses higher than 25 μg (previous studies had shown that, in our hands, a dose of 50 μg was lethal after 4 weeks of treatment; with the same dosing schedule, up to 400 μg of 13-*cis*-retinoic acid was tolerated for up to 10 weeks).

After 6 to 10 weeks of treatment with topical retinoic acids, the line pattern is much less conspicuous, and the skin reverts to a normal appearance. A dose–response effect can be demonstrated in the wrinkle effacement by quantitating the intensity of the line pattern. A monotonic response is observed, with significant differences produced by 25 μg of all-*trans*-retinoic acid and by 30 and 100 μg doses of 13-*cis*-retinoic acid. As an index of relative potency, the doses which give 50% reduction in wrinkling can be estimated to be approximately 8 and 15 μg, respectively, in the range in which effects on dermal damage were observed.

In some experiments surface impressions were recorded after 6, 8, or 10 weeks of treatment. At the lower doses of retinoid, namely, 10 μg of 13-*cis*-retinoic acid, a definite time dependence is observed in the extent of wrinkle effacement. Moreover, the rate of decrease in the wrinkle index is more than 3-fold greater for treated skin than for control. When the animals are examined 10 weeks after retinoic acid treatment has been discontinued, the skin surface impressions show that the wrinkles have remained effaced, and no return to pretreatment appearance is evident.

Summary

The effects of retinoid treatment on wrinkling in the hairless mouse can be understood in the context of the repair of the dermal elastosis. The two isomers of retinoic acid do not differ qualitatively in their effects on the histological appearance of the tissue or on the wrinkling patterns produced. The all-trans isomer is slightly more potent in this system than the 13-cis isomer but substantially more irritating, which may limit the maximum degree of repair attainable. The "reconstructed" dermis is thickened, it contains new collagen as a result of the stimulation of gene expression, and the tangled, disorganized elastin is packed into a thin layer in the lower dermis. Thus, the framework within which a wrinkle had been established is eliminated, and the skin assumes a normal state, as observed. That the effacement is apparently permanent is additional evidence of the relationship between the integrity of the elastic fiber network and the surface appearance. The only difference in the repaired skin is the absence of filamentous surface features. The thickening effect of UVB and subsequent retinoid treatment on the epidermis does not contribute substantially to the overall thickness of the repaired skin. Nevertheless, despite having a minor role in the effacement of deep wrinkles, these epidermal changes evidently preclude the formation of fine surface features.

The model is a valid one for the repair of photodamaged skin. From what is known about the role of elastin in maintaining skin integrity and from the association of wrinkling with excessive sun exposure, it is encouraging to observe the dual effect of retinoic acids. The smoothed appearance of the skin is not a transient, cosmetic adjustment but rather a return to a normal state as a result of fundamental biochemical changes occurring throughout the dermis.

[39] Effects of Topical Retinoids on Photoaged Skin as Measured by Optical Profilometry

By GARY L. GROVE and MARY JO GROVE

Introduction

The extensive family of retinoids has been important in the treatment of an impressive variety of dermatologic disorders. The therapeutic effects of topical tretinoin on unrelated, diverse disorders of keratinization have been long recognized. Indeed, retinoic acid has been important in the topical treatment of acne vulgaris for more than 20 years. Recently, tretinoin has caused a good deal of public attention owing to its purported antiaging effects. That this might be so was first suggested by Kligman's group,[1] who observed that postadolescent females with persistent acne often described smoother, less wrinkled skin after several months of treatment with topical tretinoin. This anecdotal information has been followed up by more carefully controlled investigations by our group[2] and others.[3-6]

Although these reports have indicated that topical tretinoin can be helpful in modifying many of the changes associated with photoaging, some criticism of the testing methodology has been made.[7] The primary concern that has been raised is whether the assessments can be made by the expert grader in a truly blinded fashion. For example, in the original Michigan study[3] 14 of the 15 patients who applied 0.1% tretinoin cream to

[1] A. M. Kligman, G. L. Grove, R. Hirose, and J. J. Leyden, *J. Am. Acad. Dermatol.* **15,** 836 (1986).

[2] J. J. Leyden, G. L. Grove, M. J. Grove, G. Thorne, and L. Lufrano, *J. Am. Acad. Dermatol.* **21,** 638 (1989).

[3] J. S. Weiss, C. N. Ellis, J. T. Headington, T. Tincoff, T. A. Hamilton, and J. J. Voorhees, *JAMA, J. Am. Med. Assoc.* **259,** 527 (1988).

the face on a daily basis showed some degree of overall improvement. The greatest changes noted at the end of 6 months of treatment was a decrease in fine wrinkling, followed by sallowness, tactile roughness, and coarse wrinkling. Unfortunately, more than 90% of this treated group developed a characteristic retinoid dermatitis that provides a potential bias which has led to the aforementioned criticisms. Even in situations where the incidence and severity of retinoid dermatitis in the treated group are reduced, we are very much concerned that concurrent improvements in other facial attributes which can result from effective retinoid therapy, such as decreased intensity of solar lentigines and an increased dermal blood flow, might partially unblind or unduly influence the investigator in judging the impact of the drug on wrinkles. We are also very much aware that, even under the best of conditions, clinical assessments made either in real time or from pre- versus postphotographic comparisons suffer from difficulties caused by minor changes in lighting or facial expression which can greatly influence the appearance of lines and wrinkles. Thus, there is a clear need for more objective methods for evaluating the effects of retinoids on photodamaged skin.

It has been known for some time that silicone rubber impression materials can be used to make a mold of the skin that faithfully captures the surface topography.[8] In this chapter, we review how digital image processing can be used as a quantitative method for assessing the microtopographic features in such specimens. This approach is noninvasive and allows one to serially monitor changes in skin surface features with extended periods of time passing between observation points. Although this instrumental method is the primary focus, we do not overlook the use of expert grader ratings and patient self-appraisals as effective methods for assessing such changes. Indeed, we feel that a three-pronged approach considering all three aspects represents the best strategy.[9]

---

[4] C. N. Ellis, J. S. Weiss, and J. J. Voorhees, *J. Cutaneous Aging Cosmet. Dermatol.* **1,** 33 (1989).

[5] J. S. Weiss, C. N. Ellis, J. T. Headington, and J. J. Voorhees, *J. Am. Acad. Dermatol.* **19,** 169 (1988).

[6] M. T. Goldfarb, C. N. Ellis, J. S. Weiss, and J. J. Voorhees, *J. Am. Acad. Dermatol.* **21** (Part 2), 645 (1989).

[7] C. N. Ellis, J. S. Weiss, T. A. Hamilton, and J. J. Voorhees, *JAMA, J. Am. Med. Assoc.* **259,** 3273 (1988).

[8] G. L. Grove and M. J. Grove, *in* "Cutaneous Investigation in Health and Disease" (J. L. Leveque, ed.), p. 1. Dekker, New York, 1989.

[9] G. L. Grove, *Bioeng. Skin* **3,** 359 (1987).

Procedures for Preparing Skin Surface Impressions

If one pays proper attention to details, it is a relatively simple matter to obtain high-quality impressions of the skin surface. The patients should be placed in a recumbent position (Fig. 1) so that they are comfortable while remaining still for at least 5 min. They should also be instructed to keep their eyes closed while relaxing their faces as much as possible. In the studies described below the sampled site was in the periorbital region just lateral to the commissure of the eyes and 5 mm from the external raphe. This site, which represents the "crowsfeet" region, was delineated by affixing 15-mm-diameter adhesive paper rings (Novametrix, Wallingford, CT). Because of the anisotropic nature of the skin surface markings, it is extremely important to carefully orient the tab to face outward, toward the ear. To facilitate relocating this site for subsequent serial samples, closeup 35 mm color photographs should be taken of the region with the adhesive ring properly placed for each individual subject.

Negative skin surface impressions can be obtained using a variety of materials. For our purposes we have found Silflo silicon rubber dental impression material (Flexico Developments Ltd., Potters Bar, England) to be quite satisfactory. A thin layer of freshly prepared Silflo is gently spread

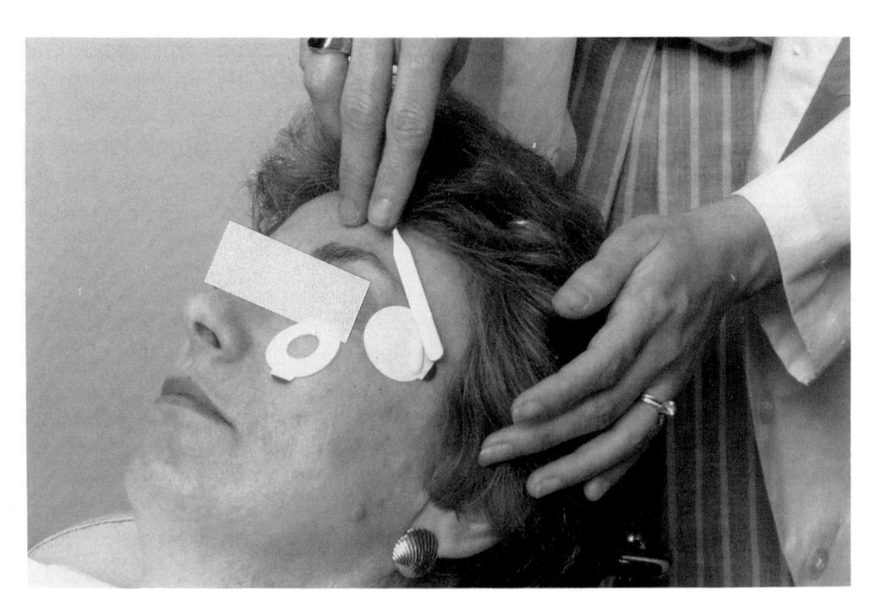

FIG. 1. Subject with crowsfeet site delineated with adhesive paper rings with tab oriented to face outward.

over the bounded area of the ring with a spatula and allowed to polymerize. This should occur within 3 to 4 min, after which time the ring can be lifted from the skin, carrying away the replica with it. Each specimen should be coded and stored in individual glassine envelopes until measured.

Optical Profilometric Methods

The Magiscan digital image processing system employed in this study has been previously described.[10] This instrument consists of a high-resolution, black and white video camera which is interfaced with a computer that contains specially designed image-processing hardware and software. In this application, we use a program entitled Optical Profilometry that is based on the remote sensing approach to lunar landscape mapping utilized during the Ranger missions of NASA.[11,12] Instead of using the sun to sidelight the moon craters and crevices, we employ a fiber optic illuminator set at a fixed angle to bring out the skin surface details. The resulting image is then digitized into a 256×512 pixel matrix with 64 gray levels for brightness.

If the gray level values across a horizontal segment of this digitized image are plotted, a profile that reflects the surface features at that specific location is created (Fig. 2). This graphic display is similar to those achieved through mechanical profilometry with mechanical stylus devices, and we extract numerical information describing the microtopographic attributes in much the same way.[8,10] Of the many parameters available for assessing skin surface topography, both R_z and R_a (Fig. 3) have proved to be especially useful. To compute R_z, the profile is first divided into five equal segments along the x axis. The min–max differences within each segment are then determined, and R_z is calculated as the average of these five local values. To compute R_a, an average line is generated to run through the center of the profile, and the area that the profile deviates above and below this line is determined and then divided by the scan length.

In analyzing a sample, 10 parallel scans are run and the average R_a and R_z computed. If any of the values from any individual scan are more than 2 standard deviations away from the group average, this outlying value is eliminated and the group average recomputed. In addition, because of the anisotropy of the skin, the R_a and R_z determinations are made with the specimens being illuminated from different orientations. To accomplish

[10] G. L. Grove, *in* "Bioengineering and the Skin" (R. Marks and P. A. Payne, eds.), p. 173. MTP Press, New York, 1981.

[11] H. M. Schurmeier, R. L. Heacock, and A. E. Wolfe, *Sci. Am.* **214,** 52 (1966).

[12] G. L. Grove, M. J. Grove, and J. J. Leyden, *J. Am. Acad. Dermatol.* **21,** 631 (1989).

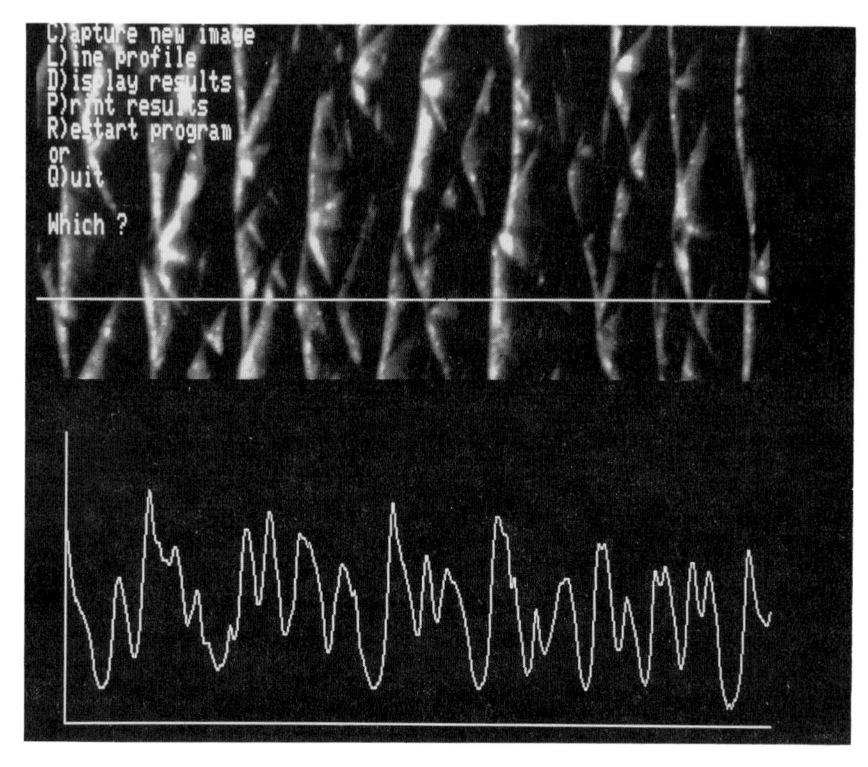

FIG. 2. Optical profilometry profile of skin surface topography. Analysis of samples involved 10 parallel scans.

this, the fiber optic illuminator remains at a fixed angle and distance while the specimen is axially moved in 90° steps on a turntable that serves as the specimen holder. Measurements referred to as North–South are obtained with the illumination running perpendicular across the major line axis, whereas East–West are with parallel lighting.

Representative data from a single specimen are shown in Table I. In general, the values are higher when taken in either the North or South orientation because the major lines are best highlighted with cross-illumination. As only small differences are to be expected from measurements taken at 180° from one another, the North and South as well as the East and West values have been respectively pooled to derive four measurements (R_a and R_z in both axes) from each specimen.

Serial assessments of the group of controlled patients indicate that, with repeat sampling of an individual, the coefficient of variation for the overall method is approximately 5% for both R_a and R_z. Of this, less than 2% can

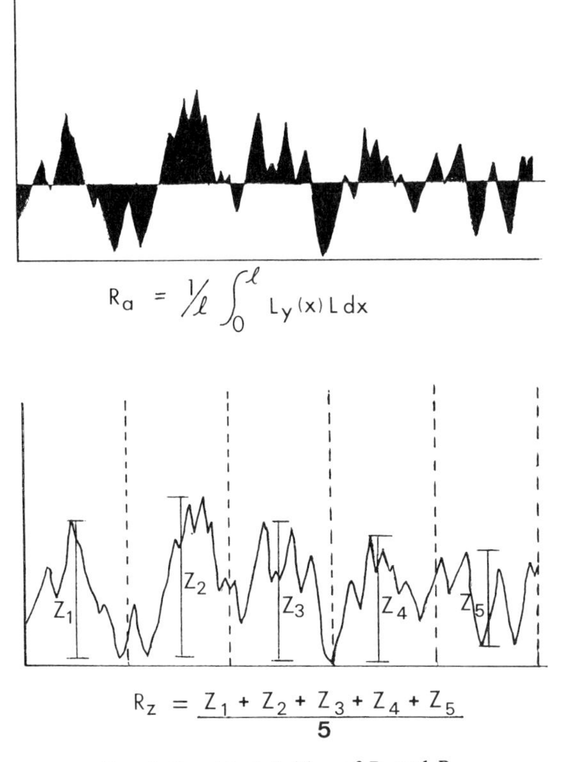

FIG. 3. Graphic definition of R_z and R_a.

TABLE I
REPRESENTATIVE DATA FROM IMAGE ANALYSIS OF SKIN SURFACE IMPRESSIONS

Direction	R_a (mean ± S.D.)	R_z (mean ± S.D.)	Shadow (%)
N	7.80 ± 0.42	31.82 ± 1.34	18.1
E	3.20 ± 0.31	15.82 ± 1.72	3.0
S	7.96 ± 0.74	32.54 ± 2.84	17.4
W	3.26 ± 0.51	16.36 ± 2.11	2.1
N–S	7.88 ± 0.58	32.18 ± 2.09	17.8
E–W	3.23 ± 0.41	16.09 ± 1.92	2.5

be directly attributable to instrumental variations in specimen placement and lighting angles. The residual variation most likely reflects real biological variations in the skin surface topography arising from changes in ambient conditions, menstrual cycle, etc., which are expected to occur over extended periods of time. Of course, failure to take care to relocate the sample site and improper mixing of the silicone rubber material can drastically decrease the reproducibility of this method.

Correlation of Image Analysis Results with Clinical Gradings

Previous studies[13-17] have established that a strong association exists between the evolution of wrinkles in the crowsfeet region and the degree of past sun exposure. Indeed, Daniell,[15] on the basis of a survey of 900 individuals, has calculated that (after controlling for age and smoking history) subjects who averaged more than 4 hr of current daily outdoor exposure were 2.5 times more likely to have prominent "crow's feet" than persons who spent less than 2 hr per day outdoors. In a collaborative study with Corcuff's group, we found that crowsfeet patterns in American women were formed by age 35 and continuously deepened with advancing age.[16]

In an earlier study,[12] 25 healthy, white women aged 24 to 64 were examined by an expert grader and the degree of photoaging as reflected by their crowsfeet patterns scored using a visual analog scale. The six-grade scoring system of Daniell[15] for visual grading of the facial skin wrinkling in the paraocular region was also employed. We found an excellent agreement between these clinical rating schemes and either R_a and R_z, especially those that were obtained in the North–South orientation. This is to be expected as the major lines and features are best highlighted under these conditions of illumination.

More recently, we have had the opportunity to do a more extensive survey involving more than 150 healthy, white women ranging from 21 to 75 years in age. Graphic displays of the correlation between the expert grader's rating of photoaging and the North–South measures of R_a and R_z

[13] P. Corcuff, J. J. deRigal, J. L. Leveque, S. Makki, and P. Agache, *J. Soc. Cosmet. Chem.* **34,** 177 (1983).

[14] C. D. J. Holman, B. K. Armstrong, P. R. Evans, G. J. Lumsden, K. J. Dallimore, C. J. Meehan, J. Beagley, and I. M. Gibson, *Br. J. Dermatol.* **110,** 129 (1984).

[15] H. W. Daniell, *Ann. Intern. Med.* **75,** 873 (1971).

[16] P. Corcuff, J. L. Leveque, G. L. Grove, and A. M. Kligman, *J. Soc. Cosmet. Chem.* **82,** 145 (1987).

[17] A. M. Kligman, *in* "Cutaneous Aging" (A. M. Kligman and Y. Takase, eds.), p. 547. Univ. of Tokyo Press, Tokyo, 1989.

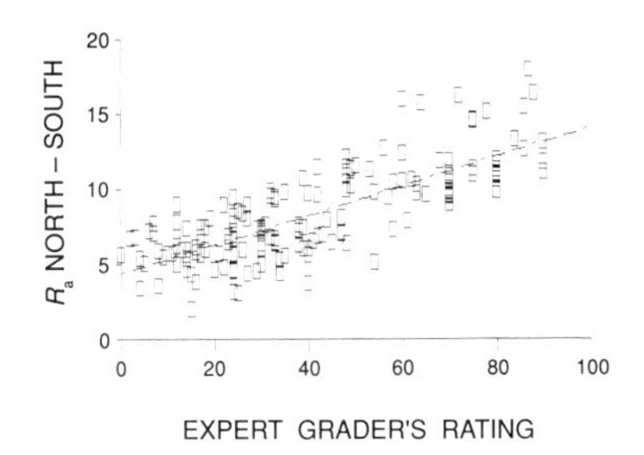

FIG. 4. Correlation of degree of facial wrinkling in the crowsfeet area with optical profilometry parameter R_a ($r = 0.77$).

are presented in Figs. 4 and 5, respectively. Close inspection of both the R_a and R_z plots indicates that neither value shows any clear-cut changes over the lower ranges (below 40) of the expert-grading rating scale, suggesting that optical profilometry is less sensitive than the well-trained eye in evaluating very fine lines.

Assessing Effects of Retinoids on Skin Surface Topography

Clinical studies in which cosmetics or drugs have been evaluated for their effects in photoaging have relied on clinical judgment and photography for assessing changes due to therapy. We feel that optical profilometry of skin surface replicas is an objective method which adds a dimension of quantification which can complement the clinical evaluation of treatments of photoaged skin. We have previously reported on a double-blind, randomized, vehicle-controlled trial to assess the effectiveness of 0.05% tretinoin cream in treating middle-aged patients with mild to moderate signs of

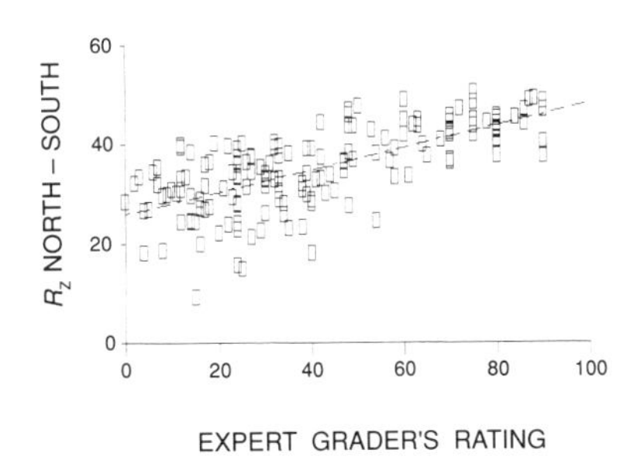

FIG. 5. Correlation of degree of facial wrinkling in the crowsfeet area with optical profilometry parameter R_z ($r = 0.68$).

facial photodamage. The reported side effects were generally mild and usually occurred within the first 2 months of therapy. The predominant complaint made by 70% of the patients on the active drug was a mild, transient, patchy scaling dermatitis sometimes associated with mild erythema. When any such irritation occurred, the usage regimen was changed to every other or every third day, until accommodation was achieved. In no case, except one patient who withdrew, were topical corticosteroids required.

In the study, we found that treatment with 0.05% tretinoin resulted in a significant amelioration of some of the changes associated with photodamage. These included a reduction in fine wrinkling and, to a lesser extent, coarse wrinkling, a lightening of hyperpigmented lesions, the replacement of a sallow complexion with a "healthy, rosy glow," and a general increase in skin firmness. These changes were gradual, occurred after any initial retinoid irritation had subsided, and became progressively more evident as treatment was continued.

The expert grader's assessment of overall improvement shown by both treatment groups after 6 months of therapy are graphically summarized in Fig. 6. The differences in the distribution of scores are highly significant ($p < .001$) in that all patients in the tretinoin-treated group showed some degree of improvement, whereas most of the vehicle-treated patients showed little or no change. Overall, 53% of the tretinoin-treated patients showed moderate changes, whereas the remainder had at least slight improvement.

We also gave each patient a self-appraisal questionnaire to rate the degree of overall improvement as either much improved, somewhat improved, the same or worse to determine whether any changes in facial skin

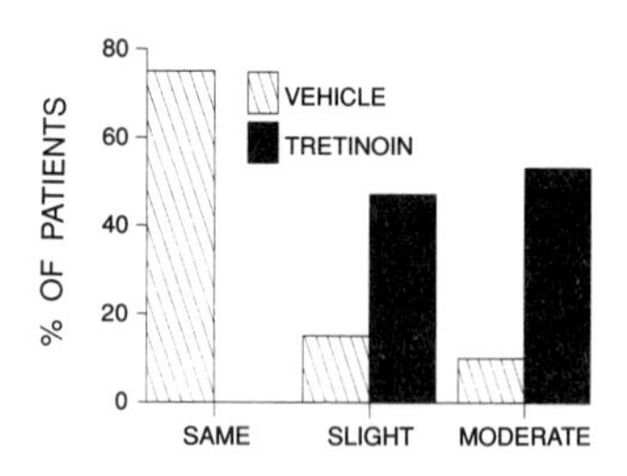

FIG. 6. Assessment by an expert grader of overall performance based on clinical examinations after 6 months of therapy with 0.05% tretinoin emollient cream or vehicle control. Results were statistically significant based on Mantel's mean score χ^2 test on the distribution of the individual ratings for each treatment group.

condition were perceptible to our patients. The different degrees of overall improvement perceived by the patients after 6 months of therapy in the tretinoin-treated and vehicle-control groups are displayed in Fig. 7. Approximately one-half of the vehicle-treated group patients indicated only slight improvement, whereas the other half indicated no improvement in facial skin appearance. In contrast, all the tretinoin-treated patients judged their skin as somewhat improved or much improved. These differences in the distribution of the degree of improvement are significant ($p < .001$) when evaluated by the chi-square (χ^2) test.

The optical profilometry findings were consistent with the ratings of an expert grader and patient self-appraisals. Significant improvement was noted in the pre–post comparisons for both treatment groups in the North–South axis where the illumination runs across the major facial lines and highlights them. When the net changes measured for each patient were compared (Fig. 8), we found that the improvement observed in the tretinoin-treated group significantly exceeded that in the vehicle-control group as far as R_a ($p < .024$). This means that the overall topography of the skin surface of the tretinoin-treated group is smoother and less wrinkled than the vehicle-control group. In the case of R_z, the values approached significance ($p < .052$). Since R_z is largely a measure of the deep wrinkles, this suggests that the major furrows and creases were less affected by tretinoin therapy than the fine lines which contribute to a great extent to the R_a and shadows measurements.

Our most recent studies have been multicentered and have been based on replica specimens taken from a number of different research centers at various geographic locations both within the United States and abroad. We have found that, with proper training and planning, good quality speci-

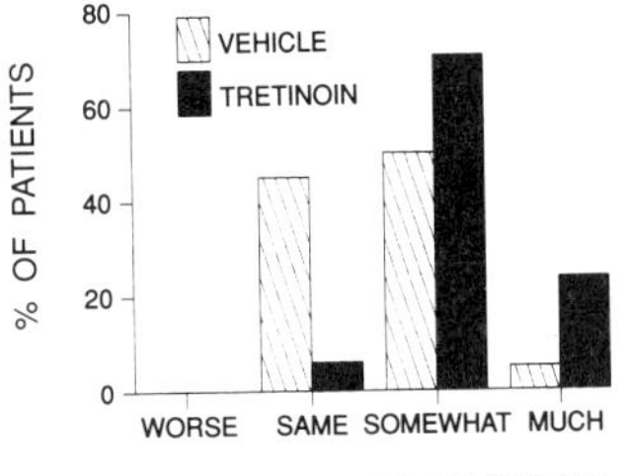

FIG. 7. Ratings of overall improvements as perceived by the patients after 6 months of therapy with 0.05% tretinoin emollient cream or vehicle control. Results were statistically significant based on Mantel's mean score χ^2 test on the distribution of the individual ratings for each treatment group.

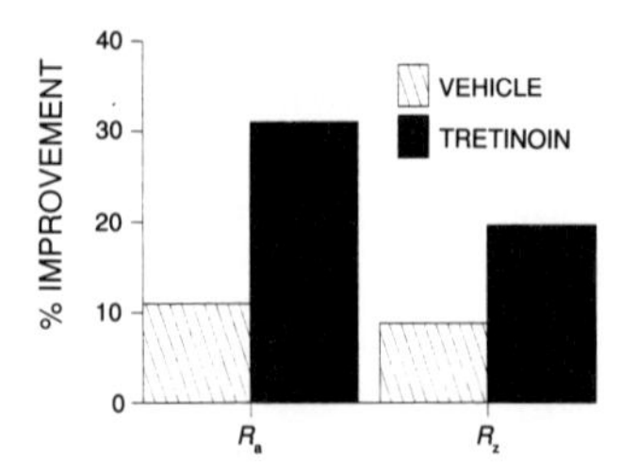

FIG. 8. Results of computerized image analysis of skin surface impressions after 6 months of therapy with 0.05% tretinoin emollient cream or vehicle control. Results were statistically significant based on Student's t test of the degree of individual change from baseline in each treatment group.

mens can be secured from patients at remote sites and forwarded via the mail to our central facility for final processing by image analysis. This is the ultimate in blinded assessments since we have no direct or indirect access to the patients or their records.

In one series of clinical trials involving approximately 300 patients and four cooperating centers, we examined the efficacy of three different doses of tretinoin emollient cream in the treatment of mild to moderately photodamaged skin. An in-depth presentation of the data is in preparation, but in general we have found certain dose-related changes in skin topography such as shown in Fig. 9 in both the crowsfeet and cheek at the end of 24 weeks of therapy. Although the magnitude of change varied across the four centers for most of the optical profilometric parameters (i.e., significant investigator effect), the results were generally consistent across the centers

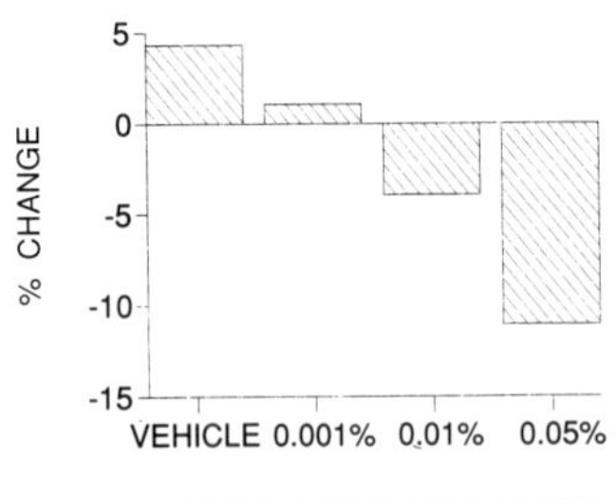

FIG. 9. Results of computerized image analysis of skin surface impressions (crowsfeet region, R_a East–West axis) after 6 months of therapy with varying doses of tretinoin emollient cream. Results with both 0.05 and 0.01% doses were significantly different from vehicle controls based on the rejective Bonferroni procedure to adjust for three multiple comparisons with a familywise error rate of 0.05.

in terms of the relative differences among the treatment groups (i.e., no significant treatment by investigator interaction). The use of computerized image analysis of skin surface replicas added a dimension of objectivity to evaluation of improvement in fine and coarse wrinkling. This mathematical analysis demonstrated a significantly greater decrease in fine lines and wrinkles in the treatment group as opposed to the vehicle controls. The image analysis results were also consistent with the clinical evaluations of the investigators and subject self-assessments in that they all indicated that the superficial fine lines were the most responsive to retinoid therapy.

Although this chapter has not addressed the underlying mechanisms that account for the observed changes, we agree with Kligman[17] that the surface topography most probably reflects alterations in the elastic network of the upper dermis instead of the morphological organization of the collagen bundles. We should also realize that an expansion of the viable epidermal compartment, which is known to occur in tretinoin-treated skin,[1] might be sufficient by itself to account for the observed fine-line changes.

Although the mode of action of tretinoin in partially reversing the signs of photodamage still needs to be resolved, related studies have shown that enhanced blood flow,[1,18] increased epidermal cell renewal,[1,19] and dermal restructuring[1,20] do occur as a result of long-term treatment with tretinoin. Additional studies are currently underway that are directed toward providing a better understanding of the mechanisms of retinoid action and maximizing its effectiveness in treating photodamaged skin while minimizing the undesirable side effects of retinoid dermatitis.

Acknowledgments

This work was partially supported by the R. W. Johnson Research Institute at Ortho Pharmaceutical Corporation, Raritan, New Jersey, and the Simon Greenberg Foundation, Philadelphia, Pennsylvania. The authors are also deeply appreciative of the technical support provided by Mrs. Trisha Alfano and Dr. Charles Zerweck in executing this investigation and the statistical analysis provided by Dr. Barry Schwab.

[18] G. L. Grove, M. J. Grove, C. R. Zerweck, and J. J. Leyden, *J. Cutaneous Aging Cosmet. Dermatol.* **1,** 27 (1988).

[19] G. L. Grove and A. M. Kligman, *in* "The Stratum Corneum" (R. Marks and G. Plewig, eds.), p. 191. Springer-Verlag, Berlin, 1982.

[20] A. S. Zelickson, J. H. Mottaz, J. S. Weiss, C. N. Ellis, and J. J. Voorhees, *J. Cutaneous Aging Cosmet. Dermatol.* **1,** 41 (1988).

[40] Retinoid Effects on Photodamaged Skin

By LORRAINE H. KLIGMAN

Introduction

Investigations into the effects of solar radiation on skin multiplied in the 1970s and 1980s for a number of reasons. Among these are the identification of several sun-sensitive diseases and drug-related phototoxic reactions. Additionally, as humans attain an increasingly vigorous old age, a greater concern for the health and appearance of skin has emerged. With morphological, biological, and biochemical techniques, researchers are identifying the hallmarks of photoaged skin and how they differ from those of chronological aging.[1] Underlying the leathery, wrinkled, and sagging clinical appearance of photoaged skin is severe damage to the connective tissue. An excessive production of thickened, twisted elastic fibers culminates in elastosis and fiber degradation. Collagen, too, is eventually degraded and displaced by the encroaching elastosis. The normally sparse glycosaminoglycans and proteoglycans comprising the ground substance become overabundant. These changes, which in humans occur during decades of chronic sun exposure, have been thought to be irreversible. Recently, in experiments to develop an animal model for photoaging, we observed that significant repair could occur even in severely photodamaged skin.[2] Within several weeks after stopping the ultraviolet (UV) exposures, a band of new, normal dermis was laid down subepidermally, displacing the old elastotic tissue downward.

We thought that retinoic acid might stimulate this repair process. Although retinoids have been reported to inhibit collagen synthesis *in vitro*,[3,4] a few *in vivo* studies suggest an opposite effect. For example, retinoids appear to aid in wound healing by stimulating collagen deposition in granulation tissue.[5] We have developed a photoaged hairless mouse model which shows that retinoids significantly enhance the repair of photodamaged skin by stimulating collagen synthesis. The repair has been mea-

[1] B. Gilchrest (ed.), "Dermatologic Clinics: The Aging Skin," Vol. 4(3). Saunders, Philadelphia, Pennsylvania, 1986.

[2] L. H. Kligman, F. J. Akin, and A. M. Kligman, *J. Invest. Dermatol.* **78,** 181 (1982).

[3] R. P. Abergel, C. A. Meeker, H. Oikarinen, A. I. Oikarinen, and J. Uitto, *Arch. Dermatol.* **121,** 632 (1985).

[4] D. L. Nelson and G. Balian, *Coll. Relat. Res.* **4,** 119 (1984).

[5] K. H. Lee and T. G. Tong, *J. Pharm. Sci.* **59,** 1195 (1970).

sured qualitatively and quantitatively with histochemistry and electron microscopy,[6] as well as with immunochemistry and biochemistry.[7]

Methods

Animals and Irradiation Schedule

Skh:hairless-1 albino female mice, aged 6–8 weeks (Temple University Health Sciences Center, Philadelphia, PA) have been used by our laboratory. The assay has also been repeated in other laboratories with the Jackson[8] and Charles River hairless mice (A. Fourtanier, personal communication). Animals are housed and irradiated in separate compartments of cages holding 12 mice and specially designed to avoid traumatic damage to the skin.[9] Because of the variability in the response of these animals to all aspects of UV radiation, it is suggested that 8–12 mice be used for each experimental group.

Lighting in the room, on a 12-hr on–off cycle, is provided by General Electric F40 GO gold fluorescent bulbs which emit no UV radiation. The UV source is a bank of eight Westinghouse FS20 or FS40 sunlamps which emit mainly UVB (290–320 nm, peak at 313 nm). Attenuation of the shorter wavelength UVC component with a cellulose triacetate film has no effect on the results. The UVA emission is too low, energetically, to affect the connective tissue. The flux of the lamp is measured weekly with an International Light 700 A Radiometer (Newburyport, MA) and a UVB sensor with peak sensitivity at 290 nm.

From prior experience with these mice, it was determined that a total UVB dose of approximately 3.5 J/cm^2 from the FS20 bulbs would produce mild connective tissue damage. Because of differences in photobiological equipment, it is extremely difficult to duplicate doses in joules from one laboratory to another. The biologic end point of a minimal erythema dose (MED) was therefore chosen as a working measurement. The MED for these mice is determined as follows. With strips of opaque adhesive tape, a six-opening template is fashioned on the dorsal trunk of each of six sedated mice. Each opening is approximately 0.7 mm^2. A sunscreen is applied to nontemplated dorsal surfaces, and the mice are placed beneath the lamp which in our laboratory is 16 cm above the mice. Every 30 sec,

[6] L. H. Kligman, H. D. Chen, and A. M. Kligman, *Connect. Tissue Res.* **12,** 139 (1984).

[7] L. Kligman, F. Cruickshank, J. Mezick, and E. Schwartz, *J. Invest. Dermatol.* **92A,** 460 (1989).

[8] G. Bryce, N. J. Bogdan, and C. C. Brown, *J. Invest. Dermatol.* **91,** 175 (1988).

[9] P. D. Forbes and F. Urbach, *in* "The Biologic Effects of Ultraviolet Radiation" (F. Urbach, ed.), p. 279. Pergamon, London, 1969.

one opening on the template is covered with a piece of the tape. Thus, small squares of skin are exposed for 0.5 to 3 min of UV radiation. Under these conditions, we found that 1 min and about 0.014 J/cm^2 of UVB produced a light, uniform pinkness or 1 MED after 24 hr. From this we devised a 10-week irradiation course with thrice weekly exposures that were gradually raised, 0.5 MED at a time, until by the end of week 4 and for the remainder of the 10 weeks, the dose per exposure was 4.5 MED.

Postirradiation Treatment

Seventy-two hours following the final irradiation, 5 to 8 mice are sacrificed, by cervical dislocation, to determine the baseline of connective tissue damage. The remaining animals are divided, randomly, into the designated treatment groups, with 8 to 12 mice per group.

Retinoids in either cream, ethanol, or ethanol–propylene glycol vehicles are applied in 50 to 100-μl amounts to the entire dorsal trunk, 5 times

FIG. 1. Results of UV treatment for 10 weeks, followed by 0.05% retinoic acid for 10 weeks in hairless mice. A zone of new collagen is present subepidermally. Elastic fibers (arrows) which immediately postirradiation were present throughout the upper dermis, are now compressed downward approximately 100 μm below the dermal–epidermal junction (bar). Note the abundance of large fibroblasts in the zone.

a week for the next 10 weeks. Skin was examined weekly for signs of irritation. In the case of some experimental retinoids, it was necessary to reduce the concentration or frequency of application because of excessive scaling, small erosions, or inflammation. With all-*trans*-retinoic acid, concentrations ranging from 0.005 to 0.05% (w/v) were well tolerated.

Histology

Central dorsal skin biopsies ($2 \times \frac{1}{2}$ cm) are placed on cardboard strips to keep them flat while in buffered formalin. Skin is then processed routinely for light microscopy. Sectioning is at 10 μm. Stains used for global evaluation are hematoxylin/eosin, Van Gieson's for collagen, Gordon and Sweet's stain for "reticulin," and Mowry's colloidal iron for ground substance. After gaining familiarity with the model, these stains can be dispensed with as they are not used in the assay. Luna's aldehyde fuchsin,[10] which stains the relatively sparse mouse elastic fibers deep purple, is routinely used for repair zone measurements. Areas of new collagen are identified by a number of criteria. With highly active retinoids, the lower border of the repair zone is delineated by a band of compressed elastic fibers (Fig. 1). With less active retinoids, vehicles, or untreated skin (Fig.

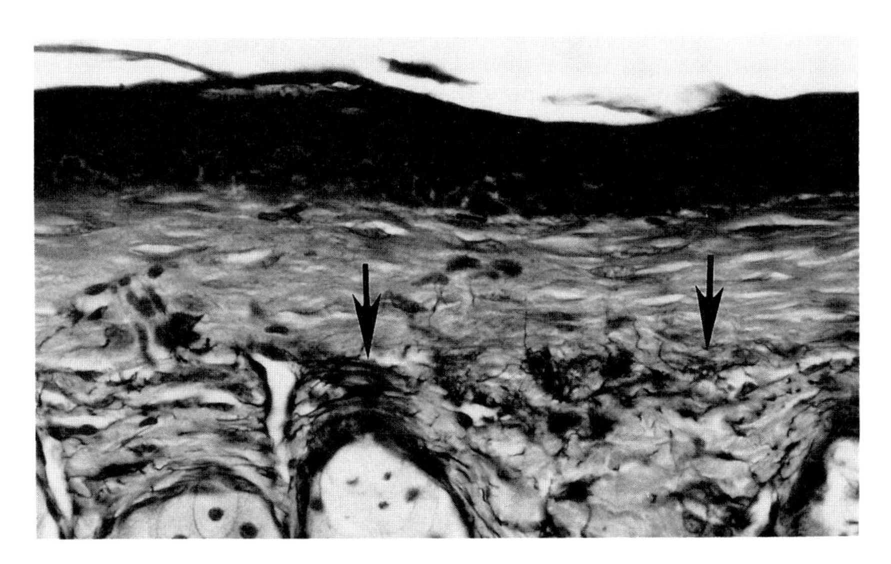

FIG. 2. Results of UV treatment for 10 weeks, followed by ethanol–propylene glycol vehicle for 10 weeks. The repair zone is shallow and less well delineated at the lower border by elastic fibers (arrows). The zone is nevertheless identifiable by the orientation of collagen bundles parallel to the dermal–epidermal junction and the absence of elastic fibers within the zone.

2), additional features such as the absence of elastic fibers in the repair zones or a slightly pinkish color and more parallel orientation of the new collagen bundles as opposed to the lavender color and random orientation of the old must be relied on. It should be noted that, for optimum results with Luna's stain, it is important to use aldehyde fuchsin with color index C.I.42510 (Harleco, Gibbstown, NJ) and stabilized paraldehyde (USP: J. T. Baker Chemical Co., Phillipsburg, NJ).

Repair Zone Measurement

Prior to the measurement of repair zones, each Luna-stained slide is microscopically scanned at a magnification of ×100. A 1.2 cm length of typical tissue is marked with ink, providing 16 to 18 fields at the magnification used for measurements (×200). Then, with an American Optical eyepiece micrometer, the depth of the repair zones is measured in two areas each in 12 of these fields for a total of 24 measurements per slide. The extra fields are allowed for sites excluded because of a high density of hair follicles or other anomalies that impede measurements. Measurements are converted to microns in accordance with instructions for the micrometer. In a field with no repair, a zero is recorded. Repair zone data consist of the average of all measurements from all skin specimens in a particular treatment group. Compared to controls, 0.05% all-*trans*-retinoic acid doubles the depth of the repair zone (Table I). The model is sensitive enough to demonstrate a retinoid dose response. Another measure of efficacy is the percentage of tissue stimulated to repair. This is calculated by dividing the number of nonzero measurements by the total number of measurements.

TABLE I
EFFECT OF RETINOIC ACID CONCENTRATION ON REPAIR ZONE MEASUREMENTS

Treatment	Mean thickness (μm)[a]	NZ/T[b]	S.D.
0.050% RA	86.5**	158/168	31.9
0.025% RA	80.4**	163/168	29.3
0.010% RA	63.3*	154/168	27.8
0.005% RA	73.5**	168/168	21.8
Vehicle	46.5	100/168	20.8
Untreated	35.6	44/84	14.6

[a]Statistical analysis with Duncan's Multiple Range Test: *$p < .05$; **$p < .01$. Mean thickness was determined on the basis of nonzero measurements.
[b]Nonzero measurements/total number of measurements.

Retinoic acid, at the concentrations used, produces repair zones in 95–100% of the tissue, whereas in controls it is 50–60% (Table I).

Enhancement of repair appears to be retinoid-specific and not related to irritancy. This assay has been repeated with 5.0% salicylic acid, 5.0% croton oil, 100% propylene glycol, 2.5% Hyamine, and 20% sodium lauryl sulfate, all of which produce scaling similar to that of 0.05% retinoic acid. In no case is the repair better than that seen in mice remaining untreated in the post-UV period.[11]

Comments

Although time-consuming, the photoaged mouse model for assessing the effect of retinoids on dermal connective tissue repair is singularly relevant to humans. Similar repair zones have been observed in retinoic acid-treated human photoaged skin.[12] The assay is sensitive enough to detect effects related to time, retinoid dose, and comparative efficacy of different retinoids. Repair zone measurements with the light microscope are labor-intensive, but at present there is no alternative. Image analysis can be used successfully only with potent retinoids where the lower border of the zone is well delineated with compressed elastosis. No doubt, with better color and image resolution, the computerized method will become completely feasible.

Electron Microscopy of Repair Zone

Standard techniques are used with fixation in Karnovsky's solution, followed by 1% osmium tetroxide. Ultrathin sections (70 nm) are stained with a tannic acid–uranyl acetate solution[13] followed by Reynolds lead citrate for 10 min. This stain renders the thin, sparse elastic fibers a deep black,[14] facilitating identification of the lower border of the repair zone. Ultrastructural analysis confirms the disruption of the collagen matrix by UV radiation (Fig. 3) and its restoration to normal architecture after retinoid treatment (Fig. 4). Fibroblasts in the repair zone are unusually numerous and have morphologic features characteristic of high metabolic activity.[6,11]

[10] L. H. Kligman, *Am. J. Dermatopathol.* **3,** 199 (1981).
[11] L. H. Kligman, *J. Am. Acad. Dermatol.* **15,** 799 (1986).
[12] A. M. Kligman, G. L. Grove, R. Hirose, and J. J. Leyden, *J. Am. Acad. Dermatol.* **15,** 836 (1986).
[13] K. Kajikawa, T. Yamaguchi, S. Katsuda, and A. Miwa, *J. Electron Microsc.* **24,** 287 (1975).
[14] R. Hirose and L. H. Kligman, *J. Invest. Dermatol.* **90,** 697 (1988).

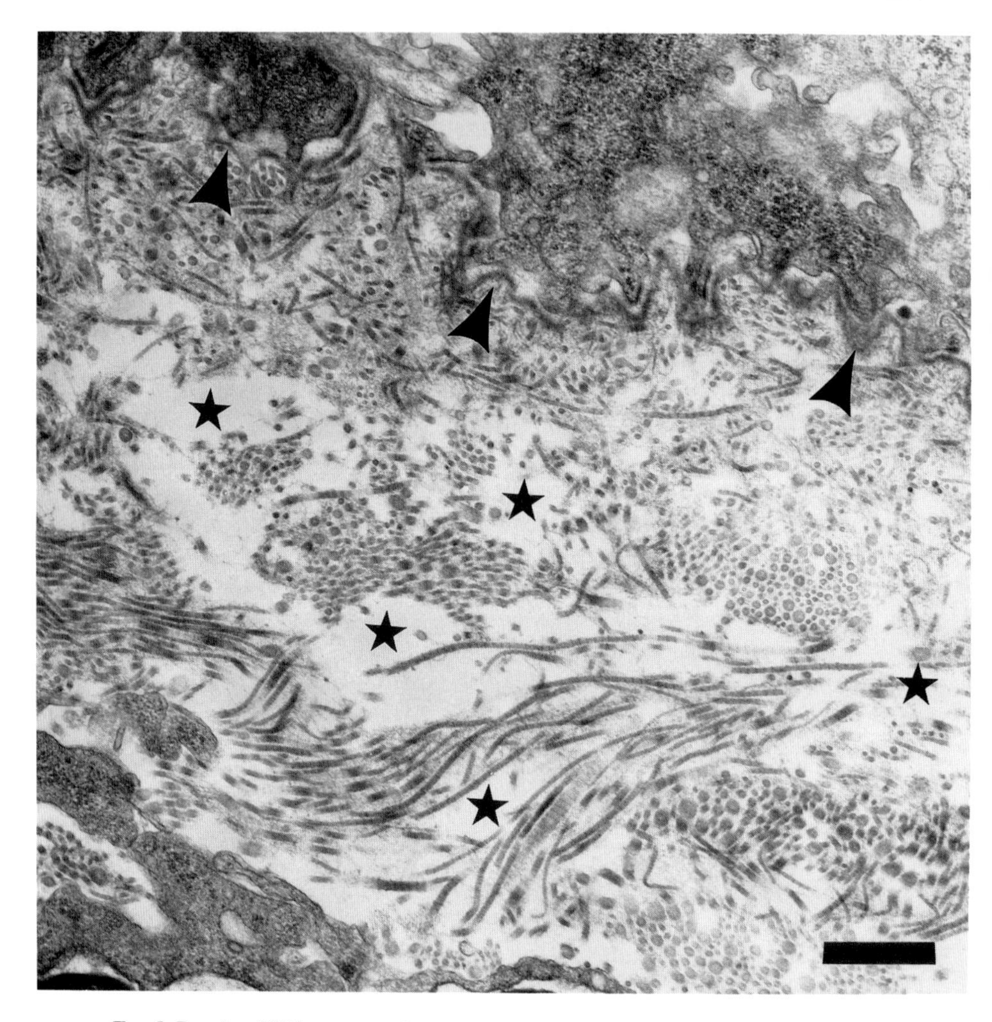

FIG. 3. Results of UV treatment for 10 weeks. Collagen bundle architecture is disrupted by an increase in electron-lucent ground substance (stars). Arrowheads mark the dermal–epidermal junction. (Original magnification: ×15,456; bar, 1 μm.)

Immunological and Biochemical Techniques

Recently, the evidence for retinoic acid-enhanced collagen synthesis has been confirmed by other techniques.[7] Using the same treatment protocol, frozen skin sections are examined with immunofluorescently tagged antibodies against the amino propeptide (AP) of type III procollagen. The

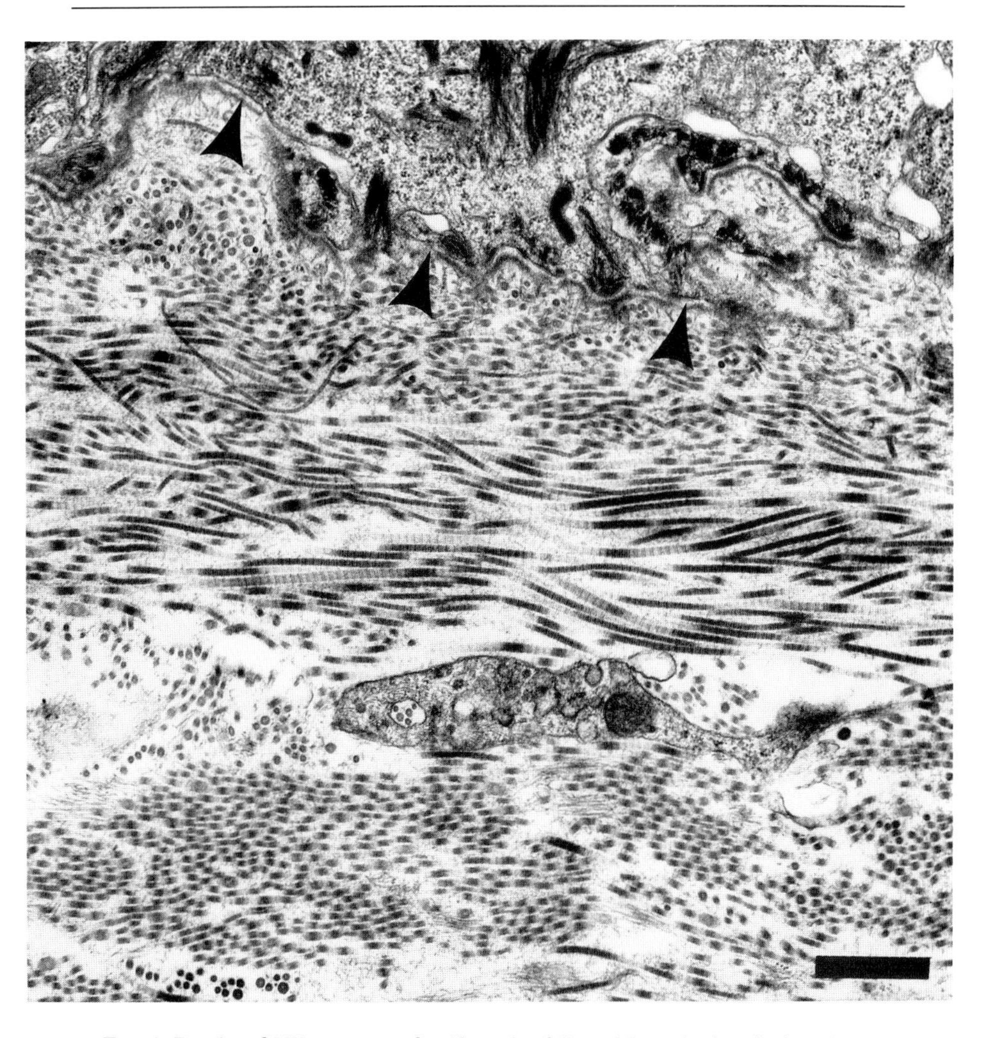

FIG. 4. Results of UV treatment for 10 weeks, followed by retinoic acid for 10 weeks. Collagen bundles are returning to a more normal, parallel arrangement. Arrowheads mark the dermal–epidermal junction. (Original magnification: ×15,456; bar, 1 μm.)

repair zone of retinoic acid treatment skins shows intense staining for this moiety. Quantification by radioimmunoassay shows that retinoic acid produces a 2-fold increase in the AP of type III collagen. Ratios of type III to type I collagen remain unchanged, suggesting that type I collagen is also stimulated proportionally.

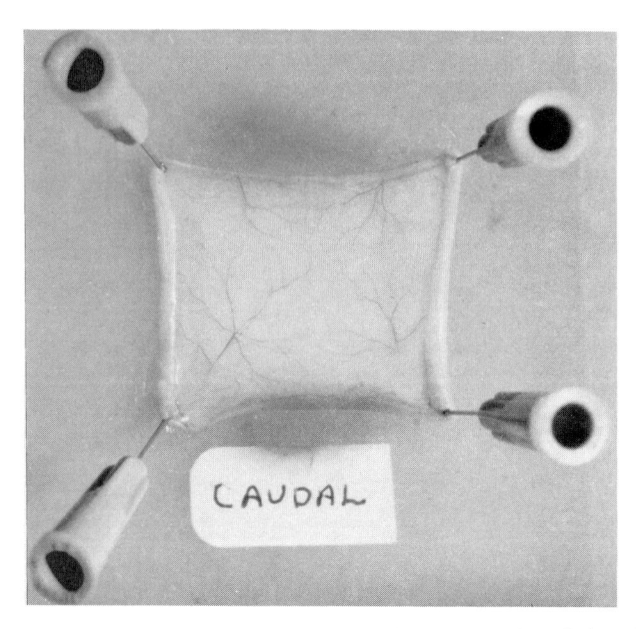

FIG. 5. Dorsal skin affixed to a wax template for photography of the subcutaneous vasculature.

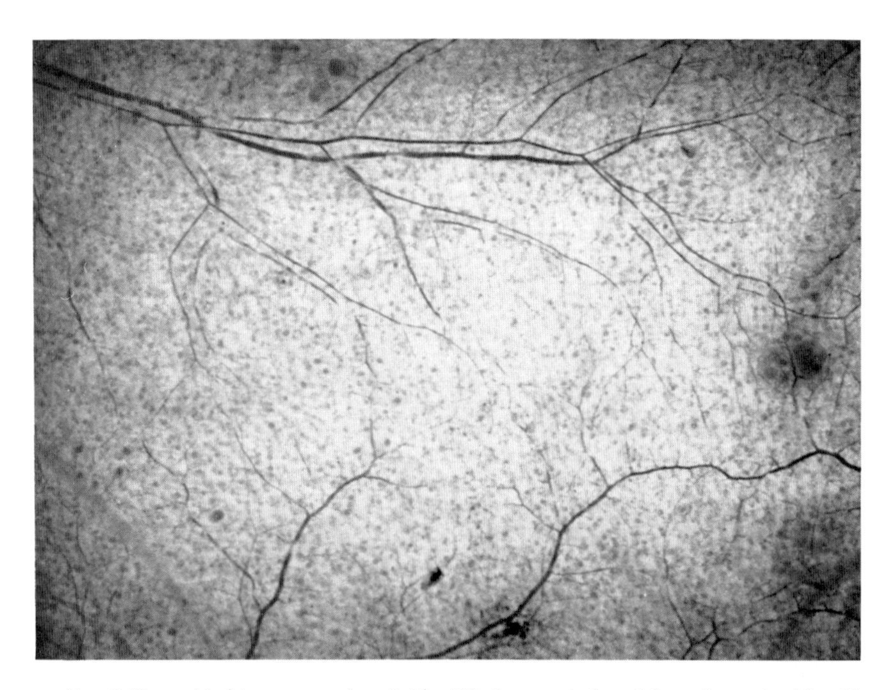

FIG. 6. Normal hairless mouse dorsal skin. Whole mount viewed from the underside with transmitted light. The vascular network is composed of fine delicate vessels. (Original magnification: ×1.5)

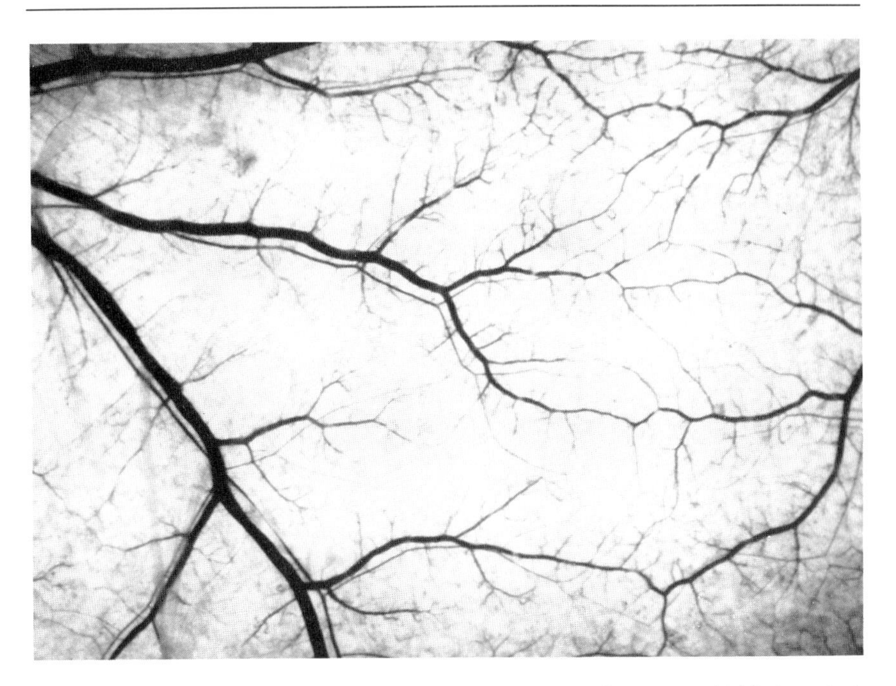

FIG. 7. Retinoic acid-treated skin. Whole mount. The vessels are more highly branched, with wider diameters. (Original magnification: ×1.5)

Retinoic Acid-Induced Angiogenesis

Another feature of photoaged human skin is the destruction and deletion of portions of the cutaneous vasculature. The hairless mouse has proved amenable for assessing the angiogenic capabilities of retinoids. In this case, normal, young (6–8 weeks old) unirradiated mice are used. Topical agents are applied each weekday for 4–6 weeks. For these experiments, mice are sacrificed by CO_2 inhalation as cervical dislocation results in drainage of blood away from skin vasculature. A 2 cm^2 section of dorsal skin is excised and affixed, dermal side up, to a wax plate with a 1.5 cm^2 opening (Fig. 5). This is placed on a light box and photographed at a magnification of ×2, allowing the subcutaneous vasculature of the thin mouse skin to be seen clearly.[11] Compared to controls (Fig. 6), retinoic acid (0.05%)-treated skins consistently have enhanced vascularity (Fig. 7), whereas enhanced vascularity is only occasional in those treated with nonspecific irritants. Similar results are seen in the superficial vasculature with sections of frozen skin utilizing the alkaline phosphatase reaction.[15]

[15] L. H. Kligman, *in* "The Dermis—From Biology to Disease" (G. E. Pierard and C. Pierard-Franchimont, eds.), p. 114. Monographies Dermatopathologiques Liegeoises, Liege, Belgium, 1989.

Finally, injection of low dose retinoic acid (0.01%) into the relatively avascular dermis of newborn mice induces striking angiogenesis in 24 hr.[16] As with repair zone induction, the angiogenic property of retinoic acid has also been demonstrated in humans.[12]

[16] L. H. Kligman, *J. Am. Acad. Dermatol.* **21,** 623 (1989).

[41] Testing of Retinoids for Systemic and Topical Use in Human Psoriasis and Other Disorders of Keratinization: Mouse Papilloma Test

By Kampe Teelmann

Introduction

Retinoids are a class of compounds consisting of natural and synthetic analogs derived from or structurally related to vitamin A. The most prominent members of this class of compounds are 13-*cis*-retinoic acid (isotretinoin, Accutane, Roaccutan), which has high clinical effectiveness in severe, treatment-resistant nodulocystic acne,[1] and etretinate (Tigason, Tegison) or acitretin (Neotigason, Soriatane), which give significant clinical improvement of severe forms of psoriasis, particularly generalized pustular psoriasis.[2]

Unfortunately, none of the dermatological diseases in which retinoids have shown major effectiveness occur in laboratory animals. Furthermore, the etiology and pathophysiology of these diseases are not fully understood. Thus, no direct *in vivo* or *in vitro* preclinical tests exist for the screening and evaluation of retinoids. Fortunately, the long experience resulting from preclinical testing and clinical use of retinoids has shown the mouse papilloma model to be a useful tool for preclinical testing of vitamin A derivatives. This model certainly does not resemble human skin diseases like psoriasis, but it contains a number of features and morphological analogies to these diseases, such as elongation of epidermal rete ridges and papillae, acanthosis as a consequence of increased proliferation of the basal cell layer, and hyper- and parakeratosis.[3-5] The mouse papilloma model

[1] A. Ward, R. N. Brodgen, R. C. Heel, T. M. Speight, and G. S. Avery, *Drugs* **28,** 6 (1984).
[2] H. Gollnik, R. Bauer, C. Brindley, *et al., J. Am. Acad. Dermatol.* **19,** 458 (1988).
[3] W. Bollag, *in* "Psoriasis" (E. M. Faber *et al.*, eds.), p. 175. Grune & Stratton, New York, 1982.

has been in use since the late 1960s. In this chapter, we give an account of how papillomas are produced in our laboratory and how retinoid testing is performed. The retinoids may be administered either systemically (intraperitoneally or orally) or topically with direct delivery to the papillomas.

Chemical Induction of Epidermal Papillomas in Mice

Animal Model

Female hairless mice from the local breeding farm (BRL Ltd., Füllinsdorf, Switzerland) are used. The strain is registered as MORO/Ibm-hr/hr; it is the hairless mutant of the outbred mouse strain Ibm/MORO. The animals are bred under specified pathogen-free conditions. In our laboratory they are kept in fully air-conditioned rooms (22 ± 2°; 50% relative humidity; ~20 air cycles/hr) in plastic boxes, type III. They receive tap water from drinking bottles and a rodent maintenance diet (NAFAG 850, cubes) containing 12000 IU of vitamin A/kg feed. The average weight of the mice at induction of papillomas is 20–22 g.

Other authors have used different mouse strains, for example, HRA/skh,[6] SENCAR,[7] and various other strains and stocks of mice.[8] The susceptibility of epidermal papilloma initiation/promotion appears to vary between different mouse strains, but there appear to exist no major sex differences.[6]

Induction/Promotion of Skin Papillomas

The chemical induction/promotion of mouse skin papillomas basically follows the classic multistage skin carcinogenesis pattern as demonstrated by Berenblum and Shubik.[9] 7,12-Dimethylbenz[a]anthracene (DMBA, Sigma, St. Louis, MO) is applied twice, with an interval of 2 weeks, to the backs of the mice. Each time, 150 μg DMBA dissolved in 0.2 ml acetone is painted on the back of each mouse to a skin surface of approximately 5 cm². Three weeks later 0.5 mg croton oil [containing the phorbol ester 12-*O*-tetradecanoylphorbol 13-acetate (TPA; supplied by Fluka Ltd., Buchs, Switzerland)] dissolved in 0.2 ml acetone is sprayed on to the back

---

[4] M. Frigg and J. Torhorst, *J. Natl. Cancer Inst.* **58,** 1365 (1977).

[5] R. Stadler, G. Schaumberger-Lever, and C. E. Orfanos, *in* "Textbook of Psoriasis" (P. D. Mier and P. C. M. van de Kerkhof, eds.), p. 40. Churchill Livingstone, Edinburgh, 1986.

[6] H. H. Steinel and R. S. U. Baker, *Cancer Lett.* **41,** 63 (1988).

[7] H. Hennings, R. Shores, M. L. Wenk, *et al. Nature (London)* **304,** 67 (1983).

[8] T. J. Slaga, *Environ. Health Perspect.* **68,** 27 (1986).

[9] I. Berenblum and P. Shubik, *Br. J. Cancer* **1,** 383 (1947).

of each mouse twice per week until papillomas develop. This may last from 3 to 5 months. It appears to make no major difference if the croton oil preparation or TPA itself is used. It is important that the promoter is well administered to the skin of the animals and that the spraying or painting of croton oil is carefully performed, even when large numbers of mice have to be treated weekly. Under precisely controlled treatment conditions in sensitive mice it may even be possible to restrict administration of the promoter to a once per week schedule. At the end of this treatment regimen at least three skin papillomas will develop in up to 80% of the mice. The yield, of course, depends on the sensitivity of the mouse strain used, and in some mice 10 or more papillomas may be formed.

The induction/promotion procedure produces papillomas as well as skin carcinomas. The carcinomas rarely emerge as such, but mostly by malignant progression from benign papillomas. The carcinoma yield can be increased if an interim period of promoter treatment (e.g., TPA for 10 weeks) is followed by a second course of initiator administration. The secondary initiators may be urethane, *N*-methyl-*N'*-nitro-*N*-nitrosoguanidine (MNNG), or 4-nitroquinoline *N*-oxide (4-NQO).[7]

Protection of Laboratory Personnel

Because the chemicals used to induce and promote skin tumors in laboratory mice must be considered as carcinogenic for humans as well, effective protective measures must be taken. We use a double strategy to prevent contact with the chemicals. First, every technician has to wear microparticle- and aerosol-rejecting dust masks, protection glasses, medical gloves, and disposable lab coats with long sleeves. Instead of the dust mask and glasses, a low pressurized ventilation system containing an antidust ventilated head visor unit connected by an air hose to a battery-driven air filter unit may also be used more effectively. The second protective barrier is a movable hood connected to a separate exhaust channel of the ventilation system (Fig. 1). This hood is the same size as the plastic cage containing the mice and is placed on top of the cage during administration (spraying) of the chemicals. Access to the mice occurs via a lateral opening; the correct administration of the chemicals to the mice can be observed through the transparent front of the hood.

Systemic Treatment of Mouse Papillomas with Retinoids

A standard procedure has been set up to test retinoids for their antipapilloma activity. The compounds are either dissolved or suspended in arachis oil and administered intraperitoneally, once per week for 2 consecutive weeks. The volume injected is 40 to 80 μl per 20 g body weight and

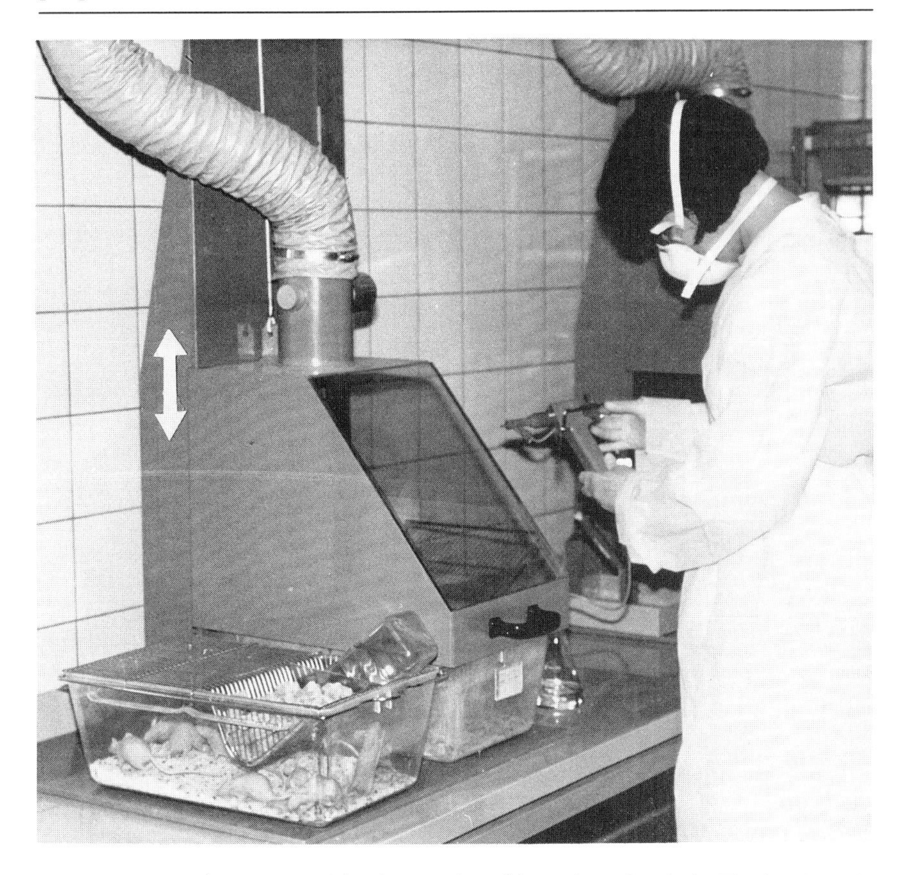

FIG. 1. Protective hood used for the spraying of hazardous chemicals. The hood can be moved up and down (white arrow) and is placed on top of the mouse cage during administration of the chemicals. Access to the mice occurs via a lateral opening on the right side of the hood. The correct administration of the chemicals can be controlled via the transparent front of the hood.

depends on the dose of the test compound to be administered. It is important to remember that retinoids are sensitive to light. In particular, solutions are easily degraded by sunlight, so protective measures have to be taken. We prepare fresh solutions each day. The stock compound is stored under N_2 gas. For routine testing we use four papilloma-bearing mice per dose of test compound. The mice are kept individually in small (type I) plastic cages, and they receive tap water from drinking bottles and cubes of laboratory chow. With the most active retinoids, additional variations of the route of administration (e.g., oral) and the dosing schedule (e.g., 5 times per week) are done to gain additional information on the potency of

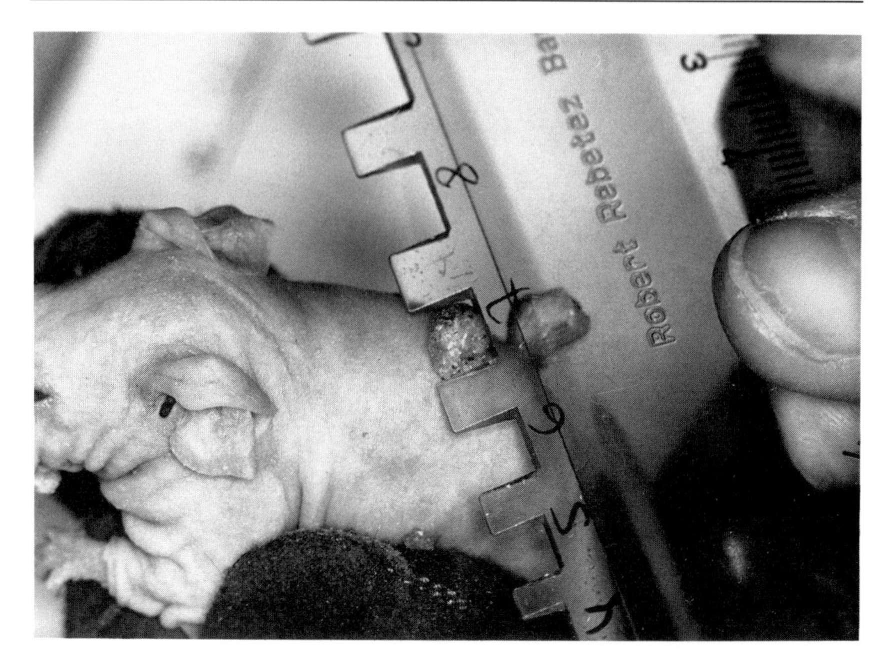

FIG. 2. Measurement of mouse papilloma diameters by means of a simple template.

the retinoid of special interest. Control mice are treated in a similar manner with the vehicle alone.

Evaluation of Retinoid Antipapilloma Activity

A retinoid is considered active when the average diameter of the papillomas per group is reduced by 50%. This information is obtained by repeated measurements of all papillomas of all mice per group. Measuring of diameters can be done by means of simple templates (Fig. 2) or callipers, and the results may be recorded manually or by electronic devices. We record the diameters of all papillomas on each mouse 3 times: before treatment and after 1 and 2 weeks of the study. Comparisons of the 2-week data to the pretreatment data are the basis for estimating the activity of the test compound. If the dose examined results in a papilloma diameter reduction of more than 50%, lower doses are used to determine the ED_{50} for that particular retinoid. This ED_{50} value then becomes one of the two variables for the calculation of the therapeutic index. An example of retinoid antipapilloma activity is given in Figs. 3 and 4.

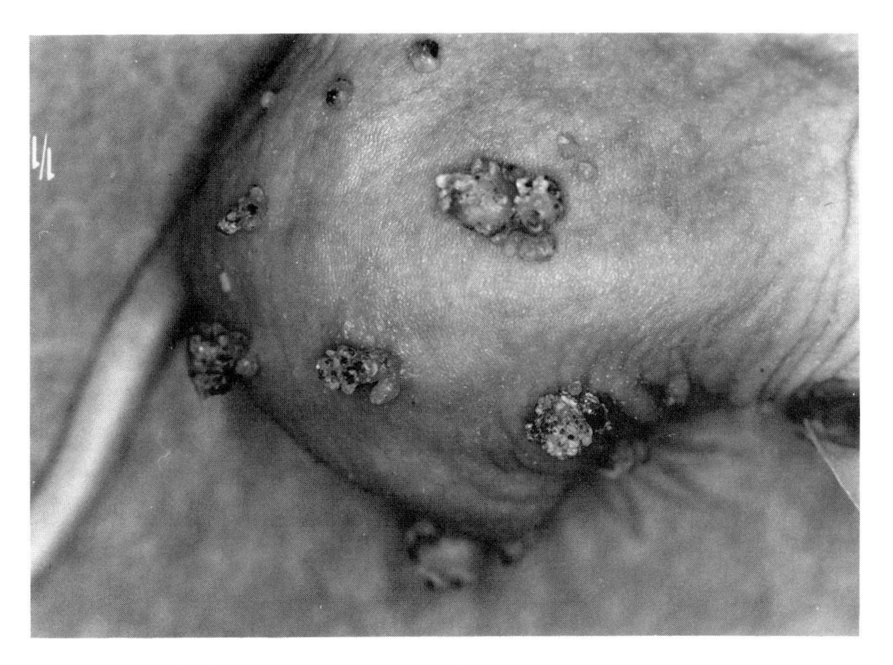

FIG. 3. Close-up view of mouse papillomas before retinoid treatment.

Test for Hypervitaminosis A Effects

In addition to testing the pharmacologic efficacy of a retinoid, experiments are performed to determine dose levels which produce a defined degree of retinoid side effects. This testing serves two purposes: it selects suitable doses for each new retinoid for use in the mouse papilloma test, and it provides a rough estimate of the range of side effects associated with the administration of the retinoid under test.

The range of side effects examined are primarily those easy to observe and selected from the variety of toxic manifestations often referred to as hypervitaminosis A syndrome.[10] The variables recorded in our laboratory are body weight development, redness and scaling of the skin, hair loss, changes in the mucosa of nose and mouth, and, finally, bone fractures. These side effects, if produced, usually become apparent after 5–7 days of treatment and are rated using a scale from 0 to 4 (see Table I).

The duration of the hypervitaminosis A test is 2 weeks. The compound is administered daily, 5 times per week, intraperitoneally. Outbred, female

[10] K. Teelmann, *Pharmacol. Ther.* **40,** 9 (1989).

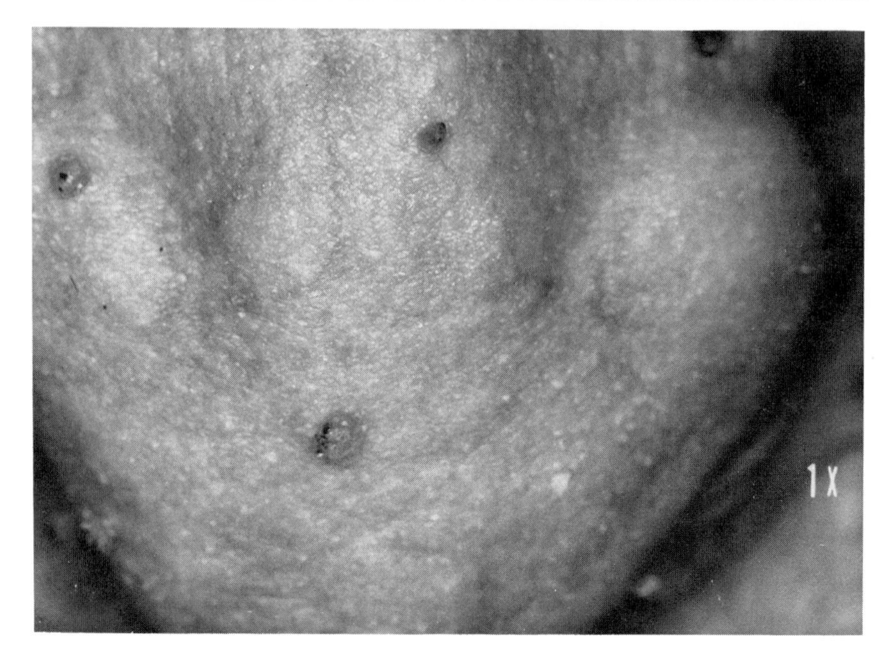

FIG. 4. Close-up view of mouse papillomas after treatment with an active retinoid (systemic treatment).

Ibm/MORO (SPF) mice are used for routine testing, 2 mice per dose group, weighing approximately 26–30 g. Hypervitaminosis A is defined as that condition of the aminals where the sum of all side effect grades yield at least 3. The lowest daily dose causing this degree of side effects serves as the second variable for the calculation of the therapeutic index.

TABLE I
RATING OF HYPERVITAMINOSIS A SYMPTOMS

	Grading				
Symptom	0	1	2	3	4
Weight loss	1 g	1–3 g	4–6 g	7–9 g	10 g
Skin scaling	None	Slight	Moderate	Severe	Very severe
Hairloss	None	Slight	Moderate	Severe	Very severe
Number of bone fractures	0	1	2	3	4

TABLE II
MOUSE PAPILLOMA TEST WITH ONCE PER WEEK INTRAPERITONEAL ADMINISTRATION OF RETINOIDS

Dose (mg/kg once/week i.p.)	Change in papilloma diameters (%)			
	Retinoic acid[a]	Etretinate[b]	Acitretin[b]	Ro 13-6298[c]
400	−48.2			
200	−36.0	−74.1	−88.8	
100		−69.6	−58.0	
50		−54.4	−47.8	
25		−48.8	−15.2	
12.5		−30.2		
0.2				−76.0
0.1				−56.2
0.05				−49.6
0.02				−42.9
0.01				−21.3

[a] From Ref. 11.
[b] From Ref. 12.
[c] From Ref. 13.

Therapeutic Index

Since various retinoids are pharmacologically active at different doses and cause side effects at different doses, it is necessary to have a parameter that allows comparison of a large number of retinoids. This parameter, originally called therapeutic ratio,[11] is now termed the therapeutic index (T.I.). It is calculated from the lowest daily dose causing hypervitaminosis A, as defined above, and the weekly dose which leads to approximately 50% regression of the diameter of chemically induced papillomas:

$$\text{T.I.} = \frac{\text{minimal toxic dose causing hypervitaminosis A (dosing: 5 times/week; i.p.)}}{\text{ED}_{50}\text{ papilloma regression (once/week dosing)}}$$

Thus, the T.I. describes the dissociation between the dose causing undesired adverse effects and the dose causing the desired antipapilloma activity. The higher the T.I., the broader the dissociation between these doses, and the more favorable are the prospects that the retinoid under test may eventually become a potent drug in dermatology and/or oncology.

Examples of test results in the mouse papilloma test with systemic

[11] W. Bollag, *Eur. J. Cancer* **10,** 731 (1974).

administration of retinoids are given in Table II.[12,13] Examples of therapeutic indices obtained by testing of retinoids are given in Table III.[14] Table III also shows the broad dose range of retinoids resulting in a 50% regression of mouse skin papillomas and causing hypervitaminosis A. Such information underlines the usefulness of the therapeutic index for comparison of different retinoids and as an indicator of the therapeutic margin in mice.

Mouse Papilloma Test as Predictor of Retinoid Activity in Human Psoriasis

To obtain information on the relevance of the mouse papilloma model as a useful predictor of retinoid activity in humans, a retrospective analysis comparing the clinical results in psoriatic patients to mouse antipapilloma activity has been performed.[15] This analysis showed that the therapeutic index in mice correctly identified 11 compounds and differentiated them into markedly active, moderately active, or inactive retinoids when subsequently used clinically in the treatment of various forms of human psoriasis. Acidic retinoids were more difficult to assess than nonacidic ones, and the therapeutic index appeared to underestimate the potency of acidic retinoids in humans. One retinoid was falsely positive in the mouse because the mouse metabolized the compound differently than humans. Thus, the therapeutic index, resulting from antipapilloma activity testing and testing for side effects in mice, may be considered a useful tool in the search for retinoids for the treatment of keratinization disorders in humans.

Skin Tumors Induced by Chemical Carcinogenesis in Mice

In order to interpret the retinoid activity on chemically induced skin tumors correctly, it must be recognized that the two-stage carcinogenic process in the mouse does not give rise to squamous cell papillomas or carcinomas exclusively. A certain percentage of nodules may arise from sebaceous glands and form sebaceous adenomas.[16] These tumors may be found on various locations over the entire body surface and are characterized as white or flesh colored nodules with a smooth or somewhat crenulated surface texture. Their size averages 1.0–1.5 mm.[16] The overlying

[12] W. Bollag, *in* "Retinoids: New Trends in Research and Therapy" (J. H. Saurat, ed.), p. 274. Karger, Basel, 1985.

[13] W. Bollag, *Cancer Chemother. Pharmacol.* **7,** 27 (1981).

[14] W. Bollag and H. R. Hartmann, *Cancer Surv.* **2,** 293 (1983).

[15] K. Teelmann and W. Bollag, *Dermatologica* **180,** 30 (1990).

[16] J. M. Rice and L. M. Anderson, *Cancer Lett.* **33,** 295 (1986).

TABLE III
THERAPEUTIC INDICES OBTAINED FOR VARIOUS RETINOIDS[a]

Compound	Chemical structure	Hypervitaminosis A (mg/kg)	Antipapilloma effect (mg/kg)	Therapeutic index
all-*trans*-Retinoic acid (tretinoin)	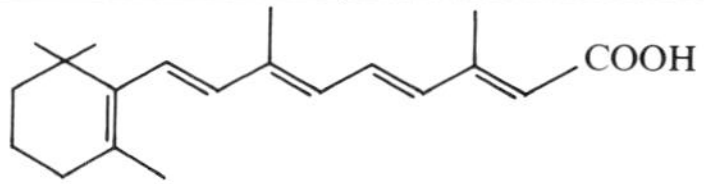 COOH	80	400	80/400 = 0.2
13-*cis*-Retinoic acid (isotretinoin), Ro 4-3780	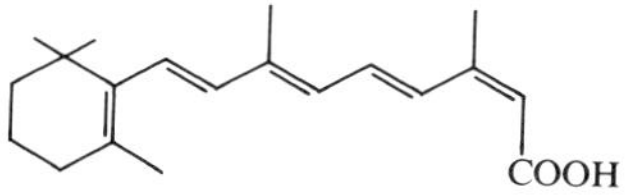 COOH	400	800	400/800 = 0.5
Trimethylmethoxyphenyl analog of retinoic acid ethylamide (motretinide), Ro 11-1430	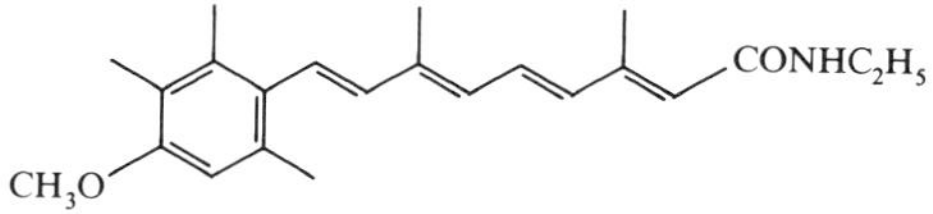 CH_3O; $CONHC_2H_5$	100	50	100/50 = 2.0
Trimethylmethoxyphenyl analog of retinoic acid ethyl ester (etretinate), Ro 10-9359	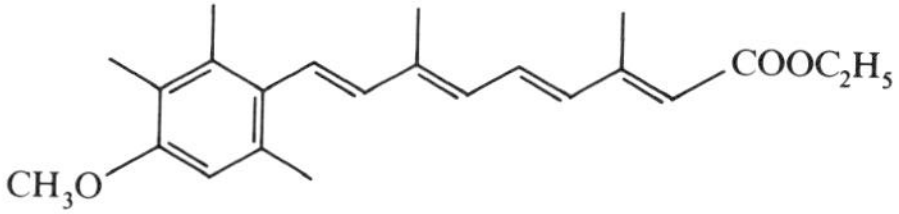 CH_3O; $COOC_2H_5$	50	25	50/25 = 2.0
Dichloromethylmethoxyphenyl analog of retinoic acid ethyl ester, Ro 12-7554	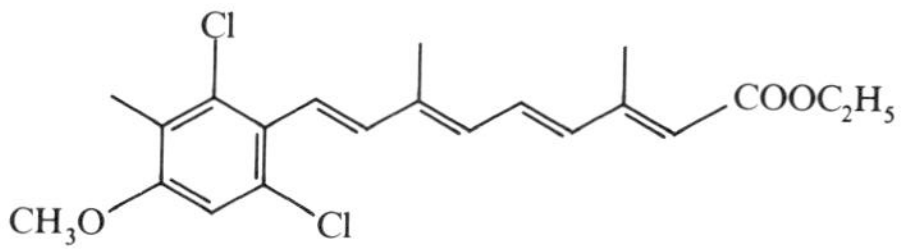 Cl; Cl; CH_3O; $COOC_2H_5$	12	6	12/6 = 2.0
Arotinoid ethyl ester, Ro 13-6298	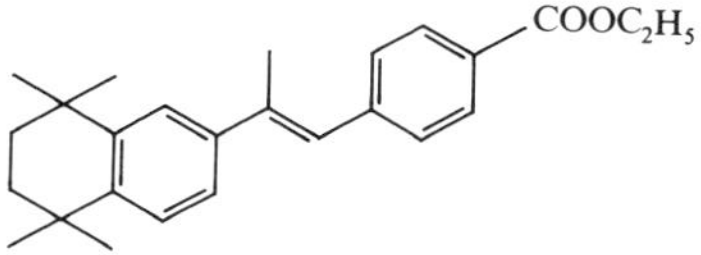 $COOC_2H_5$	0.1	0.05	0.1/0.05 = 2.0

[a] From Ref. 14.

epidermis is often moderately hyperplastic. How these nodules respond to retinoid treatment has not been examined.

Promotion of Mouse Skin Tumors by Retinoids

The efficacy of retinoids as inhibitors of carcinogenesis is well documented, but in some experimental models retinoid exposure appeared to enhance, rather than inhibit, tumor formation. all-*trans*-Retinoic acid was found to promote skin tumorigenesis in mice when administered topically for 20 weeks after tumor initiation by DMBA and before TPA tumor promotion was started.[17] In another experimental setting, all-*trans*-retinoic acid, 13-*cis*-retinoic acid, and *N*-(4-hydroxyphenyl)retinamide (4-HPR) were found capable of inducing mouse skin tumor promotion initiated with DMBA when administered topically or systemically.[18] Etretinate appeared to be negative in this respect.[19] The mechanism of this tumor-promoting activity of retinoids is not well understood.

Topical Treatment of Mouse Papillomas with Retinoids

A series of experiments has been performed to evaluate if retinoids, when applied topically to mouse papillomas, could discriminate between true topical effects and those effects mediated via systemic mechanisms after transcutaneous absorption.[20]

Methodology of Test

In these experiments papillomas are induced as described above. Approximately three papillomas per mouse are marked as "treated" papillomas and are treated with various concentrations of the retinoid under study, dissolved or suspended in acetone. Two and one-half microliters of the solution/suspension is dripped topically on each of the three treated papillomas, 5 times per week for 3 consecutive weeks. At least three additional papillomas on the same mouse are marked as "vehicle treated" and are treated with 2.5 μl of acetone alone, 5 times per week for 3 consecutive weeks (see Fig. 5). The test compounds are administered in the late afternoon under dimmed light; subsequently, the animals are kept in

[17] H. Hennings, M. L. Wenk, and R. Donahoe, *Cancer Lett.* **16,** 1 (1982).

[18] D. L. McCormick, B. J. Bagg, and T. H. Hultin, *Cancer Res.* **47,** 5989 (1987).

[19] S. M. Fischer, A. J. P. Klein-Szanto, L. M. Adams, and T. J. Slaga, *Carcinogenesis* **6,** 575 (1985).

[20] D. Hartmann, R. C. Wiegand-Chou, R. Wyss, *et al., in* "Pharmacology and the Skin" (B. Shroot and H. Schäfer, eds.), Vol. 3, p. 113. Karger, Basel, 1989.

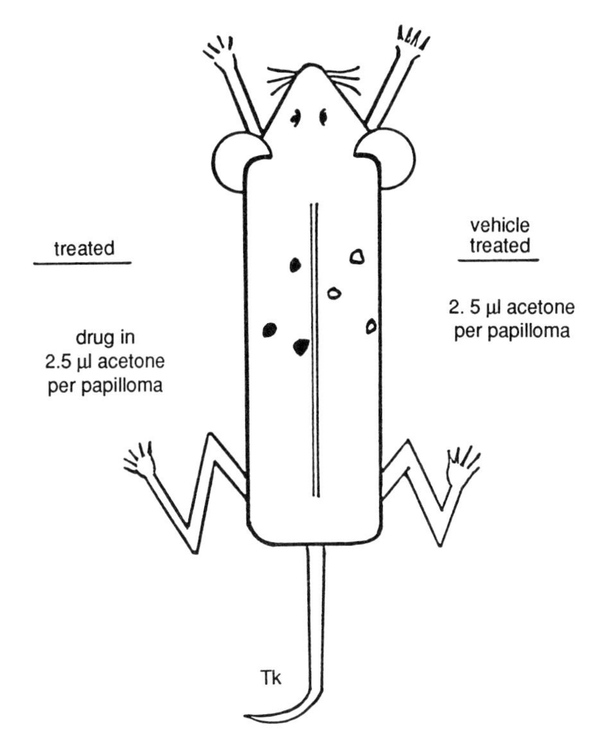

FIG. 5. Scheme of topical treatment of mouse papillomas. The test compound solution is dripped on three papillomas marked as "treated," and the vehicle is applied to at least three papillomas marked as "vehicle-treated" (open symbols).

the dark till the next morning. Four mice are used per group. The animals are housed single in separate cages.

Efficacy is evaluated by measuring two perpendicular diameters of the papillomas and by calculating the percentage change in area size. Each mouse serves as its own control in that changes of the treated papillomas are compared to changes in the vehicle-treated papillomas. Only papillomas having a diameter of at least 1 mm at the start of treatment are used. Additionally, concentrations of the various retinoids and their isomers or metabolities are determined in the plasma, in the treated and vehicle-treated papillomas, using a specific high-performance liquid chromatography (HPLC) method. Examples of test results are given in Tables IV and V.

These results suggest that the topical treatment of mouse papillomas with retinoids can discriminate between retinoid compounds or concentrations which may have a direct or true topical antipapilloma effect and

TABLE IV
CHANGE IN PAPILLOMA SIZE AFTER RETINOID TREATMENT[a]

Test compound	Concentration (%)	Change in papilloma size after 3 weeks of treatment (%)	
		Treated	Vehicle-treated
Control	Pure	+14 to +60	
all-*trans*-Retinoic acid	0.01	+46	+37
	0.1	−54[b,c]	−5
Etretinate	0.4	−32[b,c]	−25[d]
	1.2	−63[b,c]	−36[b]
	3.6	−68[b]	−53[d]

[a] From Ref. 20.
[b] Statistically significantly different from control, $p < .005$.
[c] Statistically significantly different from the vehicle-treated papillomas, $p < .05$.
[d] Statistically significantly different control, $p < .05$.

those which act via dermal absorption and activation of systemic mechanisms.[20] all-*trans*-Retinoic acid at a concentration of 0.1% appeared to act topically because only the treated papillomas responded, with 54% reduction in size, whereas the vehicle-treated papillomas showed only minimal changes (Tables IV and V). This corresponded well with the high retinoic acid concentration in the treated papillomas (113 ± 14.3 μg/g). The plasma levels were below the quantification limit (<2 ng/ml). In contrast, after topical administration of etretinate, all papillomas, either treated or vehicle-treated, showed dose-dependent reductions in size. The size reduc-

TABLE V
RETINOID CONCENTRATIONS IN MOUSE PAPILLOMA TEST[a]

Test compound	Concentration administered (%)	Plasma concentration (ng/ml)[b]	Papilloma concentration (μg/g)[b]	
			Treated	Vehicle-treated
Control	Pure	—	—	—
all-*trans*-Retinoic acid	0.01	<2	6.0 ± 8.1	0.09 ± 0.09
	0.1	<2	113.0 ± 91	13.7 ± 14.3
Etretinate	0.4	3 ± 1	240 ± 196	0.98 ± 1.4
	1.2	14 ± 7	745 ± 349	4.36 ± 2.33
	3.6	39 ± 16	2582 ± 2895	22.1 ± 9.7

[a] From Ref. 20.
[b] Mean ± S.D.

tions in the etretinate-treated papillomas were only slightly more pronounced than those in the vehicle-treated papillomas. This was accompanied by a dose-dependent increase in the plasma levels of etretinate (and even higher levels of acitretin; see Ref. 20) and a steep dose-dependent increase of etretinate in the treated papillomas (Table V). Thus, the antipapilloma activity can be interpreted as resulting from a combination of topical and systemic antipapilloma effects (treated papillomas) or a systemic effect (vehicle-treated papillomas) of etretinate.[20]

It appears that the topical papilloma test in mice is able to discriminate between true topical and systemically mediated effects, although more testing needs to be done to further validate these findings.

[42] Cancer Chemoprevention by Retinoids: Animal Models

By RICHARD C. MOON and RAJENDRA G. MEHTA

Introduction

Chemoprevention of cancer can be achieved, in most cases, either by eliminating the causative agent or by administering compounds that prevent or reduce the occurrence of cancer. Experimentally, the efficacy of a potential chemopreventive agent can be evaluated in animal models for carcinogenesis. Several target organ-specific model systems are available to study the effect of agents such as retinoids on the inhibition of carcinogenesis.[1,2] In order for a system to qualify as an effective experimental carcinogenesis model, several conditions must be satisfied. The chemical carcinogen must be specific for the target organ, that is, tumors should only arise in the organ of interest. The development of the tumor must be rapid; 6 to 9 months appears to be an optimal period for carcinogenesis studies. Histopathologically, tumors should be comparable to those found in humans. Similarly, the tumors should behave with respect to growth pattern and/or responsiveness to hormones in a manner similar to those of humans. Finally, the chemical carcinogen should cause little, if any, toxicity to the animal. Using these criteria, a number of tumor models (Table I) have been utilized for determining the chemopreventive efficacy of reti-

[1] R. C. Moon and L. Itri, *in* "The Retinoids" (M. B. Sporn, A. B. Roberts, and D. S. Goodman, eds.), Vol. 2, p. 327. Academic Press, New York, 1984.
[2] D. L. Hill and C. J. Grubbs, *Anticancer Res.* **2,** 11 (1982).

TABLE I
ANIMAL TUMOR MODELS FOR CHEMOPREVENTION[a]

Species	Target	Carcinogen[b]	Tumor Type[c]	Advantage
Rat	Mammary	MNU	Adenocarcinoma	Invasive, hormone responsive
Rat	Mammary	DMBA	Adenocarcinoma, fibroadenoma	Hormone responsive, initiation/promotion
Mouse	Urinary bladder	OH-BBN	TCC	Invasive, highly aggressive
Rat	Urinary bladder	OH-BBN	Papilloma, TCC	Slow growing, histologic stages
Hamster	Tracheobronchial	MNU	SCC	Localization of tumor
Hamster	Lung	DEN	Papilloma, SCC, adenosquamous carcinoma	Histologic types
Mouse	Colon	Azoxymethane	Adenocarcinoma	Initiation/promotion
Rat	Colon	MAM	Adenocarcinoma	Single dose
Mouse	Skin	DMBA–TPA	Papilloma	Initiation/promotion

[a] Reproduced with permission from R. C. Moon and R. G. Mehta, *Prev. Med.* (in press).
[b] MNU, *N*-Methyl-*N*-nitrosourea; DMBA, dimethylbenz[*a*]anthracene; OH-BBN, *N*-butyl-*N*-(4-hydroxybutyl)nitrosamine; DEN, diethylnitrosamine; MAM, methylazoxymethanol acetate; TPA, 12-*O*-tetradecanoylphorbol 13-acetate.
[c] TCC, Transitional cell carcinoma; SCC, squamous cell carcinoma.

noids, and, in general, these agents have shown chemopreventive activity against carcinogen-induced skin,[3] urinary bladder,[4,5] tracheobronchial,[6] pancreas,[7] and mammary[8-10] tumors.

Urinary bladder tumors are induced within 6 months in either C57BL/

[3] R. K. Boutwell, *Cancer Res.* **43,** 2465s (1983).
[4] P. J. Becci, H. J. Thompson, C. J. Grubbs, C. C. Brown, and R. C. Moon, *Cancer Res.* **39,** 3141 (1979).
[5] R. M. Hicks, *Br. Med. Bull.* **36,** 39 (1980).
[6] P. Nettesheim, M. N. Lare, and C. Snyder, *Cancer Res.* **36,** 996 (1976).
[7] D. S. Longnecker, T. J. Curphey, E. T. Kuhlmann, and B. D. Roebuck, *Cancer Res.* **42,** 19 (1982).
[8] R. C. Moon, C. J. Grubbs, and M. B. Sporn, *Cancer Res.* **36,** 2626 (1976).
[9] R. C. Moon, H. J. Thompson, P. J. Becci, C. J. Grubbs, R. J. Gander, D. L. Newton, J. M. Smith, S. R. Phillips, W. R. Henderson, L. T. Mullen, C. C. Brown, and M. B. Sporn, *Cancer Res.* **39,** 1339 (1979).
[10] D. L. McCormick, P. J. Burns, and R. E. Alberts, *JNCI, J. Natl. Cancer Inst.* **66,** 559 (1981).

6X DBA/2F1 (BDF) mice or F344 rats by *N*-butyl-*N*-hydroxybutylnitrosamine (OH-BBN). Becci *et al.*[11] showed that tumors induced by OH-BBN in BDF mice resemble the highly invasive variant of transitional carcinoma in humans. This model has proved to be effective for evaluating the chemopreventive efficacy of retinoids against urinary bladder cancer.[12]

The skin carcinogenesis model utilizes either CD-1 or SENCAR mice.[13,14] Animals are treated topically with 7,12-dimethylbenz[a]anthracene (DMBA) in acetone, at a concentration that does not induce tumors. Two weeks after DMBA treatment, mice receive 5–10 μg of the promoter 12-*O*-tetradecanoylphorbol 13-acetate (TPA) topically in 0.2 ml acetone twice weekly, and the incidence as well as growth of tumors are monitored. Animals develop a high incidence of skin papillomas by 18 weeks.

The specificity of benzo[*a*]pyrene for respiratory tract carcinogenesis in Syrian golden hamsters has been reported by Nettesheim and colleagues[6] in several studies. Recently, the effect of 4-hydroxyphenylretinamide (4-HPR) was studied on lung tumors induced by diethylnitrosamine (DEN). In DEN-treated hamsters tumors develop in the nasal cavities, larynx, trachea, and lungs regardless of the route of administration. Studies conducted in our laboratory showed that subcutaneous injections of 17.8 mg/kg DEN twice weekly for 20 weeks results in a 60–70% incidence of lung and a 90–100% incidence of tracheal tumors in the hamsters within 180 days.[15]

The study of colon carcinogenesis has been greatly facilitated by the synthesis of several carcinogens that selectively induce tumors in the large bowel. Dimethylhydrazine (DMH), methylazoxymethanol, and all other derivatives of DMH have been found to be colon carcinogens.[16] More recently, a rat colon carcinogenesis model has been used in which either *N*-methyl-*N*-nitrosourea or β-propiolactone induces colon tumors when administered intrarectally (Ratko *et al.*, unpublished).

Most of the above-mentioned animal tumor models require multiple treatments with carcinogens. Moreover, except for skin, the cancer incidence cannot be determined prior to sacrifice of the animals. Thus, an ideal model system would be one in which a single carcinogen dose would induce cancers at a select site and in which tumor incidence and/or growth

[11] P. J. Becci, H. J. Thompson, J. M. Strum, C. C. Brown, M. B. Sporn and R. C. Moon, *Cancer Res.* **41,** 927 (1981).

[12] R. C. Moon, D. L. McCormick, P. J. Becci, Y. F. Shealy, F. Frickel, J. Paust, and M. B. Sporn, *Carcinogenesis* **3,** 1469 (1982).

[13] A. K. Verma, B. G. Shapas, H. M. Rice, and R. K. Boutwell, *Cancer Res.* **39,** 419 (1979).

[14] D. L. McCormick and R. C. Moon, *Cancer Lett.* **31,** 133 (1986).

[15] R. G. Mehta, K. V. N. Rao, C. J. Detrisac, G. J. Kelloff, and R. C. Moon, *Proc. Am. Assoc. Cancer Res.* **29,** 129 (1988).

[16] B. S. Reddy, K. Watanabe, and J. J. Weisburger, *Cancer Res.* **37,** 4155 (1977).

can be monitored prior to sacrifice of the animals. Chemically induced carcinogenesis of the mammary gland offers such a luxury.

Because the majority of the retinoid chemoprevention studies conducted in our laboratory employ the rat mammary carcinogenesis model, this chapter focuses on retinoids and mammary carcinogenesis. Induction of carcinogenesis in the rat mammary gland and its chemoprevention by agents such as retinoids,[17] inhibitors of polyamine biosynthesis difluoromethyl ornithine (DFMO),[18] selenium,[19,20] and inhibitors of prostaglandin biosynthesis (indomethacin[21] or quinacrine[22]) have been successfully studied in many laboratories. In addition to ionizing radiation,[23] chemical carcinogens of such structural diversity as aromatic hydrocarbons and nitroso compounds have been used to induce mammary tumors. Generally, a single treatment with the carcinogen is sufficient to induce tumors in female rats; these tumors resemble human breast cancers histopathologically. The occurrence and growth can be monitored by palpation, and the tumors respond to hormonal modulation comparable to that of human breast tumors.

The two carcinogens used by most investigators are DMBA and *N*-methyl-*N*-nitrosourea (MNU). There are certain advantages and disadvantages in using either of the carcinogens. MNU is a direct acting agent. The MNU model was originally described by Gullino *et al.*[24] and extensively modified in our laboratory.[25] Cancers are induced in young female rats either by a single intravenous injection or by a single subcutaneous injection of the carcinogen. DMBA, on the other hand, is an aromatic hydrocarbon which requires metabolism to an active carcinogen component. DMBA is usually administered intragastrically in sesame oil.[26] Both MNU and DMBA specifically induce cancers of the mammary gland. MNU induces nearly all adenocarcinomas whereas DMBA-induced tumors are approximately 60% adenocarcinomas and 40% fibroadenomas. DMBA-induced tumors remain encapsulated and never metastasize,

[17] R. C. Moon and R. G. Mehta, *Prev. Med.* **18,** 576 (1989).

[18] H. J. Thompson, E. J. Herbst, L. D. Meeker, R. Minacha, A. M. Ronan, and R. Fite, *Carcinogenesis* **5,** 1649 (1984).

[19] D. Medina, M. Lane, and F. Shephard, *Anticancer Res.* **1,** 3771 (1981).

[20] C. Ip, *Cancer Res.* **41,** 4386 (1981).

[21] D. L. McCormick, M. Madigan, and R. C. Moon, *Cancer Res.* **45,** 1803 (1985).

[22] D. L. McCormick, *Carcinogenesis* **9,** 175 (1988).

[23] C. J. Shellaberger, D. Chmelevsky, and A. M. Kellerer, *JNCI, J. Natl. Cancer Inst.* **65,** 821 (1980).

[24] P. M. Gullino, H. M. Pettigrew, and F. H. Grantham, *JNCI, J. Natl. Cancer Inst.* **54,** 401 (1975).

[25] D. L. McCormick, C. B. Adamowski, A. Fiks, and R. C. Moon, *Cancer Res.* **41,** 1690 (1981).

[26] C. J. Grubbs, R. C. Moon, M. B. Sporn, and D. L. Newton, *Cancer Res.* **37,** 599 (1977).

whereas MNU-induced tumors invade the neighboring tissues as well as metastasize to distant sites. Both types of tumors are steroid hormone-dependent; MNU induces nearly 80–90% hormone-dependent cancers, whereas 60% of the DMBA-induced cancers are hormone-dependent. In this regard, DMBA-induced tumorigenesis is more closely related to the human situation. Histopathologically, both DMBA- and MNU-induced cancers resemble human breast adenocarcinomas; however, DMBA-induced cancers, as indicated previously, neither invade nor metastasize. A lifetime dose-dependent occurrence of MNU-induced tumors has been reported previously from our laboratory;[25] such a critical study for DMBA-induced tumors has not been reported. Procedures as presently used in our laboratory for inducing tumors with both DMBA and MNU are discussed below.

Induction of Mammary Tumors

Virgin, female Sprague-Dawley [Hsd:(SD)BR] rats are used for all mammary studies. Animals are obtained from Harlan/Sprague-Dawley (Indianapolis, IN) at 28 days of age and held in quarantine until the beginning of each study. Sprague-Dawley animals are chosen for several reasons: they are readily available, relatively healthy animals which are susceptible to the induction of mammary cancer by carcinogens, and they have been used successfully in cancer chemoprevention studies. The animals are housed in polycarbonate cages (3 rats/cage) in windowless rooms that are illuminated for 12 hr each day and maintained at a constant temperature of $22 \pm 2^\circ$ and 50% relative humidity. At 42 days of age, the animals are randomized by weight into groups, and feeding of either the basal or retinoid-supplemented diet is initiated.

N-Methyl-N-nitrosourea as Carcinogen

Breast cancer is induced by treatment of 50-day-old female Sprague-Dawley rats with a single intravenous injection of 50 mg MNU/kg body weight. Solutions of MNU (Ash-Stevens, Detroit, MI) are freshly prepared immediately prior to administration by dissolving the crystalline carcinogen in sterile physiological saline acidified with acetic acid to pH 5.0 (MNU is highly unstable at pH 6.5 or above). The MNU concentration in solution is adjusted such that a 150-g animal receives a volume of 0.6 ml. Animals are lightly anesthesized with ether, the jugular vein is exposed, and the MNU solution is administered via the jugular vein. Vehicle controls receive an injection (0.4–0.6 ml) of acidified saline only. Following MNU administration, the surgical incision is approximated with wound clips. Wound clips are removed 7 days after MNU administration. MNU can be also administered by subcutaneous injection;[27] however, a significant number of sarcomas develop at the site of injection.

7,12-Dimethylbenz[a]anthracene as Carcinogen

Crystalline DMBA (Sigma Chemical Co., St. Louis, MO) is dissolved in sesame oil (laboratory grade; Fisher Scientific Co., Pittsburgh, PA) at a dosing concentration of 12–15 mg per ml. At 50 days of age, female Sprague-Dawley rats receive a single intragastric instillation (i.g.) of 12–15 mg DMBA; vehicle controls receive a single 1.g. dose of 1 ml sesame oil only. All animals are fasted for 18 hr prior to DMBA or sesame oil administration. Seven days following DMBA or sesame oil administration, animals are randomized into groups by weight.

Observations

Beginning 4 weeks after carcinogen administration, carcinogen-treated animals are palpated weekly for the appearance of mammary tumors. Palpation is continued on a weekly basis for the duration of the study. The dates of appearance and locations of all tumors are recorded. Animals are observed twice daily and weighed weekly for the duration of the experiment.

Moribund animals are sacrificed by CO_2 narcosis. All moribund animals and any animals found dead during the course of the study are promptly necropsied. At the termination of the experiment, 180 days after carcinogen administration, all remaining animals are sacrificed by CO_2 narcosis. All mammary tumors are coded as to location, excised, individually weighed, and prepared for routine histopathological processing and classification. Any other grossly abnormal tissue is removed and processed for histologic evaluation.

Statistical Analysis

Mean body weights, mammary carcinoma multiplicities, and latencies are compared using analysis of variance (ANOVA). Mammary cancer latency curves are prepared using the method of life tables; log rank analysis is used to compare significance between latency curves. The incidences of cancer and percentages of animals surviving at termination of a study are compared by χ^2 analysis. Statistical significance is attributed to any result yielding a p value below .05.

[27] H. J. Thompson and L. D. Meeker, *Cancer Res.* **43,** 1628 (1983).

Diets

Previous studies of retinoid chemoprevention were conducted with Wayne Lab Chow or Purina Lab Chow; however, recent studies in our laboratory are conducted with semipurified AIN-76A diet. The diet, AIN-76A, is modified to contain 25% dextrose monohydrate and 40% corn starch as the carbohydrate source. Prior to mixing into the basal diet, an appropriate amount of the retinoid is dissolved in absolute ethanol–trioctanoin (1 : 3, 50 g/kg diet). This mixture is then combined with Tenox 20 (Eastman Chemicals, Kingsport, TN) to yield a final concentration of 0.5 ml/kg diet, plus basal diet to equal 500 g of agent–carrier premix. The agent–carrier premix is blended with 9.5 kg of AIN-76A diet using a P-K Liquid-Solid Blender (Patterson-Kelly, East Stroudsburg, PA). Placebo diet contains the AIN-76A diet with added carrier only. Fresh batches of diet are usually prepared weekly and stored at −20° prior to use; food supplies to the animals are changed twice a week. In cases where stability of the retinoid is questionable, food supplies are changed on a daily basis.

Analytical Methods

Diets containing retinoids are periodically analyzed for the accuracy of diet mixing and stability. Retinoids are extracted from 5 g of diet with 50 ml of methanol. Aliquots of the extract are applied to a Whatman C_{18} ODS-2 high-performance liquid chromatographic (HPLC) column (10 μm; 250 × 4.6 mm) and separated as described previously by Hultin *et al.*[28]

Subacute Toxicity and Dose Selection Studies

Prior to conducting the chemoprevention studies of rat mammary gland, each test retinoid is separately examined in a 6-week study to determine the amount which can be administered in the diet without causing suppression of body weight gain or inducing other gross manifestations of toxicity. The initial dose ranges are determined based on data obtained from the supplier or the literature. A typical protocol is given in Table II.

In all dose selection studies, animals begin treatment at the same age at which agent administration begins in the respective chemoprevention study. Prior to beginning treatment, animals are randomized by body weight into treatment groups. Animals are observed twice daily and weighed twice weekly in order to assess their general health. After at least 6

[28] T. A. Hultin, R. G. Mehta, and R. C. Moon, *J. Chromatogr.* **341,** 147 (1985).

TABLE II
PROTOCOL FOR PRELIMINARY DOSE SELECTION STUDY IN RATS[a]

Group	Number of animals	Agent dose
1	10	Vehicle control
2	10	0.25*X*
3	10	0.5*X*
4	10	*X*
5	10	2*X*
6	10	4*X*

[a] Animals: Female Sprague-Dawley rats, 43 days of age. Agent: Administered in diet with appropriate carrier. Dose *X* to be selected on information available in the literature. Route of administration given in literature to be extrapolated to dietary administration. Vehicle: Carrier in AIN-76A diet. Duration: 6 weeks.

weeks of treatment, all animals except those on doses selected for chemoprevention study are sacrificed, and necropsies are performed in order to detect possible gross toxicity of an agent. Animals on the selected doses continue to receive the compound at those doses in order to observe any potential long-term toxicity.

Chemoprevention of Breast Carcinogenesis

The mammary chemoprevention studies using the retinoids are performed according to a typical protocol in which the retinoid is mixed in the basal diet at two different concentrations, determined from the results of each dose selection study. One level is generally set at 80% and the other is 40% of the MTD (maximum tolerated dose) for each retinoid. Retinoid feeding is commenced 7 days before rats receive carcinogen and continues for an additional 180 days. For studies pertaining to the effects of retinoids on only the initiation phase of carcinogenesis, the retinoids are present prior to carcinogen treatment only. Conversely, for antipromotional studies, retinoids are present in the diet beginning 1 week postcarcinogen treatment until the end of the experiment.

At times it is necessary to include a food restriction group in the experiment, particularly if the retinoid is known to suppress body weight gain. In this case, two groups of rats receive only the basal diet at levels which are restricted to maintain a mean body weight equivalent to that of the groups which receive the retinoid at a selected dose level. Such restricted food groups makes it possible to distinguish nonspecific, anticar-

TABLE III
INFLUENCE OF *N*-METHYL-*N*-NITROSOUREA DOSE ON TUMOR INCIDENCE, TUMOR MULTIPLICITY, AND ANIMAL MORTALITY[a]

Group	Number of rats	MNU dose (mg/kg)	M_{50} (days)	T_{50} (days)	Final cancer incidence (%)	Cancers[b] per rat	Benign tumors[c] per rat
1	20	50	246	62	100	6.48 (86)[d]	1.06 (14)
2	20	45	257	68	100	4.82 (94)	0.33 (6)
3	20	40	264	79	100	3.16 (89)	0.40 (11)
4	20	35	312	96	90	3.24 (91)	0.32 (9)
5	20	30	322	105	95	2.76 (80)	0.70 (20)
6	20	25	306	238	70	1.89 (71)	0.78 (29)
7	20	20	384	393	65	2.60 (73)	0.95 (27)
8	20	15	451	434	55	1.08 (59)	0.75 (41)
9	20	10	523	—[e]	45	1.09 (42)	1.48 (58)
10	10	0	565	—	20	0.30 (50)	0.30 (50)

[a] Reproduced with permission from Ref. 25. M_{50}, time to 50% mortality; T_{50}, time to 50% cancer incidence.
[b] Includes adenocarcinomas and papillary carcinomas.
[c] Includes fibroadenomas, fibromas, and lobular hyperplasia.
[d] Numbers in parentheses represent the percentage of total tumors per rat in that group.
[e] Group never reached 50% cancer incidence.

cinogenic effects, which might result from retinoid-induced suppression of food intake or body weight gain, from those of specific, retinoid-mediated anticarcinogenic activity.

In the majority of the studies in which MNU is used as the carcinogen, the dose has been 50 mg MNU/kg body weight. However, other dose levels may be used successfully, as a lifetime dose–response relationship has been shown for the MNU-induced mammary carcinogenesis in rats. As indicated in Table III, an excellent dose-dependent induction of mammary cancers is apparent for MNU. At a 50 mg/kg dose level the appearance of the tumors is rapid, and nearly 90% of the tumors are adenocarcinomas. Such an extensive dose–response relationship for DMBA-induced carcinogenesis in a lifetime study has not been reported. DMBA at 12–15 mg dose per rat results in an 80–90% mammary tumor incidence at 180 days postcarcinogen, of which 40% are fibroadenomas.

Retinoid chemoprevention of mammary carcinogenesis has been studied in a variety of rat species, Sprague-Dawley being the most frequently used. Retinyl acetate was the first retinoid showing effectiveness in Sprague-Dawley[29] and Lewis[10] rats. However, retinyl acetate either enhanced

[29] R. C. Moon, C. J. Grubbs, M. B. Sporn, and D. S. Goodman, *Nature (London)* **267**, 620 (1977).

TABLE IV
EFFECT OF RETINOIDS ON MAMMARY CARCINOGENESIS OF RATS *in Vivo*

Carcinogen	Retinoid	Effect
DMBA	Retinyl palmitate	No effect
DMBA	Retinal Acetate Retinyl acetate *N*-(4-Hydroxyphenyl) retinamide	Inhibition of carcinogenesis
MNU	Retinyl acetate Retinyl methyl ether *N*-(4-Hydroxyphenyl) retinamide Axarophthene all-*trans*-Retinoic acid *N*-(4-Hydroxyphenyl)-13-*cis*-retinamide Temaroten	Inhibition of carcinogenesis
MMU	13-*cis*-Retinoic acid Retinyl butyl ether *S*-Ethylretinamide Retinylidene dimedone Retinylidene acetylacetone TMMP analog of retinyl ethyl ester TMMP analog of retinyl methyl ether	No inhibition
Benxo[*a*]pyrene	Retinyl acetate	Inhibition of carcinogenesis

tumorigenesis or had no effect in C3H/Avy[30] or GR[31] strains of mice. Table IV shows a list of effective and ineffective retinoids in mammary carcinogenesis. Studies from our laboratory indicate that *N*-(4-hydroxyphenyl)retinamide is the most efficacous retinoid in this system.[9,17] Effects of retinoids or any chemopreventive agent on mammary carcinogenesis can be evaluated using three principal criteria: (1) reduction in cancer incidence, with incidence being defined as percent cancer-bearing animals in a group; (2) reduction of multiplicity (average number of cancers per animal in the experimental group compared to controls); and (3) the reduction in tumor latency with latency being defined as the time to appearance of the first palpable tumor from the day of carcinogen treatment. A selected set of results from one of our recent studies using temaroten (Ro 15-0778), an arotenoid without a polar end group, is shown in Table V. Several conclusions can be derived from this study. Temaroten significantly lengthened the time to appearance of the first tumor. It clearly

[30] A. Maiorano and P. Gullino, *JNCI, J. Natl. Cancer Inst.* **64,** 655 (1980).

[31] C. W. Welsch, M. Goodrich-Smith, C. C. Brown, and N. Crowe, *JNCI, J. Natl. Cancer Inst.* **67,** 633 (1981).

TABLE V
EFFECT OF DIETARY ADMINISTRATION OF TEMAROTEN ON *N*-Methyl-*N*-NITROSOUREA-INDUCED MAMMARY CARCINOGENESIS IN FEMALE SPRAGUE-DAWLEY RATS[a]

Group	Temaroten (mg/kg body weight)	Cancer incidence (%)	Cancers per rat	Cancer latency (days)	At termination: Survival (%)	At termination: Body weight (% control)
1	Basal diet	100	7.2[b]	55[b]	60	100
2	30	93	4.5	74	75	90
3	90	89	2.9	81	76	87
4	270	41[c]	0.5[d]	75	40[e]	81
5	Basal diet FR to group 4	93	4.0	79	87	81

[a] Animals received 50 mg MMU per kg body weight intravenously at 50 days of age; each group consisted of 28 to 30 rats. Temaroten {Ro 15-0778, 1,2,3,4-tetrahydro-1,1,4,4-tetramethyl-6-[(*E*)-α-methylstryryl]naphthalene} is an arotenoid without any polar end group. Intake of rats in group 5 was matched by food restriction (FR) to that calculated for group 4. Procedures for food restriction were followed daily for the first 14 days of the study, and thereafter thrice weekly.
[b] $p < .05$ versus groups 2, 3, 4, and 5.
[c] $p < .01$ versus groups 1, 2, 3, and 5.
[d] $p < .05$ versus group 5.
[e] $p < .05$ versus groups 2, 3, and 5.

inhibited the multiplicity of cancers in a dose-related manner. Although the retinoid suppressed body weight gain, the food restriction groups of the study indicate that it is the temaroten and not suppression of body weight that affected the reduction in the multiplicity. Reduction of food intake in group 5 (basal diet) was comparable to that of group 4 and was accompanied with reduction in tumor multiplicity from 7.2 cancers to 4 cancers per rat. However it was still significantly higher than rats treated with temaroten (group 4). The data also show that retinoid toxicity may be responsible for lower survival in the rats receiving the high dose of the compound. A similar chemopreventive effectiveness of temaroten against DMBA-induced mammary tumors has been previously reported by Bollag and Hartman.[32]

In addition to the influence of retinoids on carcinogenesis as a single agent, the effectiveness of retinoids in combination with other agents such as hormone antagonists (tamoxifen)[33] or CB154[34] has also been studied.

[32] W. Bollag and H. Hartmann, *Eur. J. Cancer Clin. Oncol.* **23,** 133 (1987).
[33] T. A. Ratko, C. J. Detrisac, N. Dinger, C. F. Thomas, G. J. Kelloff, and R. C. Moon, *Cancer Res.* **49,** 4472 (1989).
[34] C. W. Welsch, C. C. Brown, M. Goodrich-Smith, J. Chiusano, and R. C. Moon, *Cancer Res.* **40,** 3095 (1980).

More recently, retinoids such as 4-HPR[35] and temaroten[36] have been shown to have possible chemotherapeutic effect as well. Finally, with a thorough understanding of the effect of retinoids in preventing mammary carcinogenesis, delaying the appearance of subsequent tumors following surgical removal of the first tumor,[37,38] their role as adjuvants in combination with antihormones,[33,34] or their chemotherapeutic action,[35,36] they could be utilized clinically as effective suppressing agents against breast cancer in humans.

[35] K. Dowlatshahi, R. G. Mehta, C. F. Thomas, N. Dinger, and R. C. Moon, *Cancer Lett.* in press (1989).
[36] H. R. Hartmann and W. Bollag, *Cancer Chemother. Pharmacol.* **15,** 141 (1985).
[37] D. L. McCormick, Z. L. Sowell, C. A. Thompson, and R. C. Moon, *Cancer* **51,** 594 (1983).
[38] R. C. Moon, J. F. Pritchard, R. G. Mehta, C. T. Nomides, C. F. Thomas, and N. Dinger, *Carcinogenesis* **10,** in press (1989).

[43] Utility of Disposition Data in Evaluating Retinoid Developmental Toxicity

By CALVIN C. WILLHITE and STEVEN A. BOOK

Introduction

The toxicity of naturally occurring and synthetic retinoids in humans and animals,[1] their dose–response parameters and biochemical actions,[2] and the requirements for retinoid chemical structure in aberrant morphogenesis have been described.[3] The retinoids are of particular interest in developmental biology as retinol is absolutely required for normal embryogenesis, and there is evidence that all-*trans*-retinoic acid functions as an endogenous morphogen, controlling local spatial relationships in the embryo. Abnormally high concentrations of acidic retinoids in avian, rodent, and human embryos result in terata, affecting the development of many of the same organ systems as those affected in retinoid deficiency.[2] Humans are 10–100 times more sensitive on a body weight basis to the teratogenic effects of certain retinoids (e.g., isotretinoin) than rodents (mice, hamsters), but humans are about equally sensitive in the case of

[1] W. B. Howard and C. C. Willhite, *J. Toxicol. Toxin Rev.* **5,** 55 (1986).
[2] C. C. Willhite, P. J. Wier, and D. L. Berry, *CRC Crit. Rev. Toxicol.* **20**(2), 113 (1989).
[3] C. C. Willhite, *in* "Chemistry and Biology of Synthetic Retinoids" (M. I. Dawson and W. H. Okamura, eds.), Chap. 21, p. 539. CRC Press, Boca Raton, Florida, 1990.

other retinoids (e.g., etretinate). These observations allow for comparison of the human and rodent teratogenic response, and they permit testing of allometric hypotheses for interspecies dose scaling in teratology.

This chapter deals with the various techniques for and calculations of retinoid pharmacokinetic parameters which may be applied to the problems of interspecies and route of administration differences in teratogenic potency. Retinol and its esters are used to illustrate the utility of metabolic fate data in the interpretation of conventional teratology bioassay results, and all-*trans*-retinoic acid is used to illustrate route-dependent differences.

Practical Considerations

Pharmacokinetic studies should be carried out using only highly purified, well-characterized retinoids. At a minimum, thin-layer[4] and standard reversed-phase high-performance liquid chromatographic (HPLC) analyses[4,5] are necessary to establish purity. The HPLC determinations should be conducted at two different wavelengths (e.g., 254 and 340 nm).[5] In addition, melting point, infrared and mass spectra,[6] ^{13}C NMR, and ^{3}H NMR[4] are used to establish retinoid identity and to confirm purity prior to structure–activity or disposition studies. Whether the retinoid to be studied is a viscous oil or crystal (generally red, orange, or yellow), factors that must be taken into account include the inherent instability of polyene retinoids to UV-induced photoisomerization,[7] oxygen, and elevated temperatures.[8] Retinoids, dosing solutions, blood, and plasma should be handled under subdued yellow or gold fluorescent light, with as little exposure to air as possible. Stock retinoids are never stored as dilute solutions, but rather are kept in the dark, under a dry inert gas (argon, nitrogen), preferably in a sealed amber glass ampoule, inside a desiccated plastic bottle, at −80°. In contrast, the conformationally restricted aromatic congeners[9] are stable at room temperature; however, we always ship these analogs under dry ice and otherwise store and handle them under conditions identical to the polyene retinoids. All manipulations of the compounds are carried out using disposable laboratory gloves. Because of their high intrinsic pharmacologic activities (particularly in the skin), the conformationally restricted

[4] C. C. Willhite and Y. F. Shealy, *J. Natl. Cancer Inst.* **72,** 689 (1984).
[5] W. B. Howard, C. C. Willhite, M. I. Dawson, and R. P. Sharma, *Toxicol. Appl. Pharmacol.* **95,** 122 (1988).
[6] W. C. Coburn, Jr., M. C. Thorpe, Y. F. Shealy, M. C. Kirk, J. L. Frye, and C. A. O'Dell, *J. Chem. Eng. Data* **28,** 422 (1983).
[7] R. W. Curley, Jr., and J. W. Fowble, *Photochem. Photobiol.* **47,** 831 (1988).
[8] D. S. Goodman and J. A. Olson, this series, Vol. 15, p. 462.
[9] J. L. Flanagan, C. C. Willhite, and V. H. Ferm, *J. Natl. Cancer Inst.* **78,** 533 (1987).

aromatic congeners (arotinoids) should be handled under containment procedures similar to those outlined in the National Cancer Institute's Guidelines for the Safe Handling of Chemical Carcinogens.

Dosing solutions for animal studies are prepared just prior to use. Retinoid solubilization is most readily accomplished by dissolving a small amount of compound in 250 μl acetone or ethanol in a 5-ml volumetric flask and subsequently bringing the solution up to volume with Tween 20. This latter vehicle is selected for experimental oral retinoid administration because it is the solvent used for retinol in over-the-counter vitamin A supplements.[10] The esters, alcohols, amides, sulfones, and dimedones disperse rapidly in acetone or other polar solvents as compared to their acidic congeners. Prior dissolution of the retinoid in the organic phase eliminates the tendency of crystalline retinoids to clump. Dimethyl sulfoxide (DMSO) should not be used as it can be teratogenic.[11] Solvent concentrations greater than 5% can induce untoward toxicity in the dams.

Use of labeled compounds permits rapid determination of total tissue distribution,[12,13] but this approach may limit the utility of the results. Radiolabeled retinoids can be handled in much the same fashion as given above; however, some derivatives (notably, retinamides) that are highly radioactive (3.5 Ci/mmol) undergo intrinsic radiodegradation,[13] making interpretation of the distribution results problematic. In contrast, aromatic retinoids of very high intrinsic radioactivity (22.8 Ci/mmol) have no such tendency.[14] Extraction and direct HPLC quantitation of parent retinoids and individual metabolites in blood, embryos, and other peripheral tissues are necessary for rigorous pharmacokinetic estimates, and automated methods have appeared[15] which reduce the tedium associated with retinoid measurement. Conventional methods for determination of total circulating and peripheral tissue residual retinoid radioactivity have been described in detail elsewhere, along with methods for extraction, isolation, and quantitation by HPLC and ultraviolet spectroscopy.[12-14] Separation by HPLC, fraction collection, and quantitation by liquid scintillation spectroscopy of the label associated with TTNPB (Ro 13-7410)[14] or all-*trans*-

[10] C. C. Willhite, *Toxicol. Lett.* **20,** 257 (1984).

[11] C. C. Willhite and P. I. Katz, *J. Appl. Toxicol.* **4,** 155 (1984).

[12] W. B. Howard, C. C. Willhite, S. T. Omaye, and R. P. Sharma, *Arch. Toxicol.* **63,** 112 (1989).

[13] W. B. Howard, C. C. Willhite, S. T. Omaye, and R. P. Sharma, *Fundam. Appl. Toxicol.* **12,** 621 (1989).

[14] W. B. Howard, C. C. Willhite, R. P. Sharma, S. T. Omaye, and A. Hatori, *Eur. J. Drug Metab. Pharmacokinet.* **14,** 153 (1989).

[15] J. Creech Kraft, C. Eckhoff, W. Kuhnz, B. Loefberg, and H. Nau, *J. Liquid Chromatogr.* **11,** 2051 (1988).

retinoic[16] acid permits an increase in the limit of sensitivity (to fmol/g plasma) compared to ultraviolet methods (nmol/g plasma).[12]

Animal Protocol

Because a pregnant animal does not necessarily handle xenobiotics in the same manner as a nonpregnant or male individual, and in order to collect embryonic and placental tissues, it is necessary to procure animals of known gestational age. Retinoid pharmacokinetic and placental permeability data have been published using rat, mouse, and hamster models. The types of congenital defects found in hamsters are remarkably similar to those reported in human infants with retinoid embryopathy.[2,3] The adult hamster body weight is 50% that of a sexually mature rat, and the retinoid dose required to induce terata can be one-tenth that required in a mouse; these characteristics are useful when only a very limited quantity of retinoid is available for study. Daily vaginal smears, detection of vaginal plugs, and determination of females in estrus are not necessary when using hamsters. Moreover, the large structure–activity series for retinoids in the hamster tracheal organ culture (a measure of intrinsic retinoid control of epithelial differentiation) can be compared to the structure–activity relationships in developmental toxicity.[3]

Hamsters should be purchased from or raised in a specific-pathogen-free, controlled entry, closed outbred colony. Fiscal considerations dictate the purchase of timed pregnant hamsters if only a small number are needed. For regulatory purposes in developmental toxicology, 20 litters per dose group are usually required, and the establishment of a standing colony as the numbers of studies increase may be advantageous. The disadvantages to shipping pregnant hamsters include the cost of express air and truck delivery, special filtered shipping crates, surcharges for specific body weights, and transportation complications; the advantages include supplier pregnancy guarantees and convenience. There is no detectable difference between pregnancy rates or incidence of congenital malformation or resorption rate in purchased pregnant or in-house colony animals.[17] For pharmacokinetic studies where interindividual variability is small, five pregnant animals per dose group is usually sufficient to derive an adequate description of drug behavior. If the interindividual variability is large, then usually three logarithmically spaced dose groups with 10 animals each are necessary.

[16] C. C. Willhite, R. P. Sharma, P. V. Allen, and D. L. Berry, *J. Invest. Dermatol.* in press (1990).

[17] R. E. Shenefelt, *Teratology* **5**, 103 (1972).

Hamsters should be free of antibody to Sendai, mouse pneumonia virus (PVM), Reo-3, lymphocytic choriomenigitis (LCM), *Encephalitozoon cuniculi,* and *Mycoplasma pulmonis.*[18,19] Additionally, the colony must be free of *Herpes simplex*[19] and "wet tail," considered to be associated with transmissible *Campylobacter jejuni*[20] or factors leading to enteric proliferation of *Escherichia coli.*[21] Clinical enteric disease is inevitably fatal, and it causes complete resorption of the litter.

Attention must be given to control of ectoparasites, helminths, and pathogenic protozoa in the colony.[21] The source colony must be free of the genetic problems leading to fetal spontaneous hemorrhagic necrosis,[22] and inbreeding can be evidenced first by an increased incidence of congenital anophthalmia. Hamsters must be separated from other species, and introduction of novel males into a closed colony can increase early embryonic death.

Prior to breeding, all hamsters, male and female, must be caged individually on a 14-hr light/10-hr dark cycle for at least 1 week. This allows establishment of the normal 4-day estrous cycle of the dominant female and cage territorial dominance by the relatively docile male. We have noted that the entire colony tends to synchronize their estrous cycle, and direct observation of copulation is all that is necessary to confirm pregnancy with a high degree of confidence. It is wise that the investigator not acquainted with hamsters perform husbandry tasks and become familiar with handling prior to attempting breeding or dosing. A single female is placed into the cage with the male; females should be 100–140 g body weight for assurance of conception. Copulation can usually be observed within minutes; if not, the female is returned to her own cage and another female is placed with the male. Animals that mate are left together during the dark period and are separated to their respective cages at the beginning of the following light period. If one assumes a 9 PM initiation of the dark period, ovulation occurs about 1 AM.[17] Within 12 hr of mating, the one cell stage with pronuclei appears in the oviduct; thus, the day following the evening of breeding is designated Day 1 of pregnancy.[23]

Our studies of the retinoid structure–activity relationships in teratology, placental permeability, and maternal pharmacokinetic parameters

[18] G. Margolis and L. Kilham, *Lab. Invest.* **21,** 189 (1969).

[19] V. H. Ferm and R. J. Low, *J. Pathol. Bacteriol.* **89,** 295 (1965).

[20] R. H. Lentsch, R. M. McLaughlin, J. E. Wagner, and T. J. Day, *Lab. Anim. Sci.* **32,** 511 (1982).

[21] T. Poole (ed.), "The UFAW Handbook on the Care and Management of Laboratory Animals," 6th Ed., p. 392. Longmans, New York, 1987.

[22] R. F. Keeler, S. Young, R. S. Spendlove, D. R. Douglas, and G. F. Stallknecht, *Teratology* **11,** 21 (1975).

[23] V. H. Ferm, *Lab. Anim. Care* **17,** 452 (1967).

have been conducted with oral intubation[12-14] or skin painting[16] on the morning of Day 8 of pregnancy, a time corresponding to the presomite[23] to early somite[17] stage of development. (The reader should note that Shenefelt[17] carried out developmental toxicity studies of all-*trans*-retinoic acid given during different stages of hamster embryogenesis and found marked differences in the types of terata that could be induced.) Firm yet gentle restraint is required during intubation to prevent pulmonary aspiration. Retinoid dosing solution or an equivalent volume of the vehicle can be given at the rate of 0.5 ml per 100 g body weight using a 1-ml tuberculin syringe affixed to a 2-inch curved feeding needle with 2.25 mm ball diameter.

In general, the time course of drug action and the pharmacologic response elicited by an active retinoid can be related to one or another aspect of the circulating concentrations of the parent drug and its active metabolite(s). Measures of plasma retinoid concentrations allow confirmation of whether the retinoid is absorbed, and, if so, to what extent and at what rate, allow comparisons of retinoid bioavailability (exposure), allow characterization of factors that enhance or retard bioavailability, allow measures of the rate of elimination, allow estimation of the influence of multiple or continuous exposures, and provide insights into the entities (parent drug, active metabolite) responsible for the response.

Pharmacokinetic studies of circulating retinoids have utilized protocols wherein rat or mouse blood is collected from the posterior vena cava or abdominal aorta at laparotomy; however, to compare animal plasma parameters to those in the human, it is preferable to collect serial blood samples just as done in clinical studies. In rats and mice, bleeding from the tail vein is an option, but because hamsters have only rudimentary tails, collection from the postorbital plexus into a herparinized glass hematocrit tube is satisfactory.[12-14] As many as 12 samples can be rapidly and easily collected, alternating eyes between time points to avoid trauma to the dam. Collection of sequential samples from the same animal, in contrast to collection of a single sample at necropsy, allows for the fit of individual plasma concentration–time values to a curve and the calculation of compartment-derived pharmacokinetic parameters. This method allows presentation of mean pharmacokinetic parameters (with attendant variation) as opposed to calculation of composite curve parameters based on mean analytical values.

The sampling times depend on the rate of enteric retinoid absorption, which in turn depends on whether the animal is fed or fasted, whether the retinoid is encapsulated or unencapsulated, the participation of intestinal carrier proteins, and the retinoid chemical structure. Sampling times depend on whether the compound accumulates in a "deep" compartment, the route (dermal application shows prolonged absorption and maintenance of elevated plasma concentrations), and the limit of analytical detection. One cannot sample too soon after dosing as oral retinoids can be absorbed with virtually no lag time;[12,14] others are delayed up to 30 min.[24] Parent retinoids can decline to less than the limit of UV detection within 4–24 hr after an oral dose, but metabolites can persist in the circulation for more than 96 hr.[12–14]

Absorption, Distribution, and Elimination

Retinoid efficacy in cancer chemoprevention and dermatologic disease prompted publication of human and rodent pharmacokinetic data for naturally occurring and synthetic derivatives. These data allow interspecies comparisons of retinoid metabolic fate and disposition, an approach which identifies those species and pharmacokinetic parameters relevant to the human teratogenic response. Dose-dependent (multiple low dose versus single high dose) and route-dependent (dermal versus oral) plasma pharmacokinetic data are measured by the area under the plasma concentration–time curve (area under curve, *AUC*). Besides *AUC*, the peak concentration (C_{max}) and the circulating plateau concentration data can assist in predicting retinoid teratogenic response, provided there is a known relationship between maternal circulating levels and the concentrations of the parent retinoid and teratogenic metabolites in the embryo.

If one uses classic compartment analysis, there are no physiological or anatomic definitions of the location of the compartments—other than that from which the data were obtained. To relate quantitatively the pharmacokinetic data to the retinoid teratogenic response, one plots the particular parameter (*AUC*, C_{max}) against the teratogenic (effective) dose ED_{10},[2] ED_{50},[3] or other measure. In general, acidic retinoid *AUC* values are proportional to dose measured on a body weight basis. The *AUC* after an oral dose is calculated to "infinite time," usually referenced as 10 times the biological half-life. However, as the plot *AUC* ("total absorbed dose") versus administered dose (mg/kg) becomes curvilinear downward with increasing dose, this implies that either biological half-life, volume of distribution, or bioavailability is changing with administered dose (dose-dependent pharmacokinetics). If the plot for the parent drug becomes curvilinear upward with dose, this implies saturation of the hepatic microsomal system responsible for retinoid oxidation and conjugation. With saturation kinetics (wherein the *AUC*/dose ratio increases with adminis-

[24] C. C. Willhite, F. I. Chow, M. J. Taylor, and S. T. Omaye, *Food Chem. Toxicol.* **23,** 51 (1985).

tered dose), as can be observed at pharmacologic doses, the retinoid plasma concentration remains elevated for prolonged periods of time.

At massive administered doses, (e.g., ≥ 150 mg/kg) even retinoids that are not teratogenic (e.g., retinamides[4])—at doses equimolar to the teratogenic ED_{100} for all-*trans*-retinoic acid—can elicit embryotoxicity. This phenomenon may not be due to parent drug pharmacokinetic parameters at all, but rather either reflects limited *in vivo* biotransformation of the parent amide or dimedone to an acidic congener or possibly reflects trace concentrations of contaminating starting materials ($<1\%$) used in their synthesis (e.g., retinal), which can assume toxicologic significance at the very high dose levels.

Rigorous data fitting to pharmacokinetic libraries requires knowledge of the plasma concentrations of individual retinoid species, as measures of total vitamin A or total retinoid radioactivity have little relevance in this context. The production, time course, and elimination of teratogenically active metabolites appear to play major roles in the relative embryonic toxicities of these compounds. Many of the oral retinoid plasma profiles can be described using a two-compartment open model with first-order (exponential) uptake (diminishing in proportion to the amount remaining to be absorbed) with a lag time to appearance in the blood, but others can best be described using a one-compartment model with first-order uptake, first-order elimination, and no lag time.

With first-order elimination, a constant fraction of retinoid in plasma is cleared in each equal interval of time; first-order processes include sequestration of retinoid in peripheral tissues, including the embryo.[25,26] In the case of plasma clearance by peripheral compartment sequestration (as in the case of lipophilic retinamide accumulation in adipose tissue[13]), the rate is dependent on blood flow to the depot and the physiochemical properties of the retinoid (as in the failure of acidic retinoids which carry a net negative charge at physiological pH to accumulate in adipose tissue[12]).

Plasma pharmacokinetic parameters can be most conveniently obtained using nonlinear regression models calculated by personal computer software (PCNONLIN).[27] The desktop software requires user justification of a model from a single dose or a multiple dose library of 23 selections each, in contrast to the mainframe pharmacokinetic decision-making programs (AUTOAN).[24] Options are provided (Nelder–Mead simplex,

[25] J. Reiners, B. Lofberg, J. Creech Kraft, D. M. Kochhar, and H. Nau, *Reprod. Toxicol.* **2,** 19 (1988).

[26] D. M. Kochhar, J. Kraft, and H. Nau, *in* "Pharmacokinetics in Teratogenesis" (H. Nau and W. J. Scott, Jr., eds.), p. 173. CRC Press, Boca Raton, Florida, 1987.

[27] Statistical Consultants, Inc., *Am. Stat.* **40,** 52 (1986).

Gauss–Newton with suppression of the magnitude of iterative parameter change) to account for situations where one has collected and analyzed as many plasma samples as possible, yet the program terminates prematurely or converges to unlikely values. The process involves selecting a mathematical relationship (written as a sum of exponential functions) which describes adequately the observed plasma concentration–time data and deriving the parameter values which describe the toxicologically relevant features of the curves. Software installation, data entry format, and program execution instructions are given in the PCNONLIN[27] and JANA[28] manuals with accompanying references to classic pharmacokinetic modeling. The least-squares program (PCNONLIN)[27] requires entry of initial estimates, derived using a technique known as curve stripping ("method of residuals"). An iterative program (JANA)[28] obtains these initial parameter estimates and calculates a "closeness of fit" as measured by an R^2 term.

As a case in point, when fitting a two-compartment open model ($C_t = A^{-\alpha t} + B^{-\beta t} + C^{-k_a t}$), PCNONLIN[27] calculates the overall rate of elimination (k_e), the rates of drug movement from the central (plasma) compartment to and from the peripheral compartment, the *AUC*, and the uptake and elimination half-times (biological half-lives). The elimination rate constant (k_e) describes the instantaneous rate of drug disappearance at time t; as the amount of retinoid available for elimination declines; the rate constant declines. For example, the plasma k_e for 13-*cis*-retinoic acid in pregnant hamsters after a single oral (35 μmol/kg) dose was 0.18 hr^{-1}, that is, about 18% per hour of the remaining 13-*cis*-retinoic acid molecules were eliminated from maternal plasma. This rate corresponds to a biological half-life of some 15 hr, accounting for a relatively large plasma *AUC* value of the parent retinoid.[12] The prolonged elimination and large plasma *AUC* are consistent with retinoid glucuronidation and enterohepatic circulation, resulting in prolonged maintenance of elevated plasma concentrations.

The extent of an acute, systemic pharmacologic response is often related to the peak concentration. The parent retinoid maximum concentration (X_{max}) for each subject can be calculated from the administered dose (D) and the first-order absorption (k_a) and elimination (k_e) rate constants supplied by PCNONLIN:

$$X_{max}/D = (k_a/k_e)^{k_e/(k_e - k_a)} \quad (1)$$

The time to maximum plasma parent retinoid concentration (t_{max}) for each subject can also be calculated from the single dose absorption and elimina-

[28] Statistical Consultants, Inc., "JANA Users Manual." Lexington, Kentucky, 1987.

tion rates:

$$t_{max} = 2.3/(k_a - k_e) \log(k_a/k_e) \qquad (2)$$

The value of t_{max} is independent of administered dose, whereas *AUC* is dose-dependent. The t_{max} value is determined by the more rapid of the two rates, and usually the rapid enteric absorption rate dictates the time to peak. Clearance, another measure used in interspecies comparisons of pharmacokinetic variation,[29] is calculated from k_e multiplied by the apparent volume of distribution (V_d) (the latter determined from the plasma level following a single intravenous injection). Retinoids are eliminated from the circulation by extensive first-pass hepatic metabolism. In the absence of an empirical V_d, the maximum intrinsic initial clearance can be calculated as dose/*AUC*. The hepatic extraction ratio is given as clearance divided by the sum of clearance and hepatic blood flow.

Applications

Individual xenobiotic biological half-times can be distributed either normally or log-normally. Log-normal distribution implies that those animals or patients at the upper end of the curve have substantially longer elimination half-times (and consequent higher sustained circulating concentrations per unit administered dose) than the population mean. The more rapid the elimination of retinoid from the maternal circulation, the lower the presumptive embryonic or fetal retinoid concentration. These variations in retinoid elimination rates (biological half-times) may explain, in part, the fact that not all rodent or human fetuses whose mothers receive pharmacologic doses of teratogenic retinoids exhibit the retinoid malformation syndrome.[1] In the cases of acidic, amide, or benzoic acid derivatives of retinoids, it has not yet been possible to correlate mean maternal peak circulating concentrations of parent drug, total plasma radioactivity (representing circulating parent retinoid and all metabolites), or other parameters (including total residual radioactivity in the fetus) with relative teratogenic potency.[12-14]

Nau and associates[26] discussed target organ (embryonic) dose (measured as embryonic *AUC*) and implicated polyene retinoid accumulation to steady-state concentrations in the embryo as "the decisive factors in retinoid teratogenesis." This latter phenomenon has implications for retinoid developmental toxicity, and it is a consequence of retinoid physical chemistry. For example, in the case of the retinoidal benzoic acid deriva-

[29] H. Boxenbaum, *J. Pharmacokinet. Biopharm.* **10,** 201 (1982).

tive Ro 13-7410 (TTNPB) (which is at least 750 times more potent a teratogen in hamsters than retinoic acid[9]), the pK_a is approximately 4.2. The low pK_a favors gastric absorption (steady state [plasma]/[stomach] = 1580), and this characteristic retards intestinal absorption. At plasma pH values, acidic retinoids are present almost entirely in the ionized state. Free charged retinoids exist initially in the extracellular space (~20% body weight); maternal–embryonic equilibrium can be reached only via penetration of protonated (electrically neutral) lipophilic species through placental membranes. The early rodent embryo maintains a relatively basic pH (7.64)[30] compared to maternal plasma (human, 7.4; hamster, 7.39; mouse, 7.26). The ratio of total parent Ro 13-7410 concentration in the embryo to that in maternal plasma is given by

$$R = \frac{1 - \text{antilog}(\text{pH embryo} - pK_a)}{1 - \text{antilog}(\text{pH plasma} - pK_a)} \tag{3}$$

Thus, the Ro 13-7410 will accumulate to levels approximately 1.5 times higher in the target tissue compared to the maternal circulating concentration. At static equilibrium, the net maternal circulating and embryonic concentrations of the nonionized moiety will be equivalent on both sides of the placenta, but there will be more total retinoid on the embryonic side, a mechanism referred to as "ion trapping." This effect is quantitatively important to retinoid teratogenic response if and only if a substantial proportion of the drug is in the ionized form in at least one of the compartments. The effect is even more pronounced where the pH differences are larger (e.g., the pH of blastocoelic fluid is 9).[30] This is not the case, for example, in the case of the parent, nonteratogenic retinamides.[13,29]

Intubation of a single large dose of all-*trans*-retinol in female hamsters or rats illustrates pronounced species differences[24] and allows comparisons to the human retinol pharmacokinetic parameters.[31] Humans show a measured plasma C_{max} of 950 ng/ml after 0.8 mg/kg retinol; hamsters and rats show C_{max} values of 1765 and 777 ng/ml, respectively, after 45 mg/kg. Humans show an *AUC* value of 6000 ng hr/ml, almost 50% that observed in hamsters (14,435 ng hr/ml) given a dose 57 times larger. Because of the relatively slow elimination, the rat *AUC* value is 50% larger (23,082 ng hr/ml) than that for the hamster. In spite of the large differences in administered dose, the human retinol C_{max} value is larger than that for the rat and about 50% that found in the hamster.

[30] H. Nau and W. J. Scott, Jr., *Nature (London)* **323,** 276 (1986).

[31] G. E. Goodman, D. S. Alberts, D. L. Earnst, and F. L. Meyskens, *J. Clin. Oncol.* **1,** 394 (1983).

These observed differences are most likely due to species-dependent intestinal esterification, as the plasma retinyl esters show massive increases following a bolus vitamin A dose.[31] Because retinol is at least 20 times less active a teratogen than retinoic acid,[32] and it is the acidic forms of the retinoids which are responsible for induction of terata,[2,3] the relative concentrations of retinoic acid and its oxidized acidic biotransformation products arising from the free plasma retinol[32] are of paramount import here.

The plasma pharmacokinetic parameters associated with a single oral[12] or single topical[16] dose of all-*trans*-retinoic acid show clear differences attributable to the route of administration. Compared to an equivalent dose given orally, the C_{max} values after dermal treatment are only 2% of those after the oral dose. The AUC values, however, are 63% of that after the oral treatment. In the case of oral all-*trans*-retinoic acid, the parent drug in plasma declines to less than the limit of detection within 4 hr, but after application to intact skin, all-*trans*-retinoic acid is detected up to 2 weeks after initial treatment. The AUC values are dose- and route-dependent.[16] These data show that there is a lower peak concentration but protracted absorption from the cutaneous depot. This latter phenomenon results in a large half-time for absorption and prolonged elevation of circulating levels. The diminished C_{max} value after topical treatment is associated with the failure to elicit a significant teratogenic response, whereas the same dose given orally causes malformations in 50% of the pups.[4] Thus, the maternal plasma AUC value is less predictive of teratogenic response than the plasma C_{max} value. The problem then becomes the delineation of the quantitative relationship between maternal circulating concentrations ("absorbed dose") and those in embryo ("effective dose").

Acknowledgments

This work was supported in part by March of Dimes Birth Defects Foundation Research Grant No. 15-130.

[32] D. M. Kochhar, J. D. Penner, and M. A. Satre, *Toxicol. Appl. Pharmacol.* **96**, 429 (1988).

[44] Teratogenicity of Retinoids: Mechanistic Studies

By ANDREAS KISTLER and W. BRIAN HOWARD

Introduction

In chick wing buds, depending on the dose applied, retinoic acid induces extra digits or causes reductions in bone development.[1] This shift of retinoic acid acting as a morphogen and/or as a teratogen happens within a doubling of the concentration; sometimes both activities are observed at the same dose.[1,2] The question therefore arises whether the mechanism(s) of action of these two effects are related.

There is general agreement that retinoids influence cellular differentiation by modifying gene expression.[3] Depending on the concentration present in the target tissue, retinoids either induce or adversely affect the normal biochemical expression of differentiation and morphogenesis.[1] At present, however, little is known of how retinoids control gene expression. It has been proposed that retinoic acid acts by binding to a cellular retinoic acid-binding protein[4] (CRABP) and, more recently, to nuclear retinoic acid receptors[5,6] (RAR). Alternative mechanisms, however, such as the interaction of retinoids at some levels of the complex signal transduction cascade involving phospholipid degradation, Ca^{2+}, arachidonic acid, prostaglandins, protein kinases, and/or other compounds[3,7] are not yet excluded.[8]

To study different aspects of retinoid teratogenicity we focus on test systems related to cartilage and bone development. All these systems have clear-cut end points, are highly reproducible, and enable investigation of retinoid pharmacokinetics, as well as their effects on morphogenesis, cellular differentiation and morphology, and biochemical aspects including molecular biology. Other test systems, such as the whole embryo culture[9] in which retinoids induce specific malformations,[10] are not described here.

[1] D. Summerbell, *J. Embryol. Exp. Morphol.* **78,** 269 (1983).
[2] C. Tickle, B. Alberts, L. Wolpert, and J. Lee, *Nature (London)* **296,** 564 (1982).
[3] M. B. Sporn and A. B. Roberts, *Cancer Res.* **43,** 3034 (1983).
[4] S. Takase, D. E. Ong, and F. Chytil, *Arch. Biochem. Biophys.* **247,** 328 (1986).
[5] M. Petkovich, N. J. Brand, A. Krust, and P. Chambon, *Nature (London)* **330,** 444 (1987).
[6] V. Giguere, E. S. Ong, P. Segni, and R. M. Evans, *Nature (London)* **330,** 624 (1987).
[7] Y. Nishizuka, *Nature (London)* **308,** 693 (1984).
[8] A. Kistler, *Biochem. Soc. Trans.* **14,** 936 (1986).
[9] D. A. T. New, *Biol. Rev.* **53,** 81 (1978).
[10] L. Cicurel and B. P. Schmid, *Experientia* **44,** 833 (1988).

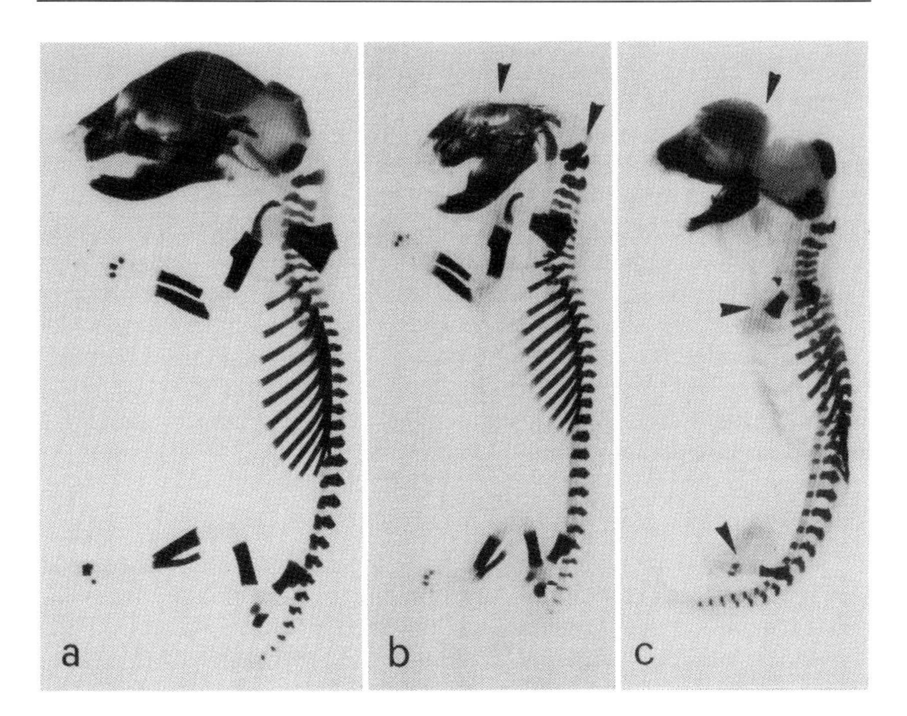

FIG. 1. Skeletal malformations induced by retinoids on Days 9 and 13 of gestation in rats. Fetuses were obtained by cesarean section on Day 20 of gestation and processed for skeletal staining with alizarin red S. (a) Control; (b) the retinoid Ro 15-1570 (arotinoid ethyl sulfone) was given on Day 9 of gestation (10 mg/kg)—note malformations of the skull and the cervical vertebrae; (c) the retinoid Ro 13-6298 (arotinoid ethyl ester) was given at Day 13 of gestation (0.3 mg/kg)—note malformations of the skull and of the fore- and hindlimbs. (Reproduced with permission from Ref. 12.)

Teratogenicity Studies *in Vivo*

Retinoids at teratogenic doses interfere with the normal development of cartilage and bone, as well as other tissues. The skeletal malformations thus induced are clearly seen and easily evaluated in near-term fetuses when the skeleton is stained with alizarin red S (Fig. 1). Stage-specific malformation patterns, depending on the day treated, are induced by retinoids in rats[11,12] (Fig. 1) as well as in mice.[13,14] For studies on the

[11] A. Kistler, *Teratology* **23,** 25 (1981).

[12] A. Kistler, B. Galli, and W. B. Howard, *Arch. Toxicol.* **64,** 43 (1990).

[13] D. M. Kochhar, *Teratology* **7,** 289 (1973).

[14] D. M. Kochhar, "Birth Defects: Original Article Series, XIII" p. 111. The National Foundation, 1977.

mechanism of action of retinoids we suggest treating animals once on a specific gestational day with a retinoid having a short elimination half-life, such as retinoic acid.[15]

For details of the dosing and recording of reproduction and malformation data, see this volume [46]. For studies on the mechanism of action, retinoid-induced limb malformations are very useful. In rats treatment on Day 12 of gestation with retinoic acid results primarily in forelimb malformations[11] ([46] in this volume). Developing limbs are a well-defined target tissue for retinoids and are easily located and removed from the embryo. They are therefore useful for pharmacokinetic, morphological, and biochemical studies. For example, the presence and distribution of retinoid-binding proteins[16] or receptors can be followed in the developing rat forelimb after a teratogenic insult on Day 12 of gestation by preparing embryos at specific time points after treatment.

Staining of Rat Fetus Skeletons with Alizarin Red S[17]

Thoracic and abdominal viscera and adipose tissue on the neck and between scapulae are removed. The fetuses are fixed in absolute ethanol for about 1 week and then macerated in 0.5% NaOH for 1 to 3 days until the skeleton is visible through the soft tissues (change NaOH if necessary). Skeletons are stained with 1 part alizarin red S stock solution [0.1% alizarin sodium sulfonate (Merck, Darmstadt, FRG) dissolved in 1% sodium hydroxide] to 35 parts 0.5% sodium hydroxide for about 24 hr. The maceration, including staining, is critical because undermaceration leads to difficulties in clearing and subsequent skeletal examination and overmaceration results in the skeleton falling apart. Fetuses are washed with tap water and cleared in a graded series of 20% (1–2 days), 50% (3–4 days), and 100% glycerol, in which they can be stored.

Examination and Evaluation of Skeletal Malformations

Depending on the intent of the study the skeletal malformations can be scored by a coarse or refined examination system (Table I). It is also possible to grade each bone of the limbs individually (Fig. 2). To study possible suppression of retinoid-induced teratogenicity by specific antagonists or inhibitors, this evaluation system is useful.[18] Using the litter as the experimental unit and determining a malformation index/litter, statistical analysis is possible.[18]

[15] A. B. Roberts and H. F. DeLuca, *Biochem. J.* **102,** 600 (1967).

[16] M. Maden, D. E. Ong, D. Summberbell, and F. Chytil, *Nature (London)* **335,** 733 (1988).

[17] A. B. Dawson, *Stain Technol.* **1,** 123 (1926).

[18] A. Kistler, *Teratog. Carcinog. Mutagen.* **6,** 93 (1986).

TABLE I
SKELETAL EXAMINATION FOR MALFORMATIONS INDUCED BY RETINOIDS[a]

Coarse examination	Refined examination
Skull	Cranium Mandible
Axial skeleton	Cervical vertebrae Thoracic vertebrae Lumbar vertebrae Sacral vertebrae
Ribs	Ribs
Forelimbs	Scapula Humerus Ulna/radius Metacarpalia/phalanges
Hindlimbs	Pelvic bones Femur Tibia/fibula Metatarsalia/phalanges

[a] Skeletal variations such as poorly ossified bones or incised neural bodies should not be included. Because in controls the sternum often shows skeletal variations which may be difficult to distinguish from abnormalities, it is excluded.

Teratogenicity Studies *in Vitro*

We propose three *in vitro* test systems related to cartilage development, more correctly, to inhibition of chondrogenesis and degradation of cartilage, by retinoids. All three test systems are related to teratogenicity of retinoids. Inhibition of chondrogenesis in limb bud cells (see this volume [45]) by retinoids correlates well with their teratogenicity *in vivo,* at least for retinoids with a free carboxylic acid end group.[19] Retinoid-induced degradation of fetal cartilage in bone organ cultures correlates well with the activity in limb bud cells (A. Kistler, unpublished data) and thus with the *in vivo* teratogenicity. The third system also uses perinatal bones as starting tissue.

[19] A. Kistler, *Arch. Toxicol.* **60,** 403 (1987).

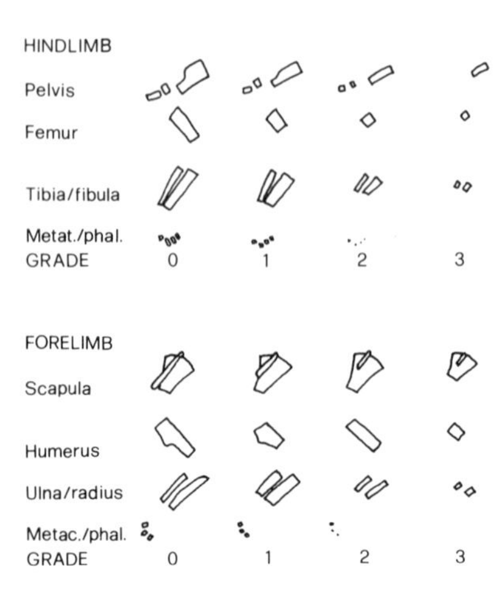

FIG. 2. Grading of skeletal malformations of fore- and hindlimbs induced by retinoids in rats. The severity of the malformation of the specific limb bones is graded from 0 (control) to 3 (severely malformed). For the sake of clarity, the shortening of the limbs in grading 2 and 3 is not shown in this scheme.

Limb Bud Cell Cultures

Details of the limb bud cell culture are described in this volume [45]. Active retinoids added on Day 1 (or 0) of culture specifically inhibit cartilage differentiation without cytotoxic effects.[19] Under these conditions retinoid treatment results in the failure of morphologically visible cartilage nodules to appear and in the lack of alcian blue-stainable proteoglycans after about 3 days. Thus, in the presence of retinoids, these two parameters remain unchanged during the culture period, whereas controls undergo marked and measurable changes. Therefore, the influence of retinoids can be judged only in comparison to controls.

Addition of retinoids on Day 5 of culture, when chondrogenesis has appeared as judged from the appearance of cartilage nodules, induces the cartilaginous nodules to disappear during the following days.[20] Thus, under these conditions retinoids induce cartilage degradation, which can be followed morphologically and biochemically (i.e., disappearance of alcian blue staining) in the treated cultures.

We believe that most of the experiments done in cell cultures to investigate the mechanism of action of retinoids can also be done at teratogenic concentrations in limb bud cell cultures. Understanding the mechanism of action of retinoid teratogenicity could help to elucidate the mechanism of action of the pharmacological effects, because it has not been possible so

[20] F. Gallandre, A. Kistler, and B. Galli, *Roux's Arch. Dev. Biol.* **189,** 25 (1980).

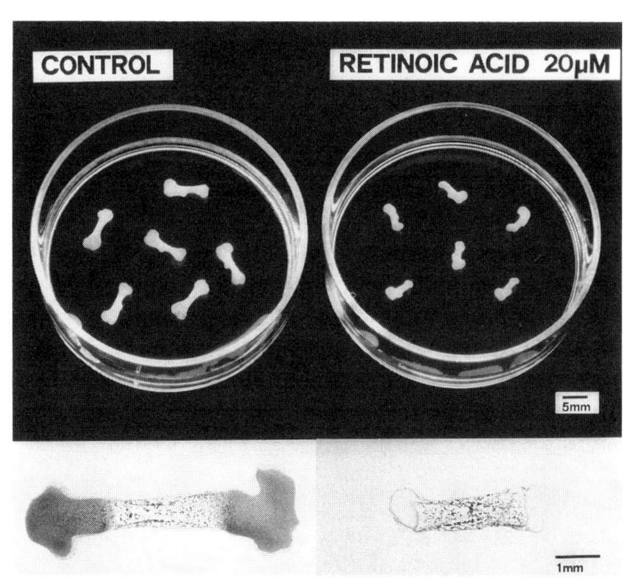

FIG. 3. Morphological changes induced by retinoic acid in cultured fetal rat bones. Humeri of fetal rats (gestation Day 19) were incubated in the absence and the presence of retinoic acid for 6 days. Note the marked shortening of the bones in the presence of the retinoid. Histological sections stained with toluidine blue demonstrate that retinoic acid induces the loss of metachromatic staining of the cartilaginous ends of the bone, which correlates with the loss of proteoglycan.[24] Furthermore, retinoic acid induces marked cartilage breakdown, represented by the very small remaining epiphyses after 6 days of incubation.

far to dissociate these two activities.[21] For mechanistic studies it would be advantageous to have an early end point (within hours after addition of the retinoid), but none has been reported so far. Limb bud cell cultures are primary cell cultures, implying that large numbers of cells are rather difficult to obtain. Limb bud cell cultures involve cell proliferation which, however, is not required for retinoid-induced cartilage degradation, as demonstrated in fetal bone organ cultures,[22] the next test system we describe.

Fetal Bone Organ Culture

Retinoids act directly on fetal bones in culture, inducing cartilage degradation.[23,24] Within 24 hr, significant release of proteoglycan[25] is in-

[21] W. Bollag, *Lancet* **1,** 860 (1983).
[22] A. Kistler, *Roux's Arch. Dev. Biol.* **193,** 121 (1984).
[23] H. B. Fell and E. Mallanby, *J. Physiol. (London)* **116,** 320 (1952).
[24] A. Kistler, *Calcif. Tissue Int.* **33,** 249 (1981).
[25] A. Kistler, *Differentiation* **21,** 168 (1982).

duced, followed by marked degradation of cartilage tissue manifested in a shortening of the bones (Fig. 3) and representing loss of DNA, RNA, and protein.[26] Depending on the age of rat humeri, retinoic acid increases the release of proteoglycan up to 6.8 times that of controls (Table II). Furthermore, the shortening of bone in response to retinoic acid is clearly expressed at gestational Days 17 to 19 but less after birth. From the practical point of view, humeri of fetuses at Days 19 to 21 are easy to prepare. These fetal humeri, however, contain less cartilage tissue than postnatal humeri. Thus, depending on the goal, fetal or postnatal humeri can be chosen.

Preparation of Humeri. Rat fetuses at the selected age (see Table II) are aseptically removed from the uterine horns and placed in physiological salt solution. After decapitation, forelimbs are dissected and collected in a new dish. Further preparation is done under a dissecting microscope in physiological salt solution. After removing the skin, the humerus is prepared free of adhering tissue with extra fine forceps. Then tendon insertions with epiphyses are cut using a pair of iridectomy scissors. For the preparation of postnatal humeri, rats are sacrificed, dipped in 70% ethanol for sterilization, decapitated, and the forelimbs dissected. Preparation of humeri is done as above.

Culturing of Humeri. Humeri are incubated in nutrient mixture F-10 (GIBCO, Glasgow, UK) supplemented with 10% fetal calf serum, 60 μg penicillin/ml, 100 μg streptomycin/ml, and 10 mM HEPES, pH 7.4 (all supplements from GIBCO) in cell culture dishes in a humidified 5% CO_2–air atmosphere at 37°. Cell culture dishes of 35 or 60 mm diameter are used with 2 or 5 ml medium, respectively.

Addition of Retinoid. The day of bone preparation is considered as Day 0 of culture. Retinoids are prepared and added as described in [45] of this volume.

Proteoglycan Determination. The amount of proteoglycan and/or glycosaminoglycans released from the bones into the medium is measured in aliquots of the medium by the alcian blue assay[27] with chondroitin sulfate as a standard. To aliquots of medium (containing 1–10 μg of proteoglycan/glycosaminoglycan) 1 ml of alcian blue complexing reagent [0.05% alcian blue (DIFCO, West Molesey, Surrey, UK), 50 mM $MgCl_2$, 50 mM sodium acetate, pH 5.8, which is freshly prepared, high shear mixed, and filtered through paper before use] is added and equilibrated for 2 hr at room temperature. After sedimentation of the alcian blue–glycosaminoglycan complex at 2000 g for 15 min, the residue is washed once with 2 ml of ethanol. The alcian blue–glycosaminoglycan complex is

[26] A. Kistler and B. Galli, *Roux's Arch. Dev. Biol.* **187,** 59 (1979).
[27] P. Whiteman, *Biochem. J.* **131,** 343 (1973).

TABLE II

AGE-DEPENDENT DIFFERENCES IN PROTEOGLYCAN RELEASE AND BONE SHORTENING INDUCED BY RETINOIC ACID IN FETAL BONE CULTURES[a]

Age	Proteoglycan release (μg/humerus/4 days)		Starting bone length (mm)	Change in length (%)	
	Control	Retinoic acid		Control	Retinoic acid
Day of gestation					
17	10.3 ± 3.8	36.0 ± 15.8	3.7 ± 0.6	$+28 \pm 7$	-46 ± 4
19	24.7 ± 1.2	168.4 ± 13.9	6.2 ± 0.1	$+19 \pm 1$	-24 ± 1
21	42.2 ± 1.6	228.9 ± 15.2	7.6 ± 0.1	$+10 \pm 1$	-1 ± 1
Day of lactation					
4	111.5 ± 10.6	498.5 ± 20.5	8.8 ± 0.2	$+12 \pm 1$	0 ± 4
8	241.5 ± 17.7	593.0 ± 52.3	9.7 ± 0.4	$+10 \pm 2$	$+6 \pm 2$
15	442.5 ± 51.6	729.0 ± 43.8	12.8 ± 0.6	$+2 \pm 0$	$+4 \pm 0$

[a] Humeri of fetal and postnatal rats (2×2 humeri/dish/group) were incubated in supplemented culture medium F-10 (final volume 2 or 5 ml depending on the size of humeri) in the absence and presence of retinoic acid (20 μM). The proteoglycan release was determined in aliquots of the medium and represents the amount released during the incubation from Days 0 to 4. The length of the bones was measured before and after 6 days of incubation and the change in length estimated. Results are means ± S.D. (Plug day = Day 0.)

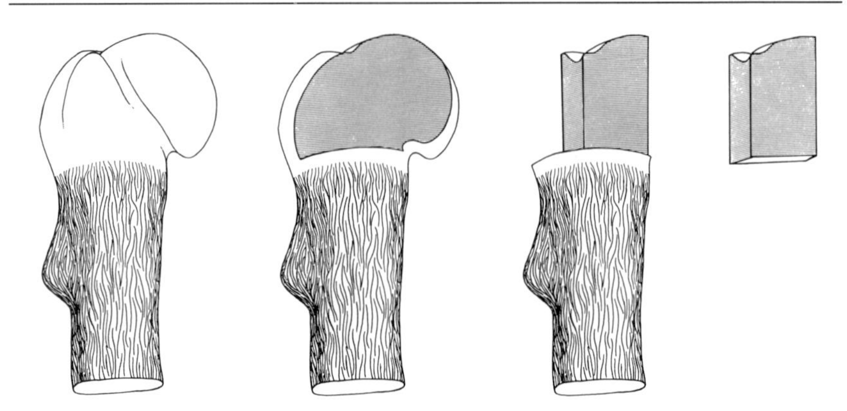

FIG. 4. Schematic representation of the preparation of cartilage pieces from the proximal epiphysis of the rat humerus. For the sake of clarity the longitudinal sections are shown in the cartilage only, but in practice the sections include part of the bony shaft. (Drawn by M. Mislin.)

dissociated with 4% sodium lauryl sulfate and measured spectrophotometrically at 620 nm.

Determination of Bone Length. To estimate the change in length during the incubation period as the parameter for tissue breakdown,[26] the length of the bones is measured under a reversed microscope and a projecting prism (Zeiss) at a final magnification of ×14.1 at the beginning and the end of the incubation.

Comment. For biochemical determinations, such as incorporation of radiolabeled precursors into macromolecules, it is possible to cut both epiphyses and thus to make the measurements in the cartilage tissue only.

Culture of Cartilage Pieces

For studies focusing on morphological (including immunohistochemical) questions, we describe here a new, not yet reported test system. Instead of preparing whole bone cultures of postnatal humeri (Table II), cartilage pieces are cut from the proximal epiphysis of the humeri (Fig. 4) and cultured. In the presence of retinoic acid, such cultures show marked cartilage degradation (Fig. 5). This system demonstrates that retinoid-induced cartilage breakdown is performed by the cartilage cells themselves and allows one to search for the cells responsible for these degradation processes.

Preparation and Culture of Cartilage Pieces. The proximal epiphysis of rat humeri from 4- to 7 day-old pups (day of birth = Day 1) are large enough to cut out a single piece of pure cartilage. Ossification of the epiphyses starts at about Day 8. Humeri are prepared as described above

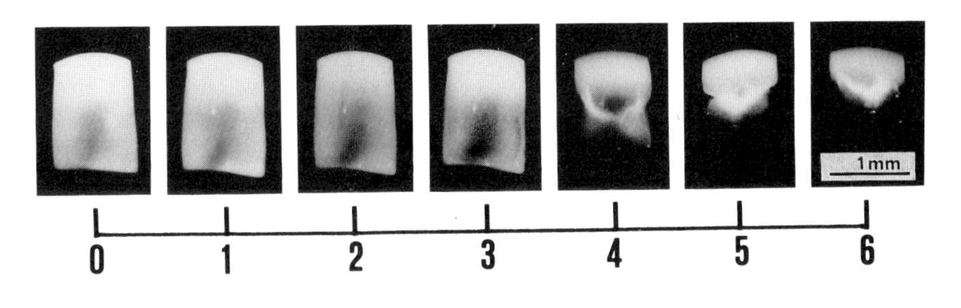

FIG. 5. Time course of retinoic acid-induced tissue degradation in a cartilage piece. A cartilage piece prepared from the proximal epiphysis of a humerus of a 7-day-old rat was incubated in supplemented nutrient mixture F-10 containing 10% fetal calf serum, antibiotics, 10 m*M* HEPES pH 7.3, and 2 μM retinoic acid. Note that after 3 days of incubation the distal part of the cartilage piece is degraded. However, the articular cartilage (uncut side, see Fig. 4) remains undegraded. In controls no degradation was observed, but the cartilage pieces increased slightly in size.

(Fetal Bone Organ Culture). With a razor blade, a section of the cartilaginous epiphysis is cut away on the four sides of the proximal joint, and then the piece is cut from the bony shaft (Fig. 4). Leaving one side uncut enables orientation of the pieces during culture. Culture conditions are the same as described in this chapter for bone organ culture. To hinder the attachment of the cartilage pieces to the plastic surface and to minimize cell outgrowth from the explants, the dishes must be shaken at least once daily.

[45] Testing of Retinoids for Teratogenicity *in Vitro:* Use of Micromass Limb Bud Cell Culture

By ANDREAS KISTLER and W. BRIAN HOWARD

Introduction

For retinoids with a free carboxylic acid end group there is a good quantitative correlation between their inhibition of chondrogenesis in limb bud cell cultures and their *in vivo* teratogenicity.[1] The limb bud cell culture system is an effective tool for a preliminary evaluation of the teratogenic potential of retinoids and has several advantages compared with other *in*

[1] A. Kistler, *Arch. Toxicol.* **60,** 403 (1987).

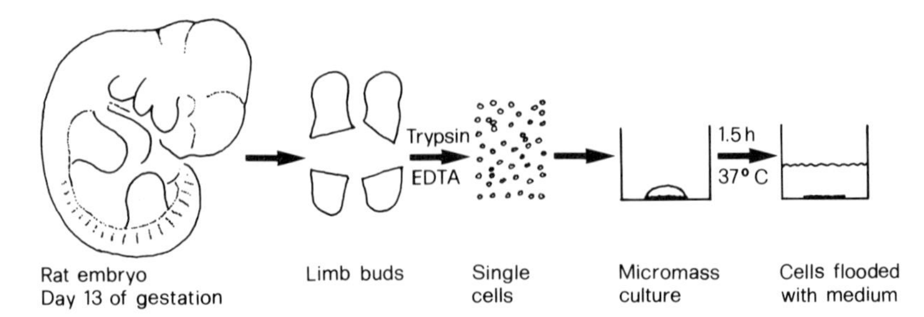

FIG. 1. Schematic representation of limb bud cell preparation. Limbs are dissected from the embryo, trypsinized to form a suspension of single cells, plated at high density (micromass), and flooded with medium.

vitro tests. It uses mammalian embryonic cells and relies on a well-established test system, namely, chondrogenesis. The test is mechanistically related to development of the skeletal system, a main target tissue for retinoids. There are quantifiable morphological and biochemical end points, enabling the estimation of IC_{50} (50% inhibition concentration) values to compare compounds with different activity. The assay is simple, easy, and requires only small amounts of retinoids. The results can be obtained rapidly, and the number of animals used is small.

In contrast to the complex responses *in vivo* which reflect absorption, distribution, metabolism, and elimination of a retinoid, responses *in vitro* are limited to the cells cultured. The comparison of effects *in vitro* with those *in vivo* may thus help to identify metabolic and pharmacokinetic factors that influence the response *in vivo.*

General Principle of Limb Bud Cell Assay

Micromass cell cultures prepared from Day 13 rat embryo fore- and hindlimbs differentiate into chondrogenic nodules and produce extracellular cartilage proteoglycans.[2,3] Limb buds are removed from the embryo, trypsinized, and cultured at high cell density (Fig. 1[4]). Biologically active retinoids inhibit both the formation of chondrogenic nodules (Figs. 2 and 3[5]) and proteoglycan production. A useful measure of biological potency, the IC_{50} can be determined spectrophotometrically using extracted alcian blue staining values of proteoglycans.

[2] J. R. Hassell and E. A. Horigan, *Teratog. Carcinog. Mutagen.* **2,** 325 (1982).
[3] F. Gallandre, A. Kistler, and B. Galli, *Roux's Arch. Dev. Biol.* **189,** 25 (1980).
[4] A. Kistler, *Concepts Toxicol.* **3,** 86 (1985).
[5] A. Kistler, M. Mislin, and A. Gehrig, *Xenobiotica* **15,** 673 (1985).

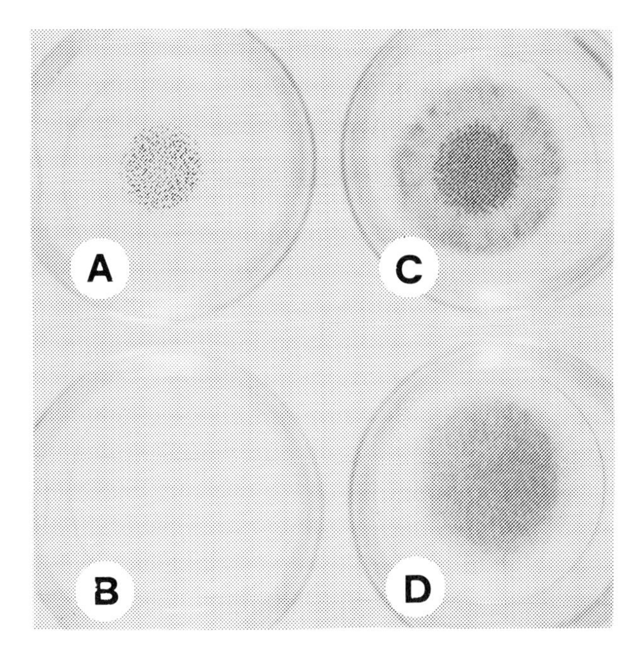

FIG. 2. Effect of retinoic acid on limb bud cell cultures. Cultures were prepared, treated with retinoic acid (1 μM), and stained as described in the text. (A) Day 8 control culture stained with alcian blue. (B) Day 8 retinoic acid-treated culture stained with alcian blue. Note the complete lack of chondrogenic nodules as evidenced by no alcian blue stain uptake. (C) Day 8 control culture stained with hematoxylin–eosin. (D) Day 8 retinoic acid-treated culture stained with hematoxylin–eosin. Note that retinoic acid inhibited chondrogenic nodule formation, but not cell growth, demonstrating that the retinoid was not cytotoxic.

Limb Bud Cell Assay

Limb Bud Cell Preparation. Female rats are mated overnight, and animals which have vaginal plugs the following morning are considered to be at Day 0 of gestation. On Day 13 of gestation the dams are sacrificed, and the uteri are aseptically removed and placed in Ringer's solution (or any other balanced salt solution). All further preparation is done in Ringer's solution. The embryos are removed from the uterine horn and collected in a new dish. The transport of the embryos and later of the prepared limb buds is done with a fine polished Pasteur pipette with the appropriate opening. Further preparation is done under a dissecting microscope. The embryonic membranes are removed with extra fine forceps. The fore- and hindlimb buds (total 100 to 200) are dissected from the embryos using a pair of iridectomy scissors. The limb buds are washed

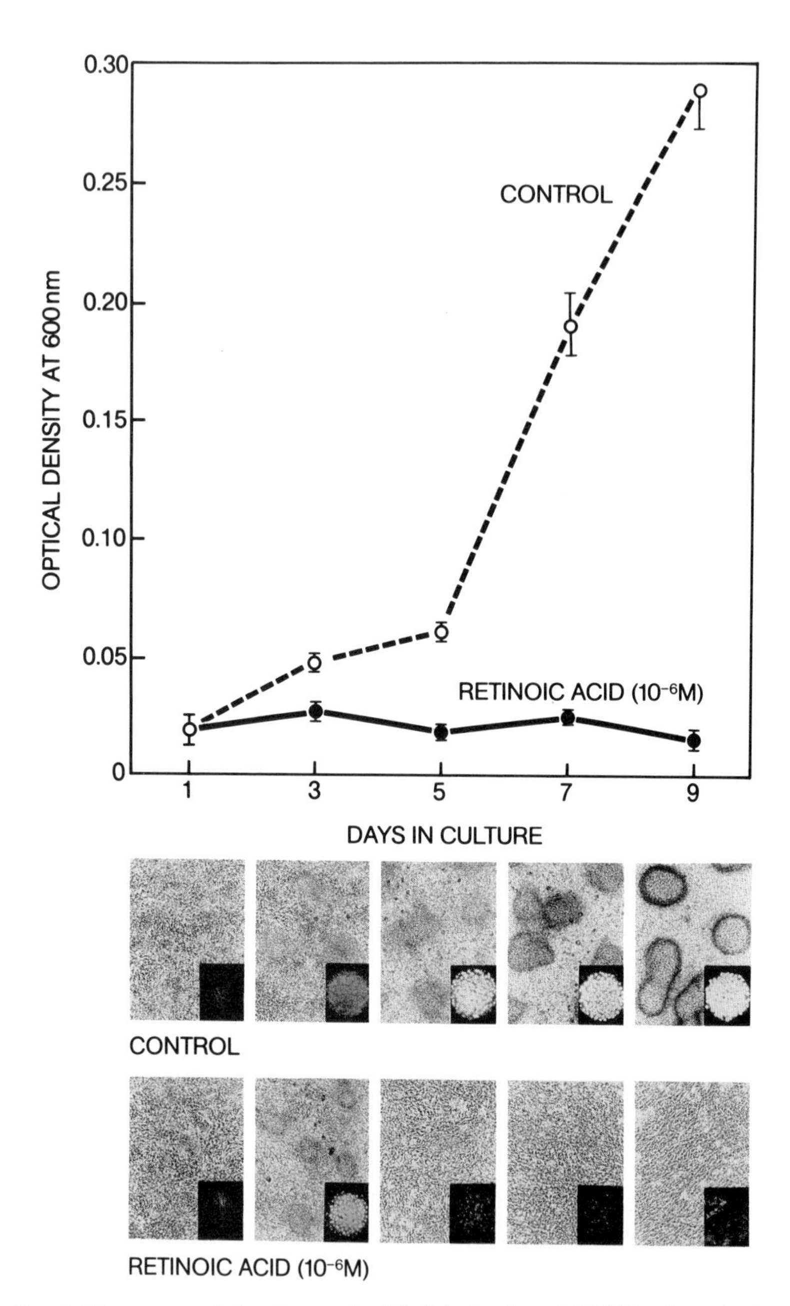

FIG. 3. Time course of chondrogenesis of limb bud cells and inhibition by retinoic acid. Retinoic acid was added on Day 1 of culture. At Days 1, 3, 5, 7, and 9, phase-contrast micrographs were taken (×10 objective; inserts show overview of the whole micromass culture photographed with a ×1.25 objective), and the cultures were stained with alcian blue. To quantify chondrogenesis, the optical density at 600 nm was measured in guanidine hydrochloride extracts of the alcian blue stain. Results are means ± S.D. of four determinations. (Reproduced with permission from Ref. 5.)

twice in sterile Ringer's, transferred to a 50-ml centrifuge tube, and dissociated in 10 ml of freshly prepared and filter-sterilized dissociation medium prepared from calcium and magnesium-free GBSS [Gey's balanced salt solution, supplemented with 100 IU penicillin/ml and 100 μg streptomycin/ml (GIBCO, Glasgow, UK)] containing 0.1% trypsin (w/v) (GIBCO), 0.1% EDTA (w/v) (GIBCO), and 50 m*M* Tris, pH 7.3.

The dissociation medium should be stirred gently in order not to damage the cells. It is best to use a hanging stirrer (a dissociation apparatus was described elsewhere[5]). Cells are dissociated for about 30 min at 37°. As soon as a single cell suspension is obtained, the dissociation is terminated with the addition of 2 ml Nu-serum (Collaborative Research Inc., Lexington, MA), a serum substitute. The cells are sedimented by centrifugation at 150 *g* for 10 min and gently resuspended in 10 ml CMRL medium (GIBCO) supplemented with 10% (v/v) Nu-serum and antibiotics. An aliquot of the cell suspension is diluted 1/10 with 0.2% (v/v) trypan blue and the number of viable and dead cells counted in a hemacytometer. Under these conditions we routinely achieve a viability of above 95%. The remaining cells are resedimented and resuspended to yield a density of 2×10^7 viable cells/ml.

Limb Bud Cell Culture. A discrete 20-μl drop of the adjusted cell suspension is placed in the center of each well in a 24-well plastic cell culture plate. The plates are placed in a humidified incubator at 37° with 5% CO_2 for 1 to 1.5 hr. It is important not to disturb the cells during this period. After the cells have attached (usually after 1 hr), they are flooded carefully with 0.5 ml supplemented CMRL medium.

Addition of Retinoid. The day of cell preparation is considered as Day 0 of culture. Retinoids, dissolved in either dimethyl sulfoxide (DMSO) or 95% ethanol, are added to the cultures on Day 1. For each retinoid tested, one 24-well cluster is used with one control group and 5 dose groups at concentrations ranging from 10^{-10} to 10^{-6} *M* in quadruplicate. Up to a concentration of 10^{-6} *M* no cytotoxic effects were noted with any retinoid tested.

The retinoids are dissolved at a 400-fold final concentration and added to the cell culture medium such that the final solvent concentration does not exceed 0.25%. An equivalent amount of solvent is added to control cultures. Culture medium is replaced on Day 4 of culture with medium containing no retinoid. This is possible because the time required for retinoids to inhibit chondrogenesis irreversibly is less than 24 hr.[3]

Alcian Blue Staining. On Day 8, the cultures are rinsed once with Ringer's and stained for about 4 hr with 0.5% alcian blue (DIFCO, West Molesey, Surrey, UK) dissolved in 3% acetic acid, pH 1.0. Unbound die is removed by subsequent rinses with 3% acetic acid, pH 1, and 3% acetic acid, unadjusted pH, followed by overnight extraction with 0.3 ml of 4 *M*

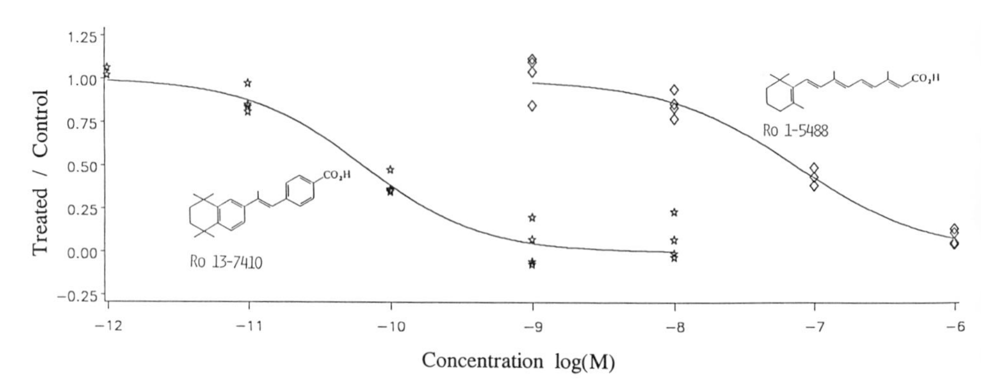

FIG. 4. Dose–response curves of two retinoids with marked differences in activity to inhibit chondrogenesis in limb bud cell cultures. The retinoids were added on Day 1 of culture. Chondrogenesis was assessed by alcian blue staining and quantified by measuring the extracted stain spectrophotometrically after 7 days of culture. Results are expressed as ratios relative to untreated controls. The IC_{50} values for Ro 13-7410 and retinoic acid (Ro 1-5488) were 6×10^{-11} and 8×10^{-8} *M*, respectively.

guanidine hydrochloride. The following day the absorbance at 600 nm is determined spectrophotometrically (Fig. 3). The IC_{50} can be determined from the dose–response curves (Fig. 4) or calculated as described elsewhere.[1]

Comments

We use rat embryos at Day 13 of gestation rather than Day 12,[6] because both fore- and hindlimbs can be used, resulting in more cells being available from the same number of embryos. Furthermore, chondrogenesis is well expressed at this developmental stage.

The pH of the dissociation solution markedly affects the time of dissociation.[5] Therefore, the pH should be controlled and the dissociation terminated as soon as a single cell suspension is obtained. Damaging cells should be avoided as cell damage leads to release of DNA; DNA causes cells to aggregate. Do not overdissociate the limb bud cells, keep centrifugal forces for sedimentation of the cells to a minimum, and resuspend the cells gently by pipetting in fresh cell culture medium.

It is possible to use fetal calf serum in place of Nu-serum. However, it should be kept in mind that there are batch-to-batch variations, and some sera may not optimally support cell growth.[4] It is our experience that attachment of the cells to the plastic surface is a critical point in the entire procedure. The following factors may affect cell attachment: cell damage, attachment time, and type or age of sera.

As an alternative method of quantifying inhibition of chondrogenesis, the number of chondrogenic foci can be determined.[6]

Some retinoids require bioactivation to the active form.[1,4,7] In preliminary experiments we were able to activate some retinoids by coincubation of limb bud cells with small amounts (< 1 μl/ml) of S9 liver fraction without cofactors. However, a metabolic activation system for routine use has not yet been established.

[6] O. P. Flint and T. C. Orton, *Toxicol. Appl. Pharmacol.* **76,** 383 (1984).
[7] D. M. Kochhar, J. D. Penner, and L. M. Minutella, *Drug Metab. Dispos.* **17,** 618 (1989).

[46] Testing of Retinoids for Teratogenicity *in Vivo*

By ANDREAS KISTLER and W. BRIAN HOWARD

Introduction

Retinoids are powerful teratogens that induce specific malformation patterns in animals[1,2] as well as in humans.[3] Representative malformed rodent fetuses at term are shown in Fig. 1.[4] Their mothers were treated with retinoids daily from Days 6 through 15 of gestation. It should be noted that some investigators consider plug day as Day 0 whereas others consider it as Day 1 of gestation; we consider plug day as Day 0 of gestation. In rats retinoic acid has a short half-life of elimination[5] and induces stage-specific malformations depending on the day of treatment[6] (Fig. 2). Thus, the malformations noted at term are unique for each day of treatment. A susceptible period regarding embryolethality and malformations of the head (including exencephaly, encephalocele, open eyes, cleft palate, and facial and cranial abnormalities) occurs between Days 8 and 10 of gestation (Fig. 2). Day 9 embryos are at the most susceptible stage of development (Table I). Therefore, for testing the teratogenic potential of a large number of new retinoids *in vivo,* we suggest treating rats once on Day

[1] J. A. G. Geelen, *Crit. Rev. Toxicol.* **6,** 351 (1979).
[2] J. J. Kamm, *J. Am. Acad. Dermatol.* **6,** 652 (1982).
[3] E. J. Lammer, D. T. Chen, R. M. Hoar, N. D. Agnish, P. J. Benke, J. T. Braun, C. J. Curry, P. M. Fernhoff, A. W. Grix, I. T. Tott, J. M. Richard, and S. C. Sun, *N. Engl. J. Med.* **313,** 837 (1985).
[4] A. Kistler, *Arch. Toxicol.* **60,** 403 (1987).
[5] A. B. Roberts and H. F. De Luca, *Biochem. J.* **102,** 600 (1967).
[6] A. Kistler, *Teratology* **23,** 25 (1981).

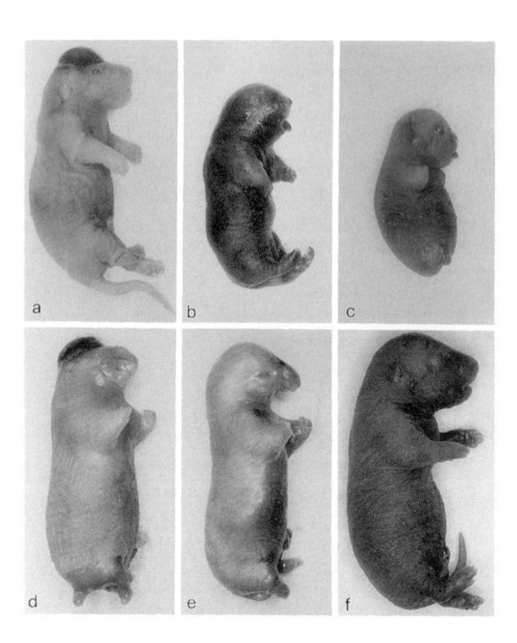

FIG. 1. Representative, severe malformations induced by retinoids in rodents. Dams were treated from Days 6 through 15 of gestation. Details regarding the various retinoids and the doses used have been reported.[4] Fetuses were obtained by cesarean section on Days 18 and 20 of gestation for mice and rats, respectively. (a–c) Mouse, (d–f) rat. The following malformations were observed: (a) exencephaly, protruding tongue, and open eyes; (b) protruding tongue and malformations of head and limbs; (c) protruding tongue, open eyes, and malformations of head, limbs, and tail; (d) exencephaly and malformations of head, limbs, and tail; (e) open eyes and malformations of head, limbs, and tail. (f) Control.

9 of gestation. After cesarean section on Day 20 of gestation, the fetuses can be examined for only externally visible abnormalities including cleft palate. This procedure enables the investigator to detect the major malformations induced by retinoids at this early susceptible stage of development without further laborious techniques such as skeletal examination after alizarin red staining ([44], this volume).

Caution

Most retinoids are teratogenic, some at such diminutive doses as 1 μg/kg. Therefore, they should be handled with utmost care, especially by women. Wearing a dust mask, working under a hood, and avoiding skin contact are recommended.

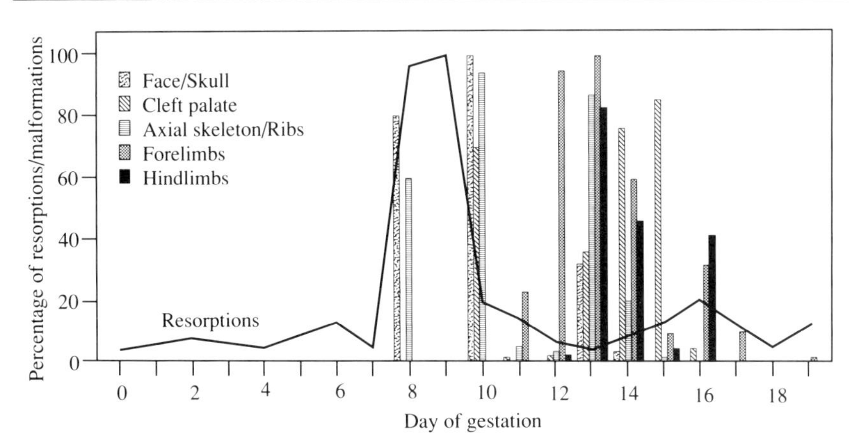

FIG. 2. Resorption rate and incidence of specific malformations induced by retinoic acid after a single treatment at specific days of gestation. Pregnant rats were treated orally with 120 mg/kg body weight at the day indicated. After cesarean section on Day 20 of gestation, fetuses were examined for external malformations, processed for visualization of the skeleton with alizarin red, and examined for skeletal malformations.

Method of Dose Preparation

Retinoids are light-sensitive, especially in solution. Stock solutions/suspensions should be kept only for a few days, protected from light and stored under argon or N_2 gas. Use high shear mixing during formulation in rapeseed oil (or other vegetable oil) and keep the suspension homogeneous during dosing with the use of a magnetic stirrer. The application volume is 5 ml/kg. Retinoids have a broad teratogenic activity range from 0.001 to over 100 mg/kg depending on the specific retinoid.[4] Therefore, for the selection of doses for exploratory studies we suggest logarithmic dose steps (i.e., 1, 3, 10 mg/kg) or even steps by one order of magnitude (i.e., 1, 10, 100 mg/kg).

Treatment of Animals and Recording of Reproduction and Malformation Data

Mated female rats on gestational Day 9 (or 8 and 9 or 8–10) are treated orally by gavage with the respective retinoid dose. Controls receive the vehicle only. On day 20 of gestation the dams are sacrificed, fetuses removed by cesarean section, and the uteri examined for the number of fetuses and sites of implantations and resorptions. The fetuses are weighed and examined for external malformations and cleft palate. The palate can

TABLE I
REPRODUCTION DATA AND EXTERNAL MALFORMATIONS INDUCED BY RETINOIC ACID AFTER A SINGLE TREATMENT AT GESTATIONAL DAYS 8, 9, AND 10 IN RATS[a]

Treatment day of gestation	Retinoic acid (mg/kg)	Number of dams	Number of resorptions/number of implantations	Percent resorptions	Number of fetuses/dam (mean ± S.D.)	Fetal body weight (mean ± S.D.)	Number of fetuses examined	Percentage of external malformations: Exencephaly encephalocele	Protruding tongue/head shape	Eyes	Cleft palate
Control	0	14	11/183	6.0	12.3 ± 2.8	3.4 ± 0.2	172	0	0	0	0
8	3	6	17/87	19.5	11.7 ± 4.9	3.2 ± 0.2	70	0	0	0	0
	10	6	7/83	8.4	12.7 ± 2.5	3.5 ± 0.2	76	0	0	0	0
	30[b]	6	12/76	15.8	10.7 ± 3.4	3.0 ± 0.3	64	5	3	2	8
	100	5	46/80	57.5	6.8 ± 6.4	2.7 ± 0.4	34	35	18	3	29
9	3	6	7/78	9.0	11.8 ± 2.5	3.4 ± 0.1	71	0	0	0	0
	10[c]	5	12/70	17.1	11.6 ± 2.9	3.3 ± 0.2	58	2	0	0	0
	30	6	69/86	80.2	2.8 ± 2.5	2.7 ± 0.4	17	94	82	18	0
	100	6	83/83	100.0	0.0 ± 0.0	0.0 ± 0.0	0				
10	3	5	6/63	9.5	11.4 ± 2.9	3.3 ± 0.6	57	0	0	0	0
	10	5	7/61	11.5	10.8 ± 2.6	3.1 ± 0.7	54	0	0	0	0
	30[d]	5	11/64	17.2	10.6 ± 6.4	2.9 ± 0.7	53	4	0	0	2
	100[e]	6	17/66	25.8	8.2 ± 6.5	3.2 ± 0.1	49	4	100	92	86

[a] Pregnant rats were treated orally with the indicated dose of retinoic acid suspended in rapeseed oil at the indicated day of gestation. After cesarean section on Day 20 of gestation, the reproduction parameters were determined and the fetuses examined for external malformations including cleft palate.

[b] One fetus with multiple malformations including ectopy of liver and intestines, and limb and genital malformations.

[c] One fetus with multiple malformations including contorted hind limbs, shortened tail, and anal and genital malformations.

[d] One fetus with spina bifida.

[e] One fetus with genital malformations.

be examined through the opened mouth of the freshly dissected fetuses. A more reliable method is the examination of the palate of formalin-fixed heads after dissection of the lower jaw.

Comments

Mice are an alternative species for teratogenicity testing. Mice have a similar susceptibility of retinoids as rats.[4] An advantage is that less drug is required. For the treatment, Days 8 and 9 of gestation are recommended, days which are very susceptible to retinoids.[7] An important advantage of a single treatment is that, in general, the drug is well tolerated by the dams,[8] in contrast to repeated dosing which induces hypervitaminosis A (weight loss, desquamation of the skin, hair loss, and alterations of the skeletal system including bone fractures).[9]

[7] D. M. Kochhar, *Acta Pathol. Microbiol. Scand.* **70,** 398 (1967).
[8] K. Teelmann, *Pharmacol. Ther.* **40,** 29 (1989).
[9] A. Kistler, H. Sterz, and K. Teelmann, *Arch. Toxicol.* **56,** 117 (1984).

[47] Correlation of Transplacental and Maternal Pharmacokinetics of Retinoids during Organogenesis with Teratogenicity

By HEINZ NAU

Introduction

Retinoids are teratogenic in humans as well as in all animal species which have been investigated.[1,2] In humans, isotretinoin[3] (13-*cis*-retinoic acid) and etretinate[4] are established teratogens; of recent concern have also been excessive doses of vitamin A[5] (retinol or esters), although a causal

[1] J. J. Kamm, *J. Am. Acad. Dermatol.* **6,** 652 (1982).
[2] B. W. Howard and C. C. Willhite, *J. Toxicol. Toxin Rev.* **5,** 55 (1986).
[3] E. J. Lammer, D. T. Chen, R. M. Hoar, N. D. Agnish, P. J. Benke, J. T. Braun, C. J. Curry, P. M. Fernhoff, A. W. Grix, I. T. Lott, J. M. Richard, and S. C. Sun, *N. Engl. J. Med.* **313,** 837 (1985).
[4] R. Happle, H. Traupe, Y. Bounameaux, and T. Fisch, *Dtsch. Med. Wochenschr.* **109,** 1476 (1984).
[5] F. W. Rosa, A. L. Wilk, and F. O. Kelsey, *Teratology* **33,** 355 (1986).

relationship between exposure and teratogenesis remains suggestive at this time.

Pharmacokinetic studies have proved useful for the interpretation of retinoid teratogenesis, as the adverse effect of a drug on embryonic development can depend on the concentration–time relationship of the drug or its active metabolite(s) in the embryo during the sensitive stages of gestation.[6,7] Of significance, therefore, are all processes which define these crucial embryonic drug concentrations: maternal absorption, distribution, elimination (metabolism as well as biliary and urinary excretion), and transfer to the embryo. The methods presented here are designed to improve our understanding of species differences, metabolic activation, and structure–activity relationships.

Transplacental Pharmacokinetics of Retinoids

Methods

Animal Treatment.[8-13] All studies are performed in NMRI:Han mice on Day 11 of gestation (day after plug detection = Day 0), as retinoid-specific limb defects and oral clefts can be produced during that gestational period[14] and the embryo has developed to a size sufficient for analytical measurements. The retinoids are dissolved in soybean oil containing 8% ethanol. Via oral intubation, a single dose of either 10 or 100 mg/kg body weight is administered in a vehicle volume of 5 ml/kg. At various periods following treatment, groups of 5 mice are anesthetized with ether, and blood is taken by heart puncture and centrifuged immediately in heparinized tubes at 4° and 3000 *g*. After the blood is collected, the mice are

[6] H. Nau, D. M. Kochhar, J. M. Creech Kraft, B. Loefberg, J. Reiners, and T. Sparenberg, *in* "Approaches to Elucidate Mechanisms in Teratogenesis" (F. Welsch, ed.), p. 1. Hemisphere, Washington, D.C., 1987.

[7] H. Nau, J. Creech Kraft, C. Eckhoff, and B. Löfberg, *in* "Pharmacology of Retinoids in the Skin" (U. Reichart and B. Shroot, eds.), Vol. III, p. 165. Karger, Basel. (1989).

[8] J. Creech Kraft, D. M. Kochhar, W. J. Scott, and H. Nau, *Toxicol. Appl. Pharmacol.* **87,** 474 (1986).

[9] J. Reiners, B. Löfberg, J. Creech Kraft, D. M. Kochhar, and H. Nau, *Reprod. Toxicol.* **2,** 19 (1988).

[10] J. Creech Kraft, B. Löfberg, I. Chahoud, G. Bochert, and H. Nau, *Toxicol. Appl. Pharmacol.* **100,** 162 (1989).

[11] C. Eckhoff, B. Löfberg, I. Chahoud, G. Bochert, and H. Nau, *Toxicol. Lett.* **48,** 171 (1989).

[12] B. Löfberg, I. Chahoud, G. Bochert, and H. Nau, *Teratology* **41,** 707 (1990).

[13] J. Creech Kraft, C. Eckhoff, W. Kuhnz, B. Löfberg, and H. Nau, *J. Liquid Chromatogr.* **11,** 2051 (1988).

[14] D. M. Kochhar, J. D. Penner, and C. I. Tellone, *Teratog. Carcinog. Mutagen.* **4,** 377 (1984).

sacrificed by cervical dislocation, and the embryo and placental tissues are collected and pooled for each litter. All samples are kept frozen at $-80°$.

Sample Preparation and Chromatographic Measurements.[8,9,11,13] The high-performance liquid chromatographic (HPLC) systems used consist of an autosampler, a precolumn switching device coupled to a gradient system, and a UV detector with integrator. The precolumn system is used for sample preconcentration to minimize sample work-up procedures. This technique proves to be especially useful for instable compounds such as retinoids, which are susceptible to decomposition during conventional sample extraction conditions. Briefly, maternal plasma is treated with an equal volume of 2-propanol; 100 mg of pooled embryos are treated with 200 μl 2-propanol. After vigorous mixing for 1 min, the samples are frozen in liquid nitrogen overnight, and the embryo samples are then homogenized for 10 sec with ultrasonication at 0°. All samples are centrifuged 20 min at 4000 *g* at 4°. Placental samples are first homogenized in aqueous buffer and ultrasonicated. These homogenates are then treated as for the embryo homogenates.

According to availability, 50–200 μl of the supernatants obtained is injected into the HPLC system. The two precolumns used are alternately flushed with the analytical buffer (2 ml/min) or with ammonium acetate (40 m*M*, pH 6.8, 1 ml/min), or methanol. The sample is injected by the autosampler, and the retinoids are enriched on the first precolumn with the ammonium acetate buffer. After 1 min, the switching module switches the solvent flows and direction so that the retinoids are flushed by the analytical buffer from the precolumns onto the analytical column. The second precolumn is cleaned with methanol to prevent carryover effects and then equilibrated for the next sample with ammonium acetate buffer. The methanol gradient used on the analytical column is adjusted according to the type of retinoid being measured. It consists of a mixture of methanol and 40 m*M* ammonium acetate, pH 7.4, which is gradually changed to 100% methanol.

External standardization is used for the calculation of the analytical results. The recoveries of the retinoids from plasma exceed 95%, and those from the embryos exceed 88%. No isomerization occurs during the analytical procedures. All handling of the retinoids is done under yellow light.

Application[6-12]

All retinoids (identified in the legend of Fig. 2 with roman numerals) are rapidly absorbed after a single oral dose of 100 mg/kg and reach peak maternal plasma levels of between 0.9 and 7 μg/ml. The concentrations then decrease rapidly within a few hours to very low levels, except for

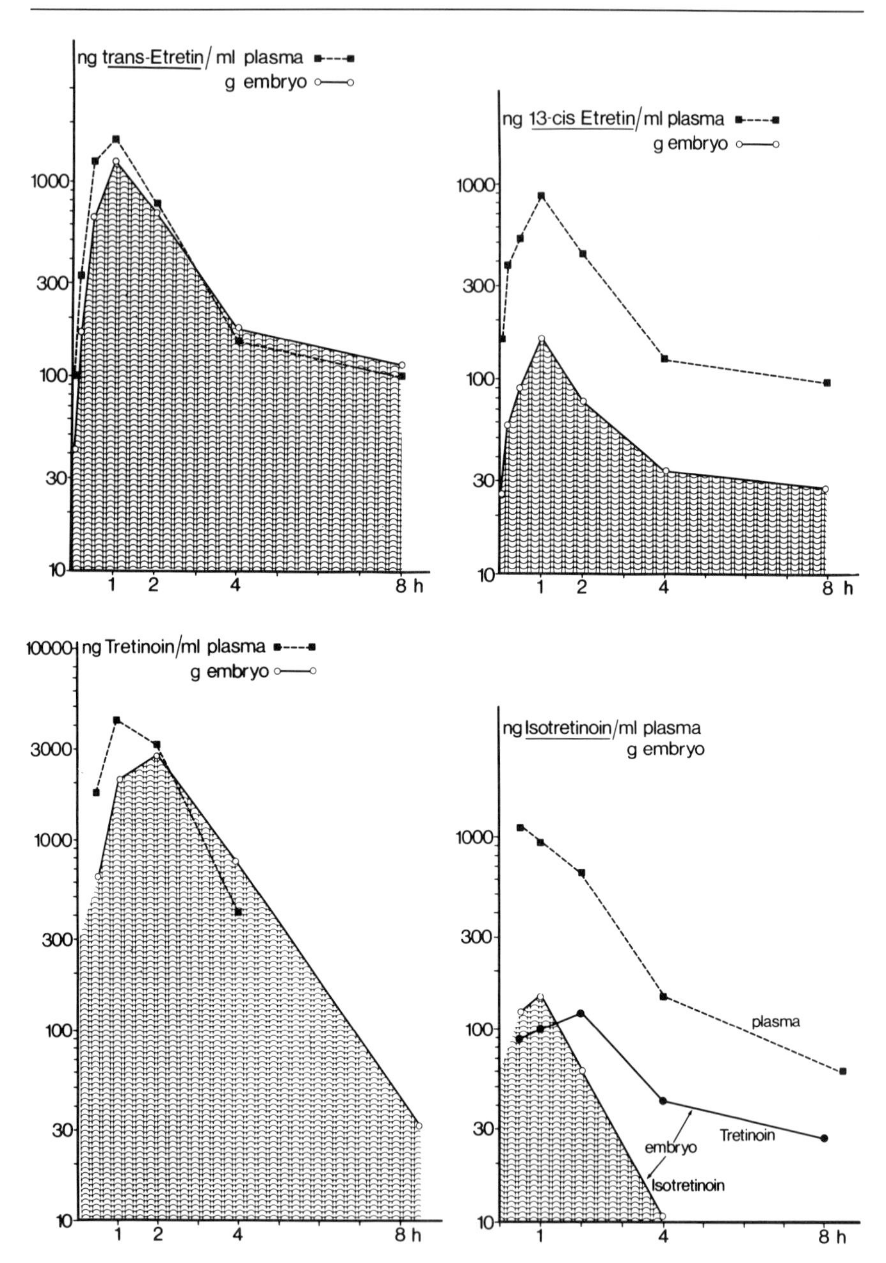

FIG. 1. Retinoid concentrations in maternal plasma and embryos following a single oral dose on Day 11 of gestation of 100 mg/kg all-*trans*-etretin (acitretin) **(V)**, 100 mg/kg 13-*cis*-etretin **(VI)**,[12] 10 mg/kg all-*trans*-retinoic acid (tretinoin) **(II)**, or 10 mg/kg 13-*cis*-retinoic acid (isotretinoin) **(I)**.[10] Note the low concentrations of the 13-*cis*-retinoids and the considerable concentrations of the corresponding all-trans isomers within the embryo.

etretinate and motretinide which exhibit terminal half-lives of about 1 day, thus indicating the presence of deep compartments for these liphophilic aromatic retinoids.

The placental transfer of retinoids varies (Figs. 1 and 2). Retinoids with the all-trans configuration **(II, III, IV, V, VII)** exhibit extensive placental transfer: during the first 2–4 hr after drug administration, embryonic

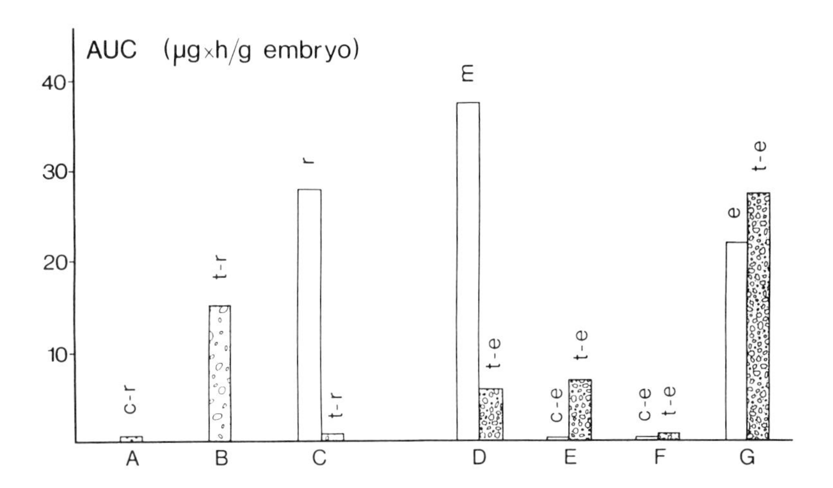

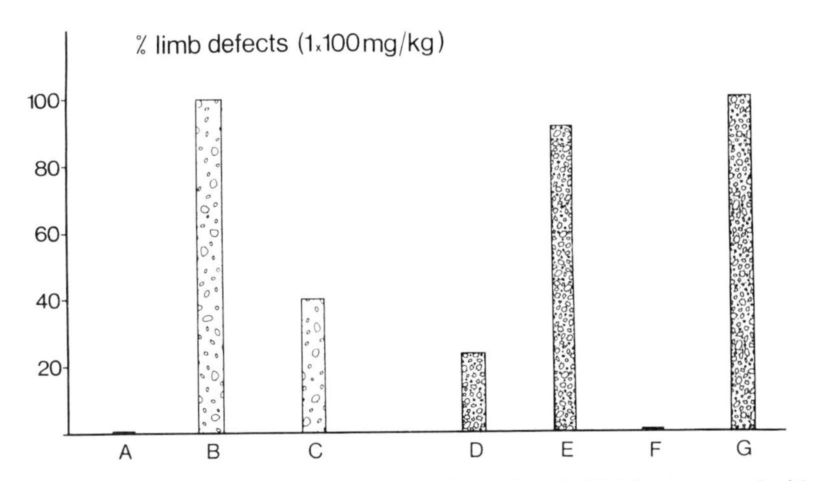

FIG. 2. Embryonic *AUC* values and rates of limb defects in NMRI mice treated with a single dose on Day 11 of gestation[6–9,11,12] of 100 mg/kg of the following retinoids: A, 13-*cis*-retinoic acid (isotretinoin) **(I)**; B, all-*trans*-retinoic acid (tretinoin) **(II)**; C, retinol **(III)**; D, motretinide **(VII)**; E, all-*trans*-etretin (acitretin) **(V)**; F, 13-*cis*-etretin **(VI)**; and G, etretinate **(IV)**. The compounds measured are as follows: c–r, **I**; t–r, **II**; r, **III**; m, **VII**; t–e, **V**; c–e, **VI**; e, **IV**.

concentrations are slightly below corresponding maternal plasma levels, whereas the reverse is found at later time intervals. Compounds with the 13-cis configuration **(I, VI)** exhibit very limited placental transfer:[8,9,10,12] concentrations in the embryo are between $\frac{1}{5}$ and $\frac{1}{50}$ lower than those in maternal plasma (Fig. 1). Both **I** and **VI** isomerize to a small extent, and because the all-trans isomers produced **(II, V)** exhibit good placental transfer, their concentration eventually exceeds those of the corresponding 13-cis isomers. This raises the possibility that isomerization of the 13-*cis*-retinoids to the all-trans form is a teratogenic activation reaction (Fig. 3). The *AUC* (area under the curve) values of the various retinoids studied correspond well with the teratogenic potency of the drugs (Fig. 2). Thus, the 13-cis isomers have very low *AUC* values as well as potency compared to the all-trans forms.[8-10,12,14]

Other metabolic activation reactions observed *in vivo* are the transformation of etretinate **(IV)** and motretinide **(VII)** to their common active metabolite all-*trans*-etretin (acitretin, **V**).[9] Both **IV** and **VII** are inactive *in vitro,* probably because the structure of these compounds contains a terminal ethyl ester or ethyl amide group, respectively, and an acidic group at the terminus of the polyene chain might be required for expression of teratogenic activity.[2,9] This may also be true *in vitro:* although motretinide **(VII)** itself reaches very high *AUC* values in the embryo, its teratogenic potency is lower than that of the other two aromatic retinoids **IV** and **V**

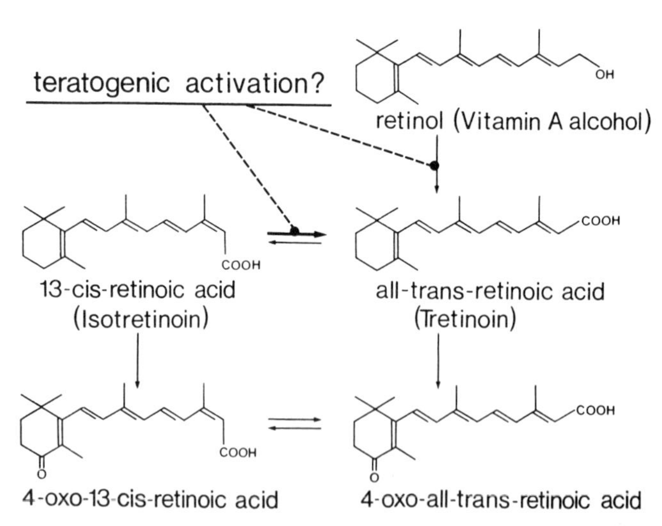

FIG. 3. Metabolic scheme of several retinoids pertinent with regard to teratogenicity. In addition, the retinoic acids **I** (13-*cis*-retinoic acid) and **II** (all-*trans*-retinoic acid) are transformed into β-D-glucuronides.

because the *AUC* of the active metabolite etretin (**V**) is lower in the embryo after administration of **VII** than after application of **IV** and **V** (Fig. 2).

Another potential teratogenic activation may be the oxidation of retinol (**III**) to all-*trans*-retinoic acid (II)[11,15] (Fig. 3). Significant amounts of **II** are produced in mice after the single high dose of **III** (*AUC* ~ 1 μg/hr/g embryo;[11] see Fig. 2). These levels of **II** in the embryo should be significant in regard to retinol-induced teratogenesis, although a comparison of the embryonic AUC values and teratogenic potencies of **II** and **III** (Fig. 2) indicates that a teratogenic effect of retinol and/or other metabolites must also be considered.

The metabolic pattern of the retinoids studied can be complex and must be studied in detail to reach valid conclusions with regard to the teratogenicity of each individual substance. As discussed above, the retinoic acids isomerize to their respective 13-cis and all-trans forms. In addition, the 4-oxo metabolites are produced (Fig. 3). The 4-oxo-all-*trans*-retinoic acid is very similar to **II** with regard to placental transfer and teratogenic potency:[10] thus, the 4-oxo metabolite, probably produced by the cytochrome *P*-450-dependent monooxygenase system, may contribute to the teratogenic potency of all-*trans*-retinoic acid. In contrast, the 4-oxo-13-*cis*-retinoic acid exhibits poor placental transfer and is not teratogenic under the same experimental conditions.[10]

Other major plasma metabolites of the retinoic acids are all-*trans*-β-D-retinoylglucuronide and 13-*cis*-retinoyl-β-D-glucuronide.[16] The latter metabolite in particular reaches very high concentrations in the maternal mouse plasma after administration of **I**. Both the all-*trans*- and the 13-*cis*-retinoyl-β-D-glucuronides reach only very low embryonic concentrations, so that their significance with regard to the teratogenicity of the parent compounds is not apparent.[16]

Based on our experimental studies it is now possible to speculate on the drastic species difference observed between the teratogenicity of isotretinoin (**I**) in mice (low potency) and humans (high potency). In mice, an extensive first-pass effect, effective glucuronidation, and high clearance (short half-life) greatly reduce the concentrations and *AUC* values of this drug in plasma;[6,17,18] in humans, longer half-lives and *AUC* values are observed for both the parent drug and its major metabolite 4-oxo-13-*cis*-

[15] D. M. Kochhar, J. D. Penner, and M. A. Satre, *Toxicol. Appl. Pharmacol.* **96,** 429 (1988).

[16] J. Creech Kraft, L. Roberts, J. R. Baily, H. Nau, and W. Slikker, Jr., *Teratology* **39**(5), 447A (1989).

[17] J. R. Kalin, M. J. Wells, and D. L. Hill, *Drug Metab. Dispos.* **10,** 391 (1982).

[18] C. C. Wang, S. Campbell, R. L. Furner, and D. L. Hill, *Drug Metab. Dispos.* **10,** 391 (1980).

[19] K. C. Khoo, D. Reik, and W. A. Colburn, *J. Clin. Pharmacol.* **22,** 395 (1982).

retinoic acid.[19] Furthermore, the very low placental transfer of **I** limits the embryonic exposure in mice;[8,10] the placental transfer in humans is not clear, although in one case we did observe considerable concentrations of **I** and **II** in human embryonic and placental tissue following interruption of pregnancy after inadvertent exposure to isotretinoin.[20]

Thus, transplacental pharmacokinetics may be used to explain species differences in retinoid teratogenesis and to extrapolate experimental results to humans. We believe that embryonic *AUC* values exceeding approximately 0.5 – 1 μg/hr/g embryo of active retinoids could elicit a significant teratogenic response.

Infusion via Implanted Osmotic Minipumps: Significance of *AUC* Values

The pharmacokinetics of retinoids following conventional administration regimens in animals is very different from those in human therapy: because of the rapid elimination in animals, drastic concentration fluctuations are usually obtained. In humans, these fluctuations are much less pronounced owing to longer half-lives. This is particularly true for etretinate following discontinuation of therapy: the drug accumulates in deep compartments such as adipose tissue and is eliminated extremely slowly, with a half-life of about 100 days.[21] Therefore, persisting concentrations of etretinate and its main metabolite are present for months or even years. We have therefore developed a novel experimental model where the long terminal half-life of etretinate is represented by intragastric infusion of the drug via implanted osmotic minipumps.[22]

Procedure[22]

Etretinate is dissolved in cottonseed oil in suitable concentrations. This solvent is not compatible with osmotic minipumps, and therefore reservoirs are constructed from polyamide tubing of 2.7 mm i.d. and 4.0 mm o.d., 35 mm long, with a volume of 200 μl. One end of the reservoir is connected by a short piece of 1 mm i.d. and 3 mm o.d. silicone tubing and an 8 cm long polyethylene tubing, 1 mm o.d. and 0.5 mm i.d., to the minipump (Model 2001, Alza Co., Palo Alto, CA, 250 μl volume, pump duration 7 days). On the other side, a 10 cm long soft silicone tubing. 0.7 mm i.d. and 0.9 mm o.d., is connected. The active device is autoclaved at 120° for 2 hr.

[20] J. C. Kraft, H. Nau, E. Lammer, and A. Olney, *N. Engl. J. Med.* **321,** 262 (1989).

[21] J. Massarella, F. Vane, C. Buggé, L. Rodriguez, W. J. Cunningham, T. Franz, and W. Colburn, *Clin. Pharmacol. Ther.* **37,** 439 (1985).

[22] B. Löfberg, J. Reiners, and H. Nau, *Dev. Pharmacol. Ther.* (submitted).

Pregnant NMRI mice are anesthetized on Day 8 of gestation (plug day = Day 0) with 375 mg/kg sodium hexobarbital. The abdominal skin (3 cm) and the peritoneum (1 cm) are opened. A small incision is made in the stomach, through which the soft silicone tubing of the reservoir is implanted and fixed with tissue glue (Histoacryl, Braun, Melsurgen, FRG). The peritoneum is closed with Histoacryl, the reservoir with the attached minipump are placed subcutaneously, and then the skin is closed with clamps and Histoacryl.

On Days 10 and 12 of gestation, blood and embryo samples are taken and analyzed for the presence of retinoids as described above (groups of 5–7 mice). On Day 17 of gestation, the number of the resorptions and live fetuses as well as the fetal weight and external malformations are recorded in further groups of animals.

Application[22]

The surgical procedures do not result in any embryotoxicity or teratogenicity. Infusion of etretinate produces dose-dependent effects, including retinoid-specific malformations such as shortening of the limb and tail structures. The lowest teratogenic dose infused is 0.84 mg/kg/day. A dose of 1 mg/kg/day results in low plasma levels of etretinate (6–11 ng/ml) and all-*trans*-etretin (40 ng/ml); the corresponding values in the embryo are higher (12 and 95 ng/g embryo, respectively). Thus, plasma *AUC* values of 0.75–1 μg/ml/hr result in retinoid-specific teratogenicity.

These results emphasize the teratogenic hazard of low concentrations of retinoids, which persist throughout the sensitive developmental stages. It is therefore highly likely that the *AUC* values of active retinoids correlate with the teratogenic effects, because concentration peaks are not produced during steady-state infusion and because low drug levels produce significant AUC levels owing to their prolonged presence. These findings should greatly assist in the extrapolation of experimental results across species, particularly to humans; the human may therefore be particularly sensitive because of persisting retinoid concentrations during therapy and relatively long sensitive stages of development—thus resulting in high AUC values as compared to experimental animals.

High-Performance Size-Exclusion Chromatography Analysis of Proteins Binding Retinoids

Retinoids are extensively bound to plasma and tissue proteins.[23,24] Retinol is bound in plasma to retinol-binding protein (RBP) which is

[23] M. Kato, K. Kato, and D. S. Goodman, *Lab Invest.* **52,** 475 (1985).
[24] D. E. Ong, *Nutr. Rev.* **43,** 225 (1985).

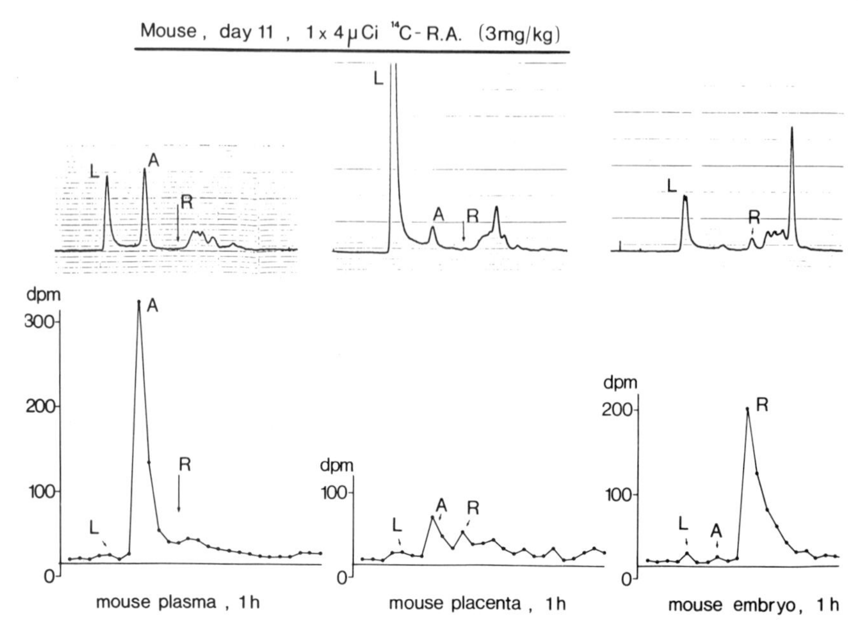

FIG. 4. Fluorescence traces indicating the elution of retinol-bound fractions (top) and radioactivity traces indicating the elution of all-*trans*-retionoic acid-bound protein fractions (bottom). Pregnant mice were injected with 4 μCi (3 mg/kg) of ^{14}C-labeled all-*trans*-retinoic ([^{14}C]RA) acid on Day 11 of gestation, and plasma, embryos, and placentas were taken 1 hr after treatment. A, Albumin; L, lipoprotein; R, retinol-binding protein.

complexed 1:1 with transthyretin (prealbumin). Retinol (and esters) are also found in lipoprotein fractions and chylomicrons (Fig. 4). In tissues, retinol and retinoic acid have their corresponding binding proteins (CRBP, CRABP, respectively). We have therefore studied the binding of some retinoids in plasma, placenta, yolk sac, and embryo in order to better understand the placental transfer of these compounds.

Procedure

The HPLC system consists of a gel-filtration column[25,26] (BioGel TSK 250, Bio-Rad, Richmond, CA) eluted with a buffer containing 20 m*M* sodium phosphate, 50 m*M* sodium sulfate, 1 m*M* EDTA, and 100 mg sodium azide/liter (pH 6.8, 0.9 ml/min). A UV detector (280 nm) and a fluorescence detector (excitation 330 nm, emission 460 nm) as well as a fraction collector (0.9-ml fractions) are coupled sequentially after the column. Plasma (20 μl) samples are directly injected. Embryo and placenta

[25] H. E. Shubeita, M. D. Patel, and A. M. McCormick, *Arch. Biochem. Biophys.* **247,** 280 (1986).

[26] H. C. Furr and J. A. Olson, *Anal. Biochem.* **171,** 360 (1988).

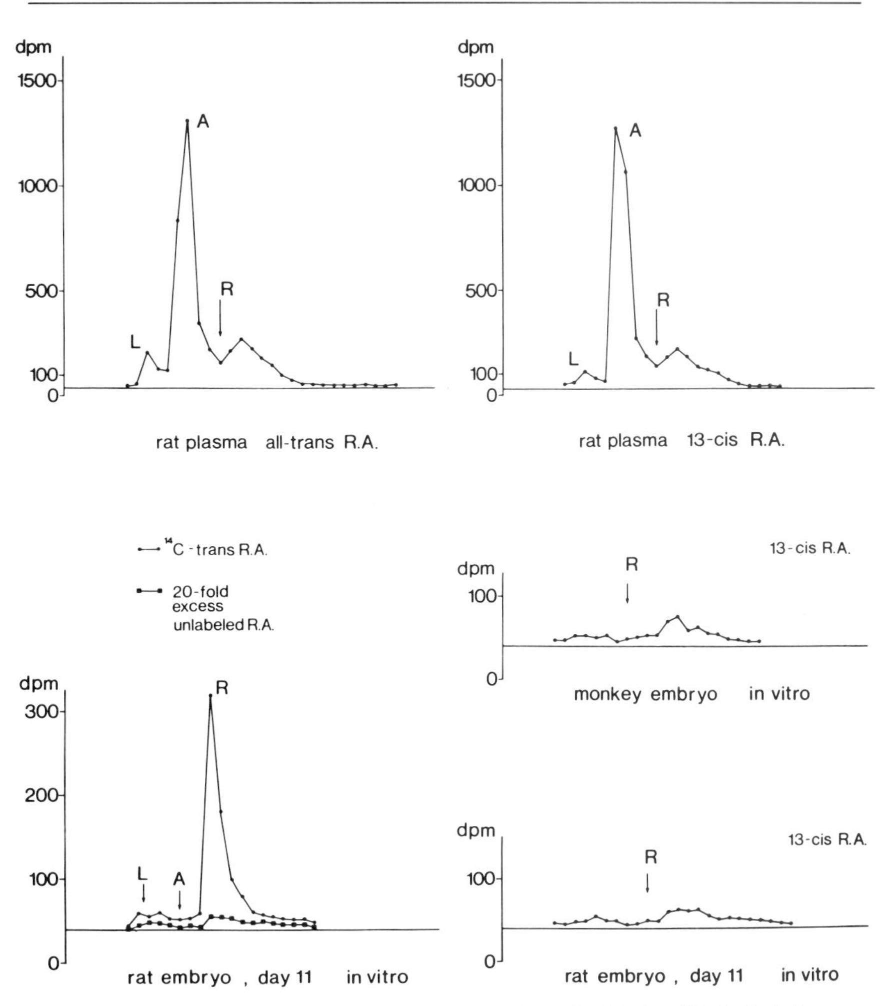

FIG. 5. Radioactivity traces indicating the elution of proteins binding ^{14}C-labeled all-*trans*- and 13-*cis*-retinoic acid in rat plasma (top) and embryo homogenates (bottom). Note the similar binding of the two isomers in plasma but the drastic difference in the embryo. The binding of all-*trans*-retinoic acid to embryo homogenates was abolished by the addition of excess unlabeled compound.

samples are homogenized at 0° with an equal amount of elution buffer and ultrasonicated; the supernatant (15,000 *g*) (20–50 μl) is then injected.

Application

Retinol bound to proteins is detected by fluorescence and three fractions are usually found to contain retinol and retinyl esters: lipoproteins

(L), albumin (A), and RBP–prealbumin complex (not separated) and RBP (R).[26] The peak of RBP in mouse plasma (Fig. 4) is very small because of low levels of retinol in pregnant mice. Retinol-related fluorescence is also found in fraction L in placentas and embryos and in fraction R in embryos.

Retinoic acid (^{14}C-radioactivity trace in the lower panels in Figs. 4 and 5) is mostly found in fraction A in plasma and in fraction R in the embryo; much less activity is bound to placenta and yolk sac proteins. The same pattern of retinoic acid binding is also found by the addition of 14 C-labeled compound to plasma *in vitro* in several species. Mouse and rat plasma show much more extensive binding in fraction A than monkey and human plasma. Also the binding of retinoic acid to embryo homogenates *in vitro* (Fig. 5) is very similar to that *in vivo* (Fig. 4): most of the radioactivity is found in fraction R. In sharp contrast, 13-*cis*-retinoic acid does not bind to mouse, rat, and monkey embryo homogenates (Fig. 5). It is therefore likely that the good placental transfer of all-*trans*-retinoic acid in the mouse is the result of extensive binding of this isomer to embryo proteins, whereas the lack of comparable binding of 13-*cis*-retinoic acid may explain the very low placental transfer of this isomer.

Comparable studies were also done with human embryos (weeks 8–12 of gestation). Extensive binding of labeled all-*trans* retinoic acid was found in homogenates of embryonic carcasses *in vitro;* similar binding was found in homogenates from human embryonic brain (week 10–12 of gestation). The HPLC fraction containing bound retinoic acid eluted at the identical position as the corresponding fractions from mouse, rat, and monkey embryos (R in Figs. 4 and 5). Binding of labeled all-*trans* retinoic acid in human embryonic homogenates, including those from brain, could be displaced only by very high concentrations of unlabeled 13-*cis*-retinoic acid; analysis of the displacement curves indicated a several hundredfold lower affinity of the 13-cis isomer as compared to the all-trans isomer in regard to human embryonic binding, which corresponds to findings in other species such as mouse, rat, and monkey. This preliminary study thus indicates that there is not a significant species difference in regard to differential embryonic binding of all-*trans*- and 13-*cis*-retinoic acids.

Acknowledgments

This work was supported by grants of the Deutsche Forschungsgemeinschaft and the Free University Berlin. The donation of reference retinoids and the many helpful discussions and suggestions of Dr. Hans Hummler, Hoffmann-La Roche, Basel, Switzerland, are gratefully acknowledged.

Author Index

Numbers in parentheses are footnote reference numbers and indicate that an author's work is referred to although the name is not cited in the text.

A

B

C

D

G

I

J

K

L

M

N

O

P

S

T

X

Y

Z

Subject Index

A

B

C

D

E

F

S

W

X

Y